Lecture Notes in Computer Science 4098

Commenced Publication in 1973
Founding and Former Series Editors:
Gerhard Goos, Juris Hartmanis, and Jan

T0216284

Frank Pfenning (Ed.)

Term Rewriting and Applications

17th International Conference, RTA 2006
Seattle, WA, USA, August 12-14, 2006
Proceedings

 Springer

Volume Editor

Frank Pfenning
Carnegie Mellon University
Department of Computer Science
Pittsburgh, PA 15213-3891, USA
E-mail: fp@cs.cmu.edu

Library of Congress Control Number: 2006929604

CR Subject Classification (1998): F.4, F.3.2, D.3, I.2.2-3, I.1

LNCS Sublibrary: SL 1 – Theoretical Computer Science and General Issues

ISSN	0302-9743
ISBN-10	3-540-36834-5 Springer Berlin Heidelberg New York
ISBN-13	978-3-540-36834-2 Springer Berlin Heidelberg New York

Springer is a part of Springer Science+Business Media

springer.com

© Springer-Verlag Berlin Heidelberg 2006
Printed in Germany

Typesetting: Camera-ready by author, data conversion by Scientific Publishing Services, Chennai, India
Printed on acid-free paper   SPIN: 11805618   06/3142   5 4 3 2 1 0

Preface

This volume contains the proceedings of the 17th International Conference on Rewriting Techniques and Applications, which was held on August 12–14, 2006 in Seattle, Washington, as part of of the 4th Federated Logic Conference (FLoC). RTA is the major forum for the presentation of research on all aspects of rewriting. Previous RTA conferences took place in Dijon (1985), Bordeaux (1987), Chapel Hill (1989), Como (1991), Montreal (1993), Kaiserslautern (1995), New Brunswick (1996), Sitges (1997), Tsukuba (1998), Trento (1999), Norwich (2000), Utrecht (2001), Copenhagen (2002), Valencia (2003), Aachen (2004), and Nara (2005).

A total of 23 regular papers and 4 system descriptions were selected for presentation from 52 submissions. Each paper was reviewed by at least 4 members of the Program Committee, with the help of 115 external referees. The committee decided to give the Best Paper Award for RTA 2006 to the contribution "Termination of String Rewriting with Matrix Interpretations" by Dieter Hofbauer and Johannes Waldmann for their original and powerful application of SAT solving in proving termination.

I would like to thank all the members of the Program Committee for their diligent, careful, and timely work and thoughtful deliberation, and Andrei Voronkov for providing the EasyChair system which greatly facilitated the reviewing process, the electronic Program Committee meeting, and the preparation of the program and the proceedings.

In addition to contributed papers, the program contained a FLoC plenary talk by Randal Bryant and two invited talks by Javier Esparza and Jürgen Giesl. I would like to thank the invited speakers not only for their presentations, but also for contributing abstracts or full papers to the proceedings.

RTA also sponsored a number of workshops held during FLoC, on the topics of Higher-Order Rewriting (HOR), Rule-Based Programming (RULE), Unification (UNIF), Reduction Strategies in Rewriting and Programming (WRS), Termination (WST), and a meeting of the IFIP Working Group 1.6 on Term Rewriting.

Many people helped to make RTA 2006 a success. I am particularly grateful to Ashish Tiwari, who took on the dual role of Conference Chair and RTA Workshop Chair, Ralf Treinen, the Publicity Chair, and Thomas Ball, Gopal Gupta, Jakob Rehof, and Moshe Vardi from the FLoC Organizing Committee who did an incredible amount of work in the arrangements for RTA and FLoC.

May 2006   Frank Pfenning

Conference Organization

Program Chair

Frank Pfenning   Carnegie Mellon University

Program Committee

Zena Ariola	University of Oregon
Franz Baader	Technical University Dresden
Gilles Dowek	École Polytechnique and INRIA
Guillem Godoy	Technical University of Catalonia
Deepak Kapur	University of New Mexico
Delia Kesner	University Paris 7
Denis Lugiez	University of Provence
Claude Marché	University Paris-Sud
Frank Pfenning	Carnegie Mellon University
Ashish Tiwari	SRI International
Yoshihito Toyama	Tohoku University
Eelco Visser	Utrecht University
Hans Zantema	Eindhoven University of Technology

Conference Chair

Ashish Tiwari   SRI International

RTA Steering Committee

Robert Nieuwenhuis	Technical University of Catalonia
Ralf Treinen	ENS Cachan, *Publicity Chair*
Jürgen Giesl	RWTH Aachen, *Chair*
Delia Kesner	University Paris 7
Vincent van Oostrom	Utrecht University
Ashish Tiwari	SRI International

FLoC Sponsors

Cadence
IBM
Microsoft Research

NEC
The John von Neumann Minerva Center
  for the Development of Reactive Systems

FLoC Organizing Committee

Thomas Ball   Stephan Kreutzer
Armin Biere   Jacob Rehof
Sandro Etalle   Nicole Schweikardt
Gopal Gupta   Ashish Tiwari
John Harrison   Mirek Truszczynski
Manuel Hermenegildo   Moshe Y. Vardi
Lydia Kavraki   Margus Veanes

External Reviewers

Cuihtlauac Alvarado   Thomas Genet   John Maraist
Takahito Aoto   Benny K. George   Luc Maranget
Albert Atserias   Silvio Ghilardi   Massimo Marchiori
Patrick Baillot   Jürgen Giesl   Ralph Matthes
Frédéric Blanqui   Guillem Godoy   Michel Mauny
Eduardo Bonelli   Bernhard Gramlich   Richard Mayr
Bernd Brassel   Thomas Hallgren   Paul-André Melliès
Martin Bravenboer   Michael Hanus   Aart Middeldorp
Pierre Castéran   Ryu Hasegawa   Oege de Moor
Taolue Chen   Sebastien Hemon   Barbara Morawska
Manuel Clavel   Joe Hendrix   Sara Negri
Thomas Colcombet   Nao Hirokawa   Monica Nesi
Evelyne Contejean   Dieter Hofbauer   Joachim Niehren
Jim Cordy   Martin Hofmann   Robert Nieuwenhuis
Solange Coupet-Grimal   Florent Jacquemard   Mizuhito Ogawa
Marcel Crabbé   Wolfram Kahl   Vincent van Oostrom
Jeremy Dawson   Kentaro Kikuchi   Vincent Padovani
Marie Duflot   Jan Willem Klop   Emir Pasalic
Irène Durand   Adam Koprowski   Adolfo Piperno
Joerg Endrullis   Masahito Kurihara   Emmanuel Polonowski
Stephan Falke   Pascal Lafourcade   Jack M. Pullikottil
Berndt Farwer   Ugo dal Lago   Femke van Raamsdonk
Jean-Christophe Filliâtre   Stéphane Lengrand   Jason Reed
Robby Findler   Jean-Jacques Lévy   Enric
Bernd Fischer   Jordi Levy   Rodríguez-Carbonell
Julien Forest   Salvador Lucas   Albert Rubio
Kim Gabarro   Bas Luttik   Michel de Rougemont
John Gallagher   Ian Mackie   Michael Rusinowitch

Table of Contents

Invited Talk

Session 4. Lambda Calculus

Session 5. Theorem Proving

Session 6. System Descriptions

Invited Talk

Session 7. Termination

Session 8. Higher-Order Rewriting and Unification

Formal Verification of Infinite State Systems Using Boolean Methods*

Randal E. Bryant

School of Computer Science, Carnegie Mellon University, Pittsburgh, PA
Randy.Bryant@cs.cmu.edu

Most successful automated formal verification tools are based on a bit-level model of computation, where a set of Boolean state variables encodes the system state. Using powerful inference engines, such as Binary Decision Diagrams (BDDs) and Boolean satisfiability (SAT) checkers, symbolic model checkers and similar tools can analyze all possible behaviors of very large, finite-state systems.

For many hardware and software systems, we would like to go beyond bit-level models to handle systems that are truly infinite state, or that are better modeled as infinite-state systems. Examples include programs manipulating integer data, concurrency protocols involving arbitrary numbers of processes, and systems containing buffers where the sizes are described parametrically.

Historically, much of the effort in verifying such systems involved automated theorem provers, requiring considerable guidance and expertise on the part of the user. We would like to devise approaches for these more expressive system models that retain the desirable features of model checking, such as the high degree of automation and the ability to generate counterexamples.

We have developed UCLID [1], a prototype verifier for infinite-state systems. The UCLID modeling language extends that of SMV [9], a bit-level model checker, to include state variables that are integers, as well as functions mapping integers to integers and integers to Booleans. Functional state variables can be used to define array and memory structures, including arrays of identical processes, FIFO buffers, and content-addressable memories.

System operation is defined in UCLID in terms of the initial values and next-state functions of the state variables. Integer operations include linear arithmetic and relational operations. Functions can be defined using uninterpreted function symbols, as well as via a restricted form of lambda expression. The underlying logic is reasonably expressive, yet it still permits a decision procedure that translates the formula into propositional logic and then uses a SAT solver [7].

UCLID supports multiple forms of verification, requiring different levels of sophistication in the handling of quantifiers. All styles verify that a safety property of the form $\forall \mathcal{X} P(s)$ holds for some set of system states s, where $\mathcal{X}$ denotes a set of integer *index variables*. Index variables can be used to express universal properties for all elements in an array of identical processes, all entries in a FIFO buffer, etc.

The simplest form of *bounded property checking* allows the user to determine that property $\forall \mathcal{X} P(s)$ holds for all states reachable within a fixed number of steps k from an

---

* This research was supported by the Semiconductor Research Corporation, Contract RID 1029.001.

F. Pfenning (Ed.): RTA 2006, LNCS 4098, pp. 1–3, 2006.

initial state. Verifying such a property can be done by direct application of the decision procedure. In practice, the effort required to verify such a property grows exponentially in k, limiting the verification to around 10–20 steps. However, it provides a useful debugging tool. In our experience, most errors are detected by this approach.

Of course, it is important to verify that properties hold for all reachable states of the system. Unfortunately, the standard fixed-point methods for bit-level model checking do not work for infinite-state systems. In many cases, the system will not reach a fixed point within a bounded number of steps. Even for those that do, checking convergence is undecidable, and our efforts to implement incomplete methods for this task have had limited success [2].

To prove that property $\forall \mathcal{X} P(s)$ holds for all reachable states s, UCLID supports *inductive invariant* checking, where the user provides an invariant Q such that Q holds for all initial states, Q implies P, and any successor for a state satisfying Q must also satisfy Q. This latter condition requires proving the validity of a formula containing existentially quantified index variables. Although this problem is undecidable for our logic, we have successfully implemented an incomplete approach using quantifier instantiation [8].

A more automated technique is to derive an inductive invariant via *predicate abstraction* [4]. Predicate abstraction operates much like the fixed-point methods of symbolic model checking, but using the concretization and abstraction operations of abstract interpretation [3] on each step. We have generalized predicate abstraction to handle the indexed predicates supported by UCLID [6]. Each step requires quantifier elimination to eliminate the current state variables, much like the relational product step of symbolic model checking. We implement this step by performing SAT enumeration on the translated Boolean formula.

As a final level of automation, we can automatically discover a set of relevant predicates for predicate abstraction based on the property P and the next-state expressions for the state variables [5].

We have successfully verified a number of systems with UCLID, including out-of-order microprocessors, distributed cache protocols, and distributed synchronization protocols.

References

1. R. E. Bryant, S. K. Lahiri, and S. A. Seshia. Modeling and verifying systems using a logic of counter arithmetic with lambda expressions and uninterpreted functions. In E. Brinksma and K. G. Larsen, editors, *Computer-Aided Verification (CAV '02)*, LNCS 2404, pages 78–92, 2002.
2. R. E. Bryant, S. K. Lahiri, and S. A. Seshia. Convergence testing in term-level bounded model checking. In *Correct Hardware Design and Verification Methods (CHARME '03)*, LNCS, September 2003.
3. P. Cousot and R. Cousot. Abstract interpretation : a unified lattice model for the static analysis of programs by construction or approximation of fixpoints. In *Principles of Programming Languages (POPL '77)*, pages 238–252, 1977.
4. S. Graf and H. Saïdi. Construction of abstract state graphs with PVS. In O. Grumberg, editor, *Computer-Aided Verification (CAV '97)*, LNCS 1254, pages 72–83, 1997.

5. S. K. Lahiri and R. E. Bryant. Indexed predicate discovery for unbounded system verification. In *Computer-Aided Verification (CAV '04)*, LNCS 3114, pages 135–147, 2004.
6. S. K. Lahiri and R. E. Bryant. Indexed predicate abstraction. *ACM Transactions on Computational Logic*, To appear.
7. S. K. Lahiri and S. A. Seshia. The UCLID decision procedure. In *Computer-Aided Verification (CAV '04)*, LNCS 3114, pages 475–478, 2004.
8. S. K. Lahiri, S. A. Seshia, and R. E. Bryant. Modeling and verification of out-of-order microprocessors in UCLID. In M. D. Aagaard and J. W. O'Leary, editors, *Formal Methods in Computer-Aided Design (FMCAD '02)*, LNCS 2517, pages 142–159, 2002.
9. K. McMillan. *Symbolic Model Checking*. Kluwer Academic Publishers, 1992.

Solving Partial Order Constraints
for LPO Termination

Michael Codish[1,*], Vitaly Lagoon[2], and Peter J. Stuckey[2,3]

[1] Department of Computer Science, Ben-Gurion University, Israel
[2] Department of Computer Science and Software Engineering
The University of Melbourne, Australia
[3] NICTA Victoria Laboratory
mcodish@cs.bgu.ac.il, {lagoon, pjs}@cs.mu.oz.au

Abstract. This paper introduces a new kind of propositional encoding for reasoning about partial orders. The symbols in an unspecified partial order are viewed as variables which take integer values and are interpreted as indices in the order. For a partial order statement on n symbols each index is represented in $\lceil \log_2 n \rceil$ propositional variables and partial order constraints between symbols are modeled on the bit representations. We illustrate the application of our approach to determine LPO termination for term rewrite systems. Experimental results are unequivocal, indicating orders of magnitude speedups in comparison with current implementations for LPO termination. The proposed encoding is general and relevant to other applications which involve propositional reasoning about partial orders.

1 Introduction

This paper formalizes a propositional logic over partial orders. Formulæ in this logic are just like usual propositional formulæ except that propositions are statements about a partial order on a finite set of symbols. For example, $(f = g) \wedge ((f > h) \vee (h > g))$ is a formula in this logic. We refer to the formulæ of this logic as *partial order constraints*. There are many applications in computer science which involve reasoning about (the satisfiability of) partial order constraints. For example, in the contexts of termination analysis, theorem proving, and planning. The main contribution of this paper is a new kind of propositional encoding of partial order constraints in propositional logic.

Contemporary propositional encodings, such as the one considered in [13], model the atoms (primitive order relations such as $f = g$ or $f > h$ on symbols) in a partial order constraint as propositional variables. Then, propositional statements are added to encode the axioms of partial orders which the atoms are subject to. For a partial order constraint on n symbols, such encodings typically introduce $O(n^2)$ propositional variables and involve $O(n^3)$ propositional connectives to express the axioms. In contrast we propose to model the symbols in a partial order constraint as integer values (in binary representation). For n

---

* Research performed while visiting the University of Melbourne.

F. Pfenning (Ed.): RTA 2006, LNCS 4098, pp. 4–18, 2006.

symbols this requires $k = \lceil \log_2 n \rceil$ propositional variables for each symbol. The integer value of a symbol reflects its index in a total order extending the partial order. Constraints of the form $(f = g)$ or $(f > h)$ are then straightforward to encode in k-bit arithmetic and involve $O(\log n)$ connectives each.

We focus on the application to termination analysis for term rewrite systems (for a survey see [6]) and in particular on LPO termination [11,5]. Experimental results are unequivocal, surpassing the performance of current termination analyzers such as TTT [10,18] and AProVE [8,2] (configured for LPO). The underlying approach is directly applicable to more powerful termination proving techniques, such as those based on dependency pairs [1], which basically involve the same kind of constraint solving.

Sections 2 and 3 introduce partial order constraints and their symbol-based propositional encoding. Section 4 introduces the LPO termination problem and its relation to partial order constraints. Section 5 describes and evaluates our implementation for LPO termination which is based on the application of a state-of-the-art propositional SAT solver [14]. Finally, we present related work and conclusions.

2   Partial Order Constraints

Informally, a partial order constraint is just like a formula in propositional logic except that propositions are atoms of the form $(f > g)$ or $(f = g)$. The semantics of a partial order constraint is a set of models. A model is an assignment of truth values to atoms which is required to satisfy both parts of the formula: the "propositional part" and the "partial order part".

Syntax: Let $\mathcal{F}$ be finite non-empty set of symbols and $\mathcal{R} = \{>, =\}$ consist of two binary relation symbols on $\mathcal{F}$. Since $\mathcal{R}$ is fixed we denote by $Atom_{\mathcal{F}}$ the set of atoms of the form $(f\ R\ g)$ where $R \in \mathcal{R}$ and $f, g \in \mathcal{F}$. A partial order constraint on $\mathcal{F}$ is a propositional formula in which the propositions are elements of $Atom_{\mathcal{F}}$. We sometimes write $(f \geq g)$ as shorthand for $(f > g) \vee (f = g)$. We denote the set of atoms occurring in a partial order constraint φ by $Atom(\varphi)$.

Semantics: The symbols in $\mathcal{R}$ are interpreted respectively as a strict partial order and as equality (both on $\mathcal{F}$). Let φ be a partial order constraint on $\mathcal{F}$. The semantics of φ is a set of models. Intuitively, a model of φ is a set of atoms from $Atom_{\mathcal{F}}$ which satisfies both parts of the formula: the propositional part and the partial order part. Before presenting a formal definition we illustrate this intuition by example.

Example 1. Let $\mathcal{F} = \{f, g, h\}$. The following are partial order constraints:

$$\varphi_1 = (f > g) \wedge ((f > h) \vee (h > f))$$
$$\varphi_2 = (f \geq g) \wedge (g \geq h) \wedge (h \geq g)$$
$$\varphi_3 = (f > g) \wedge \neg ((h > g) \vee (f > h))$$

The set of atoms $\mu_1 = \{(f > g), (f > h), (f = f), (g = g), (h = h)\}$ is a model for φ_1. It satisfies the propositional part: φ_1 evaluates to true when assigning the atoms in μ the value "true" and the others the value "false". It satisfies the partial order part: it is a partial order. The set of atoms $\{(h > f), (f > g)\}$ is not a model (for any partial order constraint) because it is not closed under transitivity (nor reflexivity). However, its extension $\mu_2 = \{(h > f), (f > g), (h > g), (f = f), (g = g), (h = h)\}$ is a model for φ_1. Formula φ_1 has additional models which are extensions of μ_1 to a total order:

$$\mu_3 = \{(f > g), (g > h), (f > h), (f = f), (g = g), (h = h)\},$$
$$\mu_4 = \{(f > h), (h > g), (f > g), (f = f), (g = g), (h = h)\}, \text{ and}$$
$$\mu_5 = \{(f > g), (g = h), (h = g), (f > h), (f = f), (g = g), (h = h)\}$$

The formula φ_2 has two models:

$$\{(f = g), (g = f), (g = h), (h = g), (f = h), (h = f), (f = f), (g = g), (h = h)\}$$

$$\{(f > g), (g = h), (h = g), (f > h), (f = f), (g = g), (h = h)\}$$

Focusing on φ_3 illustrates that there is an additional implicit condition for an assignment to satisfy a partial order constraint. We recall that a partial order can always be extended to a total order. The partial order $\mu = \{(f > g)\}$ satisfies the propositional part of φ_3 and may appear at first sight to satisfy also the partial order part (it is a partial order). However, no extension of μ to a total order satisfies the propositional part of φ_3 and hence μ will not be considered a model of φ_3. To solve this, we will restrict models to be total orders.

The following formalizes the proposed semantics for partial order constraints.

Definition 1 (assignment, model). *An assignment μ is a mapping from propositions of $Atom_{\mathcal{F}}$ to truth values, and can be identified with the set of propositions it assigns "true". Let φ be a partial order constraint on $\mathcal{F}$. We say that an assignment μ is a model for φ if: (1) it makes φ true as a propositional formula; (2) it satisfies the axioms for strict partial order and equality; and (3) it defines a total order on $\mathcal{F}$. More specifically, in (2) and in (3), an assignment μ is required to satisfy (for all $f, g, h \in \mathcal{F}$):*

reflexivity:	$(f = f) \in \mu$
symmetry:	$(f = g) \in \mu \Rightarrow (g = f) \in \mu$
asymmetry:	$\neg((f > g) \in \mu \wedge (g > f) \in \mu)$
transitivity:	$(f > g) \in \mu \wedge (g > h) \in \mu \Rightarrow (f > h) \in \mu$
	$(f = g) \in \mu \wedge (g = h) \in \mu \Rightarrow (f = h) \in \mu$
identity:	$(f > g) \in \mu \wedge (g = h) \in \mu \Rightarrow (f > h) \in \mu$
	$(f = g) \in \mu \wedge (g > h) \in \mu \Rightarrow (f > h) \in \mu$
comparability:	$(f > g) \in \mu \vee (g > f) \in \mu \vee (f = g) \in \mu$

Given that we fix the models of a partial order constraint to be total orders, we have that $\neg(f > g) \equiv (g > f) \vee (g = f)$ and that $\neg(f = g) \equiv (f > g) \vee (g > f)$. Hence we may assume without loss of generality that partial order constraints are negation free. For example, the formula φ_3 from Example 1 is equivalent to $\varphi_3' = (f > g) \wedge (g \geq h) \wedge (h \geq f)$ which is clearly unsatisfiable.

Satisfiability: In this paper we are concerned with the question of satisfiability of partial order constraints: given a partial order constraint φ does it have a model? Similarly to the general SAT problem, the satisfiability of partial order constraints is NP-complete, and the reduction from SAT is straightforward.

Solution-based interpretation: We propose a finite domain integer-based interpretation of partial order constraints. In this approach the semantics of a partial order constraint is a set of integer solutions.

Definition 2 (integer assignment and solution). *Let φ be a partial order constraint on $\mathcal{F}$ and let $|\mathcal{F}| = n$. An integer assignment for φ is a mapping $\mu : \mathcal{F} \to \{1, \ldots, n\}$. An integer solution of φ is an assignment θ which makes φ true under the natural interpretations of $>$ and $=$ on the natural numbers.*

Example 2. Consider again the partial order constraints from Example 1. The assignments mapping $\langle f, g, h \rangle$ to $\langle 3, 2, 2 \rangle$, $\langle 3, 1, 1 \rangle$ and $\langle 1, 1, 1 \rangle$ are solutions for φ_2. But only the first two are solutions to φ_1. The formula φ_3 has no solutions.

Theorem 1. *A partial order constraint φ is satisfiable if and only if it has an integer solution.*

The theorem is a direct consequence of the following lemmata.

Lemma 1. *Let θ be a solution of φ. The assignment*

$$\mu = \big\{ (f \ R \ g) \,\big|\, \{f, g\} \subseteq \mathcal{F}, \ R \in \mathcal{R}, \ (\theta(f) \ R \ \theta(g)) \big\}$$

is a model of φ.

Proof. Clearly μ satisfies both the propositional and partial order parts of φ since the integer relation $>$ is a total order. Hence μ is a model for φ by definition.

Lemma 2. *Let μ be a model of φ on $\mathcal{F}$ with n symbols. Then there exists a solution θ of φ in $\{1, \ldots, n\}$.*

Proof. Assume $\mathcal{F} = \{f_1, \ldots, f_n\}$ and let μ be a model of φ. By asymmetry, identity and comparability, for each $1 \le i < j \le n$ exactly one of $f_i > f_j$ or $f_i = f_j$ or $f_j > f_i$ hold. We can linearize the symbols in $\mathcal{F}$: $f_{k_n} \ R_{n-1} \cdots R_2 \ f_{k_2} \ R_1 \ f_{k_1}$ where for each $1 \le i < n$, $R_i \in \{>, =\}$ and $(f_{k_{i+1}} \ R_i \ f_{k_i}) \in \mu$, since μ models transitivity, symmetry, and identity. We can then construct a solution θ, using values from 1 to no more than n, where

$$\theta(f_{k_1}) = 1$$
$$\theta(f_{k_{j+1}}) = \begin{cases} \theta(f_{k_j}) & \text{where } R_{j-1} \equiv (=) \\ \theta(f_{k_j}) + 1 & \text{where } R_{j-1} \equiv (>) \end{cases} \text{ for } 1 \le j < n$$

Decomposing partial order constraint satisfaction: The atoms in a formula φ induce a graph G_φ on the symbols in $\mathcal{F}$ such that φ is satisfiable if and only if the formulae corresponding to the strongly connected components of G_φ are all satisfiable. Considering this graph facilitates the decomposition of a test for satisfiability to a set of smaller instances. This graph captures all possible cycles in the partial order and hence all potential contradictions. The following definition is inspired by [13].

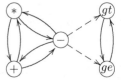

$$\varphi = ((gt > ge) \vee (- > ge)) \wedge ((ge > gt) \vee (- > gt)) \wedge$$
$$(((+ > *) \wedge (+ > -)) \vee (- > *)) \wedge$$
$$(((* > +) \wedge (* > -)) \vee (- > +)) \wedge (* > +)$$

Fig. 1. A partial order constraint (left) and its domain graph (right). The graph has two strongly connected components: $\{gt, ge\}$ and $\{-, *, +\}$. Arcs between the components are dashed.

Definition 3 (domain graph). *Let φ be a (negation free) partial order constraint on $\mathcal{F}$. The domain graph $G_\varphi = (V, E)$ is a directed graph with vertices $V = \mathcal{F}$ and edges $E = \{ (f, g) \mid \{ (f > g), (f = g), (g = f) \} \cap Atom(\varphi) \neq \emptyset \}$.*

Figure 1 illustrates a partial order constraint (a) and its domain graph (b).

Definition 4 (restricting a partial order constraint). *Let φ be a partial order constraint on $\mathcal{F}$ and let $\mathcal{F}' \subseteq \mathcal{F}$. The restriction of φ to the symbols in $\mathcal{F}'$ is the formula obtained by substituting "true" for any atom $(f \ R \ g)$ such that $(f, g) \notin \mathcal{F}' \times \mathcal{F}'$. The SCC-partition of φ is the set of graphs obtained by restricting φ to the nodes in each of the strongly connected components of G_φ.*

Example 3. Consider the partial order constraint φ and its domain graph G_φ depicted as Figure 1. The graph G_φ has two strongly connected components. The SCC-partition for φ gives:

$$\varphi_1 = ((gt > ge) \vee true) \wedge ((ge > gt) \vee true) \quad \equiv \quad true$$
$$\varphi_2 = (* > +) \wedge (((+ > *) \wedge (+ > -)) \vee (- > *)) \wedge$$
$$((((* > +) \wedge (* > -))) \vee (- > +))$$
$$\equiv (* > +) \wedge (- > *) \wedge (- > +)$$

Lemma 3. *A partial order constraint is satisfiable if and only if each of the formula in its SCC-partition is satisfiable.*

Proof. (idea) You can only get a contradiction if $x > x$ along some path in the graph. Any such path will be contained in a single SCC.

3   A Symbol-Based Propositional Encoding

This section presents a propositional encoding of partial order constraints. A partial order constraint φ on a set of symbols $\mathcal{F}$ is encoded by a propositional formula φ' such that each model of φ corresponds to a model of φ' and in particular such that φ is satisfiable if and only if φ' is. The novelty is to construct the encoding in terms of the solution-based interpretation of partial order constraints. We view the n symbols in $\mathcal{F}$ as integer variables taking finite domain values from the set $\{1, \ldots, n\}$. Each symbol is thus modeled using $k = \lceil \log_2 n \rceil$

propositional variables which encode the binary representation of its value. Constraints of the form $(f > g)$ or $(f = g)$ on $\mathcal{F}$ are interpreted as constraints on integers and it is straightforward to encode them in k-bit arithmetic.

The symbol-based propositional encoding for partial order constraints is defined as follows. For $|\mathcal{F}| = n$ we need $k = \lceil \log_2 n \rceil$ bits per symbol. We denote by $[\![a]\!]$ the propositional variable corresponding to an atom $a \in Atom_{\mathcal{F}}$ and by $[\![\varphi]\!]$ the propositional formula obtained when replacing atoms by propositional variables in partial order constraint φ.

1. For $f \in \mathcal{F}$, the k-bit representation is $f = \langle f_k, \ldots, f_1 \rangle$ with f_k the most significant bit.
2. A constraint of the form $(f = g)$ is encoded in k-bits by

$$\|(f = g)\|_k = \bigwedge_{i=1}^{k} (f_i \leftrightarrow g_i).$$

A constraint of the form $(f > g)$ is encoded in k-bits by

$$\|(f > g)\|_k = \begin{cases} (f_1 \wedge \neg g_1) & k = 1 \\ (f_k \wedge \neg g_k) \vee ((f_k \leftrightarrow g_k) \wedge \|(f > g)\|_{k-1}) & k > 1 \end{cases}$$

3. A partial order constraint φ is encoded in k bits by replacing each $[\![a]\!]$ in φ by its corresponding k-bit encoding $\|a\|_k$, which we write as:

$$\|\varphi\|_k = [\![\varphi]\!]_{[\![a]\!]/\|a\|_k} \tag{1}$$

Proposition 1. *The symbol-based encoding of partial order constraint φ with n symbols involves $O(n \log n)$ propositional variables and $O(|\varphi| \log n)$ connectives.*

Example 4. Consider the partial order constraint $\varphi_2 = (* > +) \wedge (- > *) \wedge (- > +)$ from Example 3. Each of the three symbols in φ_2 is represented in 2 bits and the propositional encoding of φ_2 is obtained as

$$\varphi_2' = ((*_2 \wedge \neg +_2) \vee (*_2 \leftrightarrow +_2 \wedge *_1 \wedge \neg +_1)) \wedge$$
$$((-_2 \wedge \neg *_2) \vee (-_2 \leftrightarrow *_2 \wedge -_1 \wedge \neg *_1)) \wedge$$
$$((-_2 \wedge \neg +_2) \vee (-_2 \leftrightarrow +_2 \wedge -_1 \wedge \neg +_1))$$

The proof of the following theorem is straightforward.

Theorem 2. *A partial order constraint φ on symbols $\mathcal{F}$ is satisfiable if and only if its symbol-based propositional encoding $encode(\varphi)$ is.*

4 LPO Termination

A term rewrite system is a set of rules of the form $\ell \to r$ where ℓ and r are terms constructed from given sets of symbols $\mathcal{F}$ and variables $\mathcal{V}$, and such that ℓ is not

$$-gt(A, B) \rightarrow ge(B, A)$$
$$-ge(A, B) \rightarrow gt(B, A)$$
$$-(A + B) \rightarrow (-A) * (-B)$$

$$-(A * B) \rightarrow (-A) + (-B)$$
$$A * (A + B) \rightarrow (A * B) + (A * C)$$
$$(B + C) * A \rightarrow (B * A) + (C * A)$$

Fig. 2. An example term rewrite system: normalizing formulæ with propositional connectives: $*, +, -$ (for: and, or, not); and partial orders: gt, ge (for: $>, \geq$)

a variable and r only contains variables also in ℓ. A rule $\ell \rightarrow r$ applies to a term t if a subterm s of t matches ℓ with some substitution σ (namely, $s = \ell\sigma$). The rule is applied by replacing the subterm s by $r\sigma$. Such an application is called a rewrite step on t. A derivation is a sequence of rewrite steps. A term rewrite system is said to be terminating if all of its derivations are finite. An example term rewrite system is depicted as Figure 2.

Termination of term rewrite systems is undecidable. However a term rewrite system terminates if there is a reduction order $\succ$ such that $\ell \succ r$ for each rule $\ell \rightarrow r$ in the system. There are many methods for defining such orders. Many of them are based on so-called simplification orders and one such order is the lexicographic path order (LPO)[11,5].

We assume an algebra of terms constructed over given sets of symbols $\mathcal{F}$ and variables $\mathcal{V}$. Let $>_{\mathcal{F}}$ denote a (strict or non-strict) partial order on $\mathcal{F}$ (a so-called *precedence*) and let $\approx_{\mathcal{F}}$ denote the corresponding equivalence relation. We denote by $\sim$ the equality of terms up to equivalence of symbols. Observe that if $>_{\mathcal{F}}$ is strict then $\approx_{\mathcal{F}}$ and $\sim$ are the identity of symbols and terms respectively. Each precedence $>_{\mathcal{F}}$ on the symbols induces a lexicographic path order $\succ_{lpo}$ on terms. If for each of the rules $\ell \rightarrow r$ in a system, $\ell \succ_{lpo} r$ then the system is LPO terminating.

Definition 5 (LPO [11]). *The lexicographic path order $\succ_{lpo}$ on terms induced by the partial order $>_{\mathcal{F}}$ is defined as $s = f(s_1, \ldots, s_n) \succ_{lpo} t$ if and only if one of the following holds:*

1. *$t = g(t_1, \ldots, t_m)$ and $s \succ_{lpo} t_j$ for all $1 \leq j \leq m$ and either*
   (i) $f >_{\mathcal{F}} g$ or   (ii) $f \approx_{\mathcal{F}} g$ and $\langle s_1, \ldots, s_n \rangle \succ_{lpo}^{lex} \langle t_1, \ldots, t_m \rangle$; or
2. *$s_i \succsim_{lpo} t$ for some $1 \leq i \leq n$.*

Here $\succ_{lpo}^{lex}$ is the lexicographic extension of $\succ_{lpo}$ to tuples of terms and $\succsim_{lpo}$ is the union of $\succ_{lpo}$ and $\sim$.

The LPO termination problem is to determine for a given term rewrite system with function symbols $\mathcal{F}$, if there exists a partial order $>_{\mathcal{F}}$ such that $\ell \succ_{lpo} r$ for each of the rules with the induced lexicographic path order. There are two variants of the problem: "strict-" and "quasi-LPO termination" depending on if we require $>_{\mathcal{F}}$ to be strict or not. The corresponding decision problems, strict- and quasi- LPO termination, are decidable and NP complete [12]. In [9], the authors observe that finding $>_{\mathcal{F}}$ such that $s \succ_{lpo} t$ is tantamount to solving a constraint obtained by unfolding the definition of $s \succ_{lpo} t$ with details depending on whether $>_{\mathcal{F}}$ is a strict or non-strict partial order. The strict- and quasi-LPO termination problems are to decide if conjunctions of these unfoldings are satisfiable — one conjunct for each rule in the given term rewrite system.

Example 5. Consider the term rewrite system of Figure 2. Unfolding Definition 5 for strict-LPO termination, we obtain the following:

$$-(gt(A,B)) >_{lpo} ge(B,A) \Longleftrightarrow (gt > ge) \vee (- > ge)$$
$$-(ge(A,B)) >_{lpo} gt(B,A) \Longleftrightarrow (ge > gt) \vee (- > gt)$$
$$-(A+B) >_{lpo} (-(A)) * (-(B)) \Longleftrightarrow ((+ > *) \wedge (+ > -)) \vee (- > *)$$
$$-(A*B) >_{lpo} (-(A)) + (-(B)) \Longleftrightarrow ((* > +) \wedge (* > -)) \vee (- > +)$$
$$A * (B+C) >_{lpo} (A*B) + (A*C) \Longleftrightarrow * > +$$
$$(B+C) * A >_{lpo} (B*A) + (C*A) \Longleftrightarrow * > +$$

The term rewrite system is LPO terminating if and only if the conjunction of the constraints on the right sides is satisfiable. This conjunction is precisely the partial order constraint φ from Figure 1 which by Lemma 3, is satisfiable if and only if the formula in its SCC-partition are. Coming back to Example 3, it is straightforward to observe that they are.

The next example illustrates a term rewrite system which is quasi-LPO terminating but not strict-LPO terminating.

Example 6. Consider the following term rewrite system.

$$div(X,e) \rightarrow i(X)$$
$$i(div(X,Y)) \rightarrow div(Y,X)$$
$$div(div(X,Y),Z) \rightarrow div(Y,div(i(X),Z))$$

Unfolding Definition 5 for strict-LPO gives

$$div(X,e) >_{lpo} i(X) \Longleftrightarrow div > i$$
$$i(div(X,Y)) >_{lpo} div(Y,X) \Longleftrightarrow i > div$$
$$div(div(X,Y),Z) >_{lpo} div(Y,div(i(X),Z)) \Longleftrightarrow div > i$$

The conjunction of the constraints on the right sides is not satisfiable indicating that there does not exist any strict partial order on $\mathcal{F}$ such that the corresponding lexicographic path order decreases on the three rules. The system is however quasi-LPO terminating. Unfolding Definition 5 for quasi-LPO gives a satisfiable partial order constraint equivalent to $(div \geq i) \wedge (i \geq div)$ which indicates that taking $div \approx i$ provides a proof of quasi-LPO termination.

5   Implementation and Experimentation

We have implemented a prototype analyzer, poSAT, for strict- and quasi- LPO termination based on the encoding proposed in Section 3. The implementation is a written primarily in SWI-Prolog [19,15] and interfaces the MiniSat solver [7,14] for solving SAT instances.

We have integrated MiniSat and SWI-Prolog through $\approx$190 lines of C-code and $\approx$140 lines of Prolog code. For details concerning this interface see [3]. SAT solvers typically consider propositional formulæ in conjunctive normal form. The transformation of a propositional formula with m connectives and n literals is performed using a (linear) Tseitin transformation [17] (for details on our implementation see [3]) and results in a conjunctive normal form with $O(m + n)$ variables and $O(m)$ clauses.

The rest of poSAT is implemented in $\approx$800 lines of Prolog code. This includes a parser (for term rewrite systems), modules to translate strict- and quasi- LPO termination problems into partial order constraints, the module converting partial order constraints into SAT instances, and finally a head module processing the command line, running the components, pretty-printing the results etc. The current implementation does not decompose partial order constraints to their SCC-components (Lemma 3). The experimental results indicate that the implementation would not benefit from that: (a) Most of the tests are very fast without this decomposition; and (b) It is typical for hard cases of LPO termination (see Table 2) to have a large strongly connected component including the majority of the symbols.

For experimentation we have taken all 751 term rewrite systems from the Termination Problem Data Base [16] which do not specify a "theory" or a "strategy". In the following, the names of term rewrite systems are indicated in typewriter font and can be found in [16]. We report on the comparison of poSAT for both strict- and quasi- LPO termination analysis with the TTT analyzer[18].

For the experiments, poSAT runs on a 1.5GHz laptop running GNU/Linux FC4. The TTT analyzer is applied via its Web interface [18] and runs on a Xeon 2.24GHz dual-CPU platform which is a considerably faster machine than ours. Experiments with AProVE running on our local (laptop) platform give results which are considerably slower than TTT (on its faster machine). For example, running AProVE configured for LPO-termination with a 10 minute timeout on the 25 examples highlighted in Table 2 takes 47 minutes and encounters three timeouts. In contrast TTT analyzes the same set of examples in about one minute (see Table 2). Hence for comparison with poSAT we provide the numbers only for TTT.

With regards to precision, as expected, both analyzers give the same results (with the exception of a single test which TTT cannot handle within the maximum timeout allocation). From the 751 example systems, 128 are LPO terminating and 132 are quasi-LPO terminating. For poSAT, run times include the complete cycle of processing each test: reading and parsing the file, translation to partial order constraints and then to propositional formula, solving by the SAT solver and printing the results. The run time of each test is computed as an average of ten identical runs.

Table 1(a) summarizes the results for strict LPO termination analysis. The columns contain times (in seconds) for our poSAT analyzer and for TTT configured to run with a timeout of 10 minutes (the maximum allowed by its Web interface). Note that the times are taken on different machines which makes the precise comparison impossible. Nevertheless, the results are indicative showing that poSAT is fast in absolute terms and scales better for hard cases. Notably, the hardest test of LPO termination for poSAT (`HM/t005.trs`) completes in

Table 1. Summary of experimental results: total, average and maximum times (sec) for 751 tests

	poSAT	TTT
Total	8.983	647.48
Average	0.012	0.86
Max	0.477	317.63

(a) strict LPO termination

	poSAT	TTT
Total	8.609	2167.44
Average	0.011	2.89
Max	0.544	600.00

(b) quasi-LPO termination

under a half second, while the hardest test for TTT (`various_14.trs`) takes more than 5 minutes.

Table 1(b) presents the results for quasi-LPO termination analysis. For this variant, poSAT completes the 751 tests in 8.6sec. The same task takes TTT over 34 *minutes* with one test (`currying/Ste92/hydra.trs`) running out of 10 minutes timeout. The next hardest test for TTT is `currying/AG01_No_3.13.trs` which completes in 182.6sec (3min). The same two tests take poSAT 0.054sec and 0.021sec respectively. The hardest quasi-LPO test for poSAT is `Zantema/z30` which takes 0.54sec in our analyzer and 5.02sec in TTT.

Once again, the timings are indicative despite the fact that the two analyzers run on different machines. By comparing the results in Table 1(a) and (b) we observe that for quasi-LPO, TTT runs about an order of magnitude slower than for strict LPO. In contrast, poSAT demonstrates similar performance for both LPO and quasi-LPO.

Table 2 presents a detailed analysis for the 25 most challenging examples for poSAT chosen by maximum total time for strict- and quasi- LPO analysis. The two parts of the table present the respective results for strict- and quasi-LPO termination analyses. The following information is provided: The columns labeled "Sym" and "CNF" characterize the partial order constraints derived from the given term rewrite systems. "Sym" indicates the number of symbols in the complete formula and in the largest component of its SCC-partition (0/0 in this column means that the partial order constraint is trivial i.e., true or false). "CNF" indicates the numbers of propositional variables and clauses in the translation of the propositional (symbol-based) encoding to conjunctive normal form. The columns labeled "poSAT" and "TTT" indicate run times (in seconds) for the poSAT and TTT solvers.

All of the tests in Table 2 are neither strict- nor quasi-LPO terminating. This is not surprising for the 25 hardest tests, as proving unsatisfiability is typically harder than finding a solution for a satisfiable formula. It is interesting to note that four examples among the hardest 25, result in trivial partial order constraints. Obviously, the challenge in these examples is not in solving the constraints but rather in obtaining them by unfolding Definition 5. Interestingly, our translation and simplification mechanisms are sometimes more powerful than those of TTT. For instance, `currying/AG01_No_3.13` is simplified to false in poSAT but not in TTT, leading to a long search for TTT. The difference is due to the fact that in the case of poSAT the generation of a partial order formula

Table 2. The 25 hardest tests for poSAT

Test	LPO				quasi-LPO			
	Sym	CNF	poSAT	TTT	Sym	CNF	poSAT	TTT
AProVE_AAECC-ring	28/10	642/2369	0.088	0.11	28/25	786/2951	0.093	0.12
Cime_mucrl1	0/0	0/1	0.298	2.56	0/0	0/1	0.248	13.88
currying_AG01_No_3.13	0/0	0/1	0.127	39.24	0/0	0/1	0.027	184.24
higher-order_Bird_H*	0/0	0/1	0.089	0.15	0/0	0/1	0.025	1.30
HM_t005	0/0	0/1	0.477	11.75	0/0	0/1	0.040	2.13
HM_t009	19/11	773/2779	0.167	0.14	19/18	1388/4880	0.175	0.16
Ex1_2_AEL03_C	19/17	630/2301	0.115	0.23	19/19	1286/4877	0.141	88.30
Ex1_2_AEL03_GM	22/17	506/1805	0.058	0.04	22/22	693/2475	0.060	19.39
Ex26_Luc03b_C	15/12	384/1307	0.055	0.08	15/15	816/2847	0.079	6.00
Ex2_Luc02a_C	15/12	390/1332	0.063	0.08	15/15	838/2939	0.086	6.09
Ex3_3_25_Bor03_C	12/10	285/945	0.050	0.06	12/12	605/2100	0.061	0.72
Ex4_7_37_Bor03_C	13/11	287/962	0.061	0.11	13/13	577/2067	0.072	0.83
Ex5_7_Luc97_C	18/15	614/2181	0.093	0.15	18/18	1341/4871	0.139	92.51
Ex6_15_AEL02_C	23/22	906/3312	0.159	0.37	23/23	1862/6756	0.215	123.47
Ex6_15_AEL02_FR	26/20	599/2146	0.060	0.05	26/26	867/3152	0.065	40.01
Ex6_15_AEL02_GM	29/25	745/2761	0.079	0.07	29/29	1074/3920	0.099	155.26
Ex6_15_AEL02_Z	26/20	587/2105	0.060	0.05	26/26	869/3196	0.061	18.31
Ex7_BLR02_C	14/11	299/1013	0.044	0.07	14/14	627/2289	0.064	1.70
Ex8_BLR02_C	12/10	280/930	0.048	0.07	12/12	546/1906	0.060	0.38
Ex9_BLR02_C	12/9	296/968	0.054	0.06	12/12	608/2071	0.065	0.37
ExAppendixB_AEL03_C	20/18	700/2576	0.121	0.29	20/20	1410/5294	0.152	109.39
ExIntrod_GM99_C	16/13	423/1428	0.080	0.11	16/16	848/3017	0.088	21.36
ExIntrod_Zan97_C	15/12	344/1167	0.051	0.08	15/15	709/2544	0.069	2.02
ExSec11_1_Luc02a_C	16/13	439/1490	0.067	0.12	16/16	985/3353	0.098	29.32
Zantema_z30	2/2	65/106	0.119	2.91	3/3	12827/18205	0.544	5.02
Total time:			2.683	58.95			2.83	922.30

never introduces trivial sub-formula ("true" or "false"), these are evaluated on-the-fly.

Another observation based on the results of Table 2 is that the partial order constraints derived from the tests typically have domain graphs with large strongly-connected components. Almost every test in the table has a "core" component including the majority of the symbols. Therefore, it is unlikely that the performance of poSAT for the presented tests can be improved by using the SCC-based decomposition of the formula.

As Table 2 shows, the maximum CNF instance solved in our tests includes 12827 propositional variables and 18205 CNF clauses. This is well below the capacity limits of MiniSat, which is reported to handle benchmarks with hundreds of thousands of variables and clauses [14].

6   Related Works

The idea of mapping LPO termination problems to a corresponding propositional formula is first addressed in [13] where the authors assume that partial order constraints contain only disjunction and conjunction of atoms of the form $(f > g)$ (no equality and no negation). This suffices for strict-LPO termination analysis. We present here a generalization of that approach which can be applied also for quasi-LPO termination and then compare it with the approach proposed in this paper.

The basic strategy is the same as in Section 3: to encode a partial order constraint φ on $\mathcal{F}$ by an equivalent propositional formula φ' such that each model of φ corresponds to a model of φ' and in particular such that φ is satisfiable if and only if φ' is. The main difference is that the approach in [13] is "atom-based". The encoding for a partial order constraint φ is obtained by: (a) viewing the atoms in φ as propositional variables, and (b) making the axioms for partial order explicit. As in Section 3, we let $[\![a]\!]$ denote the propositional variable corresponding to an atom $a \in Atom_{\mathcal{F}}$ and $[\![\varphi]\!]$ the propositional formula obtained by replacing each atom a in partial order constraint φ by the propositional variable $[\![a]\!]$. For a set of symbols $\mathcal{F}$ the following propositional formulæ make the axioms explicit:

- $R_{\mathcal{F}}^{=} = \bigwedge_{f \in \mathcal{F}} [\![f = f]\!]$

- $A_{\mathcal{F}}^{>} = \bigwedge_{f,g \in \mathcal{F}} \neg([\![f > g]\!] \wedge [\![g > f]\!])$

- $T_{\mathcal{F}}^{=} = \bigwedge_{\substack{f,g,h \in \mathcal{F} \\ f \neq g \neq h \neq f}} [\![f = g]\!] \wedge [\![g = h]\!] \to [\![f = h]\!]$

- $I_{\mathcal{F}}^{2} = \bigwedge_{\substack{f,g,h \in \mathcal{F} \\ f \neq g \neq h \neq f}} [\![f = g]\!] \wedge [\![g > h]\!] \to [\![f > h]\!]$

- $S_{\mathcal{F}}^{=} = \bigwedge_{f,g \in \mathcal{F}} [\![f = g]\!] \to [\![g = f]\!]$

- $T_{\mathcal{F}}^{>} = \bigwedge_{\substack{f,g,h \in \mathcal{F} \\ f \neq g \neq h \neq f}} [\![f > g]\!] \wedge [\![g > h]\!] \to [\![f > h]\!]$

- $I_{\mathcal{F}}^{1} = \bigwedge_{\substack{f,g,h \in \mathcal{F} \\ f \neq g \neq h}} [\![f > g]\!] \wedge [\![g = h]\!] \to [\![f > h]\!]$

- $C_{\mathcal{F}}^{\geq} = \bigwedge_{\substack{f,g \in \mathcal{F} \\ f \neq g \neq h \neq f}} [\![f > g]\!] \vee [\![g > f]\!] \vee [\![f = g]\!]$

The atom-based propositional encoding of a (negation free) partial order constraint φ on symbols $\mathcal{F}$ which does not involve equality is obtained as $encode(\varphi) = [\![\varphi]\!] \wedge T_{\mathcal{F}}^{>} \wedge A_{\mathcal{F}}^{>}$ [13]. In the general case when φ may contain also equality the encoding is obtained as

$$encode(\varphi) = [\![\varphi]\!] \wedge R_{\mathcal{F}}^{=} \wedge S_{\mathcal{F}}^{=} \wedge T_{\mathcal{F}}^{>} \wedge T_{\mathcal{F}}^{=} \wedge A_{\mathcal{F}}^{>} \wedge I_{\mathcal{F}}^{1} \wedge I_{\mathcal{F}}^{2} \wedge C_{\mathcal{F}}^{\geq} \qquad (2)$$

The two variants of atom-based propositional encodings both result in large propositional formula. For $|\mathcal{F}| = n$ they introduce $O(n^2)$ propositional variables and involve $O(n^3)$ connectives (e.g., for transitivity).

In [13] Kurihara and Kondo propose two optimizations. They note that for a given formula φ, the domain graph G_{φ} is often sparse and hence they propose to specialize the explicit representation of the axioms for those symbols from $\mathcal{F}$ actually occurring in φ. However, in view of Lemma 3 we may assume that we are testing satisfiability for partial order constraints which have strongly connected domain graphs. Moreover, as indicated by our experimental evaluation (see Table 2), the domain graphs for some of the more challenging examples have strongly connected components with up to 30 symbols.

In a second optimization Kurihara and Kondo observe that the axioms for transitivity and asymmetry can be replaced by a simpler axiom (they call it A^*)

introducing a single clause of the form $\neg((f_1 > f_2) \wedge (f_2 > f_3) \wedge \cdots \wedge (f_{k-1} > f_k) \wedge (f_k > f_1)$ for each simple cycle $(f_1 > f_2), (f_2 > f_3), \ldots, (f_{k-1} > f_k), (f_k > f_1)$ in G_φ to assert that that cycle is not present in a model. They claim correctness of the encoding and report considerable speedups when it is applied. The problem with this approach is that in general there may be an exponential number of simple cycles to consider.

Hence, the encoding described in [13] either requires $O(n^2)$ propositional variables and introduces $O(n^3)$ connectives or else relies on a potentially exponential phase of processing the simple loops in the domain graph.

It is insightful to compare the two encodings of a partial order constraint φ given as Equations (1) and (2). The common part in both encodings is the subformula $[\![\varphi]\!]$ in which atoms are viewed as propositional variables. The difference is that Equation (2) introduces explicit axioms to relate the atoms in a partial order where Equation (1) interprets the n symbols as indices represented in $\lceil \log_2 n \rceil$-bits. This is why the symbol-based encoding introduces $O(n \log n)$ propositional variables instead of $O(n^2)$ for the atom-based approach. Moreover the symbol-based encoding does not require the expensive encoding of the axioms because the encoding as integers ensures that they hold "for free". Hence the number of connectives is $O(|\varphi| \log n)$ instead of $O(n^3 + |\varphi|)$. Obviously for small n the symbol based encoding can be larger than the atom-based encoding. However, the search space is determined by the number of variables, where the $O(n \log n)$ of the symbol-based encoding is clearly superior to the $O(n^2)$ for the atom-based approach.

An implementation of the atom-based approach of [13] is described in the recent report [20] together with an experimental evaluation and comparison with our symbol-based approach. It shows the symbol-based approach is orders of magnitude faster on its benchmark set.

Testing for satisfiability of partial order constraints comes up in many other applications. First of all in the context of term rewrite systems where LPO is just one example of a simplification order and analyses based on other types of orders may also be encoded into propositional logic. Moreover, for programs which cannot be shown to terminate using these kinds of simplification orders, the dependency pairs approach [1] has proven very successful in generating sets of constraints such that the existence of a (quasi-)order satisfying them is a sufficient condition for termination. Our constraint solving technique is directly applicable and improves considerably the performance of implementations for these techniques. Initial results are described in [4].

In practice LPO termination tests are often performed in an incremental fashion, adding constraints to orient the rules in a term rewrite system one rule at a time. Methods that construct a partial order thus seek to incrementally extend that partial order if possible. In our approach, we construct a linearization of the partial order and are hence less likely to be able to extend a previous order to satisfy new constraints. However, both approaches make choices which may have to be undone to satisfy all constraints. For poSAT, the encoding (which often takes a good proportion of the analysis time) is clearly incremental. Moreover, given the raw speed advantages, and the fact that the hardest instances are unsatisfiable, where incrementality is not useful, we are confident that in an incremental context the poSAT is still superior.

7   Conclusion

We have introduced a new kind of propositional encoding for reasoning about partial orders. Previous works propose to represent the atoms in a formula as propositional variables and to explicitly encode the axioms for partial order. Our novel approach is to interpret the symbols in a formula as finite domain variables corresponding to the indices in the partial order. We illustrate the application of our approach for LPO termination analysis for term rewrite systems. Experimental results are unequivocal indicating orders of magnitude speedups in comparison with current implementations for LPO termination analysis. The proposed technique is directly applicable to more powerful termination proving techniques, such as those based on dependency pairs [1], which basically involve the same kind of constraint solving.

Acknowledgment

We are grateful to Bart Demoen for useful insights regarding the implementation and to Samir Genaim who donated the Prolog parser for term rewrite systems. Jürgen Giesl and Aart Middeldorp assisted with the use of the AProVE and TTT analyzers. Yefim Dinitz and anonymous reviewers provided many useful comments on the presentation.

References

1. T. Arts and J. Giesl. Termination of term rewriting using dependency pairs. *Theoretical Computer Science*, 236(1-2):133–178, 2000.
2. Automated program verification environment (AProVe). http://www-i2.informatik.rwth-aachen.de/AProVE. Viewed December 2005.
3. M. Codish, V. Lagoon, and P. J. Stuckey. Logic programming with satisfiability. http://www.cs.bgu.ac.il/~mcodish/Papers/Sources/lpsat.pdf, (submitted).
4. M. Codish, P. Schneider-Kamp, V. Lagoon, R. Thiemann, and J. Giesl. SAT solving for argument filterings. Available from http://arxiv.org/abs/cs.OH/0605074.
5. N. Dershowitz. Termination of rewriting. *J. Symb. Comput.*, 3(1/2):69–116, 1987.
6. N. Dershowitz and J.-P. Jouannaud. Rewrite systems. In J. van Leeuwen, editor, *Handbook of Theoretical Computer Science*, volume B: Formal Models and Semantics, pages 2435–320. Elsevier and MIT Press, 1990.
7. N. Eén and N. Sörensson. An extensible sat-solver. In E. Giunchiglia and A. Tacchella, editors, *Theory and Applications of Satisfiability Testing, 6th International Conference, SAT 2003 (Selected Revised Papers)*, volume 2919 of *LNCS*, pages 502–518. Springer, 2004.

8. J. Giesl, R. Thiemann, P. Schneider-Kamp, and S. Falke. Automated termination proofs with AProVE. In V. van Oostrom, editor, *Proc. of the 15th International Conference on Rewriting Techniques and Applications*, volume 3091 of *LNCS*, pages 210–220, Aachen, Germany, 2004. Springer.

9. N. Hirokawa and A. Middeldorp. Tsukuba termination tool. In R. Nieuwenhuis, editor, *Proc. of the 14th International Conference on Rewriting Techniques and Applications*, volume 2706 of *LNCS*, pages 311–320, Valencia, Spain, 2003.

10. N. Hirokawa and A. Middeldorp. Tyrolean termination tool. In *Proc. of the 16th International Conference on Rewriting Techniques and Applications*, volume 3467 of *LNCS*, pages 175–184, Nara, Japan, 2005. Springer.

11. S. Kamin and J.-J. Levy. Two generalizations of the recursive path ordering. Department of Computer Science, University of Illinois, Urbana, IL. Available at `http://www.ens-lyon.fr/LIP/REWRITING/OLD_PUBLICATIONS_ON_TERMINATION` (viewed December 2005), 1980.

12. M. Khrishnamoorthy and P. Narendran. On recursive path ordering. *Theoretical Computer Science*, 40:323–328, 1985.

13. M. Kurihara and H. Kondo. Efficient BDD encodings for partial order constraints with application to expert systems in software verification. In *Innovations in Applied Artificial Intelligence, 17th International Conference on Industrial and Engineering Applications of Artificial Intelligence and Expert Systems, Proceedings*, volume 3029 of *LNCS*, pages 827–837, Ottawa, Canada, 2004. Springer.

14. MiniSAT solver. `http://www.cs.chalmers.se/Cs/Research/FormalMethods/MiniSat`. Viewed December 2005.

15. Swi-prolog. `http://http://www.swi-prolog.org`. Viewed December 2005.

16. The termination problems data base. `http://www.lri.fr/~marche/tpdb/`. Viewed December 2005.

17. G. Tseitin. On the complexity of derivation in propositional calculus. In *Studies in Constructive Mathematics and Mathematical Logic*, pages 115–125. 1968. Reprinted in J. Siekmann and G. Wrightson (editors), Automation of Reasoning, vol. 2, pp. 466-483, Springer-Verlag Berlin, 1983.

18. Tyrolean termination tool. `http://cl2-informatik.uibk.ac.at/ttt`. Viewed December 2005.

19. J. Wielemaker. An overview of the SWI-Prolog programming environment. In F. Mesnard and A. Serebenik, editors, *Proceedings of the 13th International Workshop on Logic Programming Environments*, pages 1–16, Heverlee, Belgium, Dec. 2003. Katholieke Universiteit Leuven. CW 371.

20. H. Zankl. Sat techniques for lexicographic path orders. `http://arxiv.org/abs/cs.SC/0605021`, May 2006.

Computationally Equivalent Elimination of Conditions

Traian Florin Şerbănuţă and Grigore Roşu

Department of Computer Science,
University of Illinois at Urbana-Champaign
{tserban2, grosu}@cs.uiuc.edu

Abstract. An automatic and easy to implement transformation of conditional term rewrite systems into computationally equivalent unconditional term rewrite systems is presented. No special support is needed from the underlying unconditional rewrite engine. Since unconditional rewriting is more amenable to parallelization, our transformation is expected to lead to efficient concurrent implementations of rewriting.

1 Introduction

Conditional rewriting is a crucial paradigm in algebraic specification, since it provides a natural means for executing equational specifications. Many specification languages, including CafeOBJ [8], ELAN [4], Maude [6], OBJ [10], ASF/SDF [21], provide conditional rewrite engines to execute and reason about specifications. It also plays a foundational role in functional logic programming [11]. Conditional rewriting is, however, rather inconvenient to implement directly. To reduce a term, a rewrite engine needs to maintain a *control context* for each conditional rule that is tried. Due to the potential nesting of rule applications, such a control context may grow arbitrarily. The technique presented in this paper automatically translates conditional rewrite rules into unconditional rules, by *encoding the necessary control context into data context*. The obtained rules can be then executed on any unconditional rewrite engine, whose single task is to *match-and-apply* unconditional rules. Such a simplified engine can be seen as a *rewrite virtual machine*, which can be even implemented in hardware, and our transformation technique can be seen as a compiler. One can also simulate the proposed transformation as part of the implementation of a conditional engine.

Experiments performed on three fast rewrite engines, Elan[4], Maude[6] and ASF/SDF [21], show that performance increases can be obtained on current engines if one uses the proposed transformation as a front-end. However, since these rewrite engines may be optimized for conditional rewriting, we expect significant further increases in performance if one just focuses on the much simpler problem of developing optimized *unconditional* rewrite engines and use our technique. Moreover, one can focus on developing *parallel rewrite machines* without worrying about conditions, which obstruct the potential for high parallelism.

On computational equivalence. Let us formalize the informal notion of "computationally equivalent elimination of conditions". Consider a conditional term

F. Pfenning (Ed.): RTA 2006, LNCS 4098, pp. 19–34, 2006.
© Springer-Verlag Berlin Heidelberg 2006

rewriting system (CTRS) $\mathcal{R}$ over signature Σ and an (unconditional) term rewriting system (TRS) $\mathcal{R}'$ over signature Σ'. Also, assume some mapping φ from Σ-terms to Σ'-terms and some partial mapping ψ from Σ'-terms to Σ-terms that is an inverse to φ (i.e., $\psi(\varphi(s)) = s$ for any Σ-term s). Σ'-terms $\varphi(s)$ are called *initial*, while terms t' with $\varphi(s) \to_{\mathcal{R}'}^* t'$ are called *reachable* in $\mathcal{R}'$. The (partial) mapping ψ only needs to translate reachable Σ'-terms into Σ-terms. φ can be thought of as translating input terms for $\mathcal{R}$ into input terms for $\mathcal{R}'$, while ψ as taking results of rewritings in $\mathcal{R}'$ into corresponding results for $\mathcal{R}$. In other words, φ and ψ can wrap a Σ' rewrite engine into a Σ rewrite engine. Typically, φ and ψ are straightforward linear translators of syntax.

$\mathcal{R}'$ is *complete* for $\mathcal{R}$ iff any reduction in $\mathcal{R}$ has some corresponding reduction in $\mathcal{R}'$: $s \to_{\mathcal{R}}^* t$ implies $\varphi(s) \to_{\mathcal{R}'}^* \varphi(t)$. Completeness is typically easy to prove but, unfortunately, has a very limited practical use: it only allows to disprove reachability tasks in $\mathcal{R}$ by disproving corresponding tasks in $\mathcal{R}'$. $\mathcal{R}'$ is *sound* for $\mathcal{R}$ iff any reduction in $\mathcal{R}'$ of an initial term corresponds to some reduction in $\mathcal{R}$: $\varphi(s) \to_{\mathcal{R}'}^* t'$ implies $s \to_{\mathcal{R}}^* \psi(t')$. The soundness of $\mathcal{R}'$ allows to compute partial reachability sets in $\mathcal{R}$: applying ψ to all t' reached from $\varphi(s)$ in $\mathcal{R}'$, we get Σ-terms (not necessarily all) reachable from s in $\mathcal{R}$. The soundness and completeness of $\mathcal{R}'$ gives a procedure to *test* reachability in the CTRS $\mathcal{R}$ using any reachability analysis procedure for the TRS $\mathcal{R}'$: $s \to_{\mathcal{R}}^* t$ iff $\varphi(s) \to_{\mathcal{R}'}^* \varphi(t)$.

Soundness and completeness may seem the ideal properties of a transformation. Unfortunately, they do not yield the computational equivalence of the original CTRS to (the wrapping of) the resulting TRS. By *computational equivalence of $\mathcal{R}'$ to $\mathcal{R}$* we mean the following: if $\mathcal{R}$ terminates on a given term s admitting a unique normal form t, then $\mathcal{R}'$ also terminates on $\varphi(s)$ and for any of its normal forms t', we have that $\psi(t') = t$. In other words, the unconditional $\mathcal{R}'$ *can be used transparently* to perform computations for $\mathcal{R}$. Example 3 shows that the soundness and completeness of a transformation do *not* imply computational equivalence, even if the original CTRS is confluent and terminates! Note that termination of $\mathcal{R}$ is *not* required. Indeed, termination of the CTRS may be too restrictive in certain applications, e.g., in functional logic programming [2].

On Termination. Rewriting of a given term in a CTRS may not terminate for two reasons [19]: the reduction of the condition of a rule does not terminate, or there are some rules that can be applied infinitely often on the given term. In rewrite engines, the effect in both situations is the same: the system loops forever or crashes running out of memory. For this reason, we do not make any distinction between the two cases, and simply call a Σ-CTRS *operationally terminating* [13] on Σ-term s iff it always reduces s to a normal form regardless of the order rules apply. Note that this notion is different from *effective termination* [14]; Example 6 shows a system that is confluent and effectively terminating but *not* operationally terminating. Operational termination is based on the assumption that, in general, one cannot expect a rewrite engine to be "smart" enough to pick the right rewrite sequence to satisfy a condition. Formally, a CTRS $\mathcal{R}$ is operationally terminating on s if for any t, any proof tree attempting to prove that $s \to_{\mathcal{R}}^* t$ is finite. Operational termination is equivalent to decreasingness for normal CTRSs (and with quasi-decreasingness for deterministic CTRSs) [13].

We give an automatic transformation technique of CTRSs into TRSs, taking ground confluent normal CTRSs $\mathcal{R}$ into computationally equivalent TRSs $\mathcal{R}'$. This technique can be extended to more general CTRSs including ones with extra variables in conditions (see [20]). Experiments show that the resulting TRSs yield performant computational engines for the original CTRSs. On the theoretical side, our main new result is that if $\mathcal{R}$ is finite, ground confluent and operationally terminating on a term s then $\mathcal{R}'$ is ground confluent on reachable terms and terminating on $\varphi(s)$ (Theorem 6). This effectively gives computational equivalence of our transformation for (ground) confluent (finite) systems. To achieve this main result, we prove several other properties: the completeness of our transformation (Theorem 1); ground confluence (Theorem 2) *or* left linearity (Theorem 3) of $\mathcal{R}$ implies the soundness of $\mathcal{R}'$; if $\mathcal{R}$ is left linear, then ground confluence of $\mathcal{R}$ implies ground confluence of $\mathcal{R}'$ on reachable terms (Theorem 4) and operational termination of $\mathcal{R}$ on s implies termination of $\mathcal{R}'$ on $\varphi(s)$ (Theorem 5). Part of these properties recover the power of previous transformations; note however that they are not simple instances of those, due to the particularities of our transformation. Additionally, we show that left linearity and ground confluence of $\mathcal{R}'$ on reachable terms implies ground confluence of $\mathcal{R}$ (Proposition 2), and termination of $\mathcal{R}'$ on reachable terms implies operational termination of $\mathcal{R}$ (Proposition 4); these results potentially enable one to use confluence and/or termination techniques on unconditional TRSs to show confluence and/or operational termination of the original CTRS, but this was not our purpose and consequently have not experimented with this approach.

Section 2 discusses previous transformations of CTRSs into TRSs. We only focus on ones intended to be computationally equivalent and discuss their limitations. Section 3 presents our transformation. Section 4 shows it at work on several examples; some of these examples have been experimented with on the rewrite engines Elan [4], Maude [6] and ASF/SDF [21], with promising performance results. Section 5 lists theoretical results. All proofs can be found in the companion report [20], which will be published elsewhere soon. Section 6 concludes the paper.

2   Previous Transformations

Stimulated by the benefits of transforming CTRSs into equivalent TRSs, there has been much research on this topic. Despite the apparent simplicity of most transformations, they typically work for restricted CTRSs and their correctness, when true, is quite involved. We focus on transformations that generate TRSs intended to be *transparently* used to reduce terms or test reachability in the original CTRSs. Significant efforts have been dedicated to transformations preserving only certain properties, e.g., termination or confluence [18]; we do not discuss these here. We use the following two examples to illustrate the different transformations and to analyze their limitations.

Example 1. [14, 18]. The CTRS $\mathcal{R}_s$ will be used to test if a transformation is sound and $\mathcal{R}_t$ to test if it preserves termination. Let $\mathcal{R}_s$ be the CTRS

$$
\begin{array}{llll}
A \rightarrow h(f(a), f(b)) & g(d, x, x) \rightarrow B & a \rightarrow c \quad b \rightarrow c \quad c \rightarrow e \quad d \rightarrow m \\
h(x, x) \rightarrow g(x, x, f(k)) & f(x) \rightarrow x \; if \; x \rightarrow e & a \rightarrow d \quad b \rightarrow d \quad c \rightarrow l \quad k \rightarrow l \\
& & k \rightarrow m
\end{array}
$$

Let $\mathcal{R}_t$ be $\mathcal{R}_s \cup \{B \rightarrow A\}$; then $A \not\rightarrow^*_{\mathcal{R}_t} B$ and $\mathcal{R}_t$ operationally terminates.   □

Example 2. [2]. The two-rule canonical CTRS $\{f(g(x)) \rightarrow x \; if \; x \rightarrow 0, g(g(x)) \rightarrow g(x)\}$ will be used to test whether a transformation preserves confluence.   □

Bergstra&Klop. The first CTRS-to-TRS transformation appeared in [3]: start with a rule $Ix \rightarrow x$ and to each rule $\rho_i : l \rightarrow r \; if \; cl \rightarrow cr$ associate rules $\rho'_i : l \rightarrow \sigma_i(cl)r$ and $\rho''_i : \sigma_i(cr) \rightarrow I$. The transformation is proved to be complete in [3] and claimed to also be sound. Let us apply this transformation on $\mathcal{R}_s$ in Example 1. Rule $f(x) \rightarrow x \; if \; x \rightarrow e$ is replaced by $f(x) \rightarrow \sigma_1(x)x$ and $\sigma_1(e) \rightarrow I$, and rule $Ix \rightarrow x$ is added. Then:

$$
\begin{aligned}
A &\rightarrow h(f(a), f(b)) \rightarrow h(\sigma_1(a)a, f(b)) \rightarrow h(\sigma_1(a)d, f(b)) \rightarrow h(\sigma_1(c)d, f(b)) \\
&\rightarrow h(\sigma_1(c)d, \sigma_1(b)b) \rightarrow h(\sigma_1(c)d, \sigma_1(b)d) \rightarrow h(\sigma_1(c)d, \sigma_1(c)d) \\
&\rightarrow g(\sigma_1(c)d, \sigma_1(c)d, f(k)) \rightarrow g(\sigma_1(e)d, \sigma_1(c)d, f(k)) \rightarrow g(Id, \sigma_1(c)d, f(k)) \\
&\rightarrow g(d, \sigma_1(c)d, f(k)) \rightarrow g(d, \sigma_1(l)d, f(k)) \rightarrow g(d, \sigma_1(l)m, f(k)) \\
&\rightarrow g(d, \sigma_1(l)m, \sigma_1(k)k) \rightarrow g(d, \sigma_1(l)m, \sigma_1(l)k) \rightarrow g(d, \sigma_1(l)m, \sigma_1(l)m) \rightarrow B
\end{aligned}
$$

So this transformation is *not sound*. Transforming $\mathcal{R}_t$, we can see that this transformation does *not preserve termination*, because $A \rightarrow^+ A$. For the system in Example 2, $f(g(x)) \rightarrow x \; if \; x \rightarrow 0$ is replaced by $f(g(x)) \rightarrow \sigma_1(x)x$ and $\sigma_1(0) \rightarrow I$, so $f(g(g(0))) \rightarrow f(g(0)) \rightarrow \sigma_1(0)0 \rightarrow I0 \rightarrow 0$ and $f(g(g(0))) \rightarrow \sigma_1(g(0))g(0)$, both of them normal forms. Thus the resulting TRS is *not confluent*. Consequently, this transformation does not produce computationally equivalent TRSs.

Giovanetti&Moiso. The transformation in [9] (suggested in [7]) replaces each rule $\rho_i : l \rightarrow r \; if \; cl \rightarrow cr$ by $l \rightarrow if_i(Var(l), cl)$ and $if_i(Var(l), cr) \rightarrow r$ (where $Var(l)$ is the list of variables of l). However, this transformation is complete and computationally equivalent only when the original CTRS is *safely transformable* [9], that is, has no superposition, is simply terminating, and is non-overlapping on conditions. The "safely transformable" CTRSs are too restrictive; our transformation yields computationally equivalent TRSs imposing only ground confluence (safely transformable CTRSs are ground confluent) on the original CTRS.

Hintermeier [12] proposes a technique where an "interpreter" for a CTRS is specified using unconditional rewrite rules, defining the detailed steps of the application of a conditional rewrite rule including rewrite-based implementations of matching and substitution application; this result is rather expected, since unconditional term rewriting is Turing complete. Also, it has little practical relevance - this "meta" stepwise simulation leads to dramatic performance loss).

Marchiori's Unravellings. An abstract notion of transformation, called *unravelling*, and several concrete instances of it, were introduced by Marchiori in [14];

these were further studied in [15, 17, 18, 16]. An *unraveling* is a computable map U from CTRSs to TRSs over the same signature, except a special operation U_ρ for each rule ρ, such that $\downarrow_R \subseteq \downarrow_{U(R)}$ ($\downarrow$ stands for "join", i.e., $\rightarrow; \leftarrow$) and $U(T \cup R) = T \cup U(R)$ if T is a TRS. The concrete instance transformations are similar to that in [9]: each conditional rule $\rho : l \rightarrow r$ *if* $cl \rightarrow cr$ is replaced by its unravelling, rules $l \rightarrow U_\rho(cl, Var(r))$ and $U_\rho(cr, Var(l)) \rightarrow r$. Completeness holds, but soundness does not hold without auxiliary hypotheses [14] (see Example 1) such as left linearity [14, 18]. Also, (quasi) decreasingness and left linearity of the CTRS imply termination of the corresponding TRS.

Example 3. A sound and complete transformation does not necessarily yield computational equivalence even if the original CTRS is canonical. The unravelling of the system in Example 2 is $\{f(g(x)) \rightarrow U_1(x, x), U_1(0, x) \rightarrow x, g(g(x)) \rightarrow g(x)\}$. The original CTRS is left linear, so the unravelling is sound and complete, but is *not* computationally equivalent: $f(g(g(0)))$ reduces to $U_1(g(0), g(0))$, a normal form with no correspondent normal form in the original CTRS.   □

Unfortunately, no unravelling preserves confluence or termination [14] (i.e., for any unravelling U, there are confluent and/or terminating CTRSs R such that $U(R)$ is *not* a confluent and/or terminating TRS), thus they do not yield computationally equivalent TRSs. Therefore, it is not surprising that the more recent transformations discussed next that aim at computational equivalence, including ours, are *not* unravellings (they modify the original signature).

Viry. The transformation in [22] (inspired from [1]) inspired all subsequent approaches. It modifies the signature by adding to each operation as many arguments as conditional rules having it at the top of their lhs. Two unconditional rules replace each conditional rule, one for initializing the auxiliary arguments and the other for the actual rewrite step. Formally: let $\rho_{\sigma,i}$ denote the ith rule whose lhs is topped in σ; add as many arguments to σ as the number of rules $\rho_{\sigma,i}$; let $c_{\sigma,i}$ be the number $arity(\sigma) + i$, corresponding to the ith auxiliary argument added to σ; transform each rule $\rho_{\sigma,i} : l \rightarrow r$ *if* $cl \rightarrow cr$ into

$$\rho'_{\sigma,i} : \widetilde{l}[c_{\sigma,i} \leftarrow \bot] \rightarrow \widetilde{l}[c_{\sigma,i} \leftarrow [\overline{cl}, Var(l)]] \text{ and } \rho''_{\sigma,i} : l^*[c_{\sigma,i} \leftarrow [cr, Var(l)]] \rightarrow \overline{r},$$

where "$\bot$" is a special constant stating that the corresponding conditional rule has not been tried yet on the current position, $\overline{s}$ lifts a term by setting all new arguments to $\bot$, $\widetilde{s}$ lifts a term with fresh variables on the new arguments, and s^* replaces all variables in $\widetilde{s}$ with fresh variables. Structures $[u, \overrightarrow{s}]$ comprise the reduction status of conditions (u) together with corresponding substitutions ($\overrightarrow{s}$) when they were started. The substitution is used to correctly initiate the reduction of the rhs of the original conditional rule. Viry gave a wrong proof that his transformation sound and complete and that it preserves termination. We believe the completeness indeed holds, but have counterexamples for the other properties. Let us transform the CTRS $\mathcal{R}_s$ from Example 1. First, rules $h(x, x) \rightarrow g(x, x, f(k))$ and $g(d, x, x) \rightarrow B$ are replaced by $h(x, y) \rightarrow g(x, x, f(k))$ *if* $eq(x, y) \rightarrow true$ and $g(d, x, y) \rightarrow B$ *if* $eq(x, y) \rightarrow true$ to resolve non-left linearity, where $eq(x, x) \rightarrow true$ is the only non-left linear rule allowed [22]. Then these conditional rules and $f(x) \to x$ if $x \to e$ and $A \to h(f(a), f(b))$ are transformed into:

$$f(x, \bot) \to f(x, [x, x]) \qquad h(x, y, \bot) \to h(x, y, [eq(x, y), x, y]$$
$$f(y, [e, x]) \to x \qquad h(x', y', [true, x, y]) \to g(x, x, f(k, \bot), \bot)$$
$$g(d, x', y', [true, x, y]) \to B \qquad g(d, x, y, \bot) \to g(d, x, y, [eq(x, y), x, y])$$
$$A \to h(f(a, \bot), f(b, \bot), \bot)$$

The following is then a valid sequence in the generated unconditional TRS:

$$A \to h(f(a, \bot), f(b, \bot), \bot) \to h(f(a, [a, a]), f(b, \bot), \bot) \to h(f(d, [a, a]),$$
$$f(b, \bot), \bot) \to h(f(d, [c, a]), f(b, \bot), \bot) \to h(f(d, [c, c]), f(b, \bot), \bot) \to h(f(d, [c, c]),$$
$$f(b, [b, b]), \bot) \to h(f(d, [c, c]), f(d, [b, b]), \bot) \to h(f(d, [c, c]), f(d, [c, b]), \bot) \to$$
$$h(f(d, [c, c]), f(d, [c, c]), \bot) \to h(f(d, [c, c]), f(d, [c, c]), [eq(f(d, [c, c]), f(d, [c, c])),$$
$$f(d, [c, c]), f(d, [c, c])]) \to h(f(d, [c, c]), f(d, [c, c]), [true, f(d, [c, c]), f(d, [c, c])]) \to$$
$$g(f(d, [c, c]), f(d, [c, c]), f(k, \bot), \bot) \to g(f(d, [e, c]), f(d, [c, c]), f(k, \bot), \bot) \to$$
$$g(f(d, [e, e]), f(d, [c, c]), f(k, \bot), \bot) \to g(d, f(d, [c, c]), f(k, \bot), \bot) \to$$
$$g(d, f(m, [c, c]), f(k, \bot), \bot) \to g(d, f(m, [l, c]), f(k, \bot), \bot) \to g(d, f(m, [l, l]),$$
$$f(k, \bot), \bot) \to g(d, f(m, [l, l]), f(k, [k, k]), \bot) \to g(d, f(m, [l, l]), f(m, [k, k]), \bot)$$
$$\to g(d, f(m, [l, l]), f(m, [l, k]), \bot) \to g(d, f(m, [l, l]), f(m, [l, l]), \bot) \to$$
$$g(d, f(m, [l, l]), f(m, [l, l]), [eq(f(m, [l, l]), f(m, [l, l])), f(m, [l, l]), f(m, [l, l])]) \to$$
$$g(d, f(m, [l, l]), f(m, [l, l]), [true, f(m, [l, l]), f(m, [l, l])]) \to B$$

Hence, Viry's transformation is *not sound*. Using $\mathcal{R}_t$ instead of $\mathcal{R}_s$, whose corresponding TRS just adds rule $B \to A$ to that of $\mathcal{R}_s$, we can notice that it *does not preserve termination* either. Let us now transform the CTRS in Example 2 to $\{f(g(x), \bot) \to f(g(x), [x, x]), f(x, [0, y]) \to y, g(g(x)) \to g(x)\}$; note that $\mathcal{R}'$ is *not confluent* [2] (with or without Viry's conditional eagerness [22]): $f(g(g(0)), \bot)$ can be reduced to both 0 and $f(g(0), [g(0), (g(0))])$. Therefore, this transformation does not fulfill the requirements of computational equivalence.

Antoy,Brassel&Hanus proposed in [2] a simple fix to Viry's technique, namely to restrict the input CTRSs to *constructor-based* (i.e., the lhs of each rule is a term of the form $f(t_1, \ldots, t_n)$, where f is *defined* and $t_1, \ldots, t_n$ are all *constructor* terms) and *left linear* ones. Under these restrictions, they also show that the substitution needed by Viry's transformation is not necessary anymore, so they drop it and prove that the new transformation is sound and complete; moreover, if the original CTRS is additionally weakly orthogonal, then the resulting TRS is confluent on reachable terms. It is suggested in [2] that what Viry's transformation (or their optimized version of it) needs to generate computationally equivalent TRSs is to reduce its applicability to only constructor-based, weakly orthogonal and left linear CTRSs. While constructor-baseness and left linearity are common to functional logic programming and are easy to check automatically, we believe that they are, in general, an unnecessarily strong restriction on the input CTRS, which may make the translation unusable in many situations of practical interest (see, e.g., the bubble-sort algorithm in Section 4).

Roşu. The transformation in [19] is defined for join CTRSs and requires the rewrite engine to support some simple *contextual rewriting strategies*, namely an $if(\_,\_,\_)$ eager on the condition and an *equal?* eager on both arguments. As in Viry's transformation, additional arguments are added to each operation σ for each conditional rule $\rho_{\sigma,i}$, but they only need to keep truth values. The distinctive feature of this transformation is the introduction of the $\{\_\}$ operation, which allows the rewriting process to continue after a condition got stuck provided changes occur in subterms. Each conditional rule $\rho_{\sigma,i}$: $l \rightarrow r$ *if* $cl\!\downarrow\!cr$ is encoded by one unconditional rule $\overline{\rho}_{\sigma,i} : \widetilde{l}[c_{\sigma,i} \leftarrow true] \rightarrow if(equal?(\{\overline{cl}\}, \{\overline{cr}\}), \{\overline{r}\}, \widetilde{l}[c_{\sigma,i} \leftarrow false])$. The bracket clears the failed conditions on the path to the top: $\sigma(x_1, .., \{x_i\}, .., x_{arity(\sigma)}, y_1, .., y_m) \rightarrow \{\sigma(x_1, .., x_i, .., x_{arity(\sigma)}, true, .., true)\}$. It is shown in [19] that this transformation is sound and that operational termination is preserved and implies completeness and preservation of ground confluence, that is, computational equivalence. Left linearity needs *not* be assumed. Although most modern rewrite systems support the rewrite strategies required by the transformation in [19], we argue that imposing restrictions on the order of evaluation makes a rewrite engine less friendly w.r.t parallelism and more complex; in some sense, contextual strategies can be seen as some sort of conditional rules: apply the rule *if* the context permits.

Our transformation basically integrates Rosu's $\{\_\}$ operation within Viry's transformation, which allows us to also eliminate the need to carry a substitution. We recently found out[1] that a related approach was followed by Brassel in his master thesis [5], but we can't relate our results since we were unable to obtain an English translation of his results.

3   Our Transformation

Like in the last three transformations above, auxiliary arguments are added to some operators to maintain the control context information. For simplicity, we here discuss only the transformation of *normal* CTRSs, that is, ones whose conditional rules have the form $l \rightarrow r$ *if* $cl \rightarrow cr$, cr is a constant in normal form and all variables from cl and r also occur in l (and l is not a variable). In [20] we discuss extensions of our technique to more complex cases, including ones with extra variables and matching in conditions. Let $\mathcal{R}$ be any Σ-CTRS. A σ-*conditional rule* [22] is a conditional rule with σ at the top of its lhs, i.e., one of the form $\sigma(t_1, \ldots, t_n) \rightarrow r$ *if* $cl \rightarrow cr$. Let k_σ be the number of σ-conditional rules and let $\rho_{\sigma,i}$ denote the ith σ-conditional rule in $\mathcal{R}$.

The signature transformation. Let $\overline{\Sigma}$ be the signature containing: a fresh constant $\perp$; a fresh unary operator $\{\_\}$; for any $\sigma \in \Sigma_n$ (i.e., $\sigma \in \Sigma$ has n arguments), an operation $\overline{\sigma} \in \overline{\Sigma}_{n+k_\sigma}$ (the additional k_σ arguments of $\overline{\sigma}$ are written to the right of the other n arguments). An important step in our transformation is to replace Σ-terms by corresponding $\overline{\Sigma}$-terms. The reason for the additional arguments is to pass the control context (due to conditional rules) into data context: the additional i-th argument of $\overline{\sigma}$ at some position in a term maintains the status of appliance of $\rho_{\sigma,i}$; if $\perp$ then that rule was not tried, otherwise

---

[1] From a private communication with Bernd Brassel.

the condition is being under evaluation or is already evaluated. Thus, the corresponding $\overline{\Sigma}$-term of a Σ-term is obtained by replacing each operator σ by $\overline{\sigma}$ with the k_σ additional arguments all $\bot$. Formally, let $\mathcal{X}$ be an infinite set of *variables* and let $\overline{\cdot} : T_\Sigma(\mathcal{X}) \to T_{\overline{\Sigma}}(\mathcal{X})$ be defined inductively as: $\overline{x} = x$ for any $x \in \mathcal{X}$ and $\overline{\sigma(t_1,\ldots,t_n)} = \overline{\sigma}(\overline{t_1},\ldots,\overline{t_n},\bot,\ldots,\bot)$ for any $\sigma \in \Sigma_n$ and any $t_1,\ldots,t_n \in T_\Sigma(\mathcal{X})$. Let us define another map, $\widetilde{\cdot}^X : T_\Sigma(X) \to T_{\overline{\Sigma}}(\mathcal{X})$, this time indexed by a *finite* set of variables $X \subseteq \mathcal{X}$, as $\widetilde{x}^X = x$ for any $x \in X$, and as $\widetilde{\sigma(t_1,\ldots,t_n)}^X = \overline{\sigma}(\widetilde{t_1}^X,\ldots,\widetilde{t_n}^X,z_1,\ldots,z_{k_\sigma})$ for any $\sigma \in \Sigma_n$ and $t_1,\ldots,t_n \in T_\Sigma(X)$, where $z_1,\ldots,z_{k_\sigma} \in \mathcal{X} - X$ are some arbitrary but fixed different fresh variables that do not occur in X or in $\widetilde{t_1}^X,\ldots,\widetilde{t_n}^X$. Therefore, $\widetilde{t}^X$ transforms the Σ-term t into a $\overline{\Sigma}$-term by replacing each operation $\sigma \in \Sigma$ by $\overline{\sigma} \in \overline{\Sigma}$ and adding some distinct fresh variables for the additional arguments, chosen arbitrarily but deterministically.

The rewrite rules transformation. Given a Σ-CTRS $\mathcal{R}$, let $\overline{\mathcal{R}}$ be the $\overline{\Sigma}$-TRS obtained as follows. For each σ-conditional rule $\rho_{\sigma,i}: l \to r \ if \ cl \to cr$ over variables X in $\mathcal{R}$, add to $\overline{\mathcal{R}}$ two rules, namely $\overline{\rho}_{\sigma,i} : \widetilde{l}^X[c_{\sigma,i} \leftarrow \bot] \to \widetilde{l}^X[c_{\sigma,i} \leftarrow \{\widetilde{cl}\}]$ and $\overline{\rho}'_{\sigma,i} : \widetilde{l}^X[c_{\sigma,i} \leftarrow \{cr\}] \to \{\overline{r}\}$, where $c_{\sigma,i}$ is the number $arity(\sigma) + i$ corresponding to the ith conditional argument of σ. For each unconditional rule $l \to r$ in $\mathcal{R}$, add rule $\widetilde{l}^X \to \{\overline{r}\}$ to $\overline{\mathcal{R}}$. For each $\sigma \in \Sigma_n$ and each $1 \leq i \leq n$, add to $\overline{\mathcal{R}}$ a rule $\overline{\sigma}(x_1,..,x_{i-1},\{x_i\},x_{i+1},..,x_n,z_1,..,z_{k_\sigma}) \to \{\overline{\sigma}(x_1,..,x_{i-1},x_i,x_{i+1},..,x_n,\bot,..,\bot)\}$, intuitively stating that a condition tried and potentially failed in the past at some position may hold once an immediate subterm changes; the operation $\{\_\}$, symbolizing the change, also needs to be propagated bottom-up, resetting the other started conditions to $\bot$. The applicability information of an operation can be updated from several of its subterms; to keep this operation idempotent, we add $\{\{x\}\} \to \{x\}$ to $\overline{\mathcal{R}}$. The size of $\overline{\mathcal{R}}$ is $1 + u + 2 \times c + \sum_{n \geq 0} n \times |\Sigma_n|$, where u is the number of unconditional rewrite rules and c is the number of conditional rewrite rules in $\mathcal{R}$.

4   Examples and Experiments

We next illustrate our transformation on several examples.

Confluence is preserved. Let us transform the CTRS in Example 2:

$$\overline{f}(\overline{g}(x),\bot) \to \overline{f}(\overline{g}(x),\{x\}) \quad \overline{f}(\overline{g}(x),\{\overline{0}\}) \to \{x\} \quad \overline{g}(\overline{g}(x)) \to \{\overline{g}(x)\}$$
$$\overline{g}(\{x\}) \to \{\overline{g}(x)\} \quad \overline{f}(\{x\},b) \to \{\overline{f}(x,\bot)\} \quad \{\{x\}\} \to \{x\}$$

The problem that appeared in Viry's transformation is avoided in our transformation by the rules of $\{\cdot\}$, which allow the evaluation of a condition to be restarted at the top of a term once a modification occurs in a subterm. Thus, given the $\overline{\Sigma}$-term $\{\overline{f}(\overline{g}(\overline{g}(\overline{0})),\bot)\}$, even if a rewrite engine first tries to evaluate the condition at the top, a "correct" rewriting sequence is eventually obtained: $\{\overline{f}(\overline{g}(\overline{g}(\overline{0})),\bot)\} \to_{\overline{\mathcal{R}}} \{\overline{f}(\overline{g}(\overline{g}(\overline{0})),\{\overline{g}(\overline{0})\})\} \to_{\overline{\mathcal{R}}} \{\overline{f}(\{\overline{g}(\overline{0})\},\{\overline{g}(\overline{0})\})\} \to_{\overline{\mathcal{R}}} \{\{\overline{f}(\overline{g}(\overline{0}),\bot)\}\}$, and now the condition can be tried again and this time will succeed.

Odd/Even [19]. Let us consider natural numbers with 0 and successor s, constants *true* and *false* and the following on purpose inefficient conditional rules defining *odd* and *even* operators on natural numbers (here denoted as o and e):

$$o(0) \to false \quad o(s(x)) \to true \text{ if } e(x) \to true \quad o(s(x)) \to false \text{ if } e(x) \to false$$
$$e(0) \to true \quad e(s(x)) \to true \text{ if } o(x) \to true \quad e(s(x)) \to false \text{ if } o(x) \to false$$

In order to check whether a natural number n, i.e., a term consisting of n successor operations applied to 0, is odd, a conditional rewrite engine may need $\mathcal{O}(2^n)$ rewrites in the worst case. Indeed, if $n > 0$ then either the second or the third rule of *odd* can be applied at the first step; however, in order to apply any of those rules one needs to reduce the even of the predecessor of n, twice. Iteratively, the evaluation of each even involves the reduction of two odds, and so on. Moreover, the rewrite engine needs to maintain a control context data-structure, storing the status of the application of each (nested) rule that is being tried in a reduction. It is the information stored in this control context that allows the rewriting engine to backtrack and find an appropriate rewriting sequence. As shown at the end of this section, some rewrite engines perform quite poorly on this system. Let us apply it our transformation. Since there are two *odd*-conditional rules and two *even*-conditional rules, each of these operators will be enriched with two arguments. The new TRS is (for aesthetic reasons we overline only those operations that change; z_1 and z_2 are variables):

$$\overline{o}(0, z_1, z_2) \to \{false\} \qquad\qquad \overline{e}(0, z_1, z_2) \to \{true\}$$
$$\overline{o}(s(x), \{false\}, z_2) \to \{false\} \qquad \overline{e}(s(x), \{false\}, z_2) \to \{false\}$$
$$\overline{o}(s(x), z_1, \{true\}) \to \{true\} \qquad \overline{e}(s(x), z_1, \{true\}) \to \{true\}$$
$$\overline{o}(s(x), \bot, z_2) \to \overline{o}(s(x), \{\overline{e}(x, \bot, \bot)\}, z_2) \qquad \overline{e}(s(x), \bot, z_2) \to \overline{e}(s(x), \{\overline{o}(x, \bot, \bot)\}, z_2)$$
$$\overline{o}(s(x), z_1, \bot) \to \overline{o}(s(x), z_1, \{\overline{e}(x, \bot, \bot)\}) \qquad \overline{e}(s(x), z_1, \bot) \to \overline{e}(s(x), z_1, \{\overline{o}(x, \bot, \bot)\})$$
$$s(\{x\}) \to \{s(x)\} \qquad\qquad \overline{o}(\{x\}, z_1, z_2) \to \{\overline{o}(x, z_1, z_2)\}$$
$$\{\{x\}\} \to \{x\} \qquad\qquad \overline{e}(\{x\}, z_1, z_2) \to \{\overline{e}(x, z_1, z_2)\}$$

If one wants to test whether a number n, i.e., n consecutive applications of successor on 0, is odd, one should reduce the term $\{o(n, \bot, \bot)\}$.

Bubble sort. The following one-rule CTRS sorts lists of numbers (we assume appropriate rules for numbers) implementing the bubble sort algorithm:

$$\cdot(x, \cdot(y, l)) \to \cdot(y, \cdot(x, l)) \text{ if } x < y \to true$$

This CTRS is ground confluent but *not* constructor-based. Its translation is:

$$\cdot(x, \cdot(y, l, c), \bot) \to \cdot(x, \cdot(y, l, c), \{x > y\}) \qquad \{\{l\}\} \to \{l\}$$
$$\cdot(x, \cdot(y, l, c), \{true\}) \to \{\cdot(y, \cdot(x, l, \bot), \bot)\} \qquad \cdot(x, \{l\}, c) \to \{\cdot(x, l, \bot)\}$$

Experiments. Our major motivation to translate a CTRS into a computationally equivalent TRS that can run on any unrestricted unconditional rewrite engine was the potential to devise highly parallelizable rewrite engines. It was therefore an unexpected and a pleasant surprise to note that our transformation can actually bring immediate benefits if implemented as a front-end to *existing,* non-parallel rewrite engines. Note, however, that current rewrite engines are optimized for *both* conditional and unconditional rewriting; an engine optimized for just unconditional rewriting could probably be even more efficient.

We next give some numbers regarding the speed of the generated TRS. We used Elan and Maude as interpreters and ASF/SDF as a compiler - our goal was *not* to compare rewrite engines (that's why we did not use same input data for all engines) but rather to compare how our transformation performs on each of them. Besides the two examples presented above (Odd/Even and Bubble sort), we tested our transformation on two other CTRSs: Quotient/Reminder (inspired from [18]) and the evaluation of a program generating all permutations of n elements and counting them written for a rewriting based interpreter of a simple programming language with arrays (using both matching in conditions and rewriting modulo axioms - see [20] for a discussion of when our transformation is sound for rewriting modulo axioms). We have tested how long it took for a term to be rewritten to a normal form. In the table below, Cond shows the results using the original system, Ucond those using the presented transformation and Ucond* the transformation enhanced with some simple but practical optimizations described below. Times were obtained on a machine with 2 GHz Pentium 4 CPU and 1GB RAM.

Odd/Even								
Elan - odd(18)			Maude - odd(24)			ASF/SDF - odd(25)		
Cond	Uncond	Uncond*	Cond	Uncond	Uncond*	Cond	Uncond	Uncond*
85.79s	5.55s	~0s	84.97s	17.05s	~0s	0.02s	7.46s	0.01s

Bubble Sort					
Elan 100		Maude 5000		ASF/SDF 5000	
Cond	Uncond(*)	Cond	Uncond(*)	Cond	Uncond(*)
28.19s	3.46s	72.34s	43.53s	81.64s	85.71s

Quotient(Reminder)								
Elan $10^5/6$			Maude $10^7/2$			ASF/SDF $10^6/2$		
Cond	Uncond	Uncond*	Cond	Uncond	Uncond*	Cond	Uncond	Uncond*
10.85s	5.82s	5.23s	75.98s	67.59s	41.61s	13.96s	15.28s	14.53s

Rew.-based interpreter of a simple PL with arrays - *permutation generation*							
Maude				ASF/SDF			
8		9		8		9	
Cond	Uncond*	Cond	Uncond*	Cond	Uncond*	Cond	Uncond*
18.06s	12.76s	-²	144.56s	5.72s	14.20s	49.56s	-²

The optimized transformation always outperformed the original CTRS in our experiments on rewrite engine interpretors. We cannot say what exactly made it slower on some tests performed on ASF/SDF - it might be because the compiler is aimed to be efficiently executed on sequential machines; we actually expect our transformation to perform significantly better on parallel rewrite engines. Note that on the odd/even example, ASF/SDF already performs condition elimination as a stage of its compilation process.

---

² The machine ran out of memory while attempting to reduce the term.

A simple and very practical optimization is as follows: if two or more σ-conditional rules have the same lhs and their conditions also have the same lhs, then we can add only one auxiliary argument to σ in $\overline{\sigma}$ for all of these and only one rule in the TRS for starting the condition. With this, e.g., the optimized TRS generated for the odd/even CTRS is:

$$
\begin{array}{lll}
\overline{o}(0, z_1) \rightarrow \{false\} & \overline{o}(s(x), \{false\}) \rightarrow \{false\} & \overline{o}(s(x), \bot) \rightarrow \overline{o}(s(x), \{\overline{e}(x, \bot)\}) \\
\overline{e}(0, z_1) \rightarrow \{true\} & \overline{o}(s(x), \{true\}) \rightarrow \{true\} & \overline{e}(s(x), \bot) \rightarrow \overline{e}(s(x), \{\overline{o}(x, \bot)\}) \\
\{\{x\}\} \rightarrow \{x\} & \overline{e}(s(x), \{false\}) \rightarrow \{false\} & \overline{o}(\{x\}, z_1) \rightarrow \{\overline{o}(x, z_1)\} \\
s(\{x\}) \rightarrow \{s(x)\} & \overline{e}(s(x), \{true\}) \rightarrow \{true\} & \overline{e}(\{x\}, z_1) \rightarrow \{\overline{e}(x, z_1)\}
\end{array}
$$

Another optimization easy to perform statically is to restrict the number of additional conditional arguments added to each operation σ to the maximum number of overlapping rules whose lhs is rooted in σ. The intuition here is that if two conditional rules are orthogonal, their conditions can't be started at the same time for the same term. For orthogonal systems (as our language definition), for example, this means adding at most one argument per operation.

5   Theoretical Aspects

We use the terminology in [18] and, as mentioned, here only consider *normal* CTRSs. Before we formalize the relationship between CTRSs and their unconditional variants, we define and discuss several classes of $\overline{\Sigma}$-terms that will be used in the sequel. First, let $\widehat{\cdot} : T_{\overline{\Sigma}}(X) \rightarrow T_{\Sigma}(X)$ be a *partial* map, forgetting all the auxiliary arguments of operations, defined as: $\widehat{x} = x$ for any variable x, $\widehat{\{t'\}} = \widehat{t'}$ and $\widehat{\overline{\sigma}(t_1', .., t_n', z_1, .., z_{k_\sigma})} = \sigma(\widehat{t_1'}, .. \widehat{t_n'})$. In particular, $\widehat{t} = t$. The map $\widehat{\cdot}$ is partial (not defined for $\overline{\Sigma}$-terms such as, e.g., $\bot$). A $\overline{\Sigma}$-term t' is *structural* iff $\widehat{t'}$ is defined, that is, if it is "resembling" a Σ-term. Note that the lhs and rhs of any (unconditional) rule in $\overline{\mathcal{R}}$ are structural.

A position α is a string of numbers representing a path in a term seen as a tree. Let us define two mutually recursive important types of positions in structural $\overline{\Sigma}$ terms. A position α is *structural* for t' iff α has no conditional position as a prefix. A position α is *conditional* for t' iff $\alpha = \alpha' c_{\sigma,i}$ such that α' is structural for t' and $t'_{\alpha'} = \overline{\sigma}(\overline{u})$ (recall $c_{\sigma,i}$ is associated to $\rho_{\sigma,i}$'s conditions).

A rewriting step $s' \rightarrow_{\overline{\mathcal{R}}} t'$ is *structural* iff it occurs at a structural position in s' and either uses a rule of form $\overline{\rho}'_{\sigma,i}$ or one corresponding to an unconditional rule in $\mathcal{R}$. A ground $\overline{\Sigma}$-term t' is *reachable* iff there is some ground Σ-term t such that $\{\overline{t}\} \rightarrow_{\overline{\mathcal{R}}}^{*} \{t'\}$. The set of all conditions started for a reachable term s', written $cond(s')$, is defined as $\bigcup_C (\{s'|_\alpha\} \cup cond(s'|_\alpha))$ where C is the set of conditional positions α in s' such that $s'|_\alpha \neq \bot$.

Proposition 1. *(1) Any subterm of a structural term on a structural position is also structural; (2) If t' is a structural term with variables on structural positions and θ is a substitution giving structural terms for variables of t' then $\theta(t')$ is also structural; (3) Structural terms are closed under $\overline{\mathcal{R}}$; (4) Any reachable term is also structural; (5) Reachable terms are closed under $\overline{\mathcal{R}}$.*

Completeness means that rewriting that can be executed in the original CTRS can also be simulated on the corresponding TRS. We show that for any Σ-term s, "everything that can be done on s in $\mathcal{R}$ can also be done on $\{\overline{s}\}$ in $\overline{\mathcal{R}}$".

Theorem 1. *If* $s \to_{\mathcal{R}}^k t$ *then* $\{\overline{s}\} \to_{\overline{\mathcal{R}}} \{\overline{t}\}$ *with* k *structural steps.*

Although it may not seem so, $\{\overline{s}\} \to_{\overline{\mathcal{R}}}^* \{\overline{t}\}$ does not generally imply that $s \to_{\mathcal{R}}^* t$:

Example 4. Consider the transformation for $\mathcal{R}_s$ from Example 1:

$$
\begin{array}{lll}
A \to \{h(f(a, \bot), f(b, \bot))\} & f(\{x\}, y) \to \{f(x, \bot)\} & a \to \{c\} \\
h(x, x) \to \{g(x, x, f(k, \bot))\} & h(\{x\}, y) \to \{h(x, y)\} & a \to \{d\} \\
g(d, x, x) \to \{B\} & h(x, \{y\}) \to \{h(x, y)\} & b \to \{c\} \\
f(x, \bot) \to f(x, \{x\}) & g(\{x\}, y, z) \to \{g(x, y, z)\} & b \to \{d\} \\
f(x, \{e\}) \to \{x\} & g(x, \{y\}, z) \to \{g(x, y, z)\} & c \to \{e\} \\
\{\{x\}\} \to \{x\} & g(x, y, \{z\}) \to \{g(x, y, z)\} & c \to \{l\}
\end{array}
\qquad
\begin{array}{l}
k \to \{l\} \\
k \to \{m\} \\
d \to \{m\}
\end{array}
$$

Then the following rewrite sequence can be obtained in $\overline{\mathcal{R}}_s$:

$$\{A\} \to_{\overline{\mathcal{R}}}^* \{h(f(a, \bot), f(b, \bot))\} \to_{\overline{\mathcal{R}}}^* \{h(f(\{d\}, \{c\}), f(b, \bot))\}$$
$$\to_{\overline{\mathcal{R}}}^* \{h(f(\{d\}, \{c\}), f(\{d\}, \{c\}))\} \to_{\overline{\mathcal{R}}}^* \{g(f(\{d\}, \{c\}), f(\{d\}, \{c\}), f(k, \bot))\}$$
$$\to_{\overline{\mathcal{R}}}^* \{g(f(\{d\}, \{e\}), f(\{d\}, \{c\}), f(k, \bot))\} \to_{\overline{\mathcal{R}}}^* \{g(d, f(\{d\}, \{c\}), f(k, \bot))\}$$
$$\to_{\overline{\mathcal{R}}}^* \{g(d, f(\{m\}, \{l\}), f(k, \bot))\} \to_{\overline{\mathcal{R}}}^* \{g(d, f(\{m\}, \{l\}), f(\{m\}, \{l\}))\} \to_{\overline{\mathcal{R}}}^* \{B\},$$

but it is not the case that $A \to_{\mathcal{R}_s}^* B$. □

Even though Theorem 1 is too weak to give us a procedure in $\overline{\mathcal{R}}$ to test reachability in $\mathcal{R}$, it still gives us a technique to test whether a term t is *not* reachable from a term s in $\mathcal{R}$: if it is *not true* that $\{\overline{s}\} \to_{\overline{\mathcal{R}}}^* \{\overline{t}\}$ then it is also *not true* that $s \to_{\mathcal{R}}^* t$. Of course, in order for this to be mechanizable, the set of terms reachable from $\{\overline{s}\}$ must be finite. This does not give us much, but it is the most we can get without additional restrictions on $\mathcal{R}$.

Soundness means that rewrites $\{\overline{s}\} \to_{\overline{\mathcal{R}}}^* t'$ in $\overline{\mathcal{R}}$ correspond to rewrites $s \to_{\mathcal{R}}^* \widehat{t'}$ in $\mathcal{R}$. Unfortunately, as shown by Example 4, soundness does not hold without restricting $\mathcal{R}$. We show that ground confluence *or* left linearity suffices.

Theorem 2. *If* $\mathcal{R}$ *is ground confluent and* s' *is reachable such that* $s' \to_{\overline{\mathcal{R}}}^* t'$ *in* k *structural steps, then* $\widehat{s'} \to^k \widehat{t'}$. *In particular, our transformation is sound.*

The claim above may not hold if the original CTRS is not ground confluent:

Example 5. Consider the following CTRS and its corresponding TRS:

$$
(\mathcal{R}) \begin{cases} a \to true \\ a \to false \\ f(x) \to true \text{ if } x \to true \end{cases}
\qquad
(\overline{\mathcal{R}}) \begin{cases} \overline{a} \to \{\overline{true}\} & \overline{a} \to \{\overline{false}\} \\ \overline{f}(x, \bot) \to \overline{f}(x, \{x\}) & \overline{f}(x, \{\overline{true}\}) \to \{\overline{true}\} \\ \overline{f}(\{x\}, y) \to \{\overline{f}(x, \bot)\} & \{\{x\}\} \to \{x\} \end{cases}
$$

The following sequence is valid in $\overline{\mathcal{R}}$, but it is *not* the case that $f(false) \to_{\mathcal{R}} true$:

$$\{\overline{f}(\overline{a}, \bot)\} \to_{\overline{\mathcal{R}}} \{\overline{f}(\overline{a}, \{\overline{a}\})\} \to_{\overline{\mathcal{R}}} \{\overline{f}(\{\overline{false}\}, \{\overline{a}\})\} \to_{\overline{\mathcal{R}}} \boxed{\{\overline{f}(\{\overline{false}\}, \{\overline{true}\})\} \to_{\overline{\mathcal{R}}}^+ \{\overline{true}\}}$$

In Theorem 2, let $s' = \{\overline{f}(\{\overline{false}\}, \{\overline{true}\})\}$ (reachable) and $t' = \{\overline{true}\}$. □

However, the next result shows that our transformation is also sound when the original CTRS is left linear instead of ground confluent:

Theorem 3. *If $\mathcal{R}$ is left linear and $\{\overline{s}\} \to_{\overline{\mathcal{R}}}^{*} t'$ then $s \to_{\mathcal{R}}^{*} \widehat{t'}$. Moreover, if $\{\overline{s}\} \to_{\overline{\mathcal{R}}}^{*} t'$ has k structural steps, then $s \to_{\mathcal{R}}^{k'} \widehat{t'}$ with $k' \geq k$.*

Thus, our transformation is sound for Example 5. However, Example 4 shows that soundness may not hold if $\mathcal{R}$ is neither ground confluent nor left linear.

Corollary 1. *If $\mathcal{R}$ is ground confluent **or** left linear, then our transformation is sound and complete, i.e., $s \to_{\mathcal{R}}^{*} t$ iff $\{\overline{s}\} \to_{\overline{\mathcal{R}}}^{*} \{\overline{t}\}$ for any $s, t \in T_{\Sigma}$.*

This gives us a semi-decision procedure for reachability problems $s \to_{\mathcal{R}}^{*} t$ in a ground confluent or left linear CTRS: (1) transform $\mathcal{R}$ to the TRS $\overline{\mathcal{R}}$; (2) do a breadth-first search in $\overline{\mathcal{R}}$ starting with $\{\overline{s}\}$; (3) if $\{\overline{t}\}$ is reached then return true. The breadth-first search may loop forever if there is no solution for the original problem. However, it will return true *iff* the original problem has a solution. This reachability result is operationally important, since searching is very difficult in CTRSs and it can sometimes lead to defectuous implementations.

Example 6. Consider the following three-rule CTRS: $a \to c$ *if* $a \to b$ and $a \to b$ and $c \to b$. A rewrite engine sensitive to the order in which rules are given may crash when asked to verify $a \to_{\mathcal{R}}^{*} c$; indeed, Maude does so if the rules are given in the order above. The reason is that although Maude does breadth-first search in general, it chooses not to do it within conditions.
This CTRS is transformed to: $a(\bot) \to a(\{a(\bot)\})$, $c \to \{b\}$, $a(\{b\}) \to \{c\}$, $a(x) \to \{b\}$, $\{\{x\}\} \to \{x\}$. Although this TRS does not terminate either, we can use any rewrite engine which supports breadth-first searching, including Maude, to verify any reachability problem which has solutions in the original system.   □

If $\mathcal{R}$ is left linear, due to soundness and completeness, ground confluence of $\overline{\mathcal{R}}$ on reachable terms yields ground confluence of $\mathcal{R}$.

Proposition 2. *If $\overline{\mathcal{R}}$ is left linear and ground confluent on reachable terms, then $\mathcal{R}$ is ground confluent.*

Even though a transformation is sound and complete, one may not necessarily simulate $\mathcal{R}$ through $\overline{\mathcal{R}}$ (see Example 3). We show that if $\mathcal{R}$ is ground confluent and left linear then $\widehat{\cdot}$ defines a simulation relation between $\overline{\mathcal{R}}$ and $\mathcal{R}$:

Proposition 3. *(Simulation) If $\mathcal{R}$ is ground confluent and left linear then $\overline{\mathcal{R}}$ weakly simulates $\mathcal{R}$ on reachable terms: for any reachable s' with $\widehat{s'} \to_{\mathcal{R}}^{k} t$, there is a $\overline{\Sigma}$-term t' with $\widehat{t'} = t$ and $s' \to_{\overline{\mathcal{R}}}^{*} t'$ using exactly k structural steps.*

It is worthwhile noticing that the confluence of $\mathcal{R}$ does *not* imply the confluence of $\overline{\mathcal{R}}$, as the following (counter-)example shows.

Example 7. Consider the confluent one-rule CTRS $\mathcal{R}$ $f(x) \to x$ *if* $g(x) \to false$ and its corresponding TRS $\overline{\mathcal{R}}$: $\overline{f}(x, \bot) \to \overline{f}(x, \{\overline{g}(x)\})$, $\overline{f}(\{x\}, y) \to \{\overline{f}(x, \bot)\}$, $\overline{f}(x, \{\overline{false}\}) \to \{x\}$, $\{\{x\}\} \to \{x\}$. Then $\overline{f}(\{\overline{false}\}, \{\overline{false}\})$ rewrites in one step to $\{\overline{false}\}$, in normal form, and $\overline{f}(\{\overline{false}\}, \{\overline{false}\}) \to_{\overline{\mathcal{R}}} \{\overline{f}(\overline{false}, \bot)\} \to_{\overline{\mathcal{R}}} \{\overline{f}(\overline{false}, \{\overline{g}(\overline{false})\})\}$, also in normal form. Hence $\overline{\mathcal{R}}$ is not confluent.   □

In fact, for computational equivalence purposes, $\overline{\mathcal{R}}$ does *not* need to be confluent. What is needed is its confluence on *reachable* terms. The next result shows that (ground) confluence is preserved in the presence of left linearity.

Theorem 4. *If $\mathcal{R}$ is left linear and ground confluent then $\overline{\mathcal{R}}$ is ground confluent on reachable terms, or, even stronger, for any reachable terms s'_1, s'_2, if $\widehat{s'_1}$ and $\widehat{s'_2}$ are joinable in t then s'_1 and s'_2 are joinable in t' such that $\widehat{t'} = t$.*

The termination of $\overline{\mathcal{R}}$ on reachable terms implies operational termination of $\mathcal{R}$:

Proposition 4. *If $\overline{\mathcal{R}}$ terminates on $\{\overline{s}\}$, then $\mathcal{R}$ operationally terminates on s.*

The other implication does not hold without additional requirements on $\mathcal{R}$. We will show that confluence or left linearity of $\mathcal{R}$ suffices.

Example 8. Consider the system $\mathcal{R}_t$ in Example 1. Since $\mathcal{R}_t = \mathcal{R}_s \cup \{B \to A\}$, its transformed version will be the same as the one in Example 4, except adding one more rule, $B \to \{A\}$. Remember that with the system $\overline{\mathcal{R}}$ in Example 4 we have obtained that $\{A\} \to^*_{\overline{\mathcal{R}}} \{B\}$. With the new rule we therefore get that $\{A\} \to^+_{\overline{\mathcal{R}}} \{A\}$, thus the transformed version is not terminating. However, the original system is decreasing, so it is operationally terminating.   □

Confluence *or* left linearity of $\mathcal{R}$ preserves termination:

Theorem 5. (*Termination*) *If $\mathcal{R}$ operationally terminates on s and is either ground confluent or left linear, then $\overline{\mathcal{R}}$ terminates on $\{\overline{s}\}$.*

Finally, we prove that ground confluence yields computational equivalence:

Theorem 6. (*Computational equivalence*) *If $\mathcal{R}$ is finite, ground confluent and operationally terminates on s, then $\overline{\mathcal{R}}$ is ground confluent and terminates on terms reachable from $\overline{s}$. That is, $\overline{\mathcal{R}}$ is computationally equivalent to $\mathcal{R}$.*

Then one can simulate reduction in a confluent CTRS $\mathcal{R}$ by using the transformed TRS $\overline{\mathcal{R}}$. Reducing a Σ-term t to its normal form in $\mathcal{R}$ can be done as follows: start reducing $\{\overline{t}\}$ in $\overline{\mathcal{R}}$; if it does not terminate, there exists a way t might have not terminated in $\mathcal{R}$; if it terminates and $fn(\{\overline{t}\})$ is its normal form, then $\widehat{fn(\{\overline{t}\})}$ is the normal form of t in $\mathcal{R}$.

If one wants computational equivalence by means of reduction, one has to require confluence as a desired property of both the original and the transformed system, because no search is involved in the process of reaching a normal form. Instead, if one allows the underlying engine to search for normal forms, such as in logic programming paradigms, then one can replace confluence by left linearity (Theorems 3 and 5); this was also the approach taken in [2]. Note, however, that search is potentially exponential in the size of the reduction.

6   Discussion and Future Work

We presented a technique to eliminate conditional rules by replacing them with unconditional rules. The generated TRS is computationally equivalent with the original CTRS provided that the CTRS is ground confluent. Besides the theoretical results, we have also empirically shown that the proposed transformation may lead to the development of faster conditional rewrite engines. In the case of constructor-based CTRSs, the operation {_} is not needed, so our transformation becomes the same as the one in [2]; thus, our theoretical results imply that the transformation in [2] preserves ground confluence, a result not proved in [2] but approached in [5].

We believe that the results presented here can be easily generalized to conditional rewriting systems with extra variables in conditions (deterministic(D) CTRSs). In this framework, operational termination is equivalent to quasi-decreasingness and left linearity translates to *semilinearity* [15] of DCTRSs. The nontrivial proofs of the results in Section 5 (see [20]), were engineered to also work for this case. We refer the interested reader to [20] for details.

Techniques to compact the generated TRS are worthwhile investigating. Also, propagation rules for {_} can destroy useful partial reductions; can one adapt our transformation to restart only the conditions that are invalidated when a rewrite step occurred? We believe that confluence is preserved even in the absence of left linearity or termination but we have not been able to prove it. None of the transformations mentioned in this paper can handle arbitrary rewriting *modulo axioms* in the source CTRS. This seems to be a highly non-trivial problem in its entire generality; however some restricted uses of operators modulo axioms can be handled at no additional complexity (see [20]).

Acknowledgments. We warmly thank Andrei Popescu for several technical suggestions, Claude Marché for referring us to Example 1, Salvator Lucas and José Meseguer for encouraging remarks and suggestions on how to extend this work, and the careful referees for insightful comments on the draft of this paper.

References

[1] H. Aida, J. A. Goguen, and J. Meseguer. Compiling concurrent rewriting onto the rewrite rule machine. In *CTRS'90*, volume 516 of *LNCS*, pages 320–332, 1990.

[2] S. Antoy, B. Brassel, and M. Hanus. Conditional narrowing without conditions. In *PPDP'03*, pages 20–31. ACM Press, 2003.

[3] J. A. Bergstra and J. W. Klop. Conditional rewrite rules: Confluence and termination. *J. of Computer and System Sciences*, 32(3):323–362, 1986.

[4] P. Borovansky, H. Cirstea, H. Dubois, C. Kirchner, H. Kirchner, P. Moreau, C. Ringeissen, and M. Vittek. *ELAN: User Manual*, 2000. Loria, Nancy, France.

[5] B. Brassel. Bedingte narrowing-verfahren mit verzögerter auswertung. Master's thesis, RWTH Aachen, 1999. In German.

[6] M. Clavel, F. Durán, S. Eker, P. Lincoln, N. Martí-Oliet, J. Meseguer, and C. Talcott. *Maude 2.0 Manual*, 2003. http://maude.cs.uiuc.edu/manual.

[7] N. Dershowitz and D. A. Plaisted. Equational programming. In J. E. Hayes, D. Michie, and J. Richards, editors, *Machine Intelligence 11*, pages 21–56. 1988.

[8] R. Diaconescu and K. Futatsugi. *CafeOBJ Report*. World Scientific, 1998. AMAST Series in Computing, volume 6.

[9] E. Giovannetti and C. Moiso. Notes on the elimination of conditions. In *CTRS'87*, volume 308 of *LNCS*, pages 91–97. Springer, 1987.

[10] J. Goguen, T. Winkler, J. Meseguer, K. Futatsugi, and J.-P. Jouannaud. Introducing OBJ. In *Software Engineering with OBJ*, pages 3–167. Kluwer, 2000.

[11] M. Hanus. The integration of functions into logic programming: From theory to practice. *The Journal of Logic Programming*, 19 & 20:583–628, 1994.

[12] C. Hintermeier. How to transform canonical decreasing CTRSs into equivalent canonical TRSs. In *CTRS'94*, volume 968 of *LNCS*, pages 186–205, 1994.

[13] S. Lucas, C. Marché, and J. Meseguer. Operational termination of conditional term rewriting systems. *Inf. Proc. Letters*, 95(4):446–453, August 2005.

[14] M. Marchiori. Unravelings and ultra-properties. In *ALP'96*, volume 1139 of *LNCS*, pages 107–121. Springer, 1996.

[15] M. Marchiori. On deterministic conditional rewriting. Computation Structures Group, Memo 405, MIT Laboratory for Computer Science, 1997.

[16] N. Nishida, M. Sakai, and T. Sakabe. On simulation-completeness of unraveling for conditional term rewriting systems. In *LA Symposium 2004 Summer*, volume 2004-7 of *LA Symposium*, pages 1–6, 2004.

[17] E. Ohlebusch. Transforming conditional rewrite systems with extra variables into unconditional systems. In *LPAR'99*, volume 1705 of *LNCS*, pages 111–130, 1999.

[18] E. Ohlebusch. *Advanced Topics in Term Rewriting*. Springer, 2002.

[19] G. Roşu. From conditional to unconditional rewriting. In *WADT'04*, volume 3423 of *LNCS*, pages 218–233. Springer, 2004.

[20] T. F. Şerbănuţă and G. Roşu. Computationally equivalent elimination of conditions. Technical Report UIUCDCS-R-2006-2693, UIUC, February 2006.

[21] M. van den Brand, J. Heering, P. Klint, and P. A. Olivier. Compiling language definitions: the ASF+SDF compiler. *ACM TOPLAS*, 24(4):334–368, 2002.

[22] P. Viry. Elimination of conditions. *J. of Symb. Comp.*, 28:381–401, Sept. 1999.

On the Correctness of Bubbling[*]

Sergio Antoy, Daniel W. Brown, and Su-Hui Chiang

Department of Computer Science
Portland State University
P.O. Box 751
Portland, OR 97207

Abstract. Bubbling, a recently introduced graph transformation for functional logic computations, is well-suited for the reduction of redexes with distinct replacements. Unlike backtracking, bubbling preserves operational completeness; unlike copying, it avoids the up-front construction of large contexts of redexes, an expensive and frequently wasteful operation. We recall the notion of bubbling and offer the first proof of its completeness and soundness with respect to rewriting.

1 Introduction

Non-determinism is one of the most appealing features of functional logic programing. A program is *non-deterministic* when its execution may evaluate some expression that has multiple results. To better understand this concept, consider a program to color a map of the Pacific Northwest so that no pair of adjacent states shares a color. The following declarations, in Curry [15], define the well-known topology of the problem:

```
data State = WA | OR | ID | BC
states = [WA,OR,ID,BC]                                        (1)
adjacent = [(WA,OR),(WA,ID),(WA,BC),(OR,ID),(ID,BC)]
```

The colors to be used for coloring the states and a non-deterministic operation, paint, to pair its argument to a color are defined below. The library operation "?" non-deterministically selects either of its arguments.

```
data Color = Red | Green | Blue                              (2)
paint x = (x, Red ? Green ? Blue)
```

The rest of the program follows:

```
solve | all diffColor adjacent = theMap
    where theMap = map paint states
          diffColor (x,y) = colorOf x /= colorOf y           (3)
          lookup ((s,c):t) x = if s==x then c
                                       else lookup t x
          colorOf = lookup theMap
```

The evaluation of solve solves the problem. In particular, theMap associates a color to each state and so represents the map, diffColor tells whether the

---

[*] Partially supported by the NSF grant CCR-0218224.

F. Pfenning (Ed.): RTA 2006, LNCS 4098, pp. 35–49, 2006.

colors associated to two states are different, `lookup` looks up the color associated to a state in the map, `all` and `map` are well-known library functions for list manipulation, and the condition of `solve` ensures that no adjacent states have been assigned the same color.

Non-determinism reduces the effort of designing and implementing data structures and algorithms to encode this problem into a program. The simplicity of the non-deterministic solution inspires confidence in the program's correctness. The implementation of non-deterministic functional logic programs has not been studied as extensively as that of deterministic programs.

This paper addresses both theoretical and practical aspects of the implementation of non-determinism. Section 2 highlights some deficiencies of typical implementations of non-determinism and sketches our proposed solution. Section 3 discusses the background of our work. Section 4 defines a relation on graphs that is at the core of our approach. Section 5 proves the correctness of our approach. Section 6 briefly addresses related work. Section 7 offers our conclusion.

2   Motivation

We regard a functional logic program as a term rewriting system (TRS) [8,9,10,18] or a graph rewriting system (GRS) [11,21] with the constructor discipline [20]. Source-level constructs such as data declarations, currying, higher-order and anonymous functions, partial application, nested scopes, etc. can be transformed by a compilation process into ordinary rewrite rules [15]. The execution of a program is the repeated application of narrowing steps to a term until either a constructor term is reached, in which case the computation *succeeds*, or an unnarrowable term with some occurrence of a defined operation is reached, in which case the computation *fails*. Examples of the latter are an attempt to divide by zero or to return the first element of an empty list.

The instantiation of free variables in narrowing steps does not play any specific role in our discussion as well as in the program we presented in the introduction. In this paper, we are mostly concerned with rewriting. For many problems in this area, extending results from rewriting to narrowing requires only a moderate effort. We will sketch the extension of our work to narrowing in the final section.

Our focus is on the interaction of non-determinism and sharing. In a deterministic system, evaluating a shared subexpression twice is merely inefficient; in a non-deterministic system, it can lead to unsoundness. For instance, in the map coloring example, the value of `theMap` is any possible association of a color to a state. In the program, there are two occurrences of `theMap`, besides its definition. One occurrence is returned as the output of the program; the other is constrained to be a correct solution of the problem. Obviously, if the values of these occurrences were not the same, the output of the program would likely be wrong.

A TRS with non-deterministic operations is typically non-confluent. Operationally, there are two main approaches to computations in a non-confluent TRS: *backtracking* and *copying*. While the former is standard terminology, we do not know any commonly accepted name for the latter. Copying is more powerful since steps originating from distinct non-deterministic choices can be interleaved, which is essential to ensure the completeness of the results. We informally describe a computation with each approach. Let $t[u]$ be a term in which $t[\,]$ is a context and u is a subterm that non-deterministically evaluates to x or y.

With *backtracking*, the computation of $t[u]$ first requires the evaluation of $t[x]$. If this evaluation fails to produce a constructor term, the computation continues with the evaluation of $t[y]$. Otherwise, if and when the evaluation of $t[x]$ succeeds, the interpreter may give the user the option of evaluating $t[y]$.

With *copying*, the computation of $t[u]$ consists of the simultaneous (e.g., by interleaving steps), independent evaluations of $t[x]$ and $t[y]$. If either evaluation produces a constructor term, this term is a result of the computation, and the interpreter may give the user the option of continuing the evaluation of the other term. If the evaluation of one term fails to produce a constructor term, the evaluation of the other term continues unaffected.

Both backtracking and copying have been used in the implementation of FL languages. For example, PAKCS [14] and TOY [19] are based on backtracking, whereas the FLVM [7] and the interpreter of Tolmach et al. [22] are based on copying. Unfortunately, both backtracking and copying as described above have non-negligible drawbacks. Consider the following program, where div denotes the usual integer division operator and n is some positive integer.

$$\begin{aligned}
&\texttt{loop = loop} \\
&\texttt{f x = 1+(2+(...+(}n\texttt{ `div` x)...))}
\end{aligned} \tag{4}$$

We describe the evaluation of $t = \texttt{f (loop ? 1)}$ with backtracking. If the first choice for the non-deterministic expression is loop, no value of t is ever computed even though t has a value, since the evaluation of f loop does not terminate. This is a well-known problem of backtracking referred to as the loss of *completeness*.

We describe the evaluation of $t = \texttt{f (0 ? 1)}$ with copying. Both f 0 and f 1 are evaluated. Of course, the evaluation of the first one fails. The problem in this case is the construction of the term 1+(2+(...+(n `div` 0)...)). The effort to construct this term, which becomes arbitrarily large as n grows, is wasted, since the first step of the computation, which is needed, is a division by zero, and consequently the computation fails.

Thus, copying may needlessly construct terms and backtracking may fail to produce results. A recently proposed approach [5], called *bubbling*, avoids these drawbacks. The idea is to slowly "move" a choice up its context and evaluate both its arguments. Bubbling is a compromise between evaluating only one non-deterministic choice, as in backtracking, and duplicating the entire context of each non-deterministic choice, as in copying. Bubbling is free from the drawbacks of backtracking and copying discussed earlier.

The evaluation of f (loop ? 1) by bubbling produces (f loop) ? (f 1). Contrary to backtracking, no unrecoverable choice is made in this step. Both arguments of "?" can be evaluated concurrently, e.g., as in [5]. The evaluation of the first argument does not terminate; however, this does not prevent obtaining the value of the second argument.

Likewise, the evaluation of `f (0 ? 1)` goes (roughly) through the following intermediate terms, where `fail` is a distinguished symbol denoting any expression that cannot be evaluated to a constructor term:

```
f (0 ? 1)
    → 1+(2+(...+(n ʻdivʻ (0 ? 1))...))
    → 1+(2+(...+((n ʻdivʻ 0) ? (n ʻdivʻ 1))...))          (5)
    → 1+(2+(...+(fail ? (n ʻdivʻ 1))...))
    → 1+(2+(...+(n ʻdivʻ 1)...))
```

The `fail` alternative is dropped. Since `fail` occurs at a position where a constructor-rooted term is needed, it cannot lead to a successful computation.

In (5), the obvious advantages of bubbling are that no choice is left behind and no unnecessarily large context is copied. In the second step, we have distributed the parent of an occurrence of the choice operation over its arguments. Unfortunately, a "distributive property" of the kind $f(x\,?\,y) = f(x)\,?\,f(y)$ is unsound in the presence of sharing.

Consider the following operation:

$$\texttt{f x = (not x, not x)} \qquad (6)$$

and the term $t = \texttt{f (True ? False)}$. The evaluation semantics of non-right linear rewrite rules, such as (6), is called *call-time* choice [17]. Informally, the non-deterministic choice for the argument of `f` is made at the time of `f`'s application. Therefore, the instances of `x` in the right-hand side of (6) should all evaluate to `True` or all to `False`. With an eager strategy, the call-time choice is automatic, and the only available option. With a lazy strategy, the call-time choice is relatively easy to implement by "sharing" the occurrences of `x`. That is, there is only one occurrence of the term bound to `x`. All the occurrences of `x` refer to this term. The term being evaluated is the *graph* depicted in the left-hand side of the following figure:

Fig. 1. The left-hand side depicts a term graph. The right-hand side is obtained from the left-hand side by bubbling up to the parents the non-deterministic choice. The two term graphs have a different set of constructor normal forms.

The right-hand side of the above figure shows the result of bubbling up the non-deterministic choice in a way similar to (5). This term has 4 normal forms. One is `(True,False)`, which is not obtainable with either backtracking or copying and is not intended by the call-time choice semantics. Therefore, although advantageous in some situations, unrestricted bubbling can be unsound.

In the following sections we formalize a sound approach to non-deterministic computations with shared terms based on the idea of bubbling introduced in this section. This formalization is the foundation of a recently discovered strategy [5] that computes both rewriting and bubbling steps.

3   Background

TRSs have been used extensively to model FL programs. This modeling has been very successful for some problems, e.g., the discovery of efficient narrowing strategies and the study of their properties; see [4] for a survey. However, a TRS only approximates a FL program, because it does not adequately capture the sharing of subexpressions in an expression. As we discussed in the previous section, and our introductory example shows, sharing is an essential semantic component of the execution of a program.

GRSs [11,21] model FL programs more accurately than do TRSs. Unfortunately, they are also more complex than TRSs, and some variations exist in their formalization. In this paper, we follow the systemization of Echahed and Janodet [11] because the class of GRSs that they consider is a good fit for our programs. The space alloted to this paper prevents us from recalling relevant definitions and results of [11]. Luckily, this paper is easily accessible on-line at http://citeseer.ist.psu.edu/echahed97constructorbased.html.

In this paper, we assume that programs are *overlapping inductively sequential* [4,2] term graph rewriting systems, abbreviated GRSs, and computations are rewriting sequences of admissible term graphs. We recall that a graph is *admissible* [11, Def. 18] if none of its defined operations belongs to a cycle.

Our choice of programs is motivated by the expressiveness of this class (e.g., as shown by the introductory example), by the existence of a strategy that performs only steps that are needed modulo a non-deterministic choice [2], and by the fact that computations for the entire constructor based programs can be implemented by this class via a transformation [3]. Non-deterministic computations in this class are supported by the single operation defined below.

Definition 1 [Choice operation]. The *choice operation*, denoted by the infix operator "?", is defined by the following rules:

$$x \text{ ? } y = x$$
$$x \text{ ? } y = y \qquad \qquad \square$$

We assume that this is the only overlapping operation of a GRS. Any other overlapping can be eliminated, without changing the meaning of a program, using the choice operation, as discussed in [2] and shown in our introductory example.

Definition 2 [Limited overlapping]. A *limited overlapping* inductively sequential GRS, abbreviated *LOIS*, is a constructor based GRS, S, such that the signature of S contains the choice operation "?" presented in Def. 1 and every other defined operation of S is inductively sequential. □

We need an additional definition, which is crucial to our approach.

Definition 3 [Dominance]. A node d *dominates* a node n in a rooted graph t if every path from the root of t to n contains d. If d and n are distinct, then d *properly dominates* n in t. □

For example, in the left-hand side graph of Fig. 1, the occurrence of "?" is properly dominated by the root only. Every other occurrence, except the root, is properly dominated by its predecessor.

Echahed and Janodet [11] formalize rewriting, including an efficient strategy, for the inductively sequential term graph rewriting systems. This class is similar to ours, except for the presence of the choice operation. Following their lead, we always use "fresh" rules in rewrite steps. This is justified by the following example:

$$
\begin{aligned}
&\texttt{t = (ind, ind)} \\
&\texttt{ind = coin} \\
&\texttt{coin = 0 ? 1}
\end{aligned}
\tag{7}
$$

The intended semantics is that each occurrence of `ind` in `t` is evaluated independently of the other (`ind` is not a variable) and therefore `t` has four values, every pair in which each component is either `0` or `1`. To compute all the intended values of `t`, it is imperative that a rewrite step uses a *variant* [11, Def. 19] of a rewrite rule, namely a clone of the rule with fresh nodes (and variables). A consequence of using variants of rules is that the equality of graphs resulting from rewrite steps can be assessed only modulo a renaming of their nodes [11, Def. 15].

4   Bubbling

Computations that perform non-deterministic steps must preserve in some form the context of a redex when the redex has distinct replacements. Typically, some portions of the context are reconstructed, as in backtracking, or are duplicated, as in copying. An overall goal of bubbling is to limit these activities. In the following, we precisely define which portions of a context of a redex are cloned in our approach.

Definition 4 [Partial renaming]. Let $g = \langle \mathcal{N}_g, \mathcal{L}_g, \mathcal{S}_g, \mathcal{Roots}_g \rangle$ be a term graph over $\langle \Sigma, \mathcal{N}, \mathcal{X} \rangle$, $\mathcal{N}_p$ a subset of $\mathcal{N}_g$ and $\mathcal{N}_q$ a set of nodes disjoint from $\mathcal{N}_g$. A *partial renaming* of g with respect to $\mathcal{N}_p$ and $\mathcal{N}_q$ is a bijection $\Theta : \mathcal{N} \to \mathcal{N}$ such that:

$$
\Theta(n) = \begin{cases} n' & \text{where } n' \in \mathcal{N}_q, \text{ if } n \in \mathcal{N}_p; \\ n & \text{otherwise.} \end{cases}
\tag{8}
$$

Similar to substitutions, we call $\mathcal{N}_p$ and $\mathcal{N}_q$, the *domain* and *image* of Θ, respectively. We overload Θ to graphs as follows: $\Theta(g) = g'$ is a graph over $\langle \Sigma, \mathcal{N}, \mathcal{X} \rangle$ such that:

- $\mathcal{N}_{g'} = \Theta(\mathcal{N}_g)$,
- $\mathcal{L}_{g'}(m) = \mathcal{L}_g(n)$, iff $m = \Theta(n)$,
- $m_1 m_2 \ldots m_k = \mathcal{S}_{g'}(m_0)$ iff $n_1 n_2 \ldots n_k = \mathcal{S}_g(n_0)$, where for $i = 0, 1, \ldots k$, $k \geqslant 0$, $m_i = \Theta(n_i)$,
- $\mathcal{R}oots_{g'} = \mathcal{R}oots_g$. □

In simpler words, g' is equal to g in all aspects except that some nodes in $\mathcal{N}_g$, more precisely all and only those in $\mathcal{N}_p$, are consistently renamed, with a "fresh" name, in g'. Obviously, the cardinalities of the domain and image of a partial renaming are the same.

Lemma 1. *If g is a graph and g' is a partial renaming of g with respect to some $\mathcal{N}_p$ and $\mathcal{N}_q$, then g and g' are compatible.*

Proof. Immediate from the notion of compatibility [11, Def. 6] and the construction of g' in Definition 4. □

The evaluation of an admissible term graph t_0 in a GRS S is a sequence of graphs $t_0 \leadsto t_1 \leadsto t_2 \cdots$ where for every natural number i, t_{i+1} is obtained from t_i either with a rewrite step of S or with a *bubbling* step, which is defined below.

Definition 5 [Bubbling]. Let g be a graph and c a node of g such that the subgraph of g at c is of the form $x\,?\,y$, i.e., $g|_c = x\,?\,y$. Let d be a proper dominator of c in g and $\mathcal{N}_p$ the set of nodes that are on some path from d to c in g, including d and c, i.e., $\mathcal{N}_p = \{n \mid n_1 n_2 \ldots n_k \in \mathcal{P}_g(d, c)$ and $n = n_i$ for some $i\}$, where $\mathcal{P}_g(d, c)$ is the set of all paths from d to c in g. Let Θ_x and Θ_y be partial renamings of g with domain $\mathcal{N}_p$ and disjoint images. Let $g_q = \Theta_q(g|_d[c \leftarrow q])$, for $q \in \{x, y\}$. The *bubbling* relation on graphs is denoted by "$\simeq$" and defined by $g \simeq g[d \leftarrow g_x\,?\,g_y]$, where the root node of the replacement of g at d is obviously fresh. We call c and d the *origin* and *destination*, respectively, of the bubbling step, and we denote the step with "$\simeq_{cd}$" when this information is relevant. □

In simpler words, bubbling moves a choice in a graph up to a dominator node. In executing this move, some portions of the graph, more precisely those between the end points of the move, must be cloned. An example of bubbling is shown in Figure 2. In this example, the dominator is the root of the graph, but in general the destination node can be any proper dominator of the origin. In practice, it is convenient to bubble a choice only to produce a redex. The strategy introduced in [5] ensures this desirable property.

The bubbling relation entails 3 graph replacements. By Lemma 1, the graphs involved in these replacements are all compatible with each other. Therefore, the bubbling relation is well defined according to [11, Def. 9]. In particular, except for the nodes being renamed, g_x and g_y can share nodes between themselves and/or with g. Any sharing among these (sub)graphs is preserved by bubbling.

Two adjacent bubbling steps can be composed into a "bigger" step.

Theorem 1 (Transitivity of bubbling). *Let S be a GRS. For all term graphs t, u and v over the signature of S and for all c and d nodes of t and d and e nodes of u, modulo a renaming of nodes, if $t \simeq_{cd} u$ and $u \simeq_{de} v$ then e is a node of t and $t \simeq_{ce} v$.*

Fig. 2. The left-hand side depicts a term graph. The right-hand side is obtained from the left-hand side by bubbling the non-deterministic choice up to a proper dominator. The two term graphs have the same set of constructor normal forms.

Proof. If c is a node labeled by a choice operation, c_l and c_r denotes the left and right successors of c. Let w be defined by $t \simeq_{ce} w$ and consider the expressions defining u, v and w:

$$u = t[d \leftarrow (\Theta_{dc_l}(t|_d[c \leftarrow t|_{c_l}]) \, ? \, \Theta_{dc_r}(t|_d[c \leftarrow t|_{c_r}]))]$$
$$v = u[e \leftarrow (\Theta_{ed_l}(u|_e[d \leftarrow u|_{d_l}]) \, ? \, \Theta_{ed_r}(u|_e[d \leftarrow u|_{d_r}]))] \qquad (9)$$
$$w = t[e \leftarrow (\Theta_{ec_l}(t|_e[c \leftarrow t|_{c_l}]) \, ? \, \Theta_{ec_r}(t|_e[c \leftarrow t|_{c_r}]))]$$

where Θ_{xy} is a renaming whose domain is the set of the nodes in any path between x and y. Also, we assume that the images of all renamings are disjoint.

We prove that $v = w$ modulo a renaming of nodes. The portion of u at and above e is the same as in t. Using this condition *twice*, we only have to prove $\Theta_{ed_l}(t|_e[d \leftarrow u|_{d_l}]) = \Theta_{ec_l}(t|_e[c \leftarrow t|_{c_l}])$ and the analogous equation for the right-hand side argument. By construction, $u|_{d_l} = \Theta_{dc_l}(t|_d[c \leftarrow t|_{c_l}])$. Thus, $\Theta_{ed_l}(t|_e[d \leftarrow u|_{d_l}]) = \Theta_{ed_l}(t|_e[d \leftarrow \Theta_{dc_l}(t|_d[c \leftarrow t|_{c_l}])])$. Since no node is duplicated by renamings, we have that $\Theta_{ed_l}(t|_e[d \leftarrow \Theta_{dc_l}(t|_d[c \leftarrow t|_{c_l}])]) = \Theta_{ed_l} \circ \Theta_{dc_l}(t|_e[d \leftarrow t|_d[c \leftarrow t|_{c_l}]])$. Since $t|_d$ is modified only at c and c is below d, $t|_e[d \leftarrow t|_d[c \leftarrow t|_{c_l}]] = t|_e[c \leftarrow t|_{c_l}]$. Thus, by equational reasoning, $v = w$ except for the renamings of nodes, and the claim holds. $\qquad\square$

Bubbling creates a natural mapping between two graphs. If $t \simeq u$, then every node of u "comes" from a node of t. This mapping, which is instrumental in proving some of our claim, is formalized below.

Definition 6 [Natural mapping]. Let S be a *GRS*, t a graph over the signature of S and $t \simeq_{cd} u$, for some graph u and nodes c and d of t. We call *natural* the mapping $\mu : \mathcal{N}_u \rightarrow \mathcal{N}_t$ defined as follows. By construction, $u = t[d \leftarrow t']$, for some term graph t'. Let d' be the root node of t'. The construction of u involves two renamings in the sense of Def. 4; let us call them Θ_x and Θ_y. We define μ on n, a node of u, as follows:

$$\mu(n) = \begin{cases} c & \text{if } n = d'; \\ \Theta_x^{-1}(n) & \text{if } n \text{ is in the image of } \Theta_x; \\ \Theta_y^{-1}(n) & \text{if } n \text{ is in the image of } \Theta_y; \\ n & \text{otherwise.} \end{cases} \qquad (10)$$

Observe that the images of Θ_x and Θ_y are disjoint, hence the second and third cases of (10) are mutually exclusive. □

The next lemma shows that a rule of "?" applied before a bubbling step at the origin or after a bubbling step at the destination produces the same outcome.

Lemma 2 (Same rule). *Let S be a LOIS and t an admissible term graph over the signature of S. If $t \simeq_{cd} u$, $t \to_{c,R} v$ and $u \to_{d,R} w$, then $v = w$ modulo a renaming of nodes.*

Proof. R is a rule of "?". Without loss of generality, we assume that it is the rule that selects the left argument. By assumption the subgraph of t at c is of the form $x \,?\, y$. Hence $t = t[c \leftarrow x \,?\, y]$ and $t \to_{c,R} v = t[c \leftarrow x]$. By definition of bubbling, $u = t[d \leftarrow (\Theta_x(t|_d[c \leftarrow x]) \,?\, \Theta_y(t|_d[c \leftarrow y]))]$, for some renamings Θ_x and Θ_y. Therefore $u \to_{d,R} w = t[d \leftarrow \Theta_x(t|_d[c \leftarrow x])] = \Theta_x(t[d \leftarrow t|_d[c \leftarrow x]]) = \Theta_x(t[c \leftarrow x])$. □

5   Correctness

In this section we state and prove the correctness of our approach. The notion of a *redex pattern* defines the set of nodes below a node n labeled by an operation f that determines that a rule of f can be applied at n. Recall that a matcher is a function that maps the nodes of one graph to those of another, preserving the labeling and successor functions.

Definition 7 [Redex pattern]. Let t be a graph, $l \to r$ a rewrite rule, and n a node of t such that l matches t at n with matcher h. We call *redex pattern* of $l \to r$ in t at n the set of nodes of t that are images according to h of a node of l with a constructor label. □

We are convening that a node n is not in any redex pattern at n. This is just a convenient convention.

The following example shows that some pairs of bubbling and rewriting steps do not commute. This is a significant condition that prevents proof techniques based on *parallel moves* [16]. Although our GRSs are not orthogonal, some form of parallel moves is available for *LOIS* [2]. Consider the term $t = \mathtt{snd}\,(1,2\,?\,3)$, where $\mathtt{snd}$ is the function that returns the second component of a pair and, obviously, there is no sharing. Bubbling the choice to its parent (see Figure 3) produces $u = \mathtt{snd}\,((1,2)\,?\,(1,3))$. The term u cannot be obtained from t by rewriting. Furthermore, the redex at the root of t has been destroyed by the bubbling step. The following result offers a sufficient condition, namely an appropriate choice of the destination of a bubbling step, for recovering the commutativity of bubbling and rewriting. As customary, for any relation R, $R^=$ denotes the reflexive closure of R.

Lemma 3 (Parallel Bubbling Moves). *Let S be a LOIS and t an admissible term graph over the signature of S. If $t \simeq_{cd} t'$, for some graph t' and nodes c*

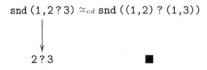

snd $(1,2\,?\,3)\simeq_{cd}$ snd $((1,2)\,?\,(1,3))$

$2\,?\,3$ ∎

Fig. 3. Bubbling and rewriting do not always commute. No *parallel moves* are available for this diagram. Note that the term on the right cannot be reached from the original term by rewriting.

and d of t, and $t \to_{p,R} u$, for some node p of t and rule R of S, and d is not in the redex pattern of R at p in t, then there exists u' such that $t' \overset{+}{\to} u'$ and $u \simeq_{cd}^= u'$ modulo a renaming of nodes.

Proof. Let $P = \mu^{-1}(p)$ be the set of nodes of t' that map to p in t. Observe that P contains either 1 or 2 nodes. We show that R can be applied to any node in P and we define u' as the result of applying R to all the nodes of P. If $p = c$, then by definition $P = \{d\}$. In this case, R is a rule of "?" and consequently $u = u'$ (modulo a renaming of nodes) and the claim immediately holds. If $p \neq c$ and $P = \{p\}$, then the reduction in t is independent of the bubbling step. The redex pattern of R at p in t is either entirely below c, since the label of c is an operation, or entirely above d, since by hypothesis d is not in the redex pattern of R at p in t. The redex pattern of R at p is the same in t and t' and the redex is equally replaced in t and t'. Hence $u \simeq_{cd} u'$. If $p \neq c$ and $P = \{p_1, p_2\}$, with $p_1 \neq p_2$, then the reduction in t is in the portion of t cloned by the bubbling step. The redex pattern of R at p in t is entirely contained in this portion. The redex pattern is entirely below d by hypothesis, and cannot include c since the the label of c is an operation. Thus the redex pattern is entirely cloned in two disjoint occurrences in t'. By reducing both occurrences, in whatever order, t' reduces to u' in two steps and $u \simeq_{cd} u'$. □

$t \simeq_{cd} t'$

$p,R \downarrow \qquad \downarrow +$

$u \simeq_{cd}^= u'$

Fig. 4. Graphical representation of Lemma 3. If the destination of the bubbling step of t is not in the redex pattern of the rewrite step of t, then, for a suitable graph u', the diagram commutes.

Definition 8 [Combined step]. We denote with "$\rightsquigarrow$", called a *combined step*, the union of the bubbling and rewriting relations in a *LOIS*, i.e., $\rightsquigarrow \; = \simeq \cup \to$. □

We now address the completeness of the combined step relation. Since this relation is an extension of the rewrite relation, a traditional proof of completeness would be trivial. Instead, we prove a more interesting claim, namely, no result of a computation is lost by the execution of bubbling steps. Therefore, an implementation of rewriting is allowed to execute bubbling steps, if it is convenient. The completeness of bubbling is not in conflict with the example of Figure 3. Although a bubbling step may destroy a redex, the redex is not irrevocably lost—there always exists a second bubbling step to recover the redex lost by the first step. In the case of Figure 3, a second bubbling step results in snd $(1,2)\,?$ snd $(1,3)$.

Theorem 2 (Completeness of bubbling). *Let S be a LOIS, t an admissible term graph and u a constructor graph such that $t \xrightarrow{*} u$. If $t \simeq_{cd} v$ for some graph v and nodes c and d of t, then $v \xrightarrow{*} u$ modulo a renaming of nodes.*

Proof. The proof is by induction on the length of $t \xrightarrow{*} u$. Base case: If $t = u$, then v does not exist, and the claim vacuously holds. Ind. case: There exist some node p, rule R and graph t_1 such that $t \rightarrow_{p,R} t_1 \xrightarrow{*} u$. We consider two exhaustive cases on d. If d is not in the redex pattern of R at p in t, then, by Lemma 3, there exists a graph v_1 such that $v \xrightarrow{*} v_1$ and $t_1 \simeq^= v_1$ modulo a renaming of nodes. By the induction hypothesis, $v_1 \xrightarrow{*} u$ modulo a renaming of nodes. If d is in the redex pattern of R at p in t, then d is neither the root of t nor the root of v. There exists a dominator e of d in v, witness the root of v, such that $v \simeq_{de} w$; by Theorem 1 $t \simeq_{ce} w$, and e is not in the redex pattern of R at p in t. By Lemma 3, there exists a graph w_1 such that $w \xrightarrow{+} w_1$ and $t_1 \simeq^= w_1$ modulo a renaming of nodes. As in the previous case $w \xrightarrow{*} u$ modulo a renaming of nodes. Since $v \simeq w$ implies $v \rightsquigarrow w$, $v \xrightarrow{*} u$ modulo a renaming of nodes. $\square$

We now turn our attention to the soundness of combined steps. This is somewhat the complement of the completeness. We prove that bubbling non-deterministic choices does not produce results that would not be obtainable without bubbling. Of course, a bubbling step of a term graph t creates a term u that is not reachable from t by rewriting, but any result (constructor normal form) obtainable from u via combined steps can be reached from t via pure rewriting. We begin by proving that a single bubbling step with destination the root node is sound.

Lemma 4 (Single copying soundness). *Let S be a LOIS and t_0 an admissible term graph over the signature of S. If $t_0 \simeq_{cd} t_1 \xrightarrow{*} t_n$, where c is a node of t_0, d is the root of t_0 and t_n is a constructor graph, then $t_0 \xrightarrow{*} t_n$ modulo a renaming of nodes.*

Proof. A diagram of the graphs and steps in the following proof are shown in Fig 5. Since in t_1 the label of the root node d is "?" and t_n is a constructor normal form, there must be an index j such that in the step $t_j \rightarrow t_{j+1}$ a rule R_j of "?" is applied at d. Without loss of generality, we assume that R_j is the rule of "?" that selects the left argument and we denote with d_l the left successor of d in t_i for $i = 1, 2, \ldots, j$. In different graphs, d_l may denote different nodes. We prove the existence of a sequence $t_0 \rightarrow u_1 \xrightarrow{*} u_n$, such that for all $i = 1, 2, \ldots, n$, $t_i \rightarrow^= u_i$ modulo a renaming of nodes. By induction on i, for all $i = 1, 2, \ldots, j$, we define u_i and we prove that $t_i|_{d_l} = u_i$ modulo a renaming of nodes. The latter implies $t_i \rightarrow_{d,R_i} u_i$. Base case: $i = 1$. The rule R_j can be applied to t_0 at c and we define u_1 as the result, i.e., $t_0 \rightarrow_{c,R_j} u_1$. By Lemma 2, $t_1|_{d_l} = u_1$ modulo a renaming of nodes. Ind. case: We assume the claim for i, where $0 < i < j$, and prove it for $i + 1$. Let $t_i \rightarrow_{p,R_i} t_{i+1}$. If p is a node of $t_i|_{d_l}$, then since $t_i|_{d_l} = u_i$ modulo a renaming of nodes, there exists a node q in u_i that renames p. We define $u_i \rightarrow_{q,R_i} u_{i+1}$ and the claim holds for $i+1$. If p is not a node of $t_i|_{d_l}$, then $t_{i+1}|_{d_l} = t_i|_{d_l}$. We define $u_{i+1} = u_i$ and the claim holds for $i + 1$ in this case too. Now, since $t_j \rightarrow_{d,R_j} t_{j+1}$, we have $t_{j+1} = t_j|_{d_l}$ and therefore $u_{i+1} = t_{j+1}$

modulo a renaming of nodes. For every i such that $j < i < n$, if $t_i \rightarrow_{p,R} t_{i+1}$, we define $u_i \rightarrow_{q,R} u_{i+1}$, where as before q renames p. Clearly, for every i, $j < i \leqslant n$, $t_i = u_i$ modulo a renaming of nodes. Thus, $t_0 \xrightarrow{+} t_n$ exists as claimed. $\qquad\square$

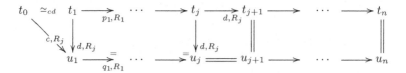

Fig. 5. Diagram of the main graphs and steps involved in the proof of Lemma 4. d is the root node of t_0. R_j is the rule of "?" that selects the left argument. q_i renames p_i.

We believe that the previous proof could be generalized to any bubbling step. However, a simpler and more elegant proof is available by taking advantage of the transitivity and the completeness of bubbling. We show this proof below.

Lemma 5 (Single bubbling soundness). *Let S be a LOIS and t_0 an admissible term graph over the signature of S. If $t_0 \simeq_{cd} t_1 \xrightarrow{*} t_n$, where c and d are nodes of t_0 and t_n is a constructor graph, then $t_0 \xrightarrow{+} t_n$ modulo a renaming of nodes.*

Proof. Suppose that d is not the root of t_0; otherwise the claim is already proved by Lemma 4. Let e be the root node of t_0 and also of t_1. Let u_1 be defined by $t_1 \simeq_{de} u_1$. By the transitivity of bubbling, Th. 1, $t_0 \simeq_{ce} u_1$. By the completeness of bubbling, Th. 2, there exists a sequence $u_1 \xrightarrow{+} t_n$ modulo a renaming of nodes. Therefore, $t_0 \simeq_{ce} u_1 \xrightarrow{*} t_n$. Since e is the root node of t_0, by Lemma 4, $t_0 \xrightarrow{+} t_n$ modulo a renaming of nodes. $\qquad\square$

Theorem 3 (Soundness of bubbling). *Let S be a LOIS and t an admissible term graph over the signature of S. If $t \xrightarrow{*} u$, for some constructor graph u, then $t \xrightarrow{*} u$ modulo a renaming of nodes.*

Proof. By induction on the number of bubbling steps in $t \xrightarrow{*} u$. $\qquad\square$

6   Related Work

Bubbling is introduced in [5] with a rewriting strategy for the overlapping inductively sequential GRSs. This strategy determines, in theory very efficiently, when to execute ordinary rewrite steps and/or bubbling steps. A bubbling step is computed only if it promotes a needed (modulo a non-deterministic choice) rewrite step. Our work proves that the execution of the bubbling steps computed by this strategy preserves all and only the constructor normal forms reachable from a term by pure rewriting. The use of bubbling in the strategy eliminates the incompleteness of backtracking and the inefficiency of copying.

Although strategies for functional logic computations [4] and term graph rewriting [21] have been extensively investigated, the work on strategies for term graph rewriting systems as models of functional logic programs has been relatively scarce. The line of work closest to ours is [11,12]. A substantial difference of our work with this line is the class of programs we consider, namely non-deterministic ones. Non-determinism is a major element of functional logic programming. Hence, our work fills a major conceptual and practical gap in this area. The attempt to minimize the cost of non-deterministic steps by limiting the copying of the context of a redex by bubbling is original.

Other efforts on handling non-determinism in functional and functional logic computations with shared subexpressions include [17], which introduces the *call-time choice semantics* to ensure that shared terms are evaluated to the same result; [13], which defines a rewriting logic that among other properties provides the call-time choice; and [1] and [22], which define operational semantics based on *heaps* and *stores* specifically for the interaction of non-determinism and sharing.

These efforts, prompted by implementations, abstract the interactions between non-determinism and sharing. In practice, all these implementations adopt strategies, summarized in [4], that have been designed and proved correct for *term* rather than *graph* rewriting or narrowing. Although for a strategy this difference is small, addressing sharing indirectly through computational data structures such as heaps and stores rather than directly prevents graph operations, such as bubbling, which are potentially beneficial.

7   Conclusion and Future Work

Bubbling, with interleaving steps on the arguments of an occurrence of the choice operation, ensures the soundness and completeness of computations without incurring the cost of copying the contexts of redexes with distinct replacements. Programs in which *don't know* non-determinism is appropriately used are likely to produce some terms that fail to evaluate to constructor normal forms. Hence, avoiding the construction of the contexts of these terms can improve the efficiency of these programs.

For example, this situation can be seen in our program for coloring a map. In finding the first solution of the problem, the operation `paint` is called 10 times. Since only four calls are needed, six choices of some color for some state eventually fail. Saving the partial construction of six contexts of `paint` can potentially improve the efficiency of execution. We are working on an implementation, within the FLVM [7], to quantify the expected improvements. The results of this paper ensure the theoretical correctness of a component of our implementation.

Bubbling steps can be executed any time a choice operation occurs at a non-root node. The problem of determining when it is appropriate to execute a bubbling step and the destination of this step is elegantly solved in [5]. A strategy similar in intent to [11] and [2] determines when a bubbling step promotes a needed (modulo a non-deterministic choice) rewrite step. Thus, bubbling steps are executed only when they are necessary to keep a computation going. This result complements quite nicely several optimality properties known for strategies for functional logic computations [4].

The focus of continued work on this topic is to extend the theory and the implementation to cover narrowing. Narrowing steps are inherently non-deterministic and therefore naturally expressed using the choice operation [6]. For example, to narrow `not x`, where `x` is a free variable, we bind `x` to `True ? False`—the patterns in the definition of `not`—and continue the evaluation of the instantiated term. In our framework, this would require a bubbling step.

Variables are singletons in their contexts. This is a key reason to represent expressions with graphs. However, in our framework, expressions with choice operations represent *sets* of ordinary expressions. Therefore, a variable that has an ancestor node labeled by a choice operation must be handled with care. For example, consider the following contrived program:

$$
\begin{aligned}
&\texttt{f x = g x ? h x} \\
&\texttt{g 0 = 0} \qquad\qquad\qquad\qquad (11) \\
&\texttt{h \_ = 1}
\end{aligned}
$$

The expression `f x`, where `x` is free, evaluates to two different terms with two different bindings. In evaluating the right-hand side of `f`, before instantiating `x` in a narrowing step, `x` must be "standardized apart" as if evaluating (`g u ? h v`) where `u` and `v` are distinct and free. The situation exemplified in (11) is characterized by a variable x that belongs to two terms encoded within a single expression of our framework. The standardization apart of a variable is accomplished by a graph transformation similar to a bubbling step.

References

1. M. Alpuente, M. Hanus, S. Lucas, and G. Vidal. Specialization of functional logic programs based on needed narrowing. *Theory and Practice of Logic Programming*, 5(3):273–303, 2005.
2. S. Antoy. Optimal non-deterministic functional logic computations. In *Proceedings of the Sixth International Conference on Algebraic and Logic Programming (ALP'97)*, pages 16–30, Southampton, UK, September 1997. Springer LNCS 1298.
3. S. Antoy. Constructor-based conditional narrowing. In *Proceedings of the Third ACM SIGPLAN International Conference on Principles and Practice of Declarative Programming*, pages 199–206. ACM Press, 2001.
4. S. Antoy. Evaluation strategies for functional logic programming. *Journal of Symbolic Computation*, 40(1):875–903, 2005.
5. S. Antoy, D. Brown, and S. Chiang. Lazy context cloning for non-deterministic graph rewriting. In *Proc. of the 3rd International Workshop on Term Graph Rewriting, Termgraph'06*, pages 61–70, Vienna, Austria, April 2006.
6. S. Antoy and M. Hanus. Overlapping rules and logic variables in functional logic programs. In *Proceedings of the 22nd International Conference on Logic Programming (ICLP'06)*, Seattle, WA, August 2006. To appear.
7. S. Antoy, M. Hanus, J. Liu, and A. Tolmach. A virtual machine for functional logic computations. In *Proc. of the 16th International Workshop on Implementation and Application of Functional Languages (IFL 2004)*, pages 108–125, Lubeck, Germany, Sept. 2005. Springer LNCS 3474.

8. F. Baader and T. Nipkow. *Term Rewriting and All That.* Cambridge University Press, 1998.
9. M. Bezem, J. W. Klop, and R. de Vrijer (eds.). *Term Rewriting Systems.* Cambridge University Press, 2003.
10. N. Dershowitz and J.-P. Jouannaud. Rewrite systems. In J. van Leeuwen, editor, *Handbook of Theoretical Computer Science, Vol. B*, pages 243–320. Elsevier, 1990.
11. R. Echahed and J.-C. Janodet. On constructor-based graph rewriting systems. Research Report 985-I, IMAG, 1997.
12. R. Echahed and J.-C. Janodet. Admissible graph rewriting and narrowing. In *Proceedings of the Joint International Conference and Symposium on Logic Programming*, pages 325 – 340, Manchester, June 1998. MIT Press.
13. J. C. González Moreno, F. J. López Fraguas, M. T. Hortalá González, and M. Rodríguez Artalejo. An approach to declarative programming based on a rewriting logic. *The Journal of Logic Programming*, 40:47–87, 1999.
14. M. Hanus (ed.). PAKCS 1.7.1: The Portland Aachen Kiel Curry System. Available at http://www.informatik.uni-kiel.de/~pakcs, March 27, 2006.
15. M. Hanus (ed.). Curry: An integrated functional logic language (vers. 0.8.2). Available at http://www.informatik.uni-kiel.de/~curry, March 28, 2006.
16. G. Huet and J.-J. Lévy. Computations in orthogonal term rewriting systems. In J.-L. Lassez and G. Plotkin, editors, *Computational logic: essays in honour of Alan Robinson.* MIT Press, Cambridge, MA, 1991.
17. H. Hussmann. Nondeterministic algebraic specifications and nonconfluent rewriting. *Journal of Logic Programming*, 12:237–255, 1992.
18. J.W. Klop. Term rewriting systems. In S. Abramsky, D. Gabbay, and T. Maibaum, editors, *Handbook of Logic in Computer Science*, volume II. Oxford University Press, 1992.
19. F. López-Fraguas and J. Sánchez-Hernández. TOY: A multiparadigm declarative system. In *Proceedings of RTA '99*, pages 244–247. Springer LNCS 1631, 1999.
20. M. J. O'Donnell. *Equational Logic as a Programming Language.* MIT Press, 1985.
21. D. Plump. Term graph rewriting. In H.-J. Kreowski H. Ehrig, G. Engels and G. Rozenberg, editors, *Handbook of Graph Grammars*, volume 2, pages 3–61. World Scientific, 1999.
22. A. Tolmach, S. Antoy, and M. Nita. Implementing functional logic languages using multiple threads and stores. In *Proc. of the Ninth International Conference on Functional Programming (ICFP 2004)*, pages 90–102, Snowbird, Utah, USA, Sept. 2004. ACM Press.

Propositional Tree Automata[*]

Joe Hendrix[1], Hitoshi Ohsaki[2], and Mahesh Viswanathan[1]

[1] University of Illinois at Urbana-Champaign
{jhendrix, vmahesh}@uiuc.edu
[2] National Institute of Advanced Industrial Science and Technology
ohsaki@ni.aist.go.jp

Abstract. In the paper, we introduce a new tree automata framework, called *propositional tree automata*, capturing the class of tree languages that are closed under an equational theory and Boolean operations. This framework originates in work on developing a sufficient completeness checker for specifications with rewriting modulo an equational theory. Propositional tree automata recognize regular equational tree languages. However, unlike regular equational tree automata, the class of propositional tree automata is closed under Boolean operations. This extra expressiveness does not affect the decidability of the membership problem. This paper also analyzes in detail the emptiness problem for propositional tree automata with associative theories. Though undecidable in general, we present a semi-algorithm for checking emptiness based on machine learning that we have found useful in practice.

1   Introduction

Tree automata techniques have been commonly used in checking consistency of tree structures. Typical examples include checking sufficient completeness of algebraic specifications [6] and the consistency of semi-structured documents [17]. These applications benefit from the good closure properties and positive decidability results for tree automata. Recently, there are more advanced applications including protocol verification [2,11], type inference [8,10], querying in databases [27,28] and theorem proving [19].

One limitation of tree automata in these applications is that the regularity of languages is not preserved when closed with respect to congruences. In other words, when some algebraic laws such as associativity and commutativity are taken into account, the congruence closure of a regular tree language may no longer be regular. In applications, this lack of closure has required users of tree automata techniques to use complicated and specialized ways of encoding protocols [5]. Many extensions of tree automata have been suggested to address this problem, including multitree automata by Lugiez [20], two-way alternating tree automata by Verma [29], and equational tree automata by Ohsaki [25].

Equational tree automata are a natural mathematical extension of tree automata that recognize tree languages modulo an equational theory. Equational

---

[*] Research supported by ONR Grant N00014-02-1-0715, NSF CAREER CCF-0448178, and NSF CCF-0429639.

tree automata enjoy several nice properties. In particular, they are weakest extensions to tree automata that are closed under congruences. More precisely, when the equational theory is induced by only linear equations (i.e equations whose left- and right-hand sides are linear terms), such automata recognize exactly the congruence closure of regular languages [25, Lemma 2].

However, checking properties of tree structures often additionally requires that the modeling language be closed under boolean operations and have efficient algorithms to check emptiness and inclusion. For example, when checking sufficient completeness, the main task is to check if the language of terms with defined functions is contained in the language of reducible terms. Thus, a sufficient completeness checker relies on a modeling language for trees for which checking inclusion is decidable. Since inclusion tests are most often implemented by complementation, intersection and a test for emptiness, these properties also are relevant for this problem. It is known that for *regular* equational tree automata with only associativity equations, the inclusion problem is undecidable. Moreover, this class of languages is not closed under intersection and complementation [24].

Motivated by this inadequacy in equational tree automata, Hendrix *et al.* proposed in [13] a further extension of tree automata, called *propositional tree automata*. These automata define a class of languages that is immediately closed under all the boolean operations via a straightforward, effective procedure for each operation. More importantly, they are the mathematically minimal extension in that the class of propositional tree automata accept the Boolean closure of languages recognizable by equational tree automata. The conservativeness of our extension leads to another desirable property: if the equational tree automata membership problem is decidable for a theory $\mathcal{E}$, then the membership problem for the propositional tree automata with $\mathcal{E}$ is decidable as well.

In [13], Hendrix *et al.* showed that the sufficient completeness problem for unconditional and left-linear membership rewrite systems modulo an equational theory can be reduced to the emptiness problem of propositional tree automata. Hence, one of the problems we investigate here is the emptiness problem modulo A- and AC-theories. Based on results for equational tree automata, we know that the problem is undecidable for propositional automata modulo A-theories. In this paper, we present a machine learning based semi-decision procedure, that is also a complete decision procedure under certain regularity conditions. We have found this algorithm effective in practice. Our algorithm has been implemented in a tree automata software library, called *CETA* [15], that can check the emptiness of propositional tree automata modulo associativity, commutativity, and identity. CETA is currently used for a next-generation sufficient completeness checker for Maude, and has already found a subtle bug in the built-in Maude specifications that can not be verified using the current checker.

This paper is organized as follows. In the next section, we define propositional tree automata. We show how this framework is closed under Boolean operations, and also investigate the recognition power relative to equational tree automata. In Section 3, we consider the membership decision problem, and analyze the complexity results with the comparison to equational tree automata. In

Sections 4 and 5, we explain our approach to the emptiness problem in detail. In Section 6, we show how our approach can be improved using ideas from machine learning. Finally, in Section 7, we conclude the paper by addressing the current software development project.

2   Preliminaries

We assume the reader is familiar with equational logic [6] and tree automata [7]. We use basic notations of rewriting from [4]. An *equational theory* is a pair $\mathcal{E} = (F, E)$ in which F is a finite set of function symbols, each with an associated *arity*, and E is a set of equations over the function symbols in F.

In the paper we are mainly interested in associative and/or commutative theory ($A \cup C$-theories for short), that is equational theories whose equations in E are associativity and/or commutativity axioms for some of the binary function symbols. Given a binary function symbol $f \in F$, $f(f(x, y), z) = f(x, f(y, z))$ is an associativity (A) axiom, and $f(x, y) = f(y, x)$ is a commutativity (C) axiom. We use F_A to denote the symbols in F with an associativity axiom in E, and F_C to be the symbols with a commutativity axiom. Since commutativity alone does not essentially affect the expressive power of the languages [25, Theorem 3], we assume that each commutative symbol is associative, i.e. $F_C \subseteq F_A$. Furthermore we write AC to denote the set E consisting of *both* A *and* C axioms from $F_A \cap F_C$.

A *propositional tree automaton* (PTA) $\mathcal{A}$ is a tuple $(\mathcal{E}, Q, \phi, \Delta)$, consisting of the equational theory $\mathcal{E} = (F, E)$, a finite set Q of states disjoint from the symbols in F (i.e. $F \cap Q = \varnothing$), a propositional formula ϕ over Q, and a finite set Δ of transition rules whose shapes are in one of the following forms:

$$\text{(Regular)} \qquad\qquad\qquad \text{(Monotone)}$$
$$f(\alpha_1, \ldots, \alpha_n) \to \beta \qquad\qquad f(\alpha_1, \ldots, \alpha_n) \to f(\beta_1, \ldots, \beta_n)$$

for some $f \in F$ with $\mathsf{arity}(f) = n$ and $p_1, \ldots, p_n, q, q_1, \ldots, q_n \in Q$. If a PTA only has regular rules, we say the PTA is *regular*; otherwise, it is *monotone*.

A move relation of $\mathcal{A} = (\mathcal{E}, Q, \phi, \Delta)$ is a rewrite relation over the set $\mathcal{T}(F \cup Q)$ of terms with respect to $\to_\Delta$ modulo $=_\mathcal{E}$, i.e. $s \to_\mathcal{A} t$ if there is a transition rule $l \to \alpha \in \Delta$ and a context $C \in \mathcal{C}(F \cup Q)$ such that $s =_\mathcal{E} C[l]$ and $t =_\mathcal{E} C[\alpha]$. The reflexive-transitive closure of $\to_\mathcal{A}$ is denoted by $\to_\mathcal{A}^*$.

A term t is *accepted* by $\mathcal{A}$ if $t \in \mathcal{T}(F)$ and the complete set of states reachable from t, $\mathsf{reach}_\mathcal{A}(t) = \{\alpha \in Q \mid t \to_\mathcal{A}^* \alpha\}$, is a model of ϕ. Boolean formulas are evaluated using their standard interpretations:

$$P \models \alpha \;\text{ if } \alpha \in P, \;\; P \models \phi_1 \vee \phi_2 \;\text{ if } P \models \phi_1 \text{ or } P \models \phi_2, \;\; P \models \neg\phi \;\text{ if not}(P \not\models \phi)$$

As an example, we consider the PTA $\mathcal{A}$ with the propositional formula $\phi = \alpha \wedge \neg\beta$ and the transition rules

$$\mathsf{a} \to \alpha \qquad \mathsf{b} \to \beta \qquad \mathsf{f}(\alpha) \to \alpha \qquad \mathsf{f}(\beta) \to \beta \qquad \mathsf{f}(\alpha) \to \gamma \qquad \mathsf{f}(\beta) \to \gamma.$$

Then a is accepted by $\mathcal{A}$, because $\mathsf{reach}_\mathcal{A}(\mathsf{a}) = \{\alpha\}$ and $\{\alpha\} \models \alpha \wedge \neg\beta$. Similarly, $\mathsf{f}(\mathsf{a})$ is accepted as $\mathsf{reach}_\mathcal{A}(\mathsf{f}(\mathsf{a})) = \{\alpha, \gamma\}$ and $\{\alpha, \gamma\} \models \alpha \wedge \neg\beta$. However, b and $\mathsf{f}(\mathsf{b})$ are not accepted, because $\mathsf{reach}_\mathcal{A}(\mathsf{b}) = \{\beta\}$ and $\{\beta\} \not\models \alpha \wedge \neg\beta$, and

$\text{reach}_\mathcal{A}(f(b)) = \{\beta, \gamma\}$ and $\{\beta, \gamma\} \not\models \alpha \wedge \neg\beta$. Intuitively, the formula $\alpha \wedge \neg\beta$ means that $\mathcal{A}$ accepts terms that rewrite to the state α and do not rewrite to β.

Propositional tree automata are closed under Boolean operations: given $\mathcal{A} = (\mathcal{E}, Q_1, \phi_1, \Delta_1)$ and $\mathcal{B} = (\mathcal{E}, Q_2, \phi_2, \Delta_2)$, then by assuming $Q_1 \cap Q_2 = \varnothing$, the intersection $\mathcal{L}(\mathcal{A}) \cap \mathcal{L}(\mathcal{B})$ is accepted by the PTA $(\mathcal{E}, Q_1 \cup Q_2, \phi_1 \wedge \phi_2, \Delta_1 \cup \Delta_2)$. The complement of $\mathcal{L}(\mathcal{A})$ is accepted by $\mathcal{A}' = (\mathcal{E}, Q_1, \neg\phi_1, \Delta_1)$, where the formula ϕ_1 of $\mathcal{A}$ has been replaced by $\neg\phi_1$. Therefore we have the following property for propositional tree automata.

Lemma 1. *The class of propositional tree automata is effectively closed under Boolean operations.* □

In the standard tree automata framework, the intersection of two tree automata may have the product of states, which is $|Q_1| \times |Q_2|$ state symbols, while the intersection of PTA $\mathcal{A}$ and $\mathcal{B}$ needs $|Q_1| + |Q_2|$ state symbols. In complementing the PTA $\mathcal{A}$, the set of states is unchanged, so the number of state symbols is $|Q_1|$, but constructing the complement of a tree automaton may require an exponential number of state symbols relative to the original.

It is also an easy lemma to show that the class of languages accepted by propositional tree automata under a certain equational theory is the smallest class of languages containing languages accepted by standard equational tree automata with the same equational theory and closed with respect to Boolean operations over the languages.

Lemma 2. *The class of tree languages accepted by PTA with an equational theory $\mathcal{E}$ corresponds precisely to the Boolean closure of tree languages accepted by equational tree automata sharing the equational theory $\mathcal{E}$.* □

One can observe that, given a term $t \in \mathcal{T}(F)$ and a propositional tree automaton $\mathcal{A}$, when $t \rightarrow_\mathcal{A}^* \alpha$ is decidable for any state α of $\mathcal{A}$, $\text{reach}_\mathcal{A}(t)$ is effectively computable. This leads to the observation:

Lemma 3. *The membership problem for equational tree automata under an equational theory $\mathcal{E}$ is decidable if and only if the membership problem for propositional tree automata with $\mathcal{E}$ is decidable.* □

3 Decidability Results

As we showed in the previous section, the decidability of the membership problem of propositional tree automata depends upon that of equational tree automata with the usual definition of acceptance in terms of final states. From previous work [22,24], we have the complexity results (in the next table) for regular and monotone cases with AC- or A-theory:

	regular A-TA	regular AC-TA	monotone A-TA	monotone AC-TA
complexity of membership	P-time	NP-complete	PSPACE-compl.	PSPACE-compl.

As an obvious observation, the membership problem for *propositional* regular AC-tree automata (abbreviated by MEM-PROP-REG-ACTA) seems harder than the problem for regular AC-tree automata. Here a propositional regular AC-tree automaton is a regular PTA over AC-theory, and a regular AC-tree automaton (regular AC-TA for short) corresponds to a regular PTA over AC-theory with a disjunction ϕ over atomic states as its propositional formula, i.e. $\phi = \alpha_1 \vee \cdots \vee \alpha_n$ for some $\alpha_1, \ldots, \alpha_n \in Q$.

As the AC-TA membership problem is NP-complete and the AC-TA non-membership problem can be converted in linear-time to the PTA membership problem, the PTA membership problem cannot be in NP unless NP equals coNP. We can show that MEM-PROP-REG-ACTA is in a higher complexity class.

Lemma 4. MEM-PROP-REG-ACTA *is in* Δ_2^P.   □

In the following, we write $A \leqslant_T^P B$ if there is an algorithm M running polynomial-time for a problem A which can ask, during its computation, some membership questions about B, where each query for B is answered in a unit time. The relation $A \leqslant_m^P B$ is polynomial-time *many-to-one* reducibility, and it is defined as follows: $A \leqslant_m^P B$ if there exists a polynomial-time function $f \colon \Sigma^* \to \Gamma^*$ such that for each $x \in \Sigma^*$, $x \in A$ if and only if $f(x) \in B$.

Proof of Lemma 4. Let $\mathcal{A} = (\mathcal{E}, Q, \phi, \Delta)$ with $\mathcal{E} = (F, E)$. We define the regular AC-tree automaton $\mathcal{B}_\mathcal{A}$ associated to $\mathcal{A}$. By assuming $\langle , \rangle$ is a fresh binary symbol, we let $\mathcal{B}_\mathcal{A} = (\mathcal{E}', P, \mathsf{p}_{\mathrm{acc}}, \Delta_\mathcal{A})$ where

$$
\begin{aligned}
\mathcal{E}' &= (F \cup Q \cup \{\langle , \rangle\}, E) \\
P &= \{\mathsf{p}_\alpha, \mathsf{q}_\alpha \mid \alpha \in Q\} \cup \{\mathsf{p}_{\mathrm{acc}}\} \\
\Delta_\mathcal{A} &= \{\alpha \to \mathsf{q}_\alpha \mid \alpha \in Q\} \\
&\quad \cup \{\langle \mathsf{p}_\alpha, \mathsf{q}_\alpha \rangle \to \mathsf{p}_{\mathrm{acc}} \mid \alpha \in Q\} \\
&\quad \cup \{f(\mathsf{p}_\alpha, \mathsf{p}_\beta) \to \mathsf{p}_\gamma \mid f(\alpha, \beta) \to \gamma \in \Delta\}.
\end{aligned}
$$

By construction, it is clear that for each $t \in \mathcal{T}(F)$ and $\alpha \in Q$, $t \to_\mathcal{A}^* q$ if and only if $\langle t, q \rangle \to_{\mathcal{B}_\mathcal{A}}^* \mathsf{p}_{\mathrm{acc}}$. One should note that $\mathcal{B}_\mathcal{A}$ can be constructed in quadratic-time to the size of $\mathcal{A}$, and the membership problem for regular AC-tree automata (abbreviated by MEM-REG-ACTA) is NP-complete.

For the next step, we take the set $S = \{\alpha \in Q \mid \alpha \text{ appears in } \phi\}$. The computation of S can be deterministically done in the size of ϕ, denoted by $|\phi|$, which is the number of occurrences of Boolean variables and Boolean connectives in ϕ. Then, for every $\alpha \in S$ (e.g. in the lexicographic order), we test by using the oracle $\mathcal{L}(\mathcal{B}_\mathcal{A})$, whether $\langle t, \alpha \rangle \in \mathcal{L}(\mathcal{B}_\mathcal{A})$. If $\langle t, \alpha \rangle \in \mathcal{L}(\mathcal{B}_\mathcal{A})$ is true, α is assigned to 1; otherwise, α is 0. By letting this Boolean assignment to be the mapping $\tau \colon S \to \{1, 0\}$, it is easy to see $\tau(\phi) = 1$ if and only if $\mathsf{reach}_\mathcal{A}(t) \models \phi$. The output value of the Boolean circuit is computable in polynomial-time relative to $|\phi|$ [9].

The above algorithm runs totally in polynomial-time with respect to the size of $\mathcal{A}$. Therefore the deterministic algorithm with an oracle set in NP solves the original membership problem in polynomially bounded time.

As a corollary of the above proof, MEM-PROP-REG-ACTA is

- $\leqslant_m^P$-hard for NP (i.e. NP-hard),
- $\leqslant_m^P$-hard for coNP (i.e. coNP-hard).

One can observe that given a term t and a regular AC-tree automaton $\mathcal{A}$, the problem of determining whether $\mathcal{A}$ *does not* accept t is coNP-complete (abbreviated by INACCEPT-REG-ACTA). Because, if $L \in$ coNP then $(L)^c \in$ NP, and thus there exists a polynomial-time function f from $(L)^c$ to MEM-REG-ACTA such that $x \in (L)^c$ if and only if $f(x)$ is accepted by a regular AC-tree automaton $(F, Q, \alpha_1 \vee \cdots \vee \alpha_n, \Delta, E)$ with $\alpha_1, \ldots, \alpha_n \in Q$. Then the reduction from L to MEM-PROP-REG-ACTA can be done by taking the propositional regular AC-tree automaton to be $(F, Q, \neg(\alpha_1) \wedge \cdots \wedge \neg(\alpha_n), \Delta, E)$.

Moreover, MEM-PROP-REG-ACTA is $\leqslant_T^P$-equivalent to MEM-REG-ACTA, because $\leqslant_m^P$ is subsumed in $\leqslant_T^P$ and $\leqslant_T^P$ is transitive. Then, $\forall A \in \Delta_2^P : A \leqslant_T^P$ MEM-REG-ACTA, and thus, MEM-PROP-REG-ACTA $\leqslant_T^P$ MEM-REG-ACTA.

In case of monotone PTA over AC-theory, using the same construction as in the proof of Lemma 4, we can show that: the membership problem for monotone PTA over AC-theory is in P^{PSPACE} ($=$ PSPACE). Then, using the fact that the membership problem for monotone TA over AC-theory is PSPACE-complete [22], we can obtain an even a stronger result that the membership problem for monotone PTA over AC-theory (indicated by monotone AC-PTA in the table) is PSPACE-complete.

The previous proof technique can also be applied to the A case. Therefore we obtain the following table of complexity results for sub-classes of propositional tree automata:

	regular A-PTA	regular AC-PTA	monotone A-PTA	monotone AC-PTA
complexity of membership	P-time	Δ_2^P	PSPACE-compl.	PSPACE-compl.

4   Emptiness Testing

We now turn our attention to the emptiness problem for PTA — given a PTA $\mathcal{A}$, does $\mathcal{L}(\mathcal{A}) = \varnothing$? This problem is computationally quite hard. Even in the free case, testing emptiness of a PTA is EXPTIME-complete. The tree automata universality problem, i.e. given a tree automaton $\mathcal{A}$ over a signature F, does $\mathcal{L}(\mathcal{A}) = \mathcal{T}(F)$?, is EXPTIME-complete [7, Theorem 14]. This problem can be converted in linear time into the PTA emptiness problem of $(\mathcal{L}(\mathcal{A}))^c$.

In AC case, regular equational tree automata are known to be closed under Boolean operations [27], and the emptiness problem is decidable [24]. It follows that the class of regular PTA over AC-theory have a decidable emptiness problem. In contrast to the above, in the A case (without commutativity axioms), the emptiness problem is undecidable:

Theorem 1. *The problem of checking whether* $\mathcal{L}(\mathcal{A}) = \varnothing$ *for regular PTA with a single associativity axiom is undecidable.*

Proof. Given a regular equational tree automaton $\mathcal{B}$ with a single associative symbol, it was shown in [25] to be undecidable whether $\mathcal{L}(\mathcal{B}) = \mathcal{T}(F)$. This problem is equivalent to checking $(\mathcal{L}(\mathcal{B}))^\mathsf{c} = \varnothing$. By Lemma 2, the language $(\mathcal{L}(\mathcal{B}))^\mathsf{c}$ is recognizable by a PTA with a single associative symbol. $\square$

Despite the lack of decidability, we nevertheless are interested in developing semi-decision algorithms that work well in practice. This is motivated by the study about the sufficient completeness checking of order-sorted equational specifications, where we have found equational tree automata techniques to be quite useful [13]. In applications, thus far we have mainly been interested in regular PTA, so we will restrict our attention to regular PTA for the remainder of this section.

Our algorithm for checking emptiness computes the set of states reachable from terms. The idea of this algorithm is similar to the use of subset construction to complement regular tree automata in [7], but with extensions to handle associative and commutative symbols. Though having no guarantee to terminate for all cases, the algorithm finds an accepting term if a language accepted by an input PTA is non-empty, and it proves the emptiness if the accepting language is empty and the PTA satisfies certain regularity conditions.

Let $\equiv_\mathcal{A}$ be the equivalence relation over terms where $s \equiv_\mathcal{A} t$ iff. $\mathsf{reach}_\mathcal{A}(s) = \mathsf{reach}_\mathcal{A}(t)$. For tree automata, the correctness of subset construction typically relies on the fact that $\equiv_\mathcal{A}$ is a congruence with respect to contexts. i.e. $s \equiv_\mathcal{A} t$ implies $C[s] \equiv_\mathcal{A} C[t]$ for all contexts C. However, this fact *does not* hold in the case when the root of s or t is an A symbol f and the context C has s or t immediately within a term labeled by f. Due to this complication, our subset construction algorithm for A and AC symbols maintains additional information.

We first define the information our subset construction algorithm for the A and AC case will eventually compute.

Definition 1. *Given a PTA* $\mathcal{A} = (\mathcal{E}, Q, \psi, \Delta)$ *over the theory* $\mathcal{E} = (F, E)$*, let* $\mathsf{det}(\mathcal{A}) \subseteq \mathcal{P}(Q) \times F$ *be the set* $\mathsf{det}(\mathcal{A}) = \{\, (\mathsf{reach}_\mathcal{A}(t), \mathsf{root}(t)) \mid t \in T_F \,\}$.

One should remark that $\mathsf{det}(\mathcal{A})$ is finite, however it is not always computable. Observe that $\mathcal{L}(\mathcal{A}) \neq \varnothing$ iff. there is a pair $(P, f) \in \mathsf{det}(\mathcal{A})$ such that $P \models \psi$. The undecidability of the emptiness problem of $\mathcal{L}(\mathcal{A})$ thus implies the membership question $(P, f) \in \mathsf{det}(\mathcal{A})$ is not decidable either.

For the remainder of this section, let $\mathcal{A}$ be a PTA with an A∪C-theory. In this case, we can obtain $\mathsf{det}(\mathcal{A})$ by an iterative computation starting from the empty set $\mathsf{d}_\mathcal{A}(0) \triangleq \varnothing$. We then expand $\mathsf{d}_\mathcal{A}(0)$ to $\mathsf{d}_\mathcal{A}(1), \mathsf{d}_\mathcal{A}(2), \ldots$ in the inference rules (defined later) until completion. Each set $\mathsf{d}_\mathcal{A}(i)$ is a subset of $\mathsf{det}(\mathcal{A})$. The mapping $\mathsf{d}_\mathcal{A}$ is simplified to d if $\mathcal{A}$ is obvious in the context.

Before describing the inference rules, we must give a few more definitions. We first extend $\mathsf{reach}_\mathcal{A}$ to allow sets of states $P_i \subseteq Q$ as constants appearing in terms. Precisely, the reachable states $\mathsf{reach}_\mathcal{A}(f(P_1, \ldots, P_n))$ for a *term with sets as constants* is the union of the reachable states for each term in $\mathcal{T}(F \cup Q)$ formed by choosing an element in each state, i.e.

$$\text{reach}_{\mathcal{A}}(f(P_1, \ldots, P_n)) = \{ \beta \in Q \mid (\exists \alpha_i \in P_i : 1 \leq i \leq n)\, f(\alpha_1, \ldots, \alpha_n) \to_{\mathcal{A}}^* \beta \}.$$

For each associative symbol $f \in F_A$, and set $\mathsf{d}(i)$, we define a context-free grammar $G_{f,\mathsf{d}(i)}$. Intuitively, the rules in the grammar are obtained from the PTA, and simulate the PTA over *flattened* terms of the form $f(P_1, \ldots, P_n)$. Each set $P_i \subset Q$ is reachable by a term whose root symbol is not f.

Definition 2. *Given a regular PTA* $\mathcal{A} = (\mathcal{E}, Q, \phi, \Delta)$ *with* $f \in F_A$ *and set* $\mathsf{d}(i)$, *we define the* flattened grammar *for* f, $G_{f,\mathsf{d}(i)}(\_\_) = (\Sigma_{f,\mathsf{d}(i)}, Q, \_\_, R)$, *where*

- $\Sigma_{f,\mathsf{d}(i)} = \{ P \mid \exists (P, g) \in \mathsf{d}(i) : g \neq f \}$,
- $R = \{ \gamma := \alpha\beta \mid f(\alpha, \beta) \to \gamma \in \Delta \} \cup \{ \gamma := P \mid P \in \Sigma_{f,\mathsf{d}(i)} \wedge \gamma \in P \}$.

In the paper, we write $\mathcal{L}(G(\alpha))$ to denote the language generated from α if G is (a mapping to) a grammar with a non-terminal symbol α. The *Parikh image* [26] of the language $\mathcal{L}(G(\alpha))$ is denoted by $\mathcal{S}(G(\alpha))$. Namely, $\mathcal{S}(G(\alpha)) = \{ \#(w) \mid w \in \mathcal{L}(G(\alpha)) \}$, where $\# : \Sigma^* \to \mathbb{N}^{|\Sigma|}$ maps each string in Σ^* to the vector counting the number of occurrences of each terminal symbol. For a subset $P\ (\subseteq Q)$ of non-terminals, let $\mathcal{L}(G(P))$ equal the strings appearing in the languages $\mathcal{L}(G(\alpha))$ generated by non-terminals $\alpha \in P$ and not in the languages $\mathcal{L}(G(\beta))$ generated by the non-terminals $\beta \in (Q - P)$. We define $\mathcal{S}(G(P))$ denote the corresponding construction from the Parikh images, i.e.,

$$\mathcal{L}(G(P)) = \bigcap_{\alpha \in P} \mathcal{L}(G(\alpha)) - \bigcup_{\beta \in (Q-P)} \mathcal{L}(G(\beta)) \qquad \mathcal{S}(G(P)) = \bigcap_{\alpha \in P} \mathcal{S}(G(\alpha)) - \bigcup_{\beta \in (Q-P)} \mathcal{S}(G(\beta))$$

As context-free grammars are not closed under intersection and complementation, $\mathcal{L}(G(P))$ is not necessarily context-free, and checking emptiness is undecidable. On the other hand, $\mathcal{S}(G(P))$ is a semi-linear set [26], because semi-linear sets are closed under Boolean operations, and moreover, have a decidable emptiness problem.

In our algorithm, we start with $\mathsf{d}(0) = \varnothing$, and then compute $\mathsf{d}(i+1)$ from $\mathsf{d}(i)$ using the inference rules below. Each step adds a pair in $\det(\mathcal{A})$ not in $\mathsf{d}(i)$.

(1) $f \notin F_A \cup F_C$:
$$\frac{(P_1, f_1), \ldots, (P_n, f_n) \in \mathsf{d}(i)}{\mathsf{d}(i+1) = \mathsf{d}(i) \uplus \{ (\,\text{reach}_{\mathcal{A}}(f(P_1, \ldots, P_n)), f\,) \}}$$

(2) $f \in F_A - F_C$:
$$\frac{P \subseteq Q \qquad \Sigma_{f,\mathsf{d}(i)}^{2+} \cap \mathcal{L}(G_{f,\mathsf{d}(i)}(P)) \neq \varnothing}{\mathsf{d}(i+1) = \mathsf{d}(i) \uplus \{ (P, f) \}}$$

(3) $f \in F_A \cap F_C$:
$$\frac{P \subseteq Q \qquad \mathbb{N}^{>1} \cap \mathcal{S}(G_{f,\mathsf{d}(i)}(P)) \neq \varnothing}{\mathsf{d}(i+1) = \mathsf{d}(i) \uplus \{ (P, f) \}}$$

In the first rule, we non-deterministically choose elements $(P_1, f_1), \ldots, (P_n, f_n)$ from $\mathsf{d}(i)$. These elements need not be distinct. In the second and third rules, we write $\Sigma_{f,\mathsf{d}(i)}^{2+}$ for the strings over $\Sigma_{f,\mathsf{d}(i)}$ containing at least two letters, and $\mathbb{N}^{>1}$ for vectors over natural numbers whose elements sum up to at least 2. We use the disjoint union operator $\uplus$ to denote that the newly added elements must be *distinct* from the other elements in $\mathsf{d}(i)$. It is relatively straightforward to show that by starting with $\mathsf{d}(0)$ and applying the rules for each operator until completion, we eventually have $\det(\mathcal{A})$. The proof is in our extended technical report [14].

Theorem 2. *Let $\mathcal{A} = (\mathcal{E}, Q, \phi, \Delta)$ be a regular PTA with $\mathcal{E} = (F, E)$ containing only associativity and commutativity axioms (A∪C-theory). Every chain $\mathsf{d}(0), \mathsf{d}(1), \dots$ obtained by applying the rules (1)–(3) until completion satisfies the following properties:*
- *the length k of the chain is $|\det(\mathcal{A})|$, and*
- *$\mathsf{d}(k) = \det(\mathcal{A})$.*

The undecidability of regular PTA with associative symbols crops up in testing the emptiness of $\Sigma^{2+}_{f,\mathsf{d}(i)} \cap \mathcal{L}(G_{f,\mathsf{d}(i)}(P))$. The focus of the next section concerns how to solve this emptiness constraint. It is worth observing that this subset construction based approach can be generalized for the monotone case as well, but in this case, the grammar $G_{f,\mathsf{d}(i)}$ must be made context-sensitive with an additional rule $\alpha\beta := \gamma\delta$ for each monotone rule $f(\gamma, \delta) \rightarrow f(\alpha, \beta) \in \Delta$.

5   Solving Language Equations for Associativity

Since at present the emptiness testing with monotone rules for associative symbols is beyond the goal of our project, we have developed an approach that is likely to work well in practice for the *regular* case with associative symbols. Our approach rests on an interactive semi-algorithm for each associative symbol $f \in F_A$ which has access to the mapping $\mathsf{d}(i)$ as it is being generated and performs two actions simultaneously: (1) recursively enumerates pairs (P, f) not in $\mathsf{d}(i)$ for which $\Sigma^{2+}_{f,\mathsf{d}(i)} \cap \mathcal{L}(G_{f,\mathsf{d}(i)}(P))$ is non-empty; and (2) applies machine learning techniques to attempt construction of a family $\{\mathcal{M}_\alpha\}_{\alpha \in Q}$ of deterministic finite automata for which $\mathcal{L}(\mathcal{M}_\alpha) = \mathcal{L}(G_{f,\det(\mathcal{A})}(\alpha))$ for all $\alpha \in Q$. If the first action succeeds, the semi-algorithm constructs the next $\mathsf{d}(i+1)$ from $\mathsf{d}(i)$. If the second action succeeds, we can decide for each subset of P states, the condition $\Sigma^{2+}_{f,\mathsf{d}(i)} \cap \mathcal{L}(G_{f,\det(\mathcal{A})}(P)) = \varnothing$ in the rule (2). We then can either obtain $\mathsf{d}(i+1)$ or prove that the conditional rule for f can no longer be applied.

A naïve approach to the first action is quite simple. We recursively enumerate the strings in $\Sigma^{2+}_{f,\mathsf{d}(i)}$ in order of increasing length to form the infinite sequence $w_1, w_2, \dots$, and parse each string w_i to get the complete set of states $P_i = \{\alpha \in Q \mid w \in \mathcal{L}(G_{f,\mathsf{d}(i)}(\alpha))\}$. If $(P_i, f) \notin \mathsf{d}(i)$, then let $\mathsf{d}(i+1) = \{(P_i, f)\} \cup \mathsf{d}(i)$. Handling the second action is more complicated. First, observe that we can enumerate the set of finite automata in order of increasing length. Because recursively enumerable sets are closed under finite products, we can even enumerate finite families of automata $\{\mathcal{M}_\alpha\}_{\alpha \in Q}$. The difficult part then lies in checking whether $\mathcal{L}(\mathcal{M}_\alpha) = \mathcal{L}(G_{f,\mathsf{d}(i)}(\alpha))$ for all $\alpha \in Q$. It is well known that given a *single* finite automaton $\mathcal{M}$ and context-free grammar G, it is *undecidable* whether $\mathcal{L}(\mathcal{M}) = \mathcal{L}(G)$ [16, Theorem 8.12(3)]. However, this result is just for a single automaton, and does not imply the undecidability of our problem. In fact, given a context-free grammar $G = (\Sigma, Q, \alpha_0, R)$ in Chomsky normal form, and a family of automata $\{\mathcal{M}_\alpha\}_{\alpha \in Q}$, the question whether $\mathcal{L}(\mathcal{M}_\alpha) = \mathcal{L}(G(\alpha))$ for all $\alpha \in Q$ is decidable.

The decidability of this problem is a direct consequence of Theorem 2.3 in [3]. Before explaining that result, however, it is necessary to shift our perspective of context-free grammars from viewing them as collections of *production rules* to viewing them as systems of *language equations*.

Definition 3. *Let $G = (\Sigma, Q, \alpha_0, R)$ be a context-free grammar. The system of equations generated by G is the family of equations $\{\alpha = P_\alpha\}_{\alpha \in Q}$ in which for each non-terminal $\alpha \in Q$, P_α is the formula $P_\alpha = w_1 \mid \cdots \mid w_n$ where $\alpha = w_1, \ldots, \alpha = w_n$ are the production rules in R whose left-hand side equals α.*

Given a system of equations with non-terminals Q and terminals Σ, a *substitution* is a mapping $\theta : Q \to \mathcal{P}(\Sigma^*)$ associating each state $\alpha \in Q$ to a language $\theta(\alpha) \subseteq \Sigma^*$. A substitution θ can be *applied* to a language formula P, yielding a language $P\theta \subseteq \Sigma^*$ which is defined using the axioms:

$$P\theta = \begin{cases} \{a\} & \text{if } P = a \text{ for some } a \in \Sigma, \\ \theta(\alpha) & \text{if } P = \alpha \text{ for some } \alpha \in Q, \\ S\theta \cup T\theta & \text{if } P = (S \mid T), \\ \{st \mid s \in S\theta \wedge t \in T\theta\} & \text{if } P = (S.T). \end{cases}$$

We may assume associativity of $\mid$ and . in the above definition. Here $S.T$ denotes the concatenation of S and T. A substitution $\theta : Q \to \mathcal{P}(\Sigma^*)$ is a *solution* to the system of equations $\{\alpha = P_\alpha\}_{\alpha \in Q}$ if and only if $\theta(\alpha) = P_\alpha\theta$ for all $\alpha \in Q$. It is known that each system of equations generated by G has a *least solution*, namely $\theta_{\mathcal{L}} : \alpha \mapsto \mathcal{L}(G(\alpha))$, and $\theta_{\mathcal{L}}(\alpha) \subseteq \psi(\alpha)$ for all solutions $\psi : Q \to \mathcal{P}(\Sigma^*)$ and $\alpha \in Q$. For grammars in Chomsky normal form, we can use the following theorem to help check whether an arbitrary solution is the least solution. Note that this is an easy consequence of Theorem 2.3 in [3].

Theorem 3. *If G is a context-free grammar in Chomsky normal form, there is a unique solution θ to the system of equations generated by G in which $\epsilon \notin \theta(\alpha)$ for any $\alpha \in Q$.* □

In the theorem ϵ denotes the empty string. The solution θ in the previous theorem is the least solution, since G does not contain a production rule of the form $\alpha := \beta$, and so $\epsilon \notin \mathcal{L}(G(\alpha))$ for any $\alpha \in Q$.

Given a context-free grammar in Chomsky normal form G and a family of finite automata $\{\mathcal{M}_\alpha\}_{\alpha \in Q}$, we can use Theorem 3 to check whether $\mathcal{L}(\mathcal{M}_\alpha) = \mathcal{L}(G(\alpha))$ for all $\alpha \in Q$.

Theorem 4. *Let G be a context-free grammar in Chomsky normal form with non-terminals Q. If $\mathcal{L}(G(\alpha))$ is regular for all $\alpha \in Q$, there is a constructable set of finite automata $\{\mathcal{M}_\alpha\}_{\alpha \in Q}$ for which $\mathcal{L}(\mathcal{M}_\alpha) = \mathcal{L}(G(\alpha))$.*

Proof. We recursively enumerate the families of finite automata $\{\mathcal{M}_\alpha\}_{\alpha \in Q}$ and check if $\mathcal{L}(\mathcal{M}_\alpha) = \mathcal{L}(G(\alpha))$ for each $\alpha \in Q$. If we let $\psi : Q \to \mathcal{P}(\Sigma^*)$ be the substitution $\alpha \mapsto \mathcal{L}(\mathcal{M}_\alpha)$, then the problem of checking whether $\mathcal{L}(\mathcal{M}_\alpha) = \mathcal{L}(G(\alpha))$ for all $\alpha \in Q$ reduces to deciding whether ψ is the unique solution satisfying Theorem 3. For each equation $\alpha = P_\alpha$, we can construct the automaton $\mathcal{M}_{P_\alpha}$ with $\mathcal{L}(\mathcal{M}_{P_\alpha}) = P_\alpha \psi$ due to the effective closure of regular languages under union and concatenation. Moreover, one can check whether $\mathcal{L}(\mathcal{M}_\alpha) = \mathcal{L}(\mathcal{M}_{P_\alpha})$ for each $\alpha \in Q$ using the standard approaches for testing the equivalence of finite automata. So clearly we can check whether ψ is a solution. But it is also trivial to check whether $\epsilon \notin \mathcal{L}(\mathcal{M}_\alpha)$ for each $\alpha \in Q$. Thus it is decidable whether ψ satisfies the conditions in Theorem 3. If it does then $\psi(\alpha)$ must equal $\mathcal{L}(G(\alpha))$ for each $\alpha \in Q$.   □

The key problem discussed in the section is determining whether the language $\mathcal{L}(G(\alpha))$ is regular for each non-terminal $\alpha \in Q$. One would expect this problem to be undecidable. Surprisingly, despite searching several texts, we could not find a decidability result for this problem. If $\mathcal{L}(G(\alpha))$ is regular for each non-terminal α, Theorem 4 shows that we can always show that by generating an equivalent family of finite automata. The other case is not so clear. Undecidability results for context-free languages such as Greibach's theorem [16, Sec. 8.7] do not apply since they concern single context-free languages and this property concerns every non-terminal in a grammar. Theorem 4's result itself relied heavily upon the assumption that every non-terminal generates a regular language. The same approach does not work to construct a finite automata corresponding to a single non-terminal in G due to the undecidability of the equivalence problem for context-free grammars and regular languages.

6   Angluin's Algorithm

Though technically sound, if one were to implement the semi-algorithm using the naïve approach outlined above, the efficiency would likely be less than desired. Enumerating finite automata in order of increasing size takes exponential time relative to the size of the automaton. Each family of finite automata would need to be checked for equivalence, and this also takes exponential time. Unfortunately, we don't see a way to improve the exponential time required to check equivalence, but by applying techniques from learning theory, we decrease the number of equivalence queries we make so that if the algorithm eventually succeeds, we will have only required a polynomial number of queries relative to the size of the accepting family of automata eventually found.

A well-known algorithm in machine learning is Angluin's algorithm [1] for learning regular languages with oracles. For an arbitrary language L, this algorithm attempts to construct a finite automaton $\mathcal{M}$ such that $\mathcal{L}(\mathcal{M}) = L$ by asking questions to two oracles: a *membership* oracle that answers whether a string $u \in \Sigma^*$ is in L; an *equivalence* oracle that answers whether $\mathcal{L}(\mathcal{M}) = L$ and if not, provides a *counterexample* string $u \in \Sigma^*$ in the symmetric difference of L and $\mathcal{L}(\mathcal{M})$, i.e. $u \in L \oplus \mathcal{L}(\mathcal{M})$ with $L \oplus \mathcal{L}(\mathcal{M}) = (L - \mathcal{L}(\mathcal{M})) \cup (\mathcal{L}(\mathcal{M}) - L)$.

Angluin's algorithm will terminate only if L is regular. However, given the appropriate oracles, one can attempt to apply it with any language, even languages not known to be regular. Due to space limitation of the paper, we roughly sketch below how Angluin's algorithm works. Readers are recommended to consult [18] for further details.

First we recall the definition of *Nerode's right congruence*: given a language $L \subseteq \Sigma^*$, the equivalence relation $\sim_L$ over Σ^* is the relation such that for $u, v \in \Sigma^*$, $u \sim_L v$ if and only if for all $w \in \Sigma^*$, $uw \in L \iff vw \in L$. It is known that a language L is regular if and only if the number of equivalence classes $|\Sigma^*/\sim_L|$ is finite. Angluin's algorithm maintains a data structure that stores two constructs: (1) a finite set $S \subseteq \Sigma^*$ of strings, each belonging to a distinct equivalence class in $\Sigma^*/\sim_L$, and (2) a finite set $D \subseteq \Sigma^*$ of distinguishing strings which in conjunction with the membership oracle, allows the algorithm to classify an arbitrary string into one of the known equivalence classes.

Initially, $S = \{\epsilon\}$ and $D = \varnothing$. Using the membership oracle in conjunction with S and D, the algorithm constructs a deterministic finite automaton $\mathcal{M}$ such that $\mathcal{L}(\mathcal{M}) = L$ when $S = \Sigma^*/\sim_L$. The algorithm then queries the equivalence oracle which either succeeds and we are done, or returns a counterexample which can be analyzed to reveal at least one additional equivalence class representative in $\Sigma^*/\sim_L$ that is not in S. If L is regular, eventually the algorithm will learn all of the equivalence classes in $\Sigma^*/\sim_L$. If L is not regular, $\Sigma^*/\sim_L$ must be infinite and so the algorithm will not terminate.

Given a finite family of regular languages $\{L_\alpha\}_{\alpha \in Q}$, Angluin's algorithm can be easily generalized to simultaneously learn a finite family of automata $\{\mathcal{M}_\alpha\}_{\alpha \in Q}$ such that $\mathcal{L}(\mathcal{M}_\alpha) = L_\alpha$ for all $\alpha \in Q$. In this version, there must be a membership oracle for each language L_α, and an equivalence oracle which given a family $\{\mathcal{M}_\alpha\}_{\alpha \in Q}$, returns true if $L_q = \mathcal{L}(\mathcal{M}_\alpha)$ for all $\alpha \in Q$, or a pair (α, u) where $\alpha \in Q$, and u is a counterexample in $L_\alpha \oplus \mathcal{L}(\mathcal{M}_\alpha)$. The generalized algorithm will terminate when L_α is regular for each $\alpha \in Q$.

In the context of this paper, we use Angluin's algorithm in conjunction with the flattened grammar $G_{f,\mathsf{d}(i)}$ with terminals $\Sigma_{f,\mathsf{d}(i)}$ and non-terminals Q. The algorithm attempts to construct of a family of finite automata $M = \{\mathcal{M}_\alpha\}_{\alpha \in Q}$ for which $\mathcal{L}(\mathcal{M}_\alpha) = \mathcal{L}(G_{f,\mathsf{d}(i)}(\alpha))$. If the process succeeds, we can easily determine whether $\Sigma^{2+} \cap \mathcal{L}(G_{f,\mathsf{d}(i)}(P)) = \varnothing$ for each pair $(P, f) \notin \mathsf{d}(i)$ using standard techniques for finite automata. If we discover that $\Sigma^{2+} \cap \mathcal{L}(G_{f,\mathsf{d}(i)}(P)) \neq \varnothing$, we set $\mathsf{d}(i+1) = \mathsf{d}(i) \uplus \{(P, f)\}$ and repeat the process for $\mathsf{d}(i+1)$.

To apply Angluin's algorithm, we need to provide the membership and equivalence oracles needed for a context-free grammar G with non-terminals Q and terminals Σ. The membership oracle for each non-terminal $\alpha \in Q$ is implemented by a context-free language parser that parses a string $u \in \Sigma^*$ and returns true if $u \in \mathcal{L}(G(\alpha))$. Given the family $\{\mathcal{M}_\alpha\}_{\alpha \in Q}$, our equivalence oracle forms the mapping $\theta : \alpha \mapsto \mathcal{L}(\mathcal{M}_\alpha)$ and checks if it is the solution to the equations generated by G satisfying Theorem 3. If θ is not the solution, the equivalence oracle must analyze the mapping to return a counterexample. The algorithm we use is presented in Fig. 1.Correctness of the oracle is shown in the following theorem:

Theorem 5. *Given a family of context-free grammars $\{\,G(\alpha)\,\}_{\alpha \in Q}$ in Chomsky normal form with non-terminals Q and terminals Σ, and a family $\{\,\mathcal{M}_\alpha\,\}_{\alpha \in Q}$ of finite automata over Σ, the algorithm* check_equiv *in Fig. 1*

- *returns* true *if $\mathcal{L}(G(\alpha)) = \mathcal{L}(\mathcal{M}_\alpha)$ for all $\alpha \in Q$; and otherwise,*
- *returns a pair (β, w) such that $w \in \mathcal{L}(G(\beta)) \oplus \mathcal{L}(\mathcal{M}_\beta)$.*

Proof sketch. Termination of this procedure is straightforward, and it is easy to verify that when a pair is returned at a return statement, it is indeed a counterexample. The non-trivial part of this theorem is that if the outer loop terminates without returning a counterexample, check_equiv returns true and $\mathcal{L}(G(\alpha)) = \mathcal{L}(\mathcal{M}_\alpha)$ is guaranteed. We obtain this property by showing that if $\mathcal{L}(\mathcal{M}_\alpha) \neq \mathcal{L}(\mathcal{G}(\alpha))$ for some $\alpha \in Q$ and the outer loop is executed, the body of the loop is guaranteed to return a pair.   □

When equipped with context-free language parsers as membership oracles and chec_equiv as an equivalence oracle, Angluin's algorithm accomplishes the same goal as the simple enumeration-based algorithm used to prove Theorem 4. However, this approach reduces the complexity from double to single exponential time. In searching for a solution, the enumeration algorithm used in Theorem 4 checks equivalence of every family of finite automata in order of increasing size. The total number of equivalence checks will be exponential relative to the size of

---

Procedure check_equiv

Input $G(\_\_) = (\Sigma, Q, \_\_, P)$: a context-free grammar
            $\{\mathcal{M}_\alpha\}_{\alpha \in Q}$: a family of finite automata over Σ

Output   true or (α, u) for some $\alpha \in Q$ and $u \in \Sigma^*$

let θ be the substitution $\alpha \mapsto \mathcal{L}(\mathcal{M}_\alpha)$;
for each $\alpha \in Q$ do
   if $\epsilon \in \mathcal{L}(\mathcal{M}_\alpha)$ then return (α, ϵ) ;

   if $\mathcal{L}(\mathcal{M}_\alpha) \neq P_\alpha \theta$ then
        choose $u \in \mathcal{L}(\mathcal{M}_\alpha) \oplus P_\alpha \theta$
        if $u \in \mathcal{L}(\mathcal{M}_\alpha) \oplus \mathcal{L}(G(\alpha))$ then return (α, u)
        else
            for each $\alpha := \beta\gamma \in P$ and $u = st$ do
                if $s \in \mathcal{L}(\mathcal{M}_\beta) \oplus \mathcal{L}(G(\beta))$ then return (β, s) ;
                if $t \in \mathcal{L}(\mathcal{M}_\gamma) \oplus \mathcal{L}(G(\gamma))$ then return (γ, t)
            od ;
od ;
return true

Fig. 1. Checking language equivalence

the final output. Since each equivalence check itself takes exponential time, the enumeration algorithm takes double exponential time relative to the size of the final output. In contrast, Angluin's algorithm makes a number of oracle queries that is polynomial [1] to the size of the final output. The equivalence oracle itself takes exponential time, and so the total time of the new algorithm is a single exponential relative to the size of the final output.

7   Concluding Remarks

The tree automata techniques developed in this paper are not only for theoretical use. The emptiness checking procedure explained in the previous two sections has been implemented in the CETA library [15]. This software provides the function for emptiness checking with not only associativity and commutativity axioms, but identity axioms as well. The identity axiom for a function symbol f with a unit symbol c is the equations of the forms $f(c, x) = x$ and $f(x, c) = x$. In CETA, identity axioms in a propositional tree automaton are converted into the rewrite rules $f(c, x) \rightarrow x$ and $f(x, c) \rightarrow x$ in conjunction with a specialized Knuth-Bendix style completion procedure modulo associativity and commutativity that preserves the set of reachable states for each term.

Though still a prototype, CETA has been integrated to work with the reachability analysis tool ACTAS [23], as well as the next generation sufficient completeness tool for Maude. In future project we plan to apply the new ACTAS for the tree automata based verification of infinite state systems including network protocols. In the Maude sufficient completeness tool, we use CETA by posing the sufficient completeness problem of an equational specification as a PTA emptiness problem. Sufficient completeness is a property of equational specifications that guarantees that enough equations have been specified so that defined operations are fully specified on all relevant data. We already experienced that CETA is useful in this context, as it allowed the checker to find a subtle bug in Maude involving lists formed from an associative operator, and also to verify the correctness of the bug-fix by proving that the language accepted by the corresponding tree automaton is empty, where the automaton often contains a theory with associativity.

References

1. D. Angluin: *Learning Regular Sets from Queries and Counterexamples*, Information and Computation 75, pp. 87–106, Elsevier, 1987.
2. A. Armando,   D. Basin,   Y. Boichut,   Y. Chevalier,   L. Compagna,   J. Cuellar, P. Hankes Drielsma, P.-C. Heám, O. Kouchnarenko, J. Mantovani, S. Mödersheim, D. von   Oheimb,   M. Rusinowitch,   J. Santiago,   M. Turuani,   L. Viganò   and L. Vigneron: *The AVISPA Tool for the Automated Validation of Internet Security Protocols and Applications*, Proc. of 17th CAV, Edinburgh (UK), LNCS 3576, pp. 281–285, Springer-Verlag, 2005.
3. J. Autebert, J. Berstel and L. Boasson: *Context-Free Languages and Push-Down Automata, Handbook of Formal Languages* 1, pp. 111–174. Springer-Verlag, 1997.

4. F. Baader and T. Nipkow: *Term Rewriting and All That*, Cambridge University Press, 1998.
5. Y. Boichut, P.-C. Heám and O. Kouchnarenko: *Automatic Verification of Security Protocols Using Approximations*, technical report RR-5727, INRIA, October 2005.
6. A. Bouhoula, J.P. Jouannaud and J. Meseguer: *Specification and Proof in Membership Equational Logic*, TCS 236, pp. 35–132, Elsevier, 2000.
7. H. Comon, M. Dauchet, R. Gilleron, F. Jacquemard, D. Lugiez, S. Tison and M. Tommasi: *Tree Automata Techniques and Applications*, incomplete draft, 2005. Available at http://www.grappa.univ-lille3.fr/tata
8. P. Devienne, J.-M. Talbot and S. Tison: *Set-Based Analysis for Logic Programming and Tree Automata*, Proc. of 4th SAS, Paris (France), LNCS 1302, pp. 127–140, Springer-Verlag, 1997.
9. D.-Z. Du and K. Ko: *Theory of Computational Complexity*, John Wiley and Sons, 2000.
10. J.P. Gallagher and G. Puebla: *Abstract Interpretation over Non-Deterministic Finite Tree Automata for Set-Based Analysis of Logic Programs*, Proc. of 4th PADL, Portland (USA), LNCS 2257, pp. 243–261, Springer-Verlag, 2002.
11. T. Genet and F. Klay: *Rewriting for Cryptographic Protocol Verification*, Proc. of 17th CADE, Pittsburgh (USA), LNCS 1831, pp. 271–290, Springer-Verlag, 2000.
12. S. Ginsburg: *The Mathematical Theory of Context-Free Languages*, McGraw-Hill, 1966.
13. J. Hendrix, H. Ohsaki and J. Meseguer: *Sufficient Completeness Checking with Propositional Tree Automata*, technical report UIUCDCS-R-2005-2635, Department of Computer Science, University of Illinois at Urbana-Champaign, 2005. Available at http://texas.cs.uiuc.edu/
14. J. Hendrix, H. Ohsaki, and M. Viswanathan *Propositional Tree Automata*, technical report UIUCDCS-R-2006-2695, Department of Computer Science, University of Illinois at Urbana-Champaign, 2006. Available at http://texas.cs.uiuc.edu/
15. J. Hendrix: *CETA: A Library for Equational Tree Automata*, Department of Computer Science, University of Illinois at Urbana-Champaign, 2006. Software available under GPL license at http://texas.cs.uiuc.edu/ceta/
16. J.E. Hopcroft and J.D. Ullman: *Introduction to Automata Theory, Languages, and Computation*, Addison-Wesley Publishing Company, 1979.
17. H. Hosoya, J. Vouillon and B.C. Pierce: *Regular Expression Types for XML*, Proc. of 5th ICFP, Montreal (Canada), SIGPLAN Notices 35(9), pp. 11–22, ACM, 2000.
18. M. Kearns and U. Vazirani: *An Introduction to Computational Learning Theory*, MIT Press, 1994.
19. N. Klarlund and A. Møller: *MONA Version 1.4 User Manual*, BRICS Notes Series NS-01-1, Department of Computer Science, University of Aarhus, 2001.
20. D. Lugiez: *Multitree Automata That Count*, TCS 333, pp. 225–263, Elsevier, 2005.
21. M. Nederhof: *Practical Experiments with Regular Approximation of Context-Free Languages*, Computational Linguistics 26(1), pp. 17–44, 2000.
22. H. Ohsaki, J.-M. Talbot, S. Tison and Y. Roos: *Monotone AC-Tree Automata*, Proc. of 12th LPAR, Montego Bay (Jamaica), LNAI 3855, pp. 337–351, Springer-Verlag, 2005.
23. H. Ohsaki and T. Takai: *ACTAS : A System Design for Associative and Commutative Tree Automata Theory*, Proc. of 5th RULE, Aachen (Germany), ENTCS 124, pp. 97–111, Elsevier, 2005.
24. H. Ohsaki and T. Takai: *Decidability and Closure Properties of Equational Tree Languages*, Proc. of 13th RTA, Copenhagen (Denmark), LNCS 2378, pp. 114–128, Springer-Verlag, 2002.
25. H. Ohsaki: *Beyond Regularity: Equational Tree Automata for Associative and Commutative Theories*, Proc. of 15th CSL, Paris (France), LNCS 2142, pp. 539–553, Springer-Verlag, 2001.

26. R. Parikh: *On Context-Free Languages*, JACM 13(4), pp. 570–581, 1966.
27. H. Seidl, T. Schwentick and A. Muscholl: *Numerical Document Queries*, Proc. of 22nd PODS, San Diego (USA), pp. 155–166, ACM, 2003.
28. I. Yagi, Y. Takata and H. Seki: *A Static Analysis Using Tree Automata for XML Access Control*, Proc. of 3rd ATVA, Taipei (Taiwan), LNCS 3707, pp. 234–247, Springer-Verlag, 2005.
29. K.N. Verma: *Two-Way Equational Tree Automata for AC-Like Theories: Decidability and Closure Properties*, Proc. of 14th RTA, Valencia (Spain), LNCS 2706, pp. 180–197, Springer-Verlag, 2003.

Generalizing Newman's Lemma for Left-Linear Rewrite Systems

Bernhard Gramlich[1] and Salvador Lucas[2]

[1] Fakultät für Informatik, Technische Universität Wien
Favoritenstr. 9 – E185/2, A-1040 Wien, Austria
gramlich@logic.at
[2] DSIC, Universidad Politécnica de Valencia
Camino de Vera s/n, 46022 Valencia, Spain
slucas@dsic.upv.es

Abstract. Confluence criteria for non-terminating rewrite systems are known to be rare and notoriously difficult to obtain. Here we prove a new result in this direction. Our main result is a generalized version of Newman's Lemma for left-linear term rewriting systems that does not need a full termination assumption. We discuss its relationships to previous confluence criteria, its restrictions, examples of application as well as open problems. The whole approach is developed in the (more general) framework of context-sensitive rewriting which thus turns out to be useful also for ordinary (context-free) rewriting.

1 Introduction and Overview

Besides termination, confluence is the most fundamental property of virtually any kind of rewrite systems (cf. e.g. [1], [2]). Newman's Lemma [19] is well-known to be the major tool for checking confluence of rewrite systems. It states that local confluence implies confluence for terminating reduction relations. However, without termination Newman's Lemma is not applicable, i.e., local confluence may be insufficient for guaranteeing confluence. In general, confluence proofs without termination are much harder. For the case of not necessarily terminating term rewriting systems (TRSs), a couple of rather restrictive criteria - mostly via strong confluence properties – are known, both for abstract rewrite systems (cf. e.g. [11], [13], [2]) as well as for TRSs (cf. e.g. [26], [11], [27], [29], [8], [23], [24], [21]). Known related decidability results include [7], [6]. Also structural and modularity properties and considerations may help in certain cases to prove confluence of non-terminating systems (cf. e.g. [25], [28], [14], [20]). The latter type of criteria is based on a *divide-and-conquer* approach, where certain sub-TRSs are shown to be confluent which in turn implies, under certain combination conditions, confluence of the whole system.

In term rewriting it is well-known that local confluence of (finite) terminating TRSs is decidable since it amounts to joinability of all *critical pairs* (*Critical Pair Lemma*, [11]). Hence, for (finite) terminating TRSs **Newman's Lemma** combined with the **Critical Pair Lemma** yields a decision procedure for confluence.

F. Pfenning (Ed.): RTA 2006, LNCS 4098, pp. 66–80, 2006.

The approach for proving confluence of (non-terminating) TRSs that we are going to present here is novel and differs methodologically from virtually all of these previous approaches in the sense that we do not consider sub-TRSs but rather certain sub-relations of the rewrite relation that are not generated by sub-TRSs.

The basic idea of our approach is as follows: Given a (non-terminating) TRS $\mathcal{R}$ with induced rewrite relation $\rightarrow_{\mathcal{R}}$, we first try to identify an appropriate terminating sub-relation $\rightarrow' \subseteq \rightarrow_{\mathcal{R}}$ (that is not induced by a sub-TRS of $\mathcal{R}$), to prove its confluence via Newman's Lemma, and then to deduce confluence of the entire TRS $\mathcal{R}$, i.e., of $\rightarrow_{\mathcal{R}}$, under some additional conditions.

The setting we are working in is *context-sensitive term rewriting* (CSR), a framework that properly extends ordinary (context-free) term rewriting by introducing context-sensitivity restrictions in the rewrite relation (cf. e.g. [17]). The necessary technical background will be provided below. CSR has turned out to be very useful for obtaining better computational properties of equational and rewrite specifications, e.g., for increased efficiency, a better termination behaviour, and an effective handling of infinite data structures (cf. e.g. [15,16,17]). Given the fact that termination is sometimes very difficult to prove and that non-termination is in many cases inherently unavoidable, CSR often provides ways for such examples to get a (restricted) terminating context-sensitive rewrite relation, while still preserving the desired computational power (e.g., for computing normal forms). Our confluence criterion will be based on such a context-sensitive view of a given ordinary TRS $\mathcal{R}$.

Let us give two simple motivating examples illustrating the problem with proving confluence.

Example 1. Consider the following TRS $\mathcal{R}$, which is a slightly modified variant of [4, Ex. (27)], and the essential part of its reduction graph:

(1)	$g(a) \rightarrow f(g(a))$
(2)	$g(b) \rightarrow c$
(3)	$a \rightarrow b$
(4)	$f(x) \rightarrow h(x,x)$
(5)	$h(x,y) \rightarrow c$

$$
\begin{array}{ccc}
g(a) & \longrightarrow & f(g(a)) \\
\downarrow & & \downarrow \\
& & h(g(a),g(a)) \\
& & \downarrow \\
g(b) & \longrightarrow & c
\end{array}
$$

Example 2. This example involves the generation of (all) natural numbers (via the constant nats and using a recursive increment operation inc) with some list destructors h(ea)d, t(ai)l and : as (infix) list constructor. The TRS $\mathcal{R}$ and the essential part of its reduction graph are as follows:

(1)	$\text{nats} \rightarrow 0 : \text{inc(nats)}$
(2)	$\text{inc}(x : y) \rightarrow \text{s}(x) : \text{inc}(y)$
(3)	$\text{hd}(x : y) \rightarrow x$
(4)	$\text{tl}(x : y) \rightarrow y$
(5)	$\text{inc}(\text{tl}(\text{nats})) \rightarrow \text{tl}(\text{inc}(\text{nats}))$

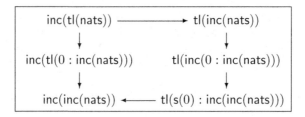

In both examples the rewrite system $\mathcal{R}$ has the following properties: It is left-linear, non-terminating and locally confluent. But as far as we know there are no known confluence criteria in the literature that would allow us to directly infer confluence in these examples. In particular, since both systems are non-terminating, Newman's Lemma is not applicable, i.e., a test for joinability of critical pairs is not sufficient. The systems are not orthogonal, since there exist critical pairs (in Ex. 1 rule (3) overlaps into (1), and in Ex. 2 rule (1) overlaps into (5)). Even though these critical pairs are joinable (cf. the reduction graphs), none of the critical pair based confluence criteria for left-linear rewrite systems in [11,8,23,24,21] is applicable here. Also, decidability results of [7], [6] are not applicable.[1] Yet, both systems are indeed confluent as we shall prove later on with our new criterion.

2   Preliminaries

We assume familiarity with the basic theory, terminology and notations in term rewriting (cf. e.g. [1], [2]). For the sake of readability some important notions and notations are recalled here.

Given a set A, $\mathcal{P}(A)$ denotes the set of all subsets of A. Given a binary relation $\to$, on a set A, we denote the transitive closure of $\to$ by $\to^+$, and its reflexive and transitive closure by $\to^*$. The inverse $\to^{-1}$ of $\to$ defined by $\{(b,a) \mid (a,b) \in \to\}$ is also denoted by $\leftarrow$. An element $a \in A$ is an $\to$-normal form, if there exists no b such that $a \to b$; $\mathrm{NF}(\to)$ is the set of $\to$-normal forms. We say that b is a $\to$-normal form of a, if $a \to^* b \in \mathrm{NF}(\to)$. We say that $\to$ is *terminating* iff there is no infinite sequence $a_1 \to a_2 \to a_3 \cdots. \to$ is *locally confluent* iff $\leftarrow \cdot \to \subseteq \to^* \cdot {}^*\!\leftarrow$, and *confluent* (or *Church-Rosser*) iff $\leftarrow^* \cdot \to^* \subseteq \to^* \cdot \leftarrow^*$. *Terms* are constructed as usual over some countable set $\mathcal{V}$ of *variables* and a *signature* $\mathcal{F}$ of *functions symbols* equipped with a fixed arity given by $ar\colon \mathcal{F} \to \mathbb{N}$. The set of all terms over $\mathcal{F}$ and $\mathcal{V}$ is denoted by $\mathcal{T}(\mathcal{F}, \mathcal{V})$. A term is *linear* if it has no multiple occurrences of a single variable. Terms are viewed as labelled trees in the usual way. *Positions* $p, q, \ldots$ in terms are

---

[1] With the *decreasing diagrams method* of [22], [2, Section 14.2], however, it is possible to prove confluence of the *linear* system of Example 2 above, by finding an appropriate well-founded labelling for rewrite steps. Yet, this powerful and general method does not directly yield easily applicable confluence criteria, but requires careful and smart design choices to become applicable. For Example 1, (practical) applicability of this method remains unclear, due to non-right-linearity.

represented by sequences of positive natural numbers. Given positions p, q, we denote their concatenation by $p.q$. Positions are ordered by the standard prefix ordering $\leq$. Two positions p and q are *parallel* (or *disjoint*), denoted $p \parallel q$, if neither $p \leq q$ nor $q \leq p$. The set of all *positions* of a term t is $Pos(t)$, the set of all its variable positions and of all its non-variable positions by $\mathcal{V}Pos(t)$ and $\mathcal{F}Pos(t)$, respectively. We denote the 'empty' root position by ϵ. The subterm of t at position p is denoted by $t|_p$ and $t[s]_p$ is the term t with the subterm at position p replaced by s. We shall also make free use of (term) *contexts* as usual. The symbol labelling the root of t is denoted as $root(t)$. For the set of all variables occurring in a term s we write $Var(s)$.

A rewrite rule is an ordered pair (l, r), written $l \to r$, with $l, r \in \mathcal{T}(\mathcal{F}, \mathcal{V}), l \notin \mathcal{V}$ and $Var(r) \subseteq Var(l)$. A *term rewriting system* (TRS) is a pair $\mathcal{R} = (\mathcal{F}, R)$ where $\mathcal{F}$ is a signature and R is a set of rewrite rules over $\mathcal{T}(\mathcal{F}, \mathcal{V})$. We will often omit the signature when it is implicitly given by the set of rules, and identify $\mathcal{R}$ and R. A TRS $\mathcal{R}$ is left-linear if for all $l \to r \in \mathcal{R}$, l is linear. The *rewrite relation* induced by a TRS $\mathcal{R}$ is defined by $s \to_{\mathcal{R}} t$ if $s|_p = l\sigma$, $t = s[r\sigma]$ for some $l \to r \in \mathcal{R}$, some $p \in Pos(s)$ and some substitution σ. Instead of $s \to_{\mathcal{R}} t$ we also write $s \to t$ if $\mathcal{R}$ is clear from the context, and $s \to_{\mathcal{R},p} t$ or $s \to_p t$ to indicate the position of the *redex contraction*. *Critical pairs* and *critical peaks* of rewrite rules and systems are defined as usual. A TRS $\mathcal{R}$ is terminating, confluent, locally confluent, etc. if $\to$ has the respective property.

Next we need some additional notions and notations for context-sensitive rewriting. Given a signature $\mathcal{F}$, a mapping $\mu \colon \mathcal{F} \to \mathcal{P}(\mathbb{N})$ is a *replacement map* (or $\mathcal{F}$-map) if for all $f \in \mathcal{F}$, $\mu(f) \subseteq \{1, \ldots, ar(f)\}$ ([15]). The set $Pos^\mu(t)$ of *(μ-)replacing* or *active* positions of $t \in \mathcal{T}(\mathcal{F}, \mathcal{V})$ is given by $Pos^\mu(t) = \{\epsilon\}$, if $t \in \mathcal{V}$ or $t \in \mathcal{F}$ with $ar(f) = 0$, and $Pos^\mu(t) = \{\epsilon\} \cup \bigcup_{i \in \mu(root(t))} \{i.q \mid q \in Pos^\mu(t|_i)\}$, otherwise. The set $\overline{Pos^\mu(t)}$ of *non-(μ-)replacing* or *inactive* positions of $t \in \mathcal{T}(\mathcal{F}, \mathcal{V})$ is just the complement of the former, i.e., $\overline{Pos^\mu(t)} = Pos(t) \setminus Pos^\mu(t)$. Replacement maps are ordered by $\sqsubseteq$, with $\mu \sqsubseteq \mu'$ if for all $f \in \mathcal{F}$, $\mu(f) \subseteq \mu'(f)$. Thus, $\mu \sqsubseteq \mu'$ means that μ considers less positions than μ' (for reduction). If $\mu \sqsubseteq \mu'$, we also say that μ is *more restrictive than* μ'. A context-sensitive rewrite system (CSRS) is a pair $(\mathcal{R}, \mu)$ (also denoted by $\mathcal{R}_\mu$), where $\mathcal{R}$ is a TRS and μ is a replacement map (over the signature of $\mathcal{R}$). In *context-sensitive rewriting* (CSR [15]), only replacing redexes are contracted. s μ-rewrites to t, denoted by $s \to_{\mathcal{R},\mu} t$ or just $s \to_\mu t$, if $s \to_{\mathcal{R},p} t$ and $p \in Pos^\mu(t)$. Note that this means that $\to_{\mathcal{R},\mu} t$ is stable under substitutions, but in general not under contexts, i.e., the monotonicity property (of $\to_\mathcal{R}$) is lost. Slightly abusing notation, we denote rewriting at non-replacing positions by $\to_{\overline{\mu}}$, i.e., $s \to_{\overline{\mu}} t$ if $s \to_p t$ for some $p \in \overline{Pos^\mu(s)}$. Observe that in general $\to_\mu \cup \to_{\overline{\mu}}$ need not be a disjoint union. A simple example illustrating this is the TRS consisting of the two rules $f(x) \to f(b)$, $a \to b$ with $\mu(f) = \emptyset$. Here we have both $f(a) \to_\mu f(b)$ and $f(a) \to_{\overline{\mu}} f(b)$. A CSRS $\mathcal{R}_\mu$ is terminating, confluent, locally confluent etc. if $\to_\mu$ has the respective property. Finally, for a given CSRS $\mathcal{R}_\mu$, we will need certain replacement maps that are not very restrictive. More precisely, all positions of non-variable subterms in the left-hand sides of the rules should be replacing (this will guarantee in particular that rewrite steps that are involved in critical overlaps of $\mathcal{R}$ are also $\mathcal{R}_\mu$-steps). The *canonical* replacement map $\mu_\mathcal{R}^{can} : \mathcal{F} \to \mathcal{P}(\mathbb{N})$ is defined by $i \in \mu_\mathcal{R}^{can}(f) \iff \exists l \to r \in \mathcal{R}, p \in \mathcal{F}Pos(l) : root(l|_p) = f, p.i \in \mathcal{F}Pos(l)$. The set $CM_\mathcal{R}$ of replacement maps (for $\mathcal{R}$) that are at most as restrictive as $\mu_\mathcal{R}^{can}$ is given by $CM_\mathcal{R} = \{\mu \in M_\mathcal{R} \mid \mu_\mathcal{R}^{can} \sqsubseteq \mu\}$. The most liberal replacement map (for $\mathcal{R}$) $\mu_\top$ is the greatest element of $M_\mathcal{R}$, i.e., with $\mu_\top(f) = \{1, \ldots, ar(f)\}$ for all $f \in \mathcal{F}$.

3 Weakening the Termination Assumption in Newman's Lemma for Left-Linear Rewrite Systems

Suppose $\mathcal{R}$ is a locally confluent non-terminating TRS. In order to try to prove confluence of $\mathcal{R}$, we will impose context-sensitivity restrictions on $\mathcal{R}$, i.e., a replacement map μ such that $\to_{\mathcal{R},\mu}$ (hopefully) becomes terminating and such that confluence of $\to_{\mathcal{R},\mu}$ implies confluence of $\to_\mathcal{R}$ (hence of $\mathcal{R}$).

3.1 Confluence Via Context-Sensitive Confluence

For reasons that will become clear later on (cf. Lemmas 3, 4, 5) we need as general assumptions, besides termination of $\to_{\mathcal{R},\mu}$, that $\mathcal{R}$ is left-linear, and that μ is at most as restrictive as the canonical replacement map $\mu_\mathcal{R}^{can}$.

Remark 1. To see why requiring $\mu \in CM_\mathcal{R}$ makes sense, consider the rewrite system $\mathcal{R}$ consisting of the rules $c \to b$, $b \to c$ and $h(b) \to a$, together with $\mu(h) = \emptyset$. Here we have one critical overlap $h(c) \leftarrow h(b) \to a$ (which is joinable via $h(c) \to h(b) \to a$). However, we cannot deduce this using only $\to -\mu$-reduction (in $\mathcal{R}_m u$ this critical peak does not even exist, since $h(c) \leftarrow h(b)$ is not a $\to -\mu$-step; moreover, for joinability we need the $h(c) \to h(b)$ which is also not a $\to -\mu$-step).

- $\mathcal{R}$ is left-linear.   (1)
- $\mu \in CM_\mathcal{R}$.   (2)
- $\mathcal{R}_\mu$ is terminating.   (3)

A first question is whether, for any such $\mathcal{R}_\mu$, its context-free version $\mathcal{R}$ is already (necessarily) confluent if $\mathcal{R}$ is locally confluent? Actually, when looking at examples in the literature, especially in papers on CSR, we have not found a single counterexample to this tempting conjecture. However, it turns out that conditions (1)–(3), together with local confluence of $\mathcal{R}$, are not yet sufficient for concluding confluence of $\mathcal{R}$. A simple counterexample is the following modified and extended version of the basic counterexample to the equivalence of local confluence and confluence (cf. [10]).

Example 3. Suppose the TRS $\mathcal{R}$ is given as follows, again with the relevant part of its reduction graph on the right.

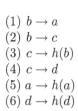

(1) $b \to a$
(2) $b \to c$
(3) $c \to h(b)$
(4) $c \to d$
(5) $a \to h(a)$
(6) $d \to h(d)$

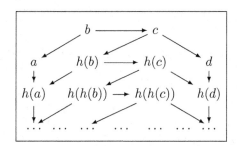

Clearly, $\mathcal{R}$ is not confluent, since for instance for $a \leftarrow b \to c \to d$ there is no common successor of a and d. However, $\mathcal{R}$ is obviously locally confluent. The two critical peaks $a \leftarrow b \to c$ and $h(b) \leftarrow c \to d$ are joinable via $a \to h(a) \leftarrow h(b) \leftarrow c$ and $h(b) \to h(c) \to h(d) \leftarrow c$, respectively. Moreover, choosing μ with $\mu(h) = \emptyset$ we have $\mu = \mu_{\mathcal{R}}^{can} \in CM_{\mathcal{R}}$. For this choice of μ, $\to_{\mathcal{R},\mu}$ is easily seen (and proved) to be terminating. However, what goes wrong in this example is the fact that $\to_{\mu}$ is not (locally) confluent. To see this, consider again the critical peaks. For $a \leftarrow_{\mu} b \to_{\mu} c$, reduction of a and c to a common successor is not possible by $\to_{\mu}$-steps only: $a \to_{\mu} h(a) \leftarrow_{\overline{\mu}} h(b) \leftarrow_{\mu} c$. Similarly, for $h(b) \leftarrow_{\mu} c \to_{\mu} d$ we only get $h(b) \to_{\overline{\mu}} h(c) \to_{\overline{\mu}} h(d) \leftarrow_{\mu} d$. In other words, although $\to$ (hence $\mathcal{R}$) is locally confluent, $\to_{\mu}$ is not. Thus we cannot argue using Newman's Lemma for $\to_{\mu}$.

Example 3 and Remark 1 suggest that in order to be able to use Newman's Lemma for the context-sensitive restriction $\to_{\mu}$ of $\to$ (in proofs of confluence of $\to$), we have to additionally require the following property of $\mathcal{R}_{\mu}$.

– Every critical peak $t_1 \leftarrow s \to t_2$ of $\mathcal{R}$ is joinable with $\to_{\mu}$-steps.   (4)

For locally diverging μ-steps (i.e., of the form $t_1 \leftarrow_{\mu} s \to_{\mu} t_2$) that correspond to a variable overlap we also have to ensure $\to_{\mu}$-joinability.[2] Actually, for proof-technical reasons we need a stronger property. To describe this formally, we first need some additional terminology.

Definition 1 (level of subterms). *Given $\mathcal{R}_{\mu}$ and a term t, the level of a subterm $t|_p$ of t (and of p), denoted by $level(t, p)$ is the number of all non-replacing positions $q = q'.i$ on the path in t from ϵ to p with $i \notin \mu(root(t|_{q'}))$. Formally:*

$$level(t, p) = \begin{cases} 0, & if\ p = \epsilon \\ level(t_i, p'), & if\ p = i.p', t = f(t_1, \ldots, t_n), i \in \mu(f) \\ 1 + level(t_i, p'), & if\ p = i.p', t = f(t_1, \ldots, t_n), i \notin \mu(f) \end{cases}$$

If $x \in Var(t)$ (for an arbitrary term t), we define – slightly abusing notation – $level(t, x) = \max\{level(t, p) \mid t|_p = x\}$ if $x \in Var(t)$, and $level(t, x) = 0$ if $x \notin Var(t)$.

---

[2] E.g., consider $f(x) \to g(x)$ and $a \to b$ with $\mu(f) = \{1\}$, $\mu(g) = \emptyset$. Then we have $f(b) \leftarrow_{\mu} f(a) \to_{\mu} g(a)$, which is $\to$-joinable via $f(b) \to g(b) \leftarrow g(a)$, but not $\to_{\mu}$-joinable, because the latter step is not a $\to_{\mu}$-step.

Intuitively, $level(t, p)$ describes 'the degree of how forbidden' it is to reduce the subterm $t|_p$ of t.

Definition 2 (level-decreasingness). *Given $\mathcal{R}_\mu$, a rule $l \to r \in \mathcal{R}$ is said to be* level-decreasing, *if for every variable $x \in Var(l)$ we have $level(l, x) \geq level(r, x)$. $\mathcal{R}_\mu$ is* level-decreasing *if every rule of $\mathcal{R}_\mu$ is level-decreasing.*

Note that ground TRSs are trivially level-decreasing.

Our last condition for the announced confluence criterion now reads as follows.

$$- \; \mathcal{R}_\mu \text{ is level-decreasing.} \tag{5}$$

Definition 3 (level of reduction steps). *Given $\mathcal{R}_\mu$ we define binary relations $\to_{\mu,i}$ and $\to_{\mu,\leq i}$, for all $i \geq 0$ as follows:*

$$s \to_{\mu,i} t \iff s \to_{\mathcal{R},p} t, \; level(s, p) = i \; .$$

$$s \to_{\mu,\leq i} t \iff s \to_{\mu,k} t \; \text{ for some } \; k \leq i \; .$$

For the sake of readability, if μ is clear from the context, we also write $\to_i$ and $\to_{\leq i}$ instead of $\to_{\mu,i}$ and $\to_{\mu,\leq i}$, respectively.

From the definitions of $\to_{\overline{\mu}}$ and $\to_k$ it is obvious that $\to_{\overline{\mu}} = \bigcup_{k \geq 1} \to_k$ holds. Clearly, $s \to_k t$ ($s \to_{\leq k} t$) means that t can be obtained from s by contracting some redex at level k (at some level $\leq k$). And $s \to_{\overline{\mu}} t$ says that we can get t from s by contracting a redex $s|_p$ of s at some non-replacing position p of s (i.e., such that the level of $s|_p$ of s is equal to some $k \geq 1$).

Example 4 (Example 3 continued). Adding levels to reduction steps, we have here e.g. $a \leftarrow_0 b \to_0 c$ and $a \to_0 h(a) \leftarrow_1 h(b) \leftarrow_0 c$ as well as $h(b) \leftarrow_0 c \to_0 d$ and $h(b) \to_1 h(c) \to_1 h(d) \leftarrow_0 d$.

Proposition 1. *Given $\mathcal{R}_\mu$ the following properties hold:*

(a) $\to_k \; \subseteq \; \to_{\leq k} \; \subseteq \; \to_{\leq k+1}$ *for all $k \geq 0$.*
(b) $\to_\mu \; = \; \to_0$, $\to_{\overline{\mu}} = \bigcup_{k \geq 1} \to_k$.
(c) $\to \; = \; \bigcup_{k \geq 0} \to_{\leq k} \; = \; \bigcup_{k \geq 0} \to_k \; = \; \to_\mu \cup \to_{\overline{\mu}}$.

Proof. Straightforward by the respective definitions.

Lemma 1 (confluence criterion for $\to_\mu$, cf. [15]). *Let $\mathcal{R}_\mu$ be a CSRS satisfying (1), (3), (4) and (5). Then $\to_\mu$ is confluent.*

Proof. Local confluence of $\to_\mu$ can be easily directly shown by considering all cases of local divergences and exploiting properties (1), (4) and (5) which together with Newman's Lemma yields confluence because of (3). In particular, our conditions (1) and (5) imply the property that $\to_\mu$ has *left-homogeneous replacing variables* (cf. [15, Def. 5]) which guarantees that variable overlaps in $\mathcal{R}_\mu$ are uncritical.[3]

---

[3] Actually, condition (5) could still be weakened a bit here by requiring only that level-decreasingness need only hold for rules with variables of level 0 in left-hand sides. However, Lemma 1 is anyway a special case of [15, Theorem 5] and we will not use the above slight generalization later on.

Lemma 2 (extraction lemma). *Given $\mathcal{R}_\mu$, suppose $s \to_{\leq k+1} t$ at some level ≥ 1. Then s has the form $s = C[s_1, \ldots, s_n]_{p_1, \ldots, p_n}$ where the p_i's are all minimal non-replacing positions in s, and $t = C[t_1, \ldots, t_n]_{p_1, \ldots, p_n}$, with $s_i \to_{\leq k} t_i$ for some $i \in \{1, \ldots, n\}$ and $s_j = t_j$ for all $j \in \{1, \ldots, n\} \setminus \{i\}$.*

Proof. Straightforward by definition of $\to_{\leq i}$.

The next result gives conditions under which reduction sequences can be re-arranged such that $\to_0$-steps are done first.[4]

Lemma 3 (exchange lemma). *Let $\mathcal{R}_\mu$ be given with (1), (2) and (5), i.e., such that $\mathcal{R}$ is left-linear, $\mu \in CM_\mathcal{R}$ and $\mathcal{R}_\mu$ is level-decreasing. Then the following rearrangement property holds for every $k \geq 0$:*[5] $\to_{\leq k} \cdot \to_0 \subseteq \to_0 \cdot \to_{\leq k}^*$.

Proof. First we observe that for $k = 0$ the inclusion holds trivially. Thus suppose $k \geq 1$. Furthermore let $s \to_{k'} t$ at position p with $0 \leq k' \leq k$ and $t \to_0 u$ at position q. Again, if $k' = 0$, the inclusion holds trivially, hence we may assume $k' \geq 1$. Now, p and q must be distinct (otherwise we have a contradiction, because a position cannot be both replacing and non-replacing). We distinguish the following cases.

(a) $p \parallel q$: Then the two steps commute and we get $s \to_0 s[u|_q]_q \to_{k'} s[u|_q][t|_p]_p = u$.

(b) $p < q$: This is impossible, since $s \to_{k'} t$ at p with $k' \geq 1$ implies $p \in \overline{Pos_\mu(s)}$ and $p \in \overline{Pos_\mu(t)}$, hence also $q \in \overline{Pos_\mu(t)}$ (because of $p < q$). But on the other hand, due to $t \to_0 u$ at position q, we have $q \in Pos_\mu(t)$, hence a contradiction.

(c) $p > q$: In this case we have $s = s[l\sigma]_p \to_{k'} s[r\sigma]_p = t = t[l'\tau]_q \to_0 t[r'\tau]_q = u$ (for some $l \to r, l' \to r' \in \mathcal{R}$ and substitutions σ, τ). If we can show that position p in t is below the pattern of l' in $t[l'\tau]_q$ we are done, because then we have – by left-linearity of $\mathcal{R}$ – $s = s[l'\tau']_q \to_{k'} t = t[l'\tau]_q \to_0 t[r'\tau]_q = u$ (for some rewrite rule $l' \to r' \in \mathcal{R}$ and some substitution τ') which commutes via $s = s[l'\tau']_q \to_0 s[r'\tau']_q \to_{\leq k'}^* s[r'\tau]_q = t[r'\tau]_q = u$. Note that for the reduction $s[r'\tau']_q \to_{\leq k'}^* s[r'\tau]_q$ in the variable parts of the right-hand side r' of $l' \to r'$, more precisely for the bound k on the level of the reduction steps) we need assumption (5). Now suppose p in t were in the pattern of l'. This would imply by (2) that $p \in Pos_\mu(t)$ and $p \in Pos_\mu(s)$, hence $k' = 0$. But this is a contradiction to our assumption $k' \geq 1$ from above. Hence we are done.

The next commutation result will be needed to prove a kind of backward preservation of $\to_\mu$-normal forms along non-replacing reduction steps (cf. Lemma 5).

---

[4] This is similar to *standardization* in left-linear TRSs (cf. e.g. [2]), except for the fact that we need additional information about the individual steps.

[5] Actually, from the proof it is clear that we even have the stronger statement $\to_{\leq k} \cdot \to_0 \subseteq \to_0 \cdot \dashrightarrow_{\leq k}$ where $\dashrightarrow_{\leq k}$ denotes a parallel reduction step with each contraction being at a level at most k. However, we don't need this stronger version later on.

Lemma 4 (commutation lemma). *Let $\mathcal{R}_\mu$ be given with (1), (2) and (5), i.e., such that $\mathcal{R}$ is left-linear, μ is at most as restrictive as $\mu_{\mathcal{R}}^{can}$ and $\mathcal{R}_\mu$ is level-decreasing. Then the following commutation property holds for every $k \geq 1$.*[6]

Proof. Suppose $t \leftarrow_k s \rightarrow_0 u$ at positions p and q, respectively. Hence, $t = s[t|_p]_p \leftarrow_k s[s|_p]_p = s = s[s|_q]_q \rightarrow_0 s[u|_q]_q = u$. We proceed by case analysis.

(a) $p \leq q$: Due to $k \geq 1$ we have $p \in \overline{Pos_\mu(s)}$ and thus also $q \in \overline{Pos_\mu(s)}$. However, $s \rightarrow_0 u$ at q implies $q \in Pos_\mu(s)$, hence a contradiction.

(b) $p \parallel q$: In this case the reductions commute as usual: $t = s[t|_p]_p \rightarrow_0 s[t|_p][u|_q]_q \leftarrow_k s[u|_q]_q = u$ where the left step is at position q and the right one at p.

(c) $p > q$: Because of assumption (2), $\mu \in CM_\mathcal{R}$, and since $k \geq 1$, this case corresponds to a variable overlap. Moreover, due to assumption (1), left-linearity of $\mathcal{R}$, we have $t = s[l\sigma']_q \leftarrow_k s = s[l\sigma]_q \rightarrow_0 s[r\sigma]_q = u$ for some $l \rightarrow r \in \mathcal{R}$ and substitutions σ, σ'. Hence, t and u are joinable via $t = s[l\sigma'] \rightarrow_0 s[r\sigma'] \leftarrow_{\leq k}^* s[r\sigma]_q = u$. Note that the (parallel) reduction $u = s[r\sigma]_q \rightarrow^* s[r\sigma']$ is of level at most k because of assumption (5), i.e., level-decreasingness of $\mathcal{R}$.

The next lemma states conditions under which reduction of some term to a $\rightarrow_\mu$-normal form implies that the original term is already a $\rightarrow_\mu$-normal form.

Lemma 5 (a condition for backward invariance of $\rightarrow_\mu$-normal forms). *Let $\mathcal{R}_\mu$ be given with (1), (2) and (5), i.e., such that $\mathcal{R}$ is left-linear and level-decreasing, and $\mu \in CM_\mathcal{R}$. Then $s \rightarrow_{\overline{\mu}}^* t \in NF(\rightarrow_\mu)$ implies $s \in NF(\rightarrow_\mu)$.*

Proof. We prove the statement for one step, i.e., $s \rightarrow_{\overline{\mu}} t \in NF(\rightarrow_\mu)$ implies $s \in NF(\rightarrow_\mu)$. The result then follows by transitivity (that is by induction on the number of steps in $s \rightarrow_{\overline{\mu}}^* t$). Suppose, for a proof by contradiction, that $s \notin NF(\rightarrow_\mu)$. Hence, there exists some s' with $s \rightarrow_0 s'$. Moreover, $s \rightarrow_{\overline{\mu}} t$ means $s \rightarrow_{k_1} t$ for some $k_1 \geq 1$. By Commutation Lemma 4 this implies that there exists some t' with $s' \rightarrow_{\leq k_1}^* t' \leftarrow_0 t$. But this is a contradiction to $t \in NF(\rightarrow_\mu)$.[7]

3.2   Main Results

Now we are ready to prove the main results of the paper. The first one is a 'level confluence' criterion for CSRSs.

Theorem 1 (level confluence criterion / technical key lemma). *Let $\mathcal{R}_\mu$ be given satisfying (1)-(5). Then $\rightarrow_{\leq k}$ is confluent for every $k \geq 0$.*

Proof. We prove confluence of $\rightarrow_{\leq k}$ for all k by induction on k.

(o) Base case $k = 0$: Confluence of $\rightarrow_{\leq 0} = \rightarrow_0 = \rightarrow_\mu$ follows from Lemma 1.

---

[6] Again, from the proof it follows that even the stronger property $\leftarrow_{\leq k} \cdot \rightarrow_0 \subseteq \rightarrow_0 \cdot \Vdash_{\leq k}$ holds.

[7] Note that the assumptions (1), (2), (5) in the lemma are needed to enable applicability of Lemma 4.

(i) Induction step "$k \implies k + 1$": Consider an arbitrary $\to_{\leq k+1}$-divergence $t_1' \leftarrow^*_{\leq k+1} s \to^*_{\leq k+1} t_2'$. Due to assumption (3) we may $\to_\mu$-normalize t_1' and t_2' yielding $D_1 : s \to^*_{\leq k+1} t_1' \to^*_0 t_1 \in \mathrm{NF}(\to_\mu)$ and $D_2 : s \to^*_{\leq k+1} t_2' \to^*_0 t_2 \in \mathrm{NF}(\to_\mu)$. Now, by repeated application of Exchange Lemma 3 and termination of $\to_\mu$ we can rearrange these derivations into $D_1' : s \to^*_0 s_1 \to^*_{\leq k+1} t_1 \in \mathrm{NF}(\to_\mu)$ and $D_2' : s \to^*_0 s_2 \to^*_{\leq k+1} t_2 \in \mathrm{NF}(\to_\mu)$, where the $\to_{\leq k+1}$-reduction steps are all non-replacing, i.e., of level ≥ 1. From Lemma 5 we infer that $s_1, s_2 \in \mathrm{NF}(\to_\mu)$. Together with confluence of $\to_0$ (see base case) this implies $s_1 = s_2$. Hence the divergence diagram collapses to $t_1 \leftarrow^*_{\leq k+1} s_1 = s_2 \to^*_{\leq k+1} t_2$. Now, let $s' = s_1 = s_2$. Repeated applications of the Extraction Lemma 2 yield $s' = C[u_1, \ldots, u_m] \to^*_{\leq k+1} C[w_1, \ldots, w_m] = t_2$ and $s' = C[u_1, \ldots, u_m] \to^*_{\leq k+1} C[v_1, \ldots, v_m] = t_1$ for some context $C[\ldots]$ such that $v_i \leftarrow^*_{\leq k} u_i \to^*_{\leq k} w_i$ for $1 \leq i \leq m$. Applying the induction hypothesis (for k) to all i, $1 \leq i \leq m$, we conclude that there exist $\overline{u_i}$ for all i with $v_i \to^*_{\leq k} \overline{u_i} \leftarrow^*_{\leq k} w_i$. Putting back these reductions in the non-extracted version, we get $t_1 = C[v_1, \ldots, v_m] \to^*_{\leq k+1} C[\overline{u_1}, \ldots \overline{u_m}] \leftarrow_{\leq k+1} C[w_1, \ldots, w_m] = t_2$. Hence, $\overline{s'} = C[\overline{u_1}, \ldots \overline{u_m}]$ is a common $\to_{\leq k+1}$-reduct of both t_1 and t_2 as desired, and we are done.

As a consequence of this level confluence criterion we thus obtain our main result, a generalized version of Newman's Lemma for left-linear TRSs.[8]

Theorem 2 (main result). *Let $\mathcal{R}$ be a TRS and μ be a replacement map on the signature of $\mathcal{R}$ such that (1)-(5) are satisfied, i.e., such that $\mathcal{R}$ is left-linear, $\mu \in CM_\mathcal{R}$, $\mathcal{R}_\mu$ is terminating and level-decreasing and all critical pairs of $\mathcal{R}$ are $\mathcal{R}_\mu$-joinable. Then $\mathcal{R}$ is confluent.*

Proof. Due to Proposition 1(c) this is an immediate corollary of Theorem 1.

Note that Newman's Lemma (for left-linear TRSs) is obtained from Theorem 2 as a special case, namely by taking – for some given $\mathcal{R}$ – μ to be the most liberal replacement map $\mu = \mu_\top$. This choice of μ clearly implies (2) and (5), and also that $\to \; = \; \to_\mu$, hence termination of $\to$ is equivalent to termination of $\to_\mu$. Actually, Theorem 2 properly generalizes Newman's Lemma (for left-linear TRSs) since there are cases (cf. e.g. Examples 1, 2) where the former is applicable, but not the latter because the system (as a TRS) is not terminating.

3.3 Examples and Comparison

Let us first reconsider our Examples 3, 1 and 2. In Example 3, $\mathcal{R}$ with μ as specified satisfies all preconditions of Theorem 2 except (4). Hence the latter is not (and should not be) applicable. In Example 1, choosing $\mu_\mathcal{R}^{can}$, i.e., with $\mu(f) = \mu(h) = \emptyset$, conditions (1)-(5) are all satisfied as is easily verified. In particular, termination of $\mathcal{R}_\mu$ is not difficult to prove by some of the methods proposed in the literature (cf. e.g. [3], [9], [5], [18]). Observe that, when choosing some $\mu \in CM_\mathcal{R}$, in order to ensure termination of $\mathcal{R}_\mu$ we must obviously have

---

[8] more precisely, for the abstract reduction systems induced by left-linear TRSs.

$\mu(f) = \emptyset$ (because otherwise this entails non-termination of $\mathcal{R}_\mu$). Hence, for h the only choice is $\mu(h) = \emptyset$. Otherwise, (5) would be violated.

In Example 2, choosing $\mu = \mu_{\mathcal{R}}^{can}$ (hence $\mu(:) = \mu(s) = \mu(hd) = \emptyset$, $\mu(\mathsf{inc}) = \mu(\mathsf{tl}) = \{1\}$) it is easy to verify that (1)-(5) do indeed hold. Hence, confluence of the TRS follows by Theorem 2.

When applying Theorem 2, there is a certain flexibility in the sense that the parameter μ may be chosen differently. We require $\mu \in CM_{\mathcal{R}}$, but not necessarily $\mu = \mu_{\mathcal{R}}^{can}$ as in the above examples. In fact, in certain cases the canonical $\mu_{\mathcal{R}}^{can}$ need not be a good choice, whereas a more liberal μ can work, cf. conditions (3)-(5).

Example 5 (Example 1 modified). Consider $\mathcal{R}$ consisting of the rules (1) $\{g(a) \to f(g(a)), (2')\ g(b) \to c(a), (3)\ a \to b, (4)\ f(x) \to h(x)$ and $(5')\ h(x) \to c(b)$. Here, choosing $\mu_{\mathcal{R}}^{can}$ we cannot apply Theorem 2 to infer confluence, since property (4) is violated. However, choosing $\mu \in CM_{\mathcal{R}}$ with $\mu(g) = \mu(c) = \{1\}$ and $\mu(i) = \emptyset$ for all other function symbols i, Theorem 2 is applicable and shows indeed confluence of $\mathcal{R}$.

Comparing our new confluence criterion of Theorem 2 with other known criteria (or decision procedures, respectively) for (possibly non-terminating) TRSs, cf. in particular those of [26], [11], [8], [23], [24], [21], [7], [6] it turns out to be incomparable w.r.t. all of them. This is easy to show by exhibiting examples where our criterion is applicable whereas the other ones are not, and vice versa. This incomparability is not really surprising, because all other confluence criteria above do not rely on a (partial) termination assumption, whereas our criterion crucially does.

3.4   Discussion

Let us first discuss the preconditions for applying Theorem 2, namely, (1)-(5), the effectiveness of using it for confluence proofs, and the inherent limitations of this confluence criterion. Then we will see how these latter limitations naturally lead to some interesting open problems.

Recall that applicability of Theorem 2 requires the following properties:

- $\mathcal{R}$ is left-linear. $\hfill (1)$
- $\mu \in CM_{\mathcal{R}}$. $\hfill (2)$
- $\mathcal{R}_\mu$ is terminating. $\hfill (3)$
- Every critical peak $t_1 \leftarrow s \to t_2$ of $\mathcal{R}$ is joinable with $\to_\mu$-steps. $\hfill (4)$
- $\mathcal{R}_\mu$ is level-decreasing. $\hfill (5)$

Note that checking (1), (2) (and (4) provided (3) holds) is easy. Furthermore, in (2), there are only finitely many possibilities for choosing some $\mu \in CM_{\mathcal{R}}$ (for finite $\mathcal{R}$), hence the search for an appropriate $\mu \in CM_{\mathcal{R}}$ can also be automated. Proving termination of $\mathcal{R}_\mu$, i.e., (3), is of course undecidable in general, but nowadays numerous powerful methods and tools exist for such context-sensitive termination proofs, cf. e.g. [5], [18]. Thus, the applicability of the confluence criterion of Theorem 2 is effectively decidable, provided that (3) holds, for some $\mu \in CM_{\mathcal{R}}$.

Having a closer look at the preconditions, termination of $\rightarrow_\mu$ (3) is crucial to get a Newman style confluence criterion. The conditions (1) left-linearity of $\mathcal{R}$, and (2) $\mu \in CM_{\mathcal{R}}$, are essential for several important lemmas (especially Lemmas 3, 4 and 5) used in the proof of the main Theorem 2. Condition (4), at least in combination with (2), seems to be unavoidable to infer confluence of $\rightarrow_\mu$ using (1). The only condition which appears to be less clear and intuitive is level-decreasingness of $\rightarrow_\mu$ (5). Besides termination of $\rightarrow_\mu$, this condition is the most restrictive application condition in practical examples. It would be nice if it could be dropped or weakened. Currently we do not know any counterexample to the modified (generalized) statement of Theorem 2 where condition (5) is dropped. On the other hand, the proof of Theorem 2 (via the "level confluence" criterion of Theorem 1) as well as Lemmas 3 and 4 heavily rely on this condition. Hence we have the following

Open Problem 1 (necessity of level-decreasingness?). *Does the statement of Theorem 2 also hold if precondition (5) is omitted? In other words, is any TRS $\mathcal{R}$ s.t*

- *$\mathcal{R}$ is left-linear,*
- *$\mu \in CM_{\mathcal{R}}$,*
- *$\mathcal{R}_\mu$ is terminating, and*
- *every critical pair of $\mathcal{R}$ is $\mathcal{R}_\mu$-joinable*

necessarily confluent?

A positive solution to this open problem would be particularly nice, since there are numerous examples (cf. e.g. the literature on CSR) where level-decreasingness is not satisfied. A basic one is the following (cf. e.g.[17]).

Example 6. Consider the $\mathcal{R}$ given by

$$\mathsf{from}(x) \rightarrow x : \mathsf{from}(\mathsf{s}(x))$$
$$\mathsf{sel}(0, y : z) \rightarrow y$$
$$\mathsf{sel}(\mathsf{s}(x), y : z) \rightarrow \mathsf{sel}(x, z)$$

where from models a kind of parameterized version of generating infinite lists of natural numbers (cf. Example 2 for a non-parameterized version), and sel serves for extracting elements from a list. This system is clearly non-terminating as a TRS, but becomes terminating as $\mathcal{R}_\mu$ with e.g. $\mu = \mu_{\mathcal{R}}^{can}$ (hence, with $\mu(:) = \{1\}$). Conditions (1)-(4) of Theorem 2 are easily verified, but (5) is violated, since the first rule is not level-decreasing. Hence Theorem 2 cannot be applied, although $\mathcal{R}$ is indeed confluent, simply because it is orthogonal ([26]).

Another issue that is related to (the preconditions and the statement of) Theorem 2 is the following which we will only touch (cf. e.g. [2] for more details and background). Let us reconsider the introductory counterexample 3 that we used to motivate the requirement that all critical pairs should be $\rightarrow_\mu$-joinable. In the example this was not the case, and $\mathcal{R}$ was not confluent, because a and d with $a \leftarrow b \rightarrow c \rightarrow d$ did not have a common reduct. But, interestingly, it turns out that, when switching from *finitary* rewriting and confluence to *infinitary* rewriting and confluence (cf. e.g. [4], [12], [16], [2]), then Example 3 behaves nicely, in the sense that $\mathcal{R}$ is *infinitary confluent* (ω-confluent). Intuitively this is easy to see since the *infinitary normal forms* of both a and d are h^ω, hence the system is indeed ω-confluent. A tempting conjecture in this direction which we state as open problem is the following.

Open Problem 2 (criterion for ω-confluence?). *Is any left-linear, non-collapsing, locally confluent TRS $\mathcal{R}$, with $\mu \in CM_\mathcal{R}$ and $\rightarrow_{\mathcal{R},\mu}$ terminating, necessarily ω-confluent?*

Orthogonal systems are known to be ω-confluent (for *strongly converging reductions*) provided they are non-collapsing (cf. [12], [2]). The typical counterexample showing that the non-collapsingness in this result cannot be dropped is as follows ([12]): Let $\mathcal{R}$ consist of the rules $a(x) \rightarrow x$, $b(x) \rightarrow x$ and $c \rightarrow a(b(c))$. Then we get the reductions

$$c \rightarrow a(b(c)) \rightarrow a(c) \rightarrow a(a(b(c))) \rightarrow a(a(c)) \rightarrow^\omega a^\omega$$

$$c \rightarrow a(b(c)) \rightarrow b(c) \rightarrow b(a(b(c))) \rightarrow b(b(c)) \rightarrow^\omega b^\omega \,,$$

hence $a^\omega \leftarrow^\omega c \rightarrow^\omega b^\omega$, but there is no term t with $a^\omega \rightarrow^{\leq\omega} t \leftarrow^{\leq\omega} b^\omega$. Now, $\mathcal{R}$ is clearly non-terminating, but $\mathcal{R}_\mu$ is also non-terminating for any μ here. This phenomenon also applies to other collapsing counterexamples (to ω-confluence) in [12]. If instead of the above system we consider $\mathcal{R}$ consisting of $a(x) \rightarrow x$, $b(x) \rightarrow x$ and $c \rightarrow d(a(b(c)))$, then $\mathcal{R}_\mu$ becomes obviously terminating e.g. for $\mu \in CM_\mathcal{R}$ with $\mu(a) = \mu(b) = \{1\}$, $\mu(d) = \emptyset$. However, for $(da)^\omega \leftarrow^\omega c \rightarrow^\omega (db)^\omega$ we can now find a common reduct (in infinitary rewriting): $(da)^\omega \rightarrow^\omega d^\omega \leftarrow^\omega (db)^\omega$. Hence, in the above Open Problem 2 it could even make sense to generalize the statement by omitting the non-collapsing requirement.

   If the answer to the above open problem were "yes", then this would be a nice way to prove ω-confluence. To the best of our knowledge it would also be the first confluence criterion for non-orthogonal infinitary rewrite systems.

4   Conclusion

To conclude, we have presented a new confluence criterion for (possibly non-terminating) left-linear TRSs which properly generalizes Newman's Lemma (for left-linear TRSs). The criterion is We think that not only the result itself is interesting, but also the proof technique employed that uses the more general framework of context-sensitive rewriting to finally derive a result about ordinary (context-free) rewriting. Methodologically, the approach strongly differs from related confluence criteria. It is neither based on critical pair criteria nor on modularity properties, but rather on a combination of Newman's Lemma (for a terminating sub-relation of the rewrite relation, that is not induced by a sub-TRS) with a level-based approach that exploits rearrangement and commutation properties.

Acknowledgements. We would like to thank the anonymous referees for various useful comments and hints.

References

1. F. Baader and T. Nipkow. *Term rewriting and All That.* Cambridge Univ. Press, 1998.
2. M. Bezem, J. W. Klop, and R. de Vrijer, eds. *Term Rewriting Systems.* Cambridge Tracts in Theoretical Computer Science 55. Cambridge Univ. Press, March 2003.
3. C. Borralleras, S. Lucas, and A. Rubio. Recursive path orderings can be context-sensitive. In A. Voronkov, ed., *Proc. 18th CADE, Copenhagen, Denmark,* LNCS 2392, pages 314–331. Springer, July 2002.
4. N. Dershowitz, S. Kaplan, and D. A. Plaisted. Rewrite, rewrite, rewrite, rewrite, rewrite, ... *Theoretical Computer Science,* 83(1):71–96, June 1991.
5. J. Giesl and A. Middeldorp. Transformation techniques for context-sensitive rewrite systems. *Journal of Functional Programming,* 14(4):379–427, July 2004.
6. G. Godoy and A. Tiwari. Confluence of shallow right-linear rewrite systems. In C.-H. Luke Ong, ed., *Proc. 14th CSL, Oxford, UK,* LNCS 3634, pages 541–556. Springer, 2005.
7. G. Godoy and A. Tiwari and R. M. Verma. On the confluence of linear shallow term rewrite systems. In H. Alt and M. Habib, eds., *Proc. 20th STACS, Berlin, Germany,* LNCS 2607, pages 85–96. Springer, 2003.
8. B. Gramlich. Confluence without termination via parallel critical pairs. In H. Kirchner, ed., *Proc. 21st CAAP, Linköping, Sweden,* LNCS 1059, pages 211–225. Springer, April 1996.
9. B. Gramlich and S. Lucas. Modular termination of context-sensitive rewriting. In C. Kirchner, ed., *Proc. 4th PPDP, Pittsburgh, PA, USA,* pages 50–61, October 2002. ACM Press.
10. J.R. Hindley. An abstract Church–Rosser theorem, part ii: Applications. *Journal of Symbolic Logic,* 39:1–21, 1974.
11. G. Huet. Confluent reductions: Abstract properties and applications to term rewriting systems. *Journal of the ACM,* 27(4):797–821, October 1980.
12. R. Kennaway, J. W. Klop, R. Sleep, and F.-J de Vries. Transfinite reductions in orthogonal term rewriting systems. *Information and Computation,* 119(1):18–38, May 1995.
13. J. W. Klop. Term rewriting systems. In S. Abramsky, D. Gabbay, and T. Maibaum, eds., *Handbook of Logic in Computer Science,* volume 2, chapter 1, pages 2–117. Clarendon Press, Oxford, 1992.
14. J. W. Klop, A. Middeldorp, Y. Toyama, and R. de Vrijer. Modularity of confluence: A simplified proof. *Information Processing Letters,* 49:101–109, 1994.
15. S. Lucas. Context-sensitive computations in functional and functional logic programs. *Journal of Functional and Logic Programming,* 1998(1):1–61, January 1998.
16. S. Lucas. Transfinite rewriting semantics for term rewriting systems. In A. Middeldorp, ed., *Proc. 12th RTA, Utrecht, The Netherlands,* LNCS 2051, pages 216–230, May 2001, Springer.

17. S. Lucas. Context-sensitive rewriting strategies. *Information and Computation*, 178(1):294–343, 2002.
18. S. Lucas. Polynomials over the reals in proofs of termination: From theory to practice. *RAIRO Theoretical Informatics and Applications*, 39(3):547-586, July 2005.
19. M. H. A. Newman. On theories with a combinatorial definition of equivalence. *Annals of Mathematics*, 43(2):223–242, 1942.
20. E. Ohlebusch. Modular properties of composable term rewriting systems. *Journal of Symbolic Computation*, 20(1):1–42, July 1995.
21. S. Okui. Simultaneous critical pairs and Church-Rosser property. In T. Nipkow, ed., *Proc. 9th RTA, Tsukuba, Japan*, LNCS 1379, pages 2–16, 1998. Springer.
22. V. van Oostrom. Confluence by decreasing diagrams. *Theoretical Computer Science*, 126(2):259–280, April 1994.
23. V. van Oostrom. Developing developments. *Theoretical Computer Science*, 175(1):159–181, March 1997.
24. M. Oyamaguchi and Y. Ohta. A new parallel closed condition for Church–Rosser of left-linear term rewriting systems. In H. Comon, ed., *Proc. 8th RTA, Sitges, Spain)*, LNCS 1232, pages 187–201, June 1997. Springer.
25. J.C. Raoult and J. Vuillemin. Operational and semantic equivalence between recursive programs. *Journal of the ACM*, 27(4):772–796, October 1980.
26. B. K. Rosen. Tree-manipulating systems and Church-Rosser theorems. *Journal of the ACM*, 20:160–187, 1973.
27. Y. Toyama. On the Church–Rosser property of term rewriting systems. ECL Technical Report 17672, NTT, December 1981. In Japanese.
28. Y. Toyama. On the Church-Rosser property for the direct sum of term rewriting systems. *Journal of the ACM*, 34(1):128–143, 1987.
29. Y. Toyama and M. Oyamaguchi. Church–Rosser property and unique normal form property of non-duplicating term rewriting systems. In N. Dershowitz and N. Lindenstrauss, eds., *Proc. 4th CTRS*, LNCS 968, pages 316–331. Springer, 1995.

Unions of Equational Monadic Theories[*]

Piotr Hoffman

Institute of Informatics, Warsaw University, Poland
piotrek@mimuw.edu.pl

Abstract. We investigate the decidability of unions of decidable equational theories. We focus on monadic theories, i.e., theories over signatures with unary symbols only. This allows us to make use of the equivalence between monoid amalgams and unions of monadic theories. We show that if the intersection theory is unitary, then the decidability of the union is guaranteed by the decidability of tensor products. We prove that if the intersection theory is a group or a group with zero, then the union is decidable. Finally, we show that even if the intersection theory is a 3-element monoid and is unitary, the union may be undecidable, but that it will always be decidable if the intersection is 2-element unitary. We also show that unions of regular theories, i.e., theories recognizable by finite automata, can be undecidable. However, we prove that they are decidable if the intersection theory is unitary.

1 Introduction

We consider the following question:

> If S and T are decidable equational theories, then under what assumptions is the union theory $S \cup T$ decidable as well?

The practical importance of this question is clear: our ability to combine decision procedures for S and T into a decision procedure for the union depends on it.

It is obvious that for the union $S \cup T$ to be decidable, something more is needed than just S and T being decidable themselves. Let Σ_S be the signature of S and Σ_T the signature of T, and let $\Sigma_U = \Sigma_S \cap \Sigma_T$ be their intersection. As usual, we assume that any equation over Σ_U is a consequence of S iff it is a consequence of T. This means that an intersection theory U over Σ_U exists which is conservatively extended by S and by T.

Since 1974 it is known [1] that if Σ_U is empty, i.e., if S and T are built over disjoint signatures, then decidability of S and T implies the decidability of $S \cup T$. A few years ago it has been proved that if *effective* U-bases of S and T exist, then the decidability of S and T implies the decidability of $S \cup T$ [2,3].[1]

Following the ideas of [4], in our work we focus on *monadic* theories, that is, on theories over signatures in which all function symbols are unary (in fact, account is also taken of constants). This admittedly strong restriction makes the

---

[*] This work has been partially supported by EU project SENSORIA (no. 016004).
[1] These results are succintly presented in Sect. 3.

F. Pfenning (Ed.): RTA 2006, LNCS 4098, pp. 81–95, 2006.

research problem simpler, but far from trivial. It is therefore a perfect vehicle for testing ideas that may be later applied to theories in which symbols of arbitrary arities appear. As for theorems stating the *undecidability* of certain unions, there is of course no need to generalize them to non-monadic theories.

Three categories of results are presented in the paper. These are: results on unitary theories and links between unions and tensor products (Sect. 3), results on the decidability of unions built over group and group-like theories (Sect. 4), and results on the undecidability of certain unions (Sect. 5).

We start with Sect. 2, in which basic notions are defined. In Sect. 3 we first recall known results on theories over effective bases (Th. 1). We then show that if the intersection theory is unitary and the tensor products are decidable, then so is the union (Th. 2). In particular, unions of *regular* theories, that is, of theories recognizable by finite automata, are decidable if their intersection is unitary (Cor. 2). In Sect. 4 we show that a union of theories whose intersection is a group is decidable under certain liberal assumptions (Th. 3, Cor. 3); a similar result is proved for groups with zero (Th. 4). We start Sect. 5 by recalling [4] that a union of theories with unitary intersection may be undecidable (Th. 5). We then show that this is true even if one of those theories is regular with only 4 equivalence classes and the intersection is the theory of two boolean constants (Th. 6). We also prove that our example is minimal, i.e., if the intersection is unitary and regular with less than 3 equivalence classes, then the union is decidable (Prop. 3). Finally, we present a construction due to Sapir [5] which proves that even the union of two regular theories may be undecidable (Th. 7). We conclude by a discussion of the forms of assumptions that might guarantee decidability of theory unions and of possible generalizations of the obtained decidability results.

2   Unions of Theories and Amalgams of Monoids

In this paper we deal only with monadic theories. A *monadic signature* is a finite set Σ_S of unary function symbols. A *monadic theory* S is a set of equations of the form $t_1(x) = t_2(x)$, where t_1, t_2 are terms over Σ_S. In fact, signatures in which constants appear can be encoded as monadic signatures by adding, for all symbols f and all constants c, equations $c(f(x)) = c(x)$. Thus, our considerations are applicable to signatures with both nullary and unary symbols.

Any monadic theory S over Σ_S naturally defines a monoid S and a homomorphism $h_S : \Sigma_S^* \to S$ from the free monoid over the alphabet Σ_S onto S. The monoid S is simply the quotient of the free monoid Σ_S^* by an equivalence relation such that words w_1, w_2 over Σ_S are considered equivalent iff the terms they represent are equal in S, and h_S is the canonical homomorphism. The neutral element 1 of S represents the equivalence class of the variable "x". The word w is said to *represent* the element $h_S(w) \in S$, and we will write w instead of $h_S(w)$. When speaking about any monoid S, we will tacitly assume that it is generated by a finite set Σ_S and that a canonical homomorphism $h_S : \Sigma_S^* \to S$ is given. A construction converse to the one presented above also exists [4], taking any monoid S and homomorphism h_S from Σ_S^* onto S to a monadic theory over Σ_S.

The above equivalence between monadic theories and monoids allows us to speak about theories in the language of monoids. Any notion concerning monadic theories has its counterpart in the world of monoids, and vice versa.

A *finite presentation* of a monoid S is a finite set of equations $w = w'$ ($w, w' \in \Sigma_S^*$) such that S is the quotient of Σ_S^* by those equations. These equations are called the *defining relations* of S. A monadic theory S directly corresponds to a set of defining relations, so if S is finite, then the monoid S is finitely presented.

The *word problem* is said to be decidable in a monoid S if one can, given words w, w' over Σ_S, decide whether $w = w'$ in S (that is, formally, whether $h_S(w) = h_S(w')$). The corresponding problem for theories is deciding whether an equation $t_1(x) = t_2(x)$ is a consequence of a theory S.

The word problem is said to be *regular* in S if the language of words $e \in (\Sigma_S \cup \{=\})^*$ of the form "$w_1 = w_2$" such that $w_1 = w_2$ holds in S is regular. Regularity has a simple characterization in terms of monoids:

Lemma 1. *The word problem is regular in a monoid S iff S is finite.*   □

In the world of theories, the regularity of a theory means that a finite automaton exists which for any term t computes an index $i(t)$ such that $t_1 = t_2$ is a consequence of the theory iff $i(t_1) = i(t_2)$.

Assume theories $\mathcal{U}$, S and $\mathcal{T}$ over signatures $\Sigma_U \subseteq \Sigma_S, \Sigma_T$ are given such that S and $\mathcal{T}$ are conservative extensions of $\mathcal{U}$. This requirement means that any equation over Σ_U is a consequence of $\mathcal{U}$ iff it is a consequence of S, and also iff it is a consequence of $\mathcal{T}$. We are interested in the theory $S \cup \mathcal{T}$ over $\Sigma_S \cup \Sigma_T$.

The corresponding notions in the world of monoids are amalgam and amalgamated product.

A *monoid amalgam* $[U \subseteq S, T]$ is a triple of monoids such that $S \cap T = U$. The monoid U is called the *core* of the amalgam. The amalgam is *finitely presented* if U, S and T are. We assume that Σ_S and Σ_T extend Σ_U.

For any monoid amalgam one can consider the pushout $\mu : S \to P, \nu : T \to P$ in the category of monoids of the span of inclusions $i : U \to S$ and $j : U \to T$. This is called the *amalgamated product*. The amalgamated product can be defined as follows. Consider a (one-step) relation $\Leftrightarrow$ on $(S \cup T)^+$ defined by the rewrite rules $s \cdot s' \Leftrightarrow ss'$ and $t \cdot t' \Leftrightarrow tt'$ for all $s, s' \in S$ and $t, t' \in T$. [2] Let P be the quotient of $(S \cup T)^+$ by the reflexive-transitive closure $\Leftrightarrow^*$, and let $\mu : S \to P$ be defined by $\mu(s) = [s]_{\Leftrightarrow^*}$ and $\nu : T \to P$ by $\nu(t) = [t]_{\Leftrightarrow^*}$, for all $s \in S$ and $t \in T$. Then P, μ, ν form an amalgamated product of the amalgam $[U \subseteq S, T]$.

An equation $w = w'$ on words w, w' over $\Sigma_S \cup \Sigma_T$ holds *in the amalgam* if $h_P(w) = h_P(w')$. The *word problem in the amalgam* is the problem of deciding whether $w = w'$ in the amalgam, for given $w, w' \in (\Sigma_S \cup \Sigma_T)^*$. We have:

Lemma 2. *For any words w, w' over $\Sigma_S \cup \Sigma_T$, $w = w'$ in the amalgam iff $w \cdot 1 \Leftrightarrow^* w' \cdot 1$.*   □

---

[2] For words over $S \cup T$ we use the symbol "$\cdot$" to represent concatenation, so that $s \cdot s'$ is a two-letter word, while ss' is a one-letter word. For other free monoids concatenation is just represented by juxtaposition.

Note that the unit "1" above is introduced in case w or w' are empty.

A monoid amalgam $[U \subseteq S, T]$ precisely corresponds to a pair of theories $\mathcal{S}$ and $\mathcal{T}$ conservatively extending a theory $\mathcal{U}$. The union $\mathcal{S} \cup \mathcal{T}$ precisely corresponds to the amalgamated product P. In particular:

Proposition 1. *If $\mathcal{S}$ and $\mathcal{T}$ conservatively extend $\mathcal{U}$, then an equation $t_1 = t_2$ is a consequence of $\mathcal{S} \cup \mathcal{T}$ iff the corresponding word equation $w_1 = w_2$ holds in the amalgam $[U \subseteq S, T]$.* □

Therefore we may, instead of the original problem of theory unions, consider the following problem:

If $[U \subseteq S, T]$ is an amalgam of monoids with decidable word problems, then under what assumptions is the word problem in the amalgam decidable as well?

3   Unitary and Tensor-Defined Amalgams

The relationship between the core U of an amalgam $[U \subseteq S, T]$ and the monoids S and T has a major influence on the difficulty of the amalgam's word problem. One notion that describes this relationship and which we will use intensively is that of a *unitary* submonoid. A submonoid U of S is unitary in S, if, for all $u \in U$ and $s \in S$, $su \in U$ or $us \in U$ imply $s \in U$.

The unitariness of U in S means that if we take an element of $S \setminus U$ and multiply it, on either side, by elements of U, then this product will always fall in $S \setminus U$. Note however, that it may well happen that the product of two elements of $S \setminus U$ falls in U. On the level of theories, unitariness means that if a certain term t over Σ_S is "new", that is it is not equal to any term over Σ_U modulo the theory $\mathcal{S}$, then its composition with an "old" term, that is one over Σ_U, will be "new" as well. Again, note that it may still happen that the composition of two "new" terms is "old", that is that it can be proved equal to a term over Σ_U.

We will also need notions describing computational aspects of the relationship between the core U and S or T: if U is a submonoid of S, then U-*membership is decidable* in S if verifying whether $w \in U$ in S is decidable for all $w \in \Sigma_S^*$, and it is *computable* if additionally a word $w' \in \Sigma_U^*$ may then be computed such that $w = w'$ in S. Thus, on the level of theories, decidability of U-membership means that it is decidable whether a given term over Σ_S is "old", that is whether it is equal to some term over Σ_U; computability then means that the term over Σ_U is also computable. We have:

Lemma 3. *If S has a decidable word problem, U is a submonoid of S, and U-membership is decidable, then U-membership is computable.*

Proof. If $w \in \Sigma_S^*$ and $w \in U$, then it suffices to enumerate all $w' \in \Sigma_U^*$ and for each in turn check $w = w'$ in S, which is decidable. □

An amalgam $[U \subseteq S, T]$ with amalgamated product P, μ, ν is said to be *weakly embeddable* if μ and ν are injective (that is, if S and T embed in P). It is said to be *embeddable* if additionally for all $s \in S$, $t \in T$, we have that $\mu(s) = \nu(t)$ implies $s = t \in U$ (that is, if $S \cup T$ embeds in P). On the level of theories, weak embeddability means that both S and T are conservatively extended by the union $S \cup T$; in other words, two terms over Σ_S (Σ_T, resp.) are equal modulo $S \cup T$ iff they are equal modulo S (T, resp.). Embeddability adds to this the requirement that S and T are *jointly* conservatively extended by $S \cup T$; in other words, if two terms, t_1 over Σ_S and t_2 over Σ_T, are equal modulo $S \cup T$, then there must be a term t over Σ_U such that $t_1 = t$ modulo S and $t = t_2$ modulo T. The importance of embeddability for describing unions of monadic theories has been discussed in detail in [4]. The results [2,3] on decidability of unions require the amalgams to be embeddable (see Lem. 4 point 3 below). We now succintly present those results.

If U is a submonoid of S, then a *U-base* of S is a set $G \subseteq S$ such that all elements $s \in S$ have a unique decomposition as $s = gu$, where $g \in G$, $u \in U$. The base is *effective* if g and u (or rather their representations, i.e., words over Σ_S and Σ_U) may be computed. A *U-cobase* is defined analogously. We have:

Lemma 4. *The following implications hold:*

1. *If S has a U-base and U-cobase, then U is unitary in S.*
2. *If G is a U-base of S, then $G \cap U = \{1\}$.*
3. *If S and T have U-bases, then $[U \subseteq S,T]$ is embeddable.*
4. *If S has a decidable word problem, then any U-base G of S such that G-membership is decidable in S is effective.*
5. *If S has a decidable word problem and an effective U-base, then U-membership is decidable in S.*

Proof. Claims 1 and 2 have been proved in [4]. Claim 3 has been proved in [2] (Prop. 4.14 and Lem. 4.18). Claim 4 holds because one may simply enumerate all pairs $(w, w') \in \Sigma_S^* \times \Sigma_U^*$ and check whether $w \in G$ and $ww' = s$ in S. Claim 5 holds because given $w \in \Sigma_S^*$ one may compute $g \in G$ and $u \in U$ such that $gu = w$ and then check whether $g = 1$ in S. □

The following theorem is a translation of the main result of Baader/Tinelli and Fiorentini/Ghilardi into the language of monoids. Note however that their result is not restricted to monadic theories. A simple proof may be found in [4]:

Theorem 1. *For any amalgam $[U \subseteq S,T]$, if:*

- *S and T have decidable word problems,*
- *there exist effective U-bases (or U-cobases) of S and T,*

then the amalgam has a decidable word problem. □

We now turn to a notion stronger than embeddability, namely *tensor-definedness*.

A *stripe* is a word from $(S \cup T)^+$ which consists either of just one letter (in particular, it may be a letter from U), or of interleaved elements of $S \setminus U$ and $T \setminus U$. It is natural to try to check whether an equality $w = w'$ holds in the amalgam by reducing w to a stripe W and w' to a stripe W' and then checking whether $W = W'$. Indeed, any word can be reduced to a stripe (though this stripe need not be unique):

Lemma 5. *For any amalgam $[U \subseteq S, T]$, if U-membership is decidable in S and T, and if S and T have decidable word problems, then for any word $w \in (\Sigma_S \cup \Sigma_T)^*$ a stripe W can be computed such that $w = W$ in the amalgam.* □

However, checking whether two stripes are equal in the amalgam need not be trivial. In general, it is not true that if two stripes W and W' are equal in the amalgam, then they must be of the same length and that then the ith letters of W and W' must be equal in S or, resp., T. For example, if U is the free monoid on one generator a, S extends U with a generator b satisfying $ab = ba = 1$, and T extends U with a generator c satisfying $ac = ca = 1$, then $W = b(cc)$ has length 2 and $W' = bcb$ has length 3, but $W = b(cc) = bc1c = bc(ba)c = bcb(ac) = bcb1 = bcb = W'$ in the amalgam.

The word problem is n-*decidable* in the amalgam if the problem of checking $W = W'$ in the amalgam is decidable for all stripes $W, W' \in (S \cup T)^+$ of length $\leq n$.

The notion of n-*tensor* $S \otimes_U T \otimes_U \dots$ is intended to reflect the idea of proving equivalence by using words in which no more than n separate "blocks" from S exist. Formally, this tensor is defined as follows. Let $\sim_n$ be the least symmetric relation on *interleaved* words of length n, i.e., on words $w = x_1 \cdot \dots \cdot x_n$ such that $x_1 \in S$, $x_2 \in T$, $x_3 \in S$, etc., satisfying the conditions:

$$w \cdot su \cdot t \cdot w' \sim_n w \cdot s \cdot ut \cdot w'$$

for all $s \in S$, $t \in T$, $u \in U$ and $w \in (S \cdot T)^k$, $w' \in (S \cdot T)^l$ (for $n = 2k + 2l + 2$) or $w' \in (S \cdot T)^l \cdot S$ (for $n = 2k + 2l + 3$), and

$$w \cdot tu \cdot s \cdot w' \sim_n w \cdot t \cdot us \cdot w'$$

for all $s \in S$, $t \in T$, $u \in U$ and $w \in (S \cdot T)^k \cdot S$, $w' \in T \cdot (S \cdot T)^l$ (for $n = 2k + 2l + 4$) or $w' \in (T \cdot S)^l$ (for $n = 2k + 2l + 3$). The n-*tensor* $\sim_n^*$ is the reflexive-transitive closure of $\sim_n$. The n-tensor $T \otimes_U S \otimes_U \dots$ is defined analogously and also denoted by $\sim_n^*$. The n-tensor is *decidable* if one can decide whether $w \sim_n^* w'$ for given interleaved words $w, w' \in (S \cup T)^n$, the letters of which are given as words over Σ_S and Σ_T.

The amalgam is *tensor-defined* if for any stripes W and W' we have $W = W'$ in the amalgam iff W and W' are of the same length, say n, and $W \sim_n^* W'$. Thus, a tensor-defined amalgam is one in which the checking the equality of two stripes does not require one to use interleaved words longer than the longer of the two stripes. This is a form of bound on the space requirements for proving equality in the amalgam. By Lem. 5, a stripe equivalent to a given word may be easily computed, and therefore the word problem in tensor-defined amalgams is reduced to the problem of deciding the tensors.

The easy proof of the following lemma may be found on the author's webpage[3]:

Lemma 6. *The following implications hold for any amalgam $[U \subseteq S, T]$ of monoids with decidable word problems:*

1. *If U-membership is decidable in S and T, then the word problem in the amalgam is decidable iff it is n-decidable for all n.*
2. *If U-membership is decidable in S and the amalgam is embeddable, then it has a 1-decidable word problem.*
3. *If the amalgam is tensor-defined, then for any $n > 0$, if for all $k \leq n$ both its k-tensors are decidable, then it has an n-decidable word problem.* □

This leads us to the following corollary, which essentially says that tensor-defined unions of theories that are recognized by finite automata are decidable (though they need not be recognizable by finite automata themselves):

Corollary 1. *If an amalgam $[U \subseteq S, T]$ of finite monoids is tensor-defined, then it has a decidable word problem.*

Proof. In an amalgam of finite monoids, all n-tensors are finite relations and are therefore decidable. So is U-membership. Since the amalgam is tensor-defined, by points 1 and 3 of Lem. 6 it has a decidable word problem. □

The notion of tensor-definedness of an amalgam is abstract and it may be difficult to verify whether an amalgam is or is not tensor-defined. Fortunately, the following proposition gives us a powerful method for showing tensor-definedness.

Proposition 2. *The following implications hold for any amalgam $[U \subseteq S, T]$:*

1. *If for all defining relations $w = w'$ of S either $w, w' \in \Sigma_U^*$, or $w, w' \in \Sigma_S^* \setminus \Sigma_U^*$, then U is unitary in S.*
2. *If U is unitary in S and T, then the amalgam is tensor-defined.*
3. *If the amalgam is tensor-defined, then it is embeddable.*

Proof. In the proof of point 2, for any words $w \Leftrightarrow^* w'$, we show, by induction on the length of the proof of this equivalence, that if w reduces to a stripe W and w' to a stripe W', then W and W' must be of the same length, say k, and $W \sim_k^* W'$ holds. This proof is quite cumbersome, and may be found, together with the easy proofs of points 1 and 3, on the author's webpage. □

From the above proposition, we can infer the following corollaries:

Theorem 2. *If $[U \subseteq S, T]$ is an amalgam such that:*

- *U-membership is decidable in S and T,*
- *U is unitary in S and T,*
- *the tensors of the amalgam are decidable,*

then the amalgam has a decidable word problem.

---

[3] In the appendix of the paper at
http://www.mimuw.edu.pl/~piotrek/RTA2006-full.ps.gz

Proof. By Prop. 2, the amalgam is tensor-defined. By point 3 of Lem. 6, it thus has an n-decidable word problem for all n. By point 1 of the same lemma, it then has a decidable word problem.   □

Corollary 2. *If* $[U \subseteq S, T]$ *is an amalgam such that:*

- *U is unitary in S and T,*
- *S and T are finite,*

then the amalgam has a decidable word problem.   □

4   Amalgams with Group-Like Cores

Results presented in Sect. 5 show that even for very simple monoids S and T the amalgam $[U \subseteq S, T]$ can have an undecidable word problem. Thus forcing U to be "small" will not make the amalgam decidable. However, it turns out that if U has a sufficiently "complete" structure, then the amalgam will have a decidable word problem for arbitary (decidable) S and T. This is quite remarkable and suggests that in the quest for decidability one should, instead of ensuring that the communication between S and T via U is difficult, rather try to ensure that any multi-step communication between them can be reduced to single-step communication.

If U is a submonoid of S, then *right U-quotients may be computed in S* if for any $s, s' \in S$ the set of all $u \in U$ satisfying $s = s'u$ is finite and may be computed (strictly speaking the s, s' are represented by words in Σ_S^* and the u by words in Σ_U^*). On the level of theories, the computability of right quotients means that for any two terms t and t' over Σ_S the set of terms t_U over Σ_U satisfying $t(x) = t_U(t'(x))$ modulo $\mathcal{S}$ contains only finitely many terms which are pairwise not equal modulo $\mathcal{U}$, and, moreover, that this set is computable.

Theorem 3. *If* $[U \subseteq S, T]$ *is an amalgam such that:*

- *S and T have decidable word problems,*
- *U is a group,*
- *right U-quotients are computable in S and in T,*

then the amalgam has a decidable word problem.

Proof. Because U is a group, U is unitary in S and T. For if $su \in U$ for $s \in S$ and $u \in U$, then $s = (su)u^{-1} \in Uu^{-1} \subseteq U$. A similar argument works for us and for T. Hence, by Prop. 2, the amalgam is tensor-defined. Note that U-membership is decidable in S and T, since for any $w \in \Sigma_S^*$ or $w \in \Sigma_T^*$ we can check whether there exists $u \in U$ such that $w = 1u$; but this holds iff $w \in U$. By points 1 and 3 of Lem. 6, it thus suffices to show that the m-tensor is decidable for all $m \geq 1$.

Consider any interleaved word $W = s_1 \cdot t_1 \cdot \ldots \cdot s_n \cdot t_n$ of length m and assume $m = 2n$. We claim that any word W' equivalent to W in the tensor is of the form

$$(s_1 u_1) \cdot (u_1^{-1} t_1 u_2) \cdot (u_2^{-1} s_2 u_3) \cdot \ldots \cdot (u_{2n-1}^{-1} t_n)$$

for some $u_1, \ldots, u_{2n-1} \in U$.

Suppose to the contrary. Let k be the least number of steps used to prove that W and W' are equivalent in the tensor. Of course $k > 0$. Assume W' is obtained from some W_0 in one step and W_0 from W in $k - 1$ steps. For W and W_0 the claim holds, that is, $W_0 = (s_1 u_1) \cdot (u_1^{-1} t_1 u_2) \cdot (u_2^{-1} s_2 u_3) \cdot \ldots \cdot (u_{2n-1}^{-1} t_n)$. The step from W_0 to W' is either of the form $\ldots \cdot su \cdot t \cdot \ldots \sim_{2n} \ldots \cdot s \cdot ut \cdot \ldots$, where $su = u_{2i-2}^{-1} s_i u_{2i-1}$ and $t = u_{2i-1}^{-1} t_i u_{2i}$, or $\ldots \cdot s \cdot ut \cdot \ldots \sim_{2n} \ldots \cdot su \cdot t \cdot \ldots$, where $s = u_{2i-2}^{-1} s_i u_{2i-1}$ and $ut = u_{2i-1}^{-1} t_i u_{2i}$, or of analogical forms with the roles of s and t swapped. In the former case we have $s = u_{2i-2}^{-1} s_i (u_{2i-1} u^{-1})$ and $ut = u u_{2i-1}^{-1} t_i u_{2i} = (u_{2i-1} u^{-1})^{-1} t_i u_{2i}$. In the latter case we have $su = u_{2i-2}^{-1} t_i (u_{2i-1} u)$ and $t = u^{-1} u_{2i-1}^{-1} t_i u_{2i} = (u_{2i-1} u)^{-1} t_i u_{2i}$. This proves the claim. Analogical claims hold for $m = 2n + 1$ and for the second tensor.

Now, all we have to do in order to decide whether

$$s_1 \cdot t_1 \cdot s_2 \cdot \ldots \sim_m^* s_1' \cdot t_1' \cdot s_2' \cdot \ldots$$

is check whether there exist $u_1, u_2, \ldots \in U$ such that

$$s_1' = s_1 u_1$$

$$t_1' = u_1^{-1} t_1 u_2$$

$$s_2' = u_2^{-1} s_2 u_3$$

$$\vdots$$

By assumption, we can compute a finite set of all possible choices for u_1, that is, all $u_1 \in U$ satisfying $s_1' = s_1 u_1$. If this set is empty, then the above system of equations is unsatisfiable. Otherwise for each choice in turn we repeat the whole procedure for the (shorter) list of equations

$$u_1 t_1' = t_1 u_2$$

$$s_2' = u_2^{-1} s_2 u_3$$

$$\vdots$$

with unknowns $u_2, u_3, \ldots \in U$. $\square$

Of course, a dual version of the above theorem holds for left U-quotients.

In certain cases, the assumption that right U-quotients are computable can easily be done away with. This is the case if U is finite, since one then can simply enumerate all elements of $u \in U$ and check which of them satisfy $s = s'u$. Thus:

Corollary 3. *If $[U \subseteq S, T]$ is a monoid amalgam such that:*

– S and T have decidable word problems,
– U is a finite group,

then the amalgam has a decidable word problem. $\square$

This corollary is especially remarkable, since it guarantees the decidability of $[U \subseteq S, T]$ solely based on the form of U.

The assumption that U is a group is satisfied in many standard cases. Consider a signature with three unary symbols pred, succ and neg, satisfying the equations:

$$\text{pred}(\text{succ}(x)) = x \qquad\qquad \text{neg}(\text{succ}(x)) = \text{pred}(\text{neg}(x))$$
$$\text{succ}(\text{pred}(x)) = x \qquad\qquad \text{neg}(\text{pred}(x)) = \text{succ}(\text{neg}(x))$$
$$\text{neg}(\text{neg}(x)) = x$$

The monoid U defined by these equations is indeed a group, since all the generators can be inverted. Thus, if right quotients are computable in S and T, *any* amalgam with U as core will have a decidable word problem if S and T do.

Note however that if we add 0 to the above group, that is, if we add an unary symbol 0 and the equations:

$$0(\text{succ}(x)) = 0(\text{pred}(x)) = 0(\text{neg}(x)) = 0(0(x)) = 0(x) = \text{neg}(0(x))$$

then the monoid U_{nat} defined by these equations is no longer a group.[4] Rather, it is the group U with a right zero 0 adjoined, which aditionally satisfies $0\,\text{neg} = 0$. Therefore Th. 3 cannot be applied here. The same is true for the monoid U_{bool} defined by equations

$$\text{f}(\text{f}(x)) = \text{f}(\text{t}(x)) = \text{f}(x) \qquad\qquad \text{t}(\text{f}(x)) = \text{t}(\text{t}(x)) = \text{t}(x)$$

over the generators f and t. The monoid U_{bool} contains an identity and two right zeros, and is not a group.

Unfortunately, it is not in general true that an amalgam of decidable monoids in which the core is a group with adjoined zero or right zeros must have a decidable word problem. In fact, in the next section we prove that there exists a monoid T with decidable word problem, such that the word problem of the amalgam $[U_{bool} \subseteq U_{bool} \cup \{0\}, T]$ is undecidable!

However, a slight generalization of Th. 3 to groups with zero is possible. A *group with zero* is a monoid of the form $G \cup \{0\}$, where G is a group and 0 is a *zero*, that is, $0g = g0 = 00 = 0$ for all $g \in G$. As is proved below, the assumption that $0s \neq 0$ and $s0 \neq 0$ for all $s \in S \setminus U$ guarantees that U is unitary in S.

Theorem 4. *If $[U \subseteq S, T]$ is a monoid amalgam such that:*

- *S and T have decidable word problems,*
- *U is a group with zero (which we denote by 0),*
- *$0s \neq 0$ and $s0 \neq 0$ for all $s \in S \setminus U$, and a similar property holds for T,*
- *right U-quotients are computable in S and in T,*

then the amalgam has a decidable word problem.

Proof. The monoid U is unitary in S and T. For if $su \in U$ for some $s \in S$, $u \in U$, then either $u \neq 0$, and then $s = (su)u^{-1} \in Uu^{-1} \subseteq U$, or $u = 0$, and then we would have $s0 \in U$. But then $0 = (s0)0 = s(00) = s0$, which implies

---

[4] The idea of this example is taken from [2].

$s \in U$. Thus, for the same reasons as in the proof of Th. 3, it suffices to prove that for all $m \geq 1$ the m-tensor is decidable. Note that the element 0 may be computed, since one can enumerate all words $w \in \Sigma_U^*$, and check for which of them $w\sigma = \sigma w = w$ holds for all $\sigma \in \Sigma_U$. For any interleaved word $W = s_1 \cdot t_1 \cdot \ldots$ of length $m = 2n$, let $Z(W)$ be the word $s_1 x_1 \cdot x_1 t_1 y_1 \cdot y_1 s_2 x_2 \cdot \ldots$, where:

- $x_i = 0$ if there is some $1 \leq j \leq i$ such that $s_j 0 = s_j, 0 t_j 0 = 0 t_j, s_{j+1} 0 = s_{j+1}, \ldots, 0 t_{i-1} 0 = 0 t_{i-1}, s_i 0 = s_i$ or $1 \leq j < i$ such that $t_j 0 = t_j, s_{j+1} 0 = s_{j+1}, \ldots, 0 t_{i-1} 0 = 0 t_{i-1}, s_i 0 = s_i$,
- $x_i = 0$ if there is some $i < j \leq n$ such that $0 s_j = s_j, 0 t_{j-1} 0 = t_{j-1} 0, 0 s_{j-1} 0 = s_{j-1} 0, \ldots, 0 t_i 0 = t_i 0$ or $i \leq j \leq n$ such that $0 t_j = t_j, 0 s_j 0 = s_j 0, 0 t_{j-1} 0 = t_{j-1} 0, 0 s_{j-1} 0 = s_{j-1} 0, \ldots, 0 t_i 0 = t_i 0$,
- otherwise $x_i = 1$,
- y_i is defined analogously.

We have $W \sim_m^* Z(W)$. Moreover, $W_1 \sim_m^* W_2$ iff $Z(W_1) \sim_m^* Z(W_2)$, since the first relation implies $Z(W_1) \sim_m^* W_1 \sim_m^* W_2 \sim_m^* Z(W_2)$ and vice versa.

Now consider the least symmetric relation $\approx_m$ on interleaved words of length m such that:

$$w \cdot su \cdot t \cdot w' \approx_m w \cdot s \cdot ut \cdot w'$$

for all $s \in S$, $t \in T$ and $u \in U$ such that $su0 \neq su$ and $0t \neq t$ and all $w \in (S \cdot T)^k$, $w' \in (S \cdot T)^l$ (for $n = k + l + 1$), and:

$$w \cdot tu \cdot s \cdot w' \approx_m w \cdot t \cdot us \cdot w'$$

for all $s \in S$, $t \in T$ and $u \in U$ such that $tu0 \neq tu$ and $0s \neq s$ and all $w \in (S \cdot T)^k \cdot S$, $w' \in T \cdot (S \cdot T)^l$ (for $n = k + l + 2$).

It can be proved that $W_1 \sim_m^* W_2$ iff $Z(W_1) \approx_m^* Z(W_2)$ and that similar results hold for the other tensor and for $m = 2n + 1$; it can also be proved that $\approx_m^*$ is decidable (for a proof of these claims see the author's webpage). Now, if we are to check whether $W_1 \sim_m^* W_2$, then all we have to do is compute $Z(W_1)$ and $Z(W_2)$ and then check whether $Z(W_1) \approx_m^* Z(W_2)$, which is decidable. □

5 Undecidability Results

In this section we prove results that explain why it is so hard to come up with methods for joining decision procedures even in the simple case of monadic theories. We show that even for amalgams of very simple monoids the word problem can be undecidable.

We start by recalling a fact proven in [4]. The proof itself is recast in terms of *2-counter machines*. This will allow us to show the similarities between the proofs of Th. 5, 6 and 7. Recall that a 2-counter machine consists of a finite set of instructions of three forms:

- "increment counter 1 (resp. 2) and jump to instruction i",
- "if counter 1 (resp. 2) is non-zero, decrement it and jump to instruction i; if it is zero, jump to instruction j",
- "stop", which is always the last instruction.

Any 2-counter machine starts at instruction 1 with counter 1 set to n and counter 2 set to 0. The input n is considered accepted if "stop" is executed with both counters 0. 2-counter machines are Turing-complete [6]. By a *state* of M a triple (i, n, m) is understood, where i is the current instruction and n and m are the values of the counters. The transition relation of the machine is denoted by $\rightarrow^*$.

Theorem 5 ([4]). *A finitely presented amalgam $[U \subseteq S, T]$ exists such that:*

- *S and T have decidable word problems,*
- *U is unitary in S and T,*
- *U is the free monoid $(\mathbb{N}, +)$,*
- *S has a U-cobase,*

and the amalgam has an undecidable word problem.

Proof. Let M be an N-instruction 2-counter machine. Let $U = \square^*$, S be generated by $\square$ and $\#$ satisfying the equation $\# = \#\square$ and let T be generated by $\square$, A, B, a, b and $q_1, \ldots, q_N$, satisfying the equations:

1. $ab = ba$, $aB = Ba$, $bA = Ab$, $AB = BA$,
2. $\square q_i = q_j a$ (resp. $\square q_i = q_j b$) if instruction i of M increments counter 1 (resp. 2) and jumps to instruction j,
3. $q_i a = \square q_j$ (resp. $q_i b = \square q_j$) and $q_i A = q_k A$ (resp. $q_i B = q_k B$) if instruction i of M decrements counter 1 (resp. 2) if it is non-zero and then jumps to instruction j, and otherwise jumps to instruction k.

The generators a and b represent single units on counters 1 and 2, while A and B represent the "end" or "zero" on counters 1 and 2. Note that we do not have relations $aA = Aa$ or $bB = Bb$, which guarantees that if A or B are read, then the counter must indeed be zero.

Obviously, U and S have decidable word problems. So does T, since for any word w of length k over Σ_T there is at most $(5 + N)^k$ words equivalent to w. This is due to the fact that the defining relations of T preserve word length. By point 1 of Prop. 2, U is unitary in S and T. Finally, the U-cobase of S is $\#^*$.

We claim that $\#q_{i_1} a^{n_1} b^{m_1} AB = \#q_{i_2} a^{n_2} b^{m_2} AB$ in the amalgam iff there is a state (i, n, m) such that $(i_1, n_1, m_1) \rightarrow^* (i, n, m)$ and $(i_2, n_2, m_2) \rightarrow^* (i, n, m)$ in the machine M. This claim may be proved by induction on the number of defining relations of types 2 and 3 applied to derive the equality in the amalgam.

From the above claim we may infer that $\#q_1 a^k AB = \#q_N AB$ holds in the amalgam iff there is a state (i, n, m) such that $(1, k, 0) \rightarrow^* (i, n, m)$ and $(N, 0, 0) \rightarrow^* (i, n, m)$ in the machine M. But instruction N of the machine M is "stop", and so $(N, 0, 0) \rightarrow^* (i, n, m)$ iff $i = N$ and $n = m = 0$. Thus $\#q_1 a^k AB = \#q_N AB$ in the amalgam iff $(1, k, 0) \rightarrow^* (N, 0, 0)$, which in turn holds iff the machine M accepts the number k.   □

The amalgam produced above is, by Prop. 2, tensor-defined and embeddable. Actually, a much stronger result can be shown if one does not insist on S having a U-base or cobase. A monoid U is *right zero* if $uu' = u'$ for all $u, u' \in U$, $u' \neq 1$.

Theorem 6. *Let U be the 3-element right zero monoid and let S be the 4-element monoid obtained by adjoining a zero to U. There exists a finitely presented monoid T satisfying $S \cap T = U$ and such that:*

- *T has a decidable word problem,*
- *U is unitary in S and T,*

and the amalgam $[U \subseteq S, T]$ has an undecidable word problem.

Proof. Let M be an N-instruction 2-counter machine. Let $U = \{1, t, f\}$ with t and f right zeros, that is, with $ut = t$ and $uf = f$ for all $u \in U$. Let $S = U \cup \{0\}$ be U with a zero adjoined, that is, with $s0 = 0s = 0$ for all $s \in S$. Let T extend U with generators A, B, a, b and $q_1, \ldots, q_N$, satisfying the equations:

1. $ab = ba$, $aB = Ba$, $bA = Ab$, $AB = BA$,
2. $tq_i = fq_j a$ (resp. $tq_i = fq_j b$) if instruction i of M increments counter 1 (resp. 2) and jumps to instruction j,
3. $tq_i a = fq_j$ (resp. $tq_i b = fq_j$) and $tq_i A = fq_k A$ (resp. $tq_i B = fq_k B$) if instruction i of M decrements counter 1 (resp. 2) if it is non-zero and jumps to instruction j, and otherwise jumps to instruction k.

The monoid T has a decidable word problem, because for any word w over Σ_T, there is only a finite number of words w' over Σ_T such that $w = w'$ in T. Indeed, relations listed under point 1 just rearrange certain letters, and relations listed under points 2 and 3 cannot lead to new equivalent words if applied more than twice. As before, U is unitary in S and T by Prop. 2.

Similarily to the proof of the previous theorem, we claim that $0q_{i_1} a^{n_1} b^{m_1} AB = 0q_{i_2} a^{n_2} b^{m_2} AB$ in the amalgam iff there is a state (i, n, m) such that $(i_1, n_1, m_1) \rightarrow^* (i, n, m)$ and $(i_2, n_2, m_2) \rightarrow^* (i, n, m)$ in the machine M; this can be proved by induction on the number of defining relations of types 2 and 3 which have to be applied to derive $0q_{i_1} a^{n_1} b^{m_1} AB = 0q_{i_2} a^{n_2} b^{m_2} AB$. As before, this claim implies that $0q_1 a^k AB = 0q_N AB$ holds in the amalgam iff the machine M accepts the number k, which is undecidable. □

One can actually show that U and S in the above theorem are minimal:

Proposition 3. *If $[U \subseteq S, T]$ is an amalgam such that:*

- *S and T have decidable word problems,*
- *U is unitary in S and T,*
- *U has less than 3 elements,*

then the amalgam has a decidable word problem.

Proof. If U has 1 element, then the amalgam is certainly decidable. If U has 2 elements, then either U is the 2-element group, or it is the trivial group with an adjoined zero. In the first case, the amalgam is decidable by Cor. 3. So assume that U is the trivial group with adjoined zero, that is, $U = \{1, 0\}$.

Since U is unitary in S, we cannot have $0s = 0$ or $s0 = 0$ for any $s \in S \setminus U$, and a similar property holds for T. Also, all right U-quotients are computable, since U is finite. Thus by Th. 4 the amalgam has a decidable word problem. □

As was the case for the amalgam constructed in Th. 5, the amalgam of Th. 6 is tensor-defined and embeddable. If the first of these properties is to be preserved and the amalgam is to have an undecidable word problem, then T must be infinite (see Cor. 1). However, if we do not insist on the amalgam being tensor-defined, then it is possible to construct an amalgam of *finite* monoids which itself has an undecidable word problem. We present an outline of the proof of Sapir [5] (Th. 1.1 and remark after Lem. 4.4). Our outline is much shorter than the original, and very similar to the proofs of Th. 5 and 6.

Theorem 7. *An amalgam* $[U \subseteq S, T]$ *exists such that:*

- *S and T are finite,*
- *the amalgam is embeddable,*

and the amalgam has an undecidable word problem.

Proof. Let M be an N-instruction 2-counter machine. Let U be generated by 0, and by α_i^1 and α_i^2 for any instruction i incrementing counter 1, β_i^1 and β_i^2 for i incrementing counter 2, $\aleph_i$, $\aleph_i^>$ and $\aleph_i^=$ for instruction i testing counter 1 on zero, and $\beth_i$, $\beth^>$ and $\beth^=$ for instruction i testing counter 2 on zero. Moreover, let $uu' = 0$ for all u, u' among the above generators.

Let S be additionally generated by a, b, $q_1, \ldots, q_N$ and $Q_1, \ldots, Q_{N-1}$. Let T be additionally generated by a', b', A and B. Let both be defined by the following relations (in addition to those defining U):

1. $q_i = a\alpha_i^1 Q_i$ in S, and $\alpha_i^1 = a'\alpha_i^2$ in T, and $\alpha_i^2 Q_i = q_j$ in S, if instruction i increments counter 1 and jumps to instruction j,
2. $q_i = Q_i \beta_i^1 b$ in S, and $\beta_i^1 = \beta_i^2 b'$ in T, and $Q_i \beta_i^2 = q_j$ in S, if instruction i increments counter 2 and jumps to instruction j,
3. $q_i = \aleph_i Q_i$ in S and:
   (a) $a'\aleph_i = \aleph_i^>$ in T and $a\aleph_i^> Q_i = q_j$ in S,
   (b) $A\aleph_i = A\aleph_i^=$ in T and $\aleph_i^= Q_i = q_k$ in S,
   if instruction i decrements counter 1 if it is non-zero and then jumps to instruction j, and otherwise jumps to instruction k,
4. $q_i = Q_i \beth_i$ in S and:
   (a) $\beth_i b' = \beth_i^>$ in T and $Q_i \beth_i^> b = q_j$ in S,
   (b) $\beth_i B = \beth_i^= B$ in T and $Q_i \beth_i^= = q_k$ in S,
   if instruction i decrements counter 2 if it is non-zero and then jumps to instruction j, and otherwise jumps to instruction k,
5. $ss' = 0$ in S and $tt' = 0$ in T for all products of generators s, s' of U and S or t, t' of U and T, except for those that appear in one of the above relations.

Both S and T extend U, since no two generators of U can be equated by applying the above relations. Also, both are finite. Finally, it can easily be verified that the amalgam is embeddable.

As before, one can prove that $A(aa')^{n_1} q_{i_1} (b'b)^{m_1} B = A(aa')^{n_2} q_{i_2} (b'b)^{m_2} B$ in the amalgam iff there is a state (i, n, m) of the machine M such that $(i_1, n_1, m_1) \to^* (i, n, m)$ and $(i_2, n_2, m_2) \to^* (i, n, m)$. This implies that $A(aa')^k q_1 B = Aq_N B$ iff $(1, k, 0) \to^* (N, 0, 0)$ in M, that is, iff k is accepted by M. □

6   Conclusions

Results such as Th. 2, 3, 4 or 6 show that the approach via monoid amalgams, proposed in [4] where the weaker Th. 5 was proved, works in practice. Still, even in the realm of monadic theories, a great number of problems concerning the decidability of unions remains open. For example, unitariness is sufficient for the amalgam to be tensor-defined, but is certainly not necessary; a sufficient and necessary notion is not yet known. The decidability of n-tensors and the notion of n-decidability remain unexplored (though see [7]). Decidability of amalgams with "small" cores has been explored in this paper only in the unitary case. Decidability of amalgams with "complete" cores has been dealt with for groups and groups with zero; but certainly there are other classes of cores that are "complete" enough for amalgams over them to have decidable word problems.

Moreover, in none of the criteria employed in the context of combining theories (effective bases, unitariness, amalgams with group-like cores) does the relationship between the theories S and T appear directly; the only exception is tensor-definedness. All the other criteria say something about S, about T, and about the intersection U, but nothing about the relationship between S and T. If we want to show the decidability of complex unions of theories, then it seems inevitable that our criteria refer directly to this relationship.

Of course, work is also, and maybe foremost, needed on applying the ideas developed for monadic theories to theories with symbols of arbitrary arities. It should then be possible to rather easily generalize the obtained results to many-sorted equational logic: most results of this paper can be reformulated in a many-sorted framework without any difficulty.

Finally, in the context of combining theories a number of interesting automata-theoretic questions arise. As we have seen, a union of regular theories may be undecidable. One is thus tempted to ask: When is it decidable (regular, automatic)? What about unions of automatic theories? What is the situation for relations other than equality? All these questions may also be posed for non-monadic theories, thus leading to problems for tree-automata.

References

1. Pigozzi, D.: The join of equational theories. Coll. Math. **30**(1) (1974) 15–25
2. Baader, F., Tinelli, C.: Deciding the word problem in the union of equational theories. Inf. Comp. **178**(2) (2002) 346–390
3. Fiorentini, C., Ghilardi, S.: Combining word problems through rewriting in categories with products. Th. Comp. Sci. **294** (2003) 103–149
4. Hoffman, P.: Union of equational theories: An algebraic approach. In: RTA'05. Volume 3467 of LNCS. (2005) 61–73
5. Sapir, M.V.: Algorithmic problems for amalgams of finite semigroups. J. Algebra **229**(2) (2000) 514–531
6. Minsky, M.: Computation: Finite and Infinite Machines. Prentice-Hall (1967)
7. Birget, J.C., Margolis, S., Meakin, J.: On the word problem for tensor products and amalgams of monoids. Intnl. J. Alg. Comp. **9** (1999) 271–294

Modular Church-Rosser Modulo

Jean-Pierre Jouannaud*

École Polytechnique
LIX, UMR CNRS 7161
91400 Palaiseau, France
http://www.lix.polytechnique.fr/Labo/Jean-Pierre.Jouannaud/

Abstract. In [12], Toyama proved that the union of two confluent term-rewriting systems that share absolutely no function symbols or constants is likewise confluent, a property called modularity. The proof of this beautiful modularity result, technically based on slicing terms into an homogeneous cap and a so called alien, possibly heterogeneous substitution, was later substantially simplified in [5,11].

In this paper we present a further simplification of the proof of Toyama's result for confluence, which shows that the crux of the problem lies in two different properties: a cleaning lemma, whose goal is to anticipate the application of collapsing reductions; a modularity property of ordered completion, that allows to pairwise match the caps and alien substitutions of two equivalent terms.

We then show that Toyama's modularity result scales up to rewriting modulo equations in all considered cases.

1 Introduction

Let R and S be two rewrite systems over disjoint signatures. Our goal is to prove that confluence is a modular property of their *disjoint union*, that is that $R \cup S$ inherits the confluence properties of R and S, a result known as Toyama's theorem. In the case of rewriting modulo an equationnal theory also considered in this paper, confluence must be generalized as a Church-Rosser property. Toyama apparently anticipated this generalization by using the word Church-Rosser in his title.

A first contribution of this paper is a new comprehensive proof of Toyama's theorem, obtained by reducing modularity of the confluence property to modularity of ordered completion, the latter being a simple property of disjoint unions. It is organized around the notion of *stable equalizers*, which are heterogeneous terms in which collapsing reductions have been anticipated with respect to the rewrite system $R^\infty \cup S^\infty$ obtained by (modular) ordered completion of $R \cup S$. Confluence of $R^\infty \cup S^\infty$ implies that equivalent terms have the same stable equalizers, made of a homogeneous *cap* which cannot collapse, and an alien stable substitution. This makes it possible to prove Toyama's theorem by induction on the structure of stable equalizers.

A second contribution is a study of modularity of the Church-Rosser property when rewriting with a set of rules R modulo a set of equations E. We prove that all rewrite relations introduced in the litterature, class rewriting, plain rewriting modulo, rewriting modulo, normal rewriting and normalized rewriting enjoy a modular Church-Roser

---

* Project LogiCal, Pôle Commun de Recherche en Informatique du Plateau de Saclay, CNRS, École Polytechnique, INRIA, Université Paris-Sud.

F. Pfenning (Ed.): RTA 2006, LNCS 4098, pp. 96–107, 2006.

property. We indeed show a more general generic result which covers all these cases. The proof is again obtained by applying selected results of the previous contribution to the rewrite system $R \cup E^{\rightarrow} \cup E^{\leftarrow}$, obtained by orienting the equations in E both ways, which results in a confluent system when the original rewrite relation is conluent.

We introduce terms in Section 2, and recall the basic notions of caps and aliens in Section 3. The new proof of Toyama's theorem is carried out in Section 4. Modularity of rewriting modulo is adressed in Section 5. Concluding remarks come in Section 6. We assume familiarity with the basic concepts and notations of term rewriting systems and refer to [1,11] for supplementary definitions and examples.

2   Preliminaries

Given a *signature* $\mathcal{F}$ of *function symbols*, and a set $\mathcal{X}$ of *variables*, $T(\mathcal{F}, \mathcal{X})$ denotes the set of *terms* built up from $\mathcal{F}$ and $\mathcal{X}$.

Terms are identified with finite labelled trees as usual. *Positions* are strings of positive integers, identifying the empty string Λ with the root position. We use $\mathcal{P}os(t)$ (resp. $\mathcal{FP}os(t)$) to denote the set of positions (resp. non-variable positions) of t, $t(p)$ for the symbol at position p in t, $t|_p$ for the *subterm* of t at position p, and $t[u]_p$ for the result of replacing $t|_p$ with u at position p in t. We may sometimes omit the position p, writing $t[u]$ for simplicity. $Var(t)$ is the set of variables occuring in t.

Substitutions are sets of pairs (x, t) where x is a variable and t is a term. The *domain* of a substitution σ is the set $\mathcal{D}om(\sigma) = \{x \in \mathcal{X} \mid \sigma(x) \neq x\}$. A substitution of finite domain $\{x_1, \ldots, x_n\}$ is written as in $\sigma = \{x_1 \mapsto t_1, \ldots, x_n \mapsto t_n\}$. A substitution is *ground* if $\sigma(x)$ is a ground term for all $x \in \mathcal{X}$. We use greek letters for substitutions and postfix notation for their application to terms. Composition is denoted by juxtaposition. Bijective substitutions are called *variable renamings*.

Given two terms s, t, computing the substitution σ when it exists such that $t = s\sigma$ is called *matching*, and s is then said to be *more general* than t. This quasi-ordering is naturally extended to substitutions. Given to terms s, t their *most general unifier* whenever it exists is the most general substitution σ (unique up to variable renaming) such that $s\sigma = t\sigma$.

A *(plain) rewrite rule* is a pair of terms, written $l \rightarrow r$, such that $l \notin \mathcal{X}$ and $Var(r) \subseteq Var(l)$. Plain rewriting uses plain pattern-matching for firing rules: a term t *rewrites* to a term u at position p with the rule $l \rightarrow r \in R$ and the substitution σ, written $t \xrightarrow{p}_{l \rightarrow r} u$ if $t|_p = l\sigma$ and $u = t[r\sigma]_p$. A *(plain) term rewriting system* is a set of rewrite rules $R = \{l_i \rightarrow r_i\}_i$. An *equation* is a rule which can be used both ways. An equation $x = s$ with $x \in \mathcal{X}$ is *collapsing*. We use AC for associativity and commutativity, and $\leftrightarrow_E$ for rewriting with a set E of equations.

The reflexive transitive closure of a relation $\rightarrow$, denoted by $\rightarrow^*$, is called *derivation*, while its symmetric, reflexive, transitive closure is denoted by $\leftrightarrow^*$, or $\leftrightarrow^*_R$ or $=_R$ when the relation is generated by a rewrite system R. A term rewriting system R is *confluent* (resp. *Church-Rosser*) if $t \rightarrow^* u$ and $t \rightarrow^* v$ (resp. $u \leftrightarrow^* v$) implies $u \rightarrow^* s$ and $v \rightarrow^* s$ for some s. The Church-Rosser property shall sometimes be used for some subset $T \subset \mathcal{T}(\mathcal{F}, \mathcal{X})$, in which case u, v are assumed to belong to T.

An ordering $\succ$ on terms is *monotonic* if $s \succ t$ implies $u[s] \succ u[t]$ for all terms u, and *stable* if $s \succ t$ implies $s\sigma \succ t\sigma$ for all substitutions σ. A *rewrite ordering* is a well-founded, monotonic, stable ordering on terms.

Given a set of equations E and a rewrite ordering $\succ$ total on ground terms, *ordered rewriting* with the pair $(E, \succ)$ is defined as plain rewriting with the infinite system $R = \{l\sigma \rightarrow r\sigma \mid l = r \in E, \gamma \text{ ground and } l\gamma \succ r\gamma\}$. When R is not confluent, the pair $(E, \succ)$ can be completed into a pair $(E^\infty, \succ)$ such that the associated rewrite system R^∞ is confluent, a process called *ordered completion*: given two equations $g = d \in E$, $l = r \in E$ such that (i) the substitution σ is the most general unifier of the equation $g = l|_p$ and (ii) $g\sigma\gamma \succ d\sigma\gamma$ and $l\sigma\gamma \succ r\sigma\gamma$ for some ground substitution γ, then, the so-called *ordered critical pair* $l[d\sigma]_p = r\sigma$ is added to E if it is not already confluent.

Given two sets of equations E and S sharing absolutely no function symbol, a key observation is that $(E \cup S)^\infty = E^\infty \cup S^\infty$ for any rewrite ordering $\succeq$ total on ground terms. Because, if the signatures are disjoint, there are no critical pairs between E and S. Therefore, ordered completion is *modular* for disjoint unions. Note that the result of completion is not changed by adding an arbitrary set of free variables provided the ordering is extended so as to remain a total rewrite ordering for terms in the extended signature, which is possible with the recursive path ordering.

3   Caps and Aliens

Following Toyama, our main assumption throughout this paper is that we are given two disjoint vocabularies $\mathcal{F}_R$ and $\mathcal{F}_S$, that is, such that

$$\mathcal{F}_R \cap \mathcal{F}_S = \emptyset.$$

We also assume without loss of generality a fixed bijective mapping ξ from a denumerable set of variables $\mathcal{Y}$ disjoint from $\mathcal{X}$, to the set of terms $\mathcal{T}(\mathcal{F}_R \cup \mathcal{F}_S, \mathcal{X})$.

We proceed by slicing terms into homogeneous subparts:

Definition 1. *A term in the union $\mathcal{T}(\mathcal{F}_R \cup \mathcal{F}_S, \mathcal{X})$ is* heterogeneous *if it uses symbols of both $\mathcal{F}_R$ and $\mathcal{F}_S$, otherwise it is* homogeneous.

A heterogeneous term can be decomposed into a topmost maximal homogeneous part, its *cap*, and a multiset of remaining subterms, its *aliens*. Thanks to our assumption, there is only one way of slicing a term by separating its homogeneous cap from its aliens rooted by symbols of the other signature.

Definition 2 (Cap and alien positions). *Given a term t, a position*

(i) $q \in \mathcal{D}om(t)$ is a cap position *if and only if $\forall p \leq q\ t(p) \in \mathcal{F}_R \cup \mathcal{X}$ iff $t(\Lambda) \in \mathcal{F}_R \cup \mathcal{X}$. In particular, Λ is a cap position;*

(ii) $q \in \mathcal{D}om(t) \setminus \{\Lambda\}$ is an alien position, *and the subterm $t|_q$ is an* alien *if and only if $t(q) \in \mathcal{F}_S$ (resp. $\mathcal{F}_R$) iff $\forall p < q$, $t(p) \in \mathcal{F}_R$ (resp. $\mathcal{F}_S$).*

We use $\mathcal{CP}os(t)$ for the set of cap positions in t, $\mathcal{AP}os(t)$ for its set of alien positions, and $Aliens(t)$ for the multiset of aliens in t.

A term is its own (trivial) alien at level 0. (Non-trivial) aliens at level $i > 0$ in t are the aliens of the aliens at level $i - 1$. The rank of a term is the maximal level of its aliens.

Definition 3 (Cap term and alien substitution). *Given a term t, its cap $\widehat{t}$ and alien substitution γ_t are defined as follows:*

(i) $\mathcal{P}os(\widehat{t}) = \mathcal{C}\mathcal{P}os(t) \cup \mathcal{A}\mathcal{P}os(t)$;
(ii) $\forall p \in \mathcal{C}\mathcal{P}os(t)\ \widehat{t}(p) = t(p)$;
(iii) $\forall q \in \mathcal{A}\mathcal{P}os(t)\ \widehat{t}(q) = \xi^{-1}(t|_q)$
(iv) γ_t *is the restriction of ξ to the variables in* $\mathcal{V}ar(\widehat{t}) \cap \mathcal{Y}$.

We will often use ξ instead of γ_t. The following result is straithforward:

Lemma 1. *Given a term t, its hat $\widehat{t}$ and alien substitution γ_t are uniquely defined and satisfy* $t = \widehat{t}\gamma_t$. *Moreover* $\mathcal{A}\mathcal{P}os(t) = \emptyset$ *and* $\widehat{t} = t$ *if t is homogeneous.*

4 Plain Rewriting

Let R and S be two rewrite systems operating on sets of terms defined over the respective vocabularies $\mathcal{F}_R \cup \mathcal{X}$ and $\mathcal{F}_S \cup \mathcal{X}$. We will often write $s \longrightarrow^* t$ for $s \longrightarrow^*_{R \cup S} t$ operating on sets of terms defined over the vocabulary $\mathcal{F}_R \cup \mathcal{F}_S \cup \mathcal{X}$.

4.1 Cap Reduction, Alien Reductions, and Equalizers

The notion of an equalizer is the key original notion of this paper, which allows us to perform reductions in the cap independently of reductions in the aliens (even if the rewrite rules are not left-linear) by anticipating reductions in the aliens.

Definition 4 (Equalizer). *A term t is an* equalizer *if for any two non-trivial aliens u at level i and v at level j in t,* $u \leftrightarrow^*_{R \cup S} v$ *iff* $u = v$.
   A substitution γ is an equalizer substitution *if* $\forall x \in \mathcal{D}om(\gamma)$, $\gamma(x)$ *is an equalizer, and* $\forall x, y \in \mathcal{D}om(\gamma)$, $x = y$ *iff* $\gamma(x) \leftrightarrow^*_{R \cup S} \gamma(y)$.

Example 1. Let $\mathcal{F}_R = \{f, c, +, a, b\}$, $\mathcal{F}_S = \{g, h\}$, $R = \{+(a, b) \rightarrow +(b, a)\}$ and $S = \{g(x) \rightarrow h(x)\}$. The term $c(f(g(+(b, a))), g(+(b, a)))$ is an $R \cup S$-equalizer, while the term $c(f(g(+(a, b))), h(+(b, a)))$ is not.

Definition 5. *We define a* cap reduction $s \longrightarrow_C t$ *if* $s \xrightarrow[R \cup S]{p} t$ *with* $p \in \mathcal{C}\mathcal{P}os(s)$, *and an* alien reduction $s \longrightarrow_A t$ *if* $s \xrightarrow[R \cup S]{p} t$ *with* $p \notin \mathcal{C}\mathcal{P}os(s)$.

Alien reductions take place inside an alien term, not necessarily at an alien position.

Lemma 2. *Let* $s \longrightarrow_A t$. *Then* $\widehat{t}(p) = \widehat{s}(p)$ *for all* $p \in \mathcal{C}\mathcal{P}os(s)$ *while* $Aliens(s)$ $(\longrightarrow)_{mul} Aliens(t)$.

Proof. Since the reduction takes place in the aliens, $\mathcal{C}\mathcal{P}os(s) \subseteq \mathcal{C}\mathcal{P}os(t)$ and $\widehat{t}(p) = t(p) = s(p) = \widehat{s}(p)$ for all $p \in \mathcal{C}\mathcal{P}os(s)$.   □

Note that $\mathcal{D}om(\widehat{t})$ and $\mathcal{D}om(\widehat{s})$ become different in case the rule has collapsed the cap of an alien of s to a subterm in the other signature, hence enlarging the cap of the whole term.

Lemma 3. *Let $s \xrightarrow{p}_C t$ with rule $l \to r \in R$ and substitution σ. Then,*

(i) $\widehat{s} \xrightarrow[l \to r]{p} \widehat{t} \in T(\mathcal{F}_R, \mathcal{Y}) \setminus \mathcal{Y}$*; or*

(ii) $\widehat{s} \xrightarrow[l \to r]{p} y \in \mathcal{Y}, r \in Var(l)$ *and* $t = \xi(y)$ *is an alien of s.*

In both cases t is an equalizer if s is an equalizer.

Note that $\widehat{s}$ and $\widehat{t}$ belong to the same signature and s and t have the same rank in the first case, while they do not in the second case and the rank has decreased strictly from s to t.

Proof. Since $s = \widehat{s}\gamma_s$ by lemma 1, $p \in \mathcal{CP}os(s)$ and rules are homogeneous, $\sigma = \delta\gamma_s$ for some homogeneous substitution δ. Therefore, $s = s[s|_p]_p = \widehat{s}\gamma_s[l\delta\gamma_s]_p = (\widehat{s}[l\delta]_p)\gamma_s$ since $p \in \mathcal{FP}os(\widehat{s})$, and $t = (\widehat{s}[r\delta]_p)\gamma_s$. Assuming that s is an equalizer, then γ_s is an equalizer substitution and t is an equalizer as well.

   Case (i) $\widehat{s}[r\delta]_p \notin \mathcal{Y}$, that is, $\widehat{s}[r\delta]_p \in T(\mathcal{F}_R, \mathcal{Y}) \setminus \mathcal{Y}$. Then, $\widehat{s} \longrightarrow_{l \to r} \widehat{t} = \widehat{s}[r\delta]_p$ and $t = \widehat{t}\gamma_s$.

   Case (ii) $\widehat{s}[r\delta]_p \in \mathcal{Y}$. Necessarily, $r\delta = y \in Var(\widehat{s}), r \in Var(l)$, and $t = y\gamma_s = y\xi$. □

4.2   Stable Equalizers

Due to the possible action of collapsing reductions, the cap and the aliens may grow or change signature along derivations. In particular, the cap may change signature if the term is equivalent to one of its aliens. Before introducing a stronger notion of equalizer, let us consider an example:

Example 2. Let $R = \{f(x,x) \to x, h(x) \to x\}$ and $S = \{a \to b\}$.
Then, $f(h(a), b) \longrightarrow f(h(b), b) \longrightarrow f(b, b) \longrightarrow b$.

Collapsing the cap here needs rewriting first an alien in order to transform the starting term into an equalizer, before applying the non-linear collapsing rules according to Lemma 3 (ii). Starting from the equalizer directly would not need any alien rewrite step. This suggests a stronger notion of equalizer.

Definition 6 (Stability). *A rewrite step $s \longrightarrow t$ is* cap-stable *if $\widehat{s}$ and $\widehat{t}$ belong to the same signature. A cap-stable derivation is a sequence of cap-stable rewrite steps. An equalizer s is* cap-collapsing *if there exists a cap-stable derivation $s \longrightarrow_C^* t$ and an alien u of t such that $t \longrightarrow_C u$.*

   An equalizer s is cap-stable *if it is not cap-collapsing,* stable *if it is cap-stable and its aliens are themselves stable, and* alien-stable *if its aliens are stable.*

According to Lemma 2, alien rewrite steps are cap-stable. We proceed with a thorough investigation of the properties of stable equalizers, of which the first is straightforward.

Lemma 4. *Any alien of a stable equalizer is a stable equalizer.*

Lemma 5. *Assume that s is a stable equalizer such that $s \longrightarrow_C t$. Then s and t have their cap in the same signature and t is a stable equalizer.*

Proof. By stability assumption of s, the rewrite step $s \longrightarrow_C t$ must satisfy Lemma 3 case (i). It is therefore a cap-stable step, and $\widehat{s}, \widehat{t}$ are built from the same signature. By Lemma 3 (i) again, every alien u of t is an alien of s, hence is a stable equalizer, and therefore t is an alien-stable equalizer. We are left to show that t is cap-stable.

If it were not, then $t \longrightarrow_C^* u$ for some cap-stable derivation and $u \longrightarrow_C v$ for some alien v of u. The derivation $s \longrightarrow_C t \longrightarrow_C^* u \longrightarrow_C v$ now contradicts the stability assumption of s. $\square$

Lemma 6. *Given an alien-stable equalizer s such that $s \longrightarrow_A^* t$, there exists a substitution θ from $Var(\widehat{t}) \cap \mathcal{Y}$ to $Var(\widehat{s}) \cap \mathcal{Y}$ such that $\widehat{s} = \widehat{t}\theta$ and $\theta\gamma_s \longrightarrow^* \gamma_t$. Moreover, θ is a bijection if t is an equalizer.*

Proof. By Lemma 2, $\widehat{s}$ and $\widehat{t}$ are in the same signature, and by Lemma 5 t is alien-stable. Hence, $\mathcal{CP}os(s) = \mathcal{CP}os(t)$, and $\forall p \in \mathcal{CP}os(s)$ $s(p) = t(p)$, and $\widehat{s}$ and $\widehat{t}$ may only differ by the names of their variables in $\mathcal{Y}$. Let $p, q \in \mathcal{AP}os(t)$ such that $\widehat{t}|_p = \widehat{t}|_q \in \mathcal{Y}$. Then $t|_p = t|_q$, therefore $s|_p \leftrightarrow^* s|_q$ since $s \longrightarrow_A^* t$. Hence $s|_p = s|_q$ since s is an equalizer. Therefore $\widehat{s}|_p = \widehat{s}|_q$, and $\widehat{s} = \widehat{t}\theta$ for some θ from $Var(\widehat{t}) \cap \mathcal{Y}$ to $Var(\widehat{s}) \cap \mathcal{Y}$. Also $\theta\gamma_s \longrightarrow^* \gamma_t$ since $s \longrightarrow_A^* t$.

If t is an equalizer, then $\widehat{s}|_p = \widehat{s}|_q \in \mathcal{Y}$ implies $t|_p = t|_q$, hence $\widehat{t}|_p = \widehat{t}|_q$, and θ is bijective. $\square$

Lemma 7. *Let s be an alien-stable equalizer such that $s \longrightarrow_A^* u \longrightarrow_C v$. Then there exists a term t such that $s \longrightarrow_C t$.*

Proof. By Lemma 6, $\widehat{u} = \widehat{s}\sigma$ for some substitution σ. By Lemma 3, $\widehat{u}$ is rewritable, and therefore $\widehat{s} \longrightarrow w$ for some w, hence $s \longrightarrow_C w\gamma_s = t$. $\square$

Lemma 8. *Let e be an alien-stable equalizer. Then, e is cap-collapsing iff $\widehat{e} \longrightarrow^* y$ for some variable $y \in \mathcal{Y}$.*

Proof. The if part is clear, we show the converse. Assume that $e \longrightarrow_C^* u$ is a cap-stable derivation and that $u \longrightarrow_C v$ for some alien v of u. Since all rewrite steps from e to u are cap-stable cap-rewrite steps, they satisfy Lemma 3(i), and therefore $\widehat{e} \longrightarrow^* \widehat{u}$. Since v is an alien of u, their caps are not in the same signature, hence the rewrite step from u to v is not cap-stable. It therefore satisfies Lemma 3(ii), and $\widehat{u} \longrightarrow_{l \rightarrow x} y \in \mathcal{Y}$. It follows that $\widehat{s} \longrightarrow^* y \in \mathcal{Y}$. $\square$

4.3   Structure Lemma

The goal of this section is to show that equivalence proofs between non-homogenous stable terms can be decomposed into a proof between their caps, and a proof between their aliens.

Lemma 9 (Cleaning). *Let t be a term such that the set of all its non-trivial aliens has the Church-Rosser property with respect to $R \cup S$. Then, there exists a stable equalizer e such that $t \longrightarrow_{R \cup S}^* e$.*

Proof. By induction on the rank of $t = \hat{t}\gamma_t$. By confluence assumption on the aliens, $\gamma_t \longrightarrow^* \gamma'$ such that $\gamma_t(x) \leftrightarrow^*_{R \cup S} \gamma_t(y)$ iff $\gamma'(x) = \gamma'(y)$. Let $\mathcal{D}om(\gamma') = \{x_1, \dots, x_m\}$ and $y \in \mathcal{D}om(\gamma')$. By induction hypothesis, $y\gamma' \longrightarrow^*_{R \cup S} y\gamma''$, a stable equalizer. Let $s = \hat{t}\gamma''$, hence $t \longrightarrow^*_A s$. We now compute $\hat{s}$ and γ_s, show that γ_s is a stable equalizer substitution, and that s rewrites to a stable equalizer e.

From Lemma 2, $\mathcal{P}os(\hat{t}) \subseteq \mathcal{P}os(\hat{s})$. Let $y \in \mathcal{V}ar(\hat{t}) \setminus \mathcal{V}ar(t)$ occurring at position p in $\hat{t}$ and $\theta(y) = \hat{s}|_p$. By construction, $\hat{s} = \hat{t}\theta$ and $\gamma'' = \theta\gamma_s$. Since γ'' is a stable equalizer substitution, so is γ_s by Lemma 4, hence s is an alien-stable equalizer. If s is not cap-collapsing, it is a stable equalizer and we are done. Otherwise, $\hat{s} \longrightarrow^* y$ by Lemma 8, hence $s \longrightarrow^* y\gamma_s$, which is a stable equalizer as already shown and we are done again.   □

By property of ordered completion, let $R^\infty \cup S^\infty$ be a confluent rewrite system such that $\leftrightarrow^*_{R \cup S} = \leftrightarrow^*_{R^\infty \cup S^\infty}$. By definition, both presentations define the same notions of equalizers.

Lemma 10. *Let u be a stable equalizer with respect to $R \cup S$. Then, it is a stable equalizer with respect to $R^\infty \cup S^\infty$.*   □

Proof. Since $R \cup S$ and $R^\infty \cup S^\infty$ define the same equationnal theory, they enjoy the same set of equalizers. We now prove that u is stable with respect to $R^\infty \cup S^\infty$ by induction on the rank. By induction hypothesis, u is alien-stable. We are left to show that it is cap-stable.

Assume it does not hold. By Lemma 8, $\hat{u} \longrightarrow^*_{R^\infty \cup S^\infty} y$ for some variable $y \in \mathcal{Y}$, and therefore $\hat{u} \leftrightarrow^*_{R \cup S} y$. Since $\hat{u}$ is homogeneous, by confluence of $R \cup S$ on homogeneous terms, $\hat{u} \longrightarrow^*_{R \cup S} y$. Lemma 8 now yields a contradiction.   □

The fact that $R^\infty \cup S^\infty$ and $R \cup S$ define the same notions of stable equalizers is crucial in the coming structural property of equalizers.

Lemma 11 (Structure). *Let $R \cup S$ be a disjoint union, and v and w be stable equalizers such that $v \leftrightarrow^*_{R \cup S} w$. Then, there exists a variable renaming η such that   (i) $\hat{v} \leftrightarrow^*_{R \cup S} \hat{w}\eta$ and   (ii) $\gamma_v \leftrightarrow^*_{R \cup S} \eta^{-1}\gamma_w$.*

Proof. By assumption, v and w are stable equalizers with respect to $R \cup S$, hence to $R^\infty \cup S^\infty$ by Lemma 10. Let v' and w' be their respective normal forms with respect to cap-rewrites with $R^\infty \cup S^\infty$. By lemma 5 (applied repeatedly), v' and w' are stable equalizers with respect to $R^\infty \cup S^\infty$.

Therefore, $v' \leftrightarrow^*_{R^\infty \cup S^\infty} w'$, $v' \longrightarrow^*_{R^\infty \cup S^\infty} s$ and $w' \longrightarrow^*_{R^\infty \cup S^\infty} s$ for some s. Now, since v' and w' are in normal form for cap-rewrites, all rewrites from v' to s and w' to s must occur in the aliens by Lemma 7. Since $R^\infty \cup S^\infty$ is confluent, equivalent aliens of s are joinable, and therefore, we can assume without loss of generality than s is an equalizer.

Since v and w are stable, Lemma 3(i) shows that $\hat{v} \longrightarrow^* \hat{v'}$ and $\hat{w} \longrightarrow^* \hat{w'}$ and therefore $\mathcal{V}ar(\hat{v'}) \subseteq \mathcal{V}ar(\hat{v})$ and $\mathcal{V}ar(\hat{w'}) \subseteq \mathcal{V}ar(\hat{w})$. Since $v' \longrightarrow^*_A s$ and $w' \longrightarrow^*_A s$, Lemma 6 shows that $\hat{v'} = \hat{s}\mu$ and $\hat{w'} = \hat{s}\nu$ for some bijection μ from $\mathcal{V}ar(\hat{s}) \cap \mathcal{Y}$ to $\mathcal{V}ar(\hat{v'}) \cap \mathcal{Y}$ and ν from $\mathcal{V}ar(\hat{s}) \cap \mathcal{Y}$ to $\mathcal{V}ar(\hat{w'}) \cap \mathcal{Y}$. Therefore, $\hat{v'} = \hat{w'}\nu^{-1}\mu$, and $\hat{v} \leftrightarrow^*_{R \cup S} \hat{w}\eta$ where $\eta = \mu^{-1}\nu$ is a bijection from $\mathcal{V}ar(\hat{v'}) \cap \mathcal{Y}$ to $\mathcal{V}ar(\hat{w'}) \cap \mathcal{Y}$.

Using now Lemmas 3(i) and 6 to relate the alien substitutions of u, v, s, we get $\gamma_v = \gamma_{v'}$ and $\mu\gamma_{v'} \longrightarrow^*_{R^\infty \cup S^\infty} \gamma_s$, hence $\mu\gamma_v \leftrightarrow^* \gamma_s$. Similarly $\nu\gamma_w \leftrightarrow^* \gamma_s$, and therefore, $\mu\gamma_v \leftrightarrow^* \nu\gamma_w$ yielding (ii). $\qquad\square$

4.4 Modularity

Theorem 1. *The union of two Church-Rosser rewrite systems R, S over disjoint signatures is Church-Rosser.*

Proof. We show the Church-Rosser property for terms v, w: $v \leftrightarrow^*_{R \cup S} w$ iff $v \longrightarrow^*_{R \cup S} \longleftarrow^*_{R \cup S} w$. The if direction is straightforward. The proof of the converse proceeds by induction on the maximum of the ranks of v, w. By induction hypothesis, the Church-Rosser property is therefore satisfied for the aliens of v, w.

1. By the cleaning Lemma 9, $v \longrightarrow^*_{R \cup S} v'$, $w \longrightarrow^*_{R \cup S} w'$, v' and w' being stable equalizers.
2. By the structure Lemma 11, $\widehat{v'} \leftrightarrow^*_{R \cup S} \widehat{w'}\eta$ and $\gamma_{v'} \leftrightarrow^*_{R \cup S} \eta^{-1}\gamma_{w'}$.
3. By the Church-Rosser assumption for homogeneous terms, $\widehat{v'} \longrightarrow^* s = t \longleftarrow^* \widehat{w'}\eta$.
4. By the induction hypothesis applied to $\gamma_{v'}$ and $\eta^{-1}\gamma_{w'}$ whose ranks are strictly smaller than those of v, w, $\gamma_{v'} \longrightarrow^* \sigma = \tau \longleftarrow^* \eta^{-1}\gamma_{w'}$.
5. Conclusion:

$$v \longrightarrow^* v' = \widehat{v'}\gamma_{v'} \longrightarrow^* s\gamma_{v'} \longrightarrow^* s\sigma$$
$$=$$
$$w \longrightarrow^* w' = \widehat{w'}\gamma_{w'} = \widehat{w'}\eta\eta^{-1}\gamma_{w'} = t\eta^{-1}\gamma_{w'} \longleftarrow^* t\tau$$

$\qquad\square$

This new proof of Toyama's theorem appears to be much simpler and shorter than previous ones. We will see next that it is the key to our generalization to rewriting modulo.

5   Rewriting Modulo Equations

We assume now given a set R of rewrite rules and a set E of equations used for equational reasoning, both built over the signature $\mathcal{F}_R$. Orienting the equations of E from left-to-right and right-to-left respectively, we denote by $E^\rightarrow$ and $E^\leftarrow$ the obtained rewrite systems. the notation $E^\leftarrow$ implies the assumption that no equation $x = t$ with $x \in \mathcal{X}$ can be in E.

Note that $\leftrightarrow^*_E = \longrightarrow^*_{E^\rightarrow \cup E^\leftarrow}$, and that $E^\rightarrow \cup E^\leftarrow$ is trivially confluent.

Similarly, we are also given a set S of rewrite rules and a set D of equations built over the signature $\mathcal{F}_S$.

5.1   The Zoo of Rewrite Relations Modulo Equations

We will consider five different rewrite relations in the case of rewriting with the pair (R, E):

1. Class rewriting [6], defined as $u \longrightarrow_{RE} t$ if $\exists s$ such that $u \leftrightarrow^*_E s \longrightarrow^*_R t$;
2. Plain rewriting modulo [2], defined as plain rewriting $\longrightarrow_R$;

3. Rewriting modulo [10,3], assuming that E-matching is decidable, defined as $u \longrightarrow_{R_E}^p t$ if $u|_p =_E l\sigma$ and $t = u[r\sigma]_p$ for some $l \to r \in R$;
4. Normal rewriting [4], assuming E-matching is decidable and E admits normal forms (a modular property [9]), writing $u \downarrow_E$ for the normal form of u, defined as $u \longrightarrow_E^* u \downarrow_E \longrightarrow_{R_E}^* t$;
5. Normalized rewriting [7], for which $E = S \cup AC$ and S is AC-Church-Rosser in the sense of rewriting modulo defined at case 3, defined as $u \longrightarrow_{S_{AC}}^* u \downarrow_{S_{AC}} \longrightarrow_{R_{AC}} t$.

One step class-rewriting requires searching the equivalence class of u until an equivalent term s is found that contains a redex for plain rewriting. Being the least efficient, class-rewriting has been replaced by the other more effective definitions. Normal rewriting has been introduced for modelling higher-order rewriting (using higher-order pattern matching). But our results *do not* apply directly to the case of higher-order rewriting in the sense of Nipkow [8] and its generalizations [4], since the E-equational part is then shared.

5.2   Modularity of Class Rewriting

Modularity of class-rewriting reduces easily to modularity of plain rewriting by using the fact that $R \cup E^\rightarrow \cup E^\leftarrow$ and $S \cup D^\rightarrow \cup D^\leftarrow$ are confluent rewrite systems over disjoint signatures whenever class-rewriting with (E, R) and (S, D) are confluent.

Theorem 2. *The Church-Rosser property is modular for class rewriting.*

Proof. Class rewriting relates to plain rewriting with $R \cup E^\rightarrow \cup E^\leftarrow$ as follows: $u \longrightarrow_{RE} w$ iff $u \leftrightarrow_E^* v \longrightarrow_R w$ iff $u \longrightarrow_{E^\rightarrow \cup E^\leftarrow}^* v \longrightarrow_R w$, and therefore $u \longrightarrow_{RE}^* \leftrightarrow_E^* w$ iff $u \longrightarrow_{R \cup E^\rightarrow \cup E^\leftarrow}^* w$. As a consequence, class rewriting with (R, E) is Church-Rosser iff plain rewriting with $\longrightarrow_{R \cup E^\rightarrow \cup E^\leftarrow}^*$ is Church-Rosser. Since the former is modular by Toyama's theorem, so is the latter.   □

This proof does not scale up to the other relations for rewriting modulo, unfortunately.

5.3   Modularity of Rewriting Modulo Equations

In order to show the modularity property of all these relations at once, we adopt an abstract approach using a generic notation $\Longrightarrow_{R,E}$ for rewriting modulo with the pair (R, E). More precisely, we prove that any rewrite relation $\Longrightarrow_{R,E}$ satisfying

(i) $\longrightarrow_R \subseteq (\Longrightarrow_{R,E} \leftrightarrow_E^*)^*$
(ii) $\Longrightarrow_{R,E} \subseteq (\leftrightarrow_E^* \longrightarrow_R \leftrightarrow_E^*)^*$
(iii) Variables are in normal form for $\Longrightarrow_{R,E}$
(iv) E does not admit collapsing equations

enjoys a modular Church-Rosser property defined as

$$\forall s, t \text{ s.t. } s \underset{R \cup E}{\overset{*}{\leftrightarrow}} t \quad \exists v, w \text{ s.t. } s \underset{R,E}{\overset{*}{\Longrightarrow}} v, t \underset{R,E}{\overset{*}{\Longrightarrow}} w \text{ and } v \underset{E}{\overset{*}{\leftrightarrow}} w$$

Note that all concrete rewriting modulo relations considered in Section 5.1 satisfy conditions (i,ii), including of course class-rewriting, and moreover that any rewriting modulo relation should satisfy these conditions to make sense, since (i,ii) imply soundeness

$$\left(\underset{R,E}{\Longrightarrow} \cup \underset{R,E}{\Longleftarrow} \cup \underset{E}{\leftrightarrow}\right)^* = \left(\underset{R}{\longrightarrow} \cup \underset{R}{\longleftarrow} \cup \underset{E}{\leftrightarrow}\right)^*$$

For all these relations, one $\Longrightarrow_R$ step suffices in the righthand side of (i), while for (ii), one $\longrightarrow_R$ step suffices in the righthand side with no $\leftrightarrow_E$ step on its right. For rewriting modulo, no $\leftrightarrow_E$ steps are needed in (i). They are needed on the left of $\longrightarrow_R$ in (ii) for modulo, normal and normalized rewriting. Finally, note that (iv) implies (iii) for all our relations, but is a much stronger assumption.

Our coming generalization of Toyama's theorem makes an essential use of the rewrite system $R \cup E^{\rightarrow} \cup E^{\leftarrow}$. We therefore first need to precisely relate the rewrite relation $\Longrightarrow_{R,E}$ to the relation $\longrightarrow_{R \cup E^{\rightarrow} \cup E^{\leftarrow}}$ when the former is Church-Rosser.

Lemma 12. *Assume* $\Longrightarrow_{R,E}$ *is Church-Rosser. Then, plain rewriting with* $R \cup E^{\rightarrow} \cup E^{\leftarrow}$ *is Church-Rosser.*

Proof. Straightforward consequence of (ii). □

Form now on, we consider two sets of pairs (R, E) and (S, D), and assume that the corresponding generic relations for rewriting modulo, $\Longrightarrow_{R,E}$ and $\Longrightarrow_{S,D}$, are both Church-Rosser. We shall use the abbreviation $\Longrightarrow$ for $\Longrightarrow_{R \cup S, E \cup D}$.

Our proof that the generic relation $\Longrightarrow$ is Church-Rosser for terms in $T(\mathcal{F}_R \cup \mathcal{F}_S, \mathcal{X})$ is essentially based on the structure Lemma 11 for the rewrite systems $R \cup E^{\rightarrow} \cup E^{\leftarrow}$ and $S \cup D^{\rightarrow} \cup D^{\leftarrow}$. By Lemma 12, both are Church-Rosser under our assumption that $\Longrightarrow_{R,E}$ and $\Longrightarrow_{S,D}$, are Church-Rosser.

To this end, we first need generalizing the cleaning lemma:

Lemma 13 (Cleaning). *Let* t *be a term such that the set of its non-trivial aliens has the Church-Rosser property for* $\Longrightarrow$. *Then, there exists a stable equalizer* e *such that* $t \Longrightarrow^* e$.

The proof uses the cleaning Lemma 9 for the rewrite relation $R \cup E^{\rightarrow} \cup E^{\leftarrow} \cup S \cup D^{\rightarrow} \cup D^{\leftarrow}$, which is Church-Rosser for the aliens of t by lemma 12, in order to dispense us with showing all intermediate properties needed in a direct proof of the lemma. This is possible since a stable equalizer does not depend upon the rewrite relation in use, but upon the equational theory itself.

Proof. The proof is by induction on the rank. By our assumptions, Lemmas 12 and 9 (we actually need a more precise statement extracted from its proof), there exists an alien-stable equalizer $s = \widehat{t}\gamma$ such that $\gamma_t \longrightarrow^*_{R \cup E^{\rightarrow} \cup E^{\leftarrow} \cup S \cup D^{\rightarrow} \cup D^{\leftarrow}} \gamma$ (hence, by induction hypothesis, $\gamma_t \Longrightarrow^* \gamma$), and an equalizer e such that

1. either s is cap-stable, in which case $e = \widehat{s}\gamma$ is stable and $t \Longrightarrow^* e$,
2. or s is cap-collapsing, in which case $\widehat{s} \longrightarrow^*_{R \cup E^{\rightarrow} \cup E^{\leftarrow} \cup S \cup D^{\rightarrow} \cup D^{\leftarrow}} x$ for some variable x and $e = x\gamma$. By the Church-Rosser property of $\Longrightarrow$ for homogeneous terms and assumption (iii), $\widehat{s} \Longrightarrow^* x$, hence $s \Longrightarrow^* e$ again. □

We now obtain our main new result:

Theorem 3. *The Church-Rosser property is modular for any rewriting modulo relation satisfying assumptions (i,ii,iii,iv).*

Proof. The proof mimics the proof of Theorem 1. Let v, w satisfying $v \leftrightarrow^*_{R \cup E \cup S \cup D} w$. The proof is by induction on the maximum rank of v, w. By induction hypothesis, the Church-Rosser property is therefore satisfied for the aliens of v, w.

1. By the cleaning Lemma 13, $v \Longrightarrow^* v'$, $w \Longrightarrow^* w'$, v' and w' being stable equalizers for the theory generated by $R \cup E \cup S \cup D$.

2. By assumptions (i) and (ii), $v' \leftrightarrow^*_{R \cup E \cup S \cup D} w'$.

3. By the structure Lemma 11, $\widehat{v'} \leftrightarrow^*_{R \cup E \cup S \cup D} \widehat{w'} \eta$ and $\gamma_{v'} \leftrightarrow^*_{R \cup E \cup S \cup D} \eta^{-1} \gamma_{w'}$.

4. By the Church-Rosser assumption for homogeneous terms, $\widehat{v'} \Longrightarrow^* s \doteq_{E \cup D} t \Longleftarrow^* \widehat{w'} \eta$. Note that $E \cup D$ applies here to an homogeneous term, that is, we do not know which of E or D is used to relate s and t.

5. By the induction hypothesis applied to $\gamma_{v'}$ and $\eta^{-1} \gamma_{w'}$ whose ranks are strictly smaller than those of v, w, $\gamma_{v'} \Longrightarrow^* \sigma \doteq_{E \cup D} \tau \Longleftarrow^* \eta^{-1} \gamma_{w'}$.

6. Conclusion:

$$v \Longrightarrow^* v' = \widehat{v'} \gamma_{v'} \Longrightarrow^* s \gamma_{v'} \Longrightarrow^* s\sigma$$
$$\doteq_{E \cup D}$$
$$w \Longrightarrow^* w' = \widehat{w'} \gamma_{w'} = \widehat{w'} \eta \eta^{-1} \gamma_{w'} = t \eta^{-1} \gamma_{w'} \Longleftarrow^* t\tau$$

□

We have not investigated whether this result extends to a theory E that does not satisfy assumption (iv). We suspect it does when assumption (iii) is satisfied, by generalizing the notion of collapsing reduction to $s \longrightarrow_R \doteq_E t$ for some alien t of s, but have not tried.

6   Conclusion

We have given a comprehensive treatment of Toyama's theorem which should ease its understanding. Moreover, we have generalized Toyama's theorem to rewriting modulo equations for all rewriting relations considered in the litterature (and for those not yet considered as well, if any, since they should satisfy our conditions to make sense), under the assumption that the equations are non-collapsing.

The question arises whether our proof method scales up to the constructor sharing case. This requires extending the modularity of ordered completion to cope with constructor sharing. We have tried without succes, except for the trivial case where constructors cannot occur on top of righthand sides of rules (a rule violating this assumption is called constructor lifting in the litterature). This implies that the modularity of the Church-Rosser property of higher-order rewriting cannot be derived from our results, except when the higher-order rewrite rules do not have a binder or an application at the root of their righthand sides. This shows that extending our method to the constructor sharing case is an important direction for further research.

On the other hand, we think that our proof method should yield a simpler proof of other modularity results, in particular for the existence of a normal form. We have not tried this direction.

Acknowledgments. The author thanks Nachum Dershowitz and Maribel Fernandez for numerous discussions about modularity and Yoshito Toyama for suggesting the trick of orienting equations both ways instead of repeating the basic proof. An anonymous referee suggested a potential further simplification by stabilizing terms before equalization.

References

1. Nachum Dershowitz and Jean-Pierre Jouannaud. Rewrite systems. In J. van Leeuwen, editor, *Handbook of Theoretical Computer Science*, volume B, pages 243–309. North-Holland, 1990.
2. Gérard Huet. Confluent reductions: abstract properties and applications to term rewriting systems. *Journal of the ACM*, 27(4):797–821, October 1980.
3. Jean-Pierre Jouannaud and Hélène Kirchner. Completion of a set of rules modulo a set of equations. *SIAM Journal on Computing*, 15(4):1155–1194, 1986.
4. Jean-Pierre Jouannaud, Femke van Raasdon, and Albert Rubio. Rewriting with types and arities, 2005. available from the web.
5. Jan Willem Klop, Aart Middeldorp, Yoshihito Toyama, and Roel de Vrijer. Modularity of confluence: A simplified proof. *Information Processing Letters*, 49(2):101–109, 1994.
6. Dallas S. Lankford and A. M. Ballantyne. Decision procedures for simple equational theories with permutative axioms: Complete sets of permutative reductions. Research Report Memo ATP-37, Department of Mathematics and Computer Science, University of Texas, Austin, Texas, USA, August 1977.
7. Claude Marché. Normalised rewriting and normalised completion. In *Proc. 9th IEEE Symp. Logic in Computer Science*, pages 394–403, 1994.
8. Richard Mayr and Tobias Nipkow. Higher-order rewrite systems and their confluence. *Theoretical Computer Science*, 192(1):3–29, February 1998.
9. A. Middeldorp. Modular aspects of properties of term rewriting systems related to normal forms. In *Proc. 3rd Rewriting Techniques and Applications, Chapel Hill, LNCS 355*, pages 263–277. Springer-Verlag, 1989.
10. Gerald E. Peterson and Mark E. Stickel. Complete sets of reductions for some equational theories. *Journal of the ACM*, 28(2):233–264, April 1981.
11. M. Bezem, J.W. Kop and R. de Vrijer eds. *Term Rewriting Systems*. Cambridge Tracts in Theoretical Computer Science 55, Cambridge University Press, 2003.
12. Y. Toyama. On the Church-Rosser property for the direct sum of term rewriting systems. *Journal of the ACM*, 34(1):128–143, April 1987.

Hierarchical Combination of Intruder Theories[*]

Yannick Chevalier[1] and Michaël Rusinowitch[2]

[1] IRIT Université Paul Sabatier, France
ychevali@irit.fr
[2] LORIA-INRIA-Lorraine, France
rusi@loria.fr

Abstract. Recently automated deduction tools have proved to be very effective for detecting attacks on cryptographic protocols. These analysis can be improved, for finding more subtle weaknesses, by a more accurate modelling of operators employed by protocols. Several works have shown how to handle a single algebraic operator (associated with a fixed intruder theory) or how to combine several operators satisfying disjoint theories. However several interesting equational theories, such as exponentiation with an abelian group law for exponents remain out of the scope of these techniques. This has motivated us to introduce a new notion of hierarchical combination for intruder theories and to show decidability results for the deduction problem in these theories. Under a simple hypothesis, we were able to simplify this deduction problem. This simplification is then applied to prove the decidability of constraint systems w.r.t. an intruder relying on exponentiation theory.

1 Introduction

Recently many procedures have been proposed to decide insecurity of cryptographic protocols in the Dolev-Yao model w.r.t. a finite number of protocol sessions [1,4,24,22]. Among the different approaches the symbolic ones [22,9,3] are based on reducing the problem to constraint solving in a term algebra. While these approaches rely on a perfect encryption hypothesis, the design of some protocols (see *e.g.* [26]) rely on lower-level primitives such as exponentiation or bitwise exclusive or (xor). These specification may give rise to new attacks exploiting the underlying algebraic structure when it is not abstracted as perfect encryption. For attacks exploiting the bitwise xor equational properties in the context of mobile communications see for instance [5].

Hence several protocol decision procedures have been designed for handling equational properties [21,11,6,18] of the cryptographic primitives. A very fruitful concept in this area is the notion of locality introduced by McAllester [19] which applies to several intruder theories [12,18]. When an intruder theory is *local* then we can restrict every intruder deduction to contain only subterms of its inputs, i.e. its hypotheses and its goal and this may lead to decidability of intruder constraints. Here we extend this approach to a case where the signature can be

---

[*] Supported by ACI-SI SATIN, ACI-Jeune Chercheur JC9005.

F. Pfenning (Ed.): RTA 2006, LNCS 4098, pp. 108–122, 2006.

divided into two disjoint sets and where the term algebra can be divided into two types of terms, say 0 and 1 type, according to their root symbol. Then we give sufficient conditions so that we can restrict intruder deductions to deductions where all subterms of type 1 that occur in the deduction are subterms of the inputs (i.e. some initially given terms and the goal term). Our goal is to bound the deductions of terms of type 1 by the intruder, thus permitting subsequent analysis to focus deductions of terms of type 0.

This approach allows us to decide interesting intruder theories presented as non-disjoint combination of theories, and that were not considered before, by reducing them to simpler theories. For instance it allows one to combine the Abelian group theory of [23] with a theory of an exponential operator.

Related works. In [8] we have extended the combination algorithm for solving E-unification problems of [2] to solve intruder constraints on disjoint signatures. Here we show that we can handle some non-disjoint combinations. In [13] Delaune and Jacquemard consider theories presented by rewrite systems where the right-hand side of every rule is a ground term or a variable. Comon and Treinen [12,10] have also investigated general conditions on theories for deciding insecurity with passive intruders.

As an application, we have obtained a decidable intruder theory combining Abelian group and exponential which has less restrictions than any previous one: unlike [7] it permits the intruder to multiply terms outside exponents, which is natural with the Diffie-Hellman protocol where the prime decomposition of the module is public. The setting is also less restrictive than in [25] where bases of exponentials have to be constants and exponential terms must not appear inside exponents.

Outline. In Section 2 we will first recall basic notions about terms, substitutions, term rewriting and define a new notion of *mode*. We then derive a notion of *subterm value* from the mode, and study properties of term replacement operations. We recall the definition of intruder systems in Section 3, and define the notion of *well-moded intruders*. We also prove the existence of special sequences of deductions called *quasi well-formed derivations*. Then we define constraint systems in Section 4. In Section 5 we define for a constraint system $\mathcal{C}$ a special kind of substitutions called *bound substitutions*. We prove that whenever a constraint system $\mathcal{C}$ is satisfiable it is also satisfied by a bound substitution. We also prove that these solutions do not increase the number of subterms of $\mathcal{C}$ of type 1, *i.e.* after instanciating $\mathcal{C}$ with a bound solution, the number of subterms of type 1 in the result is lesser or equal. We then give in Section 6 an application of these results to an interesting class of security protocols.

2   Terms, Subterms and Modes

2.1   Basic Notions

We consider an infinite set of free constants C and an infinite set of variables $\mathcal{X}$. For all signatures $\mathcal{G}$ (*i.e.* sets of function symbols not in C with arities), we denote by $T(\mathcal{G})$ (resp. $T(\mathcal{G}, \mathcal{X})$) the set of terms over $\mathcal{G} \cup C$ (resp. $\mathcal{G} \cup C \cup \mathcal{X}$). The former is called the set of ground terms over $\mathcal{G}$, while the latter is simply called the set of terms over $\mathcal{G}$. The arity of a function symbol f is denoted by $\mathrm{AR}(f)$. Variables are denoted by x, y, terms are denoted by s, t, u, v, and finite sets of terms are written $E, F, ...$, and decorations thereof, respectively. We abbreviate $E \cup F$ by E, F, the union $E \cup \{t\}$ by E, t and $E \setminus \{t\}$ by $E \setminus t$.

Given a signature $\mathcal{G}$, a *constant* is either a free constant or a function symbol of arity 0 in $\mathcal{G}$. We define the set of atoms $\mathcal{A}$ to be the union of $\mathcal{X}$ and the set of constants. Given a term t we denote by $\mathrm{Var}(t)$ the set of variables occurring in t and by $\mathrm{Cons}(t)$ the set of constants occurring in t. We denote by $\mathrm{Atoms}(t)$ the set $\mathrm{Var}(t) \cup \mathrm{Cons}(t)$. A substitution σ is an involutive mapping from $\mathcal{X}$ to $T(\mathcal{G}, \mathcal{X})$ such that $\mathrm{Supp}(\sigma) = \{x | \sigma(x) \neq x\}$, the *support* of σ, is a finite set. The application of a substitution σ to a term t (resp. a set of terms E) is denoted $t\sigma$ (resp. $E\sigma$) and is equal to the term t (resp. E) where all variables x have been replaced by the term $\sigma(x)$. A substitution σ is *ground* w.r.t. $\mathcal{G}$ if the image of $\mathrm{Supp}(\sigma)$ is included in $T(\mathcal{G})$.

An *equational presentation* $\mathcal{H} = (\mathcal{G}, A)$ is defined by a set A of equations $u = v$ with $u, v \in T(\mathcal{G}, \mathcal{X})$ and u, v without free constants. For any equational presentation $\mathcal{H}$ the relation $=_{\mathcal{H}}$ denotes the equational theory generated by $(\mathcal{G}, A)$ on $T(\mathcal{G}, \mathcal{X})$, that is the smallest congruence containing all instances of axioms of A. Abusively we shall not distinguish between an equational presentation $\mathcal{H}$ over a signature $\mathcal{G}$ and a set A of equations presenting it and we denote both by $\mathcal{H}$. We will also often refer to $\mathcal{H}$ as an equational theory (meaning the equational theory presented by $\mathcal{H}$).

The *syntactic subterms* of a term t are denoted $\mathrm{Sub}_{\mathrm{syn}}(t)$ and are defined recursively as follows. If t is a variable or a constant then $\mathrm{Sub}_{\mathrm{syn}}(t) = \{t\}$. If $t = f(t_1, \ldots, t_n)$ then $\mathrm{Sub}_{\mathrm{syn}}(t) = \{t\} \cup \bigcup_{i=1}^{n} \mathrm{Sub}_{\mathrm{syn}}(t_i)$. The *positions* in a term t are sequences of integers defined recursively as follows, ϵ being the empty sequence. The term t is at position ϵ in t. We also say that ϵ is the root position. We write $p \leq q$ to denote that the position p is a prefix of position q. If u is a syntactic subterm of t at position p and if $u = f(u_1, \ldots, u_n)$ then u_i is at position $p \cdot i$ in t for $i \in \{1, \ldots, n\}$. We write $t_{|p}$ the subterm of t at position p. We denote $t(s_1, \ldots, s_m)$ a term that admits $s_1 \ldots s_m$ among its syntactic subterms. We write $t[s]$ to denote a term t where s is a syntactic subterm of t.

In this paper, we will consider two disjoint signatures $\mathcal{F}_0$ and $\mathcal{F}_1$, an equational theory $\mathcal{E}_0$ (resp. $\mathcal{E}_1$) on $\mathcal{F}_0$ (resp. $\mathcal{F}_0 \cup \mathcal{F}_1$). We denote by $\mathcal{F}$ the union of the signatures $\mathcal{F}_0$ and $\mathcal{F}_1$ and by $\mathcal{E}$ the union of the theories $\mathcal{E}_0$ and $\mathcal{E}_1$. We assume that $\mathcal{E}$ is consistent (i.e. two free constants are not equal modulo $\mathcal{E}$). A term t in $T(\mathcal{F}_0, \mathcal{X})$ (resp. $T(\mathcal{F}_1, \mathcal{X})$) is called a *pure 0-term* (resp. *pure 1-term*). We denote by $\mathrm{TOP}(\cdot)$ the function that associates to each term t its root symbol. We also partition the set of variables $\mathcal{X}$ into two infinite sets $\mathcal{X}_0$ and $\mathcal{X}_1$.

2.2   Congruences and Ordered Rewriting

In this subsection we shall introduce the notion of *ordered rewriting* [14] which has been useful (*e.g.* [2]) for proving the correctness of combination of unification algorithms. Let $<$ be a simplification ordering on $\mathrm{T}(\mathcal{G})^1$ assumed to be total on $\mathrm{T}(\mathcal{G})$ and such that the minimum for $<$ is a constant $c_{\min} \in C$ and non-free constants are smaller than any non-constant ground term.

Given a signature $\mathcal{G}$, we denote by $\mathrm{C}_{\mathrm{spe}_{\mathcal{G}}}$ the set containing the constants in $\mathcal{G}$ and $c_{\min}$. For the the signature $\mathcal{F} = \mathcal{F}_0 \cup \mathcal{F}_1$ defined earlier, we abbreviate $\mathrm{C}_{\mathrm{spe}_{\mathcal{F}}}$ by $\mathrm{C}_{\mathrm{spe}}$. Given a possibly infinite set of equations $\mathcal{O}$ on the signature $\mathrm{T}(\mathcal{G})$ we define the ordered rewriting relation $\to_{\mathcal{O}}$ by $s \to_{\mathcal{O}} s'$ iff there exists a position p in s, an equation $l = r$ in $\mathcal{O}$ and a substitution τ such that $s = s[p \leftarrow l\tau]$, $s' = s[p \leftarrow r\tau]$, and $l\tau > r\tau$. It has been shown (see [16,14]) that by applying the *unfailing completion procedure* to a set of equations $\mathcal{H}$ we can derive a (possibly infinite) set of equations $\mathcal{O}$, called *o-completion* of $\mathcal{H}$ and such that, *first*, the congruence relations $=_{\mathcal{O}}$ and $=_{\mathcal{H}}$ are equal on $\mathrm{T}(\mathcal{F})$; and *second*, the ordered rewrite relation $\to_{\mathcal{O}}$ is convergent (*i.e.* terminating and confluent) on $\mathrm{T}(\mathcal{F})$.

From now on when we will say *"the rewrite system $\to_{\mathcal{O}}$"* this will mean "the ordered rewrite relation $\to_{\mathcal{O}}$", when will say *"by convergence of $\mathcal{O}$"*, we will mean "by convergence of $\to_{\mathcal{O}}$ on ground terms". By convergence of $\mathcal{O}$ we can define $(t)\!\downarrow_{\mathcal{O}}$ as the unique normal form of the ground term t for $\to_{\mathcal{O}}$. A ground term t is in *normal form*, or *normalized*, if $t = (t)\!\downarrow_{\mathcal{O}}$. Given a ground substitution σ we denote by $(\sigma)\!\downarrow_{\mathcal{O}}$ the substitution with the same support such that for all variables $x \in \mathrm{Supp}(\sigma)$ we have $x(\sigma)\!\downarrow_{\mathcal{O}} = ((x\sigma)\!\downarrow_{\mathcal{O}})$. A substitution σ is *normal* if $\sigma = (\sigma)\!\downarrow_{\mathcal{O}}$. In the following we will denote by R an o-completion of $\mathcal{E} = \mathcal{E}_1 \cup \mathcal{E}_2$.

2.3   Modes

When one considers the union of two equational theories over two disjoint signatures, a standard processing is to decompose the terms according to the signature of their inner symbols into a set of equations whose members are pure terms (i.e. built with symbols from a single signature). The rational for this decomposition is that by construction, in the case of disjoint signatures, the rewrite system obtained by o-completion is the union of two independent rewrite systems, each one operating on pure terms. This decomposition cannot be applied *as is* in the case of non-disjoint signatures. We provide here a notion of *mode* that allows one (under some hypothesis) to decompose terms in *subterm values* such that that the left-hand sides of rules in the o-completion never overlap two terms in the decomposition of a term. This notion of *mode* is different from the standard notion of *type* that would define how terms can be built.

In the following we assume that there exists a *mode* function $\mathrm{M}(\cdot, \cdot)$ such that $\mathrm{M}(f, i)$ is defined for every symbol $f \in \mathcal{F}$ and every integer i such that $1 \leq i \leq \mathrm{AR}(f)$. For all f, i we have $\mathrm{M}(f, i) \in \{0, 1\}$ and for all $f \in \mathcal{F}_0$ and for all i, $\mathrm{M}(f, i) = 0$.

For all $f \in \mathcal{F} \cup \mathcal{X}$ we define a function that gives the *class* $\mathrm{SIG}(f)$ of a symbol:

$$\mathrm{SIG} : \mathcal{F} \cup \mathcal{X} \to \{0, 1, 2\}$$
$$\mathrm{SIG}(f) = \begin{cases} i \text{ if } f \in \mathcal{F}_i \cup \mathcal{X}_i \text{ for } i \in \{0, 1\} \\ 2 \text{ otherwise, i.e. when f is a free constant} \end{cases}$$

The function SIG is extended to terms by taking $\mathrm{SIG}(t) = \mathrm{SIG}(\mathrm{TOP}(t))$.

---

[1] By definition $<$ satisfies for all $s, t, u \in \mathrm{T}(\mathcal{G})$ $s \leq t[s]$ and $s < u$ implies $t[s] < t[u]$.

A position different from ϵ in a term t is *well-moded* if it can be written $p \cdot i$ (where p is a position and i a nonnegative integer) such that $\text{SIG}(t_{|p \cdot i}) = \text{M}(\text{TOP}(t_{|p}), i)$. In other words the position in a term is well-moded if the subterm at that position is of the expected type w.r.t. the function symbol immediately above it. A term is well-moded if all its non root positions are well-moded. If a non root position of t is not well-moded we say it is *ill-moded* in t. An equational presentation $\mathcal{H} = (\mathcal{G}, A)$ is well-moded if for all equations $u = v$ in A the terms u and v are well-moded and $\text{SIG}(u) = \text{SIG}(v)$. One can prove that if an equational theory is well-moded then its completion is also well-moded.

We call a *subterm value* of a term t a syntactic subterm of t that is either atomic or occurs at an ill-moded position of t[2]. We denote $\text{Sub}(t)$ the set of subterm values of t. By extension, for a set of terms E, the set $\text{Sub}(E)$ is defined as the union of the subterm values of the elements of E. The subset of the maximal and strict subterm values of a term t plays an important role in the sequel. We call these subterm values the *factors* of t, and denote this set $\text{Factors}(t)$.

Example 1. Consider two binary symbols f and g with $\text{SIG}(f) = \text{SIG}(g) = \text{M}(f, 1) = \text{M}(g, 1) = 1$ and $\text{M}(f, 2) = \text{M}(g, 2) = 0$, and $t = f(f(g(a, b), f(c, c)), d)$. Its subterm values are a, b, $f(c, c)$, c, d, and its factors are a, b, $f(c, c)$ and d.

In the rest of this paper and unless otherwise indicated, *the notion of subterm will refer to subterm values.* From now on we assume that $\mathcal{E}$ is a well-moded equational presentation, and thus that R is a well-moded rewrite system. Under this assumption, one can prove that rewriting never overlaps subterm values.

2.4   Normalisation and Replacement

Subterms and Normalisation. We now study the evolution of the subterms of a term t when t is being normalized. Assuming the theory is well-moded, we can prove that (ordered) rewriting by R preserves factors in normal form. Since R is convergent, this permits to prove the following lemma.

Lemma 1. *Let t be a term with all its factors in normal form. Then either $(t){\downarrow} \in \text{Factors}(t) \cup C_{\text{spe}}$ or $\text{SIG}((t){\downarrow}) = \text{SIG}(t)$. Moreover $\text{Sub}((t){\downarrow}) \subseteq (\text{Sub}(t)){\downarrow} \cup C_{\text{spe}}$.*

Replacement and Normalization. We now give conditions under which the replacement of a normal subterm s of a term t commutes with the normalisation of t. First let us define replacement with respect to the subterm value relation on terms. If Π is a set of non-comparable positions in term t we denote by $t[\Pi \leftarrow v]$ the term obtained by putting v at all positions of t that are in Π. We denote $\delta_{u,v}$ the replacement of u by v such that if u appears at positions Π_u as a subterm (*i.e.* as a subterm value) of t then $t\delta_{u,v} = t[\Pi_u \leftarrow v]$. We denote in short δ_u the replacement $\delta_{u,c_{\min}}$.

We define the notion of *free terms* to express that a term s is not in a set of terms T once a substitution σ has been applied. A term s is *free* in T with respect to a ground substitution σ if there is no $t \in T$ such that $(t\sigma){\downarrow} = (s){\downarrow}$. A

---

[2] Note that the root position of a term is *always* ill-moded.

term which is not free is said to be *bound* by σ in T. We feel free to omit σ or T when they are clear from context. Since rewriting by R never overlaps subterm values, we can prove that normalization and subterm replacement commute.

Lemma 2. *Let t be a ground term with all its factors in normal form, and let s be a ground term in normal form with $s \neq (t){\downarrow}$ and $s \notin C_{\text{spe}}$. Then we have $(t\delta_s){\downarrow} = ((t){\downarrow}\delta_s){\downarrow}$.*

Example 2. Consider the equational theory $\mathcal{E} = \{f(g(x)) = x\}$. The only valid mode functions set either the mode of the argument of f and g to 0 or to 1. Since there is no critical pairs and the right-hand side is a subterm of the left-hand side, the rewrite system obtained by unfailing completion is $f(g(x)) \rightarrow x$. Consider now the terms $t = f(g(a))$ and $s = g(a)$. In both choices of the mode function, the subterms of t are t, and thus $t\delta_s = t$. This shows how the notion of mode permits to define replacements compatible with normalization.

Let s be a normalized ground term with $\text{SIG}(s) = 1$ and σ be a ground normal substitution. Next lemma shows that under the provision that a normalized term s is free in $\text{Sub}(t)$ for a ground substitution σ, the replacement of s in $(t\sigma){\downarrow}$ yields the same result as the replacement of s in σ. This will permit to transfer a pumping argument on instantiated terms to a pumping argument on substitutions. The proof again relies on the convergence of R.

Lemma 3. *Let t be a term, σ be a normalized substitution and s be a ground term in normal form with $\text{SIG}(s) = 1$. Assume s is free in $\text{Sub}(t)$ for σ and let $\sigma' = (\sigma\delta_s){\downarrow}$. We have:*

$$((t\sigma){\downarrow}\delta_s){\downarrow} = (t\sigma'){\downarrow}$$

Example 3. Consider now the equational theory $\mathcal{E} = \{f(x, x) = 0\}$, the term $t = f(f(x, x), f(x, c_{\min}))$ and $x\sigma = a$, and consider the replacement δ_a. Using the notations of Lemma 3, we have $\sigma' = \{x \mapsto c_{\min}\}$, and thus $t\sigma' = f(f(c_{\min}, c_{\min}), f(c_{\min}, c_{\min}))$, while on the other hand $(t\sigma){\downarrow}\delta_a = f(0, f(c_{\min}, c_{\min}))$. This example shows even though s is in normal form, an extra normalization is needed after replacement. Replacing one of the occurrence of x by a also shows why we need s to be free in Lemma 3.

3   Intruder Deduction Systems

We first recall here the general definition of intruder systems, as is given in [8]. Then we define the *well-moded intruder* in which we are interested in this paper. In the context of a security protocol (see *e.g.* [20] for a brief overview), we model messages as ground terms and intruder deduction rules as rewrite rules on sets of messages representing the knowledge of an intruder. The intruder derives new messages from a given (finite) set of messages by applying intruder rules. Since we assume some equational axioms $\mathcal{H}$ are satisfied by the function symbols in the signature, all these derivations have to be considered *modulo* the equational congruence $=_{\mathcal{H}}$ generated by these axioms. An intruder deduction rule in our setting is specified by a term t in some signature $\mathcal{G}$. Given values for the variables of t the intruder is able to generate the corresponding instance of t.

Definition 1. *An* intruder system $\mathcal{I}$ *is given by a triple* $\langle \mathcal{G}, \mathcal{S}, \mathcal{H} \rangle$ *where* $\mathcal{G}$ *is a signature,* $\mathcal{S} \subseteq \mathrm{T}(\mathcal{G}, \mathcal{X})$ *and* $\mathcal{H}$ *is a set of equations between terms in* $\mathrm{T}(\mathcal{G}, \mathcal{X})$. *To each* $t \in \mathcal{S}$ *we associate a deduction rule* $\mathrm{L}^t : \mathrm{Var}(t) \to t$ *and* $\mathrm{L}^{t,\mathrm{g}}$ *denotes the set of ground instances of the rule* L^t *modulo* $\mathcal{H}$:

$$\mathrm{L}^{t,\mathrm{g}} = \{l \to r \mid \exists \sigma, \text{ ground substitution on } \mathcal{G}, \ l = \mathrm{Var}(t)\sigma \text{ and } r =_{\mathcal{H}} t\sigma\}$$

The set of rules $\mathrm{L}_{\mathcal{I}}$ *is defined as the union of the sets* $\mathrm{L}^{t,\mathrm{g}}$ *for all* $t \in \mathcal{S}$.

Each rule $l \to r$ in $\mathrm{L}_{\mathcal{I}}$ defines an intruder deduction relation $\to_{l \to r}$ between finite sets of terms. Given two finite sets of terms E and F we define $E \to_{l \to r} F$ if and only if $l \subseteq E$ and $F = E \cup \{r\}$. We denote $\to_{\mathcal{I}}$ the union of the relations $\to_{l \to r}$ for all $l \to r$ in $\mathrm{L}_{\mathcal{I}}$ and by $\to_{\mathcal{I}}^*$ the transitive closure of $\to_{\mathcal{I}}$. Note that by definition, given sets of terms E, E', F and F' such that $E =_{\mathcal{G}} E'$ and $F =_{\mathcal{G}} F'$ we have $E \to_{\mathcal{I}} F$ iff $E' \to_{\mathcal{I}} F'$. We simply denote by $\to$ the relation $\to_{\mathcal{I}}$ when there is no ambiguity about $\mathcal{I}$.

Example 4. Let $\to_{\mathcal{I}_\times}$ be the relation between ground sets of terms defined by the Abelian group intruder $\mathcal{I}_\times = \langle \{\times, \mathrm{i}, 1\}, \{x \times y, \mathrm{i}(x), 1\}, \mathcal{E}_\times \rangle$. One has:

$$a, b, c \times a \to_{\mathcal{I}_\times} a, b, c, c \times a, \mathrm{i}(a) \to_{\mathcal{I}_\times} a, b, c, c \times a, \mathrm{i}(a), c$$

The latter deduction resulting from the application of the rule $x, y \to x \times y$ with x instantiated by $\mathrm{i}(a)$, y instantiated by $c \times a$, with right-hand side c which is equal to $\mathrm{i}(a) \times (c \times a)$ modulo the equational theory.

A *derivation* D of length n, $n \geq 0$, is a sequence of steps of the form $E_0 \to_{\mathcal{I}}$ $E_0, t_1 \to_{\mathcal{I}} \cdots \to_{\mathcal{I}} E_n$ with finite sets of ground terms $E_0, \ldots E_n$, and ground terms $t_1, \ldots, t_n$, such that $E_i = E_{i-1} \cup \{t_i\}$ for every $i \in \{1, \ldots, n\}$. The term t_n is called the *goal* of the derivation. We define $\overline{E}^{\mathcal{I}}$ to be equal to the set $\{t \mid \exists F \text{ s.t. } E \to_{\mathcal{I}}^* F \text{ and } t \in F\}$ *i.e.* the set of terms that can be derived from E. If there is no ambiguity on the deduction system $\mathcal{I}$ we write $\overline{E}$ instead of $\overline{E}^{\mathcal{I}}$.

With this definition of deduction, one can easily prove that it suffices to consider deductions on sets of terms in normal form. We will thus only consider derivations on sets of terms in normal form. From now on we will consider intruder systems over the signature $\mathcal{F}_0 \cup \mathcal{F}_1$ modulo the equational theory $\mathcal{E} = \mathcal{E}_0 \cup \mathcal{E}_1$ as defined in Section 2.1. Let $\mathcal{I}_1 = \langle \mathcal{F}, \mathcal{S}, \mathcal{E}_0 \cup \mathcal{E}_1 \rangle$ be an intruder system where terms in $\mathcal{S}$ are well-moded.

In the case of a well-moded intruder it is possible to split $\mathcal{S}$ into two sets of well-moded terms $\mathcal{S}_0$ and $\mathcal{S}_1$ such that for all terms t in $\mathcal{S}_i$ we have $\mathrm{SIG}(t) = i$ for $i \in \{0, 1\}$ and such that $\mathcal{S}_0$ contains terms built from symbols of $\mathcal{F}_0$. This permits to extract from $\mathcal{I}_1$ a simpler intruder, namely $\mathcal{I}_0 = \langle \mathcal{F}_0, \mathcal{S}_0, \mathcal{E}_0 \rangle$. In the sequel, we will reduce some decision problems on $\mathcal{I}_1$ to decision problems on $\mathcal{I}_0$ under some adequate hypotheses. We define $E \to_{\mathcal{S}_0} F$ (resp. $E \to_{\mathcal{S}_1} F$, resp. $E \to_{\mathcal{S}} F$) if $E \to_{l \to r} F$ with $l \to r \in \mathrm{L}^{t,\mathrm{g}}$ for $t \in \mathcal{S}_0$ (resp. $\mathcal{S}_1$, resp. $\mathcal{S}$).

Properties of deduction rules. Under the assumption that $\mathcal{S}$ is well-moded, one can prove the following key lemmas. Lemma 4 states that when a term appears as a new subterm of a knowledge set, it has just been built by the intruder. Considering a derivation, this will permit to apply Lemma 5 iteratively in order to show that this term may be eliminated from the derivation. This is the main step of the proof that terms not appearing as instance subterms of the initial constraint systems can be replaced by smaller terms (w.r.t. $<$) in a solution to yield a smaller solution.

Lemma 4. *Assume E and F are in normal form. If $E \to {}_{\mathcal{S}} F$ and $t \in \mathrm{Sub}(F) \setminus (\mathrm{Sub}(E) \cup \mathrm{C}_{\mathrm{spe}})$, then $F \setminus E = t$ and $E \to_{L^u} F$, with $u \in \mathcal{S}$ and $\mathrm{SIG}(u) = \mathrm{SIG}(t)$.*

Proof. The hypotheses permit to apply Lemma 1. If the rule is applied with substitution τ this implies $\mathrm{Sub}((u\tau)\!\downarrow) \subseteq \{(u\tau)\!\downarrow\} \cup \mathrm{Sub}(E) \cup \mathrm{C}_{\mathrm{spe}}$. Thus $t \notin \mathrm{Sub}(E) \cup \mathrm{C}_{\mathrm{spe}}$ implies $t = (u\tau)\!\downarrow$ and $t \notin \mathrm{C}_{\mathrm{spe}} \cup \mathrm{Factors}(u\tau)$. Thus by Lemma 1 $\mathrm{SIG}(t) = \mathrm{SIG}(u\tau) = \mathrm{SIG}(u)$.

Lemma 5. *Assume E, s and t are in normal form, $s \notin (E \cup \mathrm{C}_{\mathrm{spe}})$, $s \neq t$ and $\mathrm{c}_{\min} \in E$. Then $E, s \to E, s, t$ implies $(E\delta_s)\!\downarrow, s \to ((E, t)\delta_s)\!\downarrow, s$.*

Locality hypothesis on intruder systems. The previous lemma will be used in conjunction with an extra hypothesis that is related to the locality property [15].

> HYPOTHESIS 1: If $E \to_{\mathcal{S}_1} E, r \to_{\mathcal{S}_1} E, r, t$ and $r \notin \mathrm{Sub}(E, t) \cup \mathrm{C}_{\mathrm{spe}}$ then there is a set of terms F such that $E \to_{\mathcal{S}_0}^* F \to_{\mathcal{S}_1} F, t$.

Let us define the *closure* of $\mathcal{S}_1$ as the smallest set $\langle \mathcal{S}_1 \rangle$ of terms that contains $\mathcal{S}_1$ and such that if $s, s' \in \mathcal{S}_1$ and x is a variable of s of mode 1 then $s[x \leftarrow s'] \in \langle \mathcal{S}_1 \rangle$. By construction the set $\langle \mathcal{S}_1 \rangle$ contains only terms with head in $\mathcal{F}_1$ and thus contains only well-moded terms. We can prove that for any set of terms $\mathcal{S}_1$ the set of terms $\langle \mathcal{S}_1 \rangle$ satisfies Hypothesis 1.

4   Constraint Systems

We introduce now the constraint systems to be solved for checking protocols. It is shown in [8] how these constraint systems permit to express the reachability of a state in a protocol execution.

Definition 2. *(Unification systems) Let $\mathcal{H}$ be a set of equational axioms on $\mathrm{T}(\mathcal{G}, \mathcal{X})$. An $\mathcal{H}$-Unification system $\mathcal{S}$ is a finite set of couples of terms in $\mathrm{T}(\mathcal{G}, \mathcal{X})$ denoted by $\{u_i \overset{?}{=} v_i\}_{i \in \{1,\dots,n\}}$. It is satisfied by a ground substitution σ, and we note $\sigma \models \mathcal{S}$, if for all $i \in \{1, \dots, n\}$ we have $u_i\sigma =_{\mathcal{H}} v_i\sigma$.*

Definition 3. *(Constraint systems) Let $\mathcal{I} = \langle \mathcal{G}, S, \mathcal{H} \rangle$ be an intruder system. An $\mathcal{I}$-Constraint system $\mathcal{C}$ is denoted: $((E_i \rhd v_i)_{i \in \{1,\dots,n\}}, \mathcal{S})$ and it is defined by a sequence of couples $(E_i, v_i)_{i \in \{1,\dots,n\}}$ with $v_i \in \mathcal{X}$ and $E_i \subseteq \mathrm{T}(\mathcal{G}, \mathcal{X})$ for $i \in \{1, \dots, n\}$, and $E_{i-1} \subseteq E_i$ for $i \in \{2, \dots, n\}$ and by an $\mathcal{H}$-unification system $\mathcal{S}$.*

An $\mathcal{I}$-Constraint system $\mathcal{C}$ *is satisfied by a ground substitution* σ *if for all* $i \in \{1, \ldots, n\}$ *we have* $v_i\sigma \in \overline{E_i}\sigma$ *and if* $\sigma \models_\mathcal{H} \mathcal{S}$. *If a ground substitution* σ *satisfies a constraint system* $\mathcal{C}$ *we denote it by* $\sigma \models_\mathcal{I} \mathcal{C}$.

Constraint systems are denoted by $\mathcal{C}$ and decorations thereof. Note that if a substitution σ is a solution of a constraint system $\mathcal{C}$, by definition of constraint and unification systems the substitution $(\sigma)\!\downarrow_\mathcal{O}$ is also a solution of $\mathcal{C}$. In the context of cryptographic protocols the inclusion $E_{i-1} \subseteq E_i$ means that the knowledge of an intruder does not decrease as the protocol progresses: after receiving a message a honest agent will respond to it. This response can be added to the knowledge of an intruder who listens to all communications.

We are not interested in general constraint systems but only in those related to protocols. In particular we need to express that a message to be sent at some step i should be built from previously received messages recorded in the variables $v_j, j < i$, and from the initial knowledge. To this end we define:

Definition 4. *(Deterministic Constraint Systems) We say that an $\mathcal{I}$-constraint system* $((E_i \rhd v_i)_{i \in \{1, \ldots, n\}}, \mathcal{S})$ *is* deterministic *if for all i in $\{1, \ldots, n\}$ we have* $\mathrm{Var}(E_i) \subseteq \{v_1, \ldots, v_{i-1}\}$.

In order to be able to combine solutions of constraints for the intruder theory $\mathcal{I}_1$ with solutions of constraint systems for intruders defined on a disjoint signature we have, as for unification, to introduce some ordering constraints to be satisfied by the solution. Intuitively, these ordering constraints prevent from introducing cycle when building a global solution. This motivates us to define the *Ordered Satisfiability* problem:

Ordered Satisfiability

Input:	an $\mathcal{I}$-constraint system $\mathcal{C}$, X the set of all variables and C the set of all free constants occurring in $\mathcal{C}$ and a linear ordering $\prec$ on $X \cup C$.
Output:	SAT iff there exists a substitution σ such that $\sigma \models_\mathcal{I} \mathcal{C}$ and for all $x \in X$ and $c \in C$, $x \prec c$ implies $c \notin \mathrm{Sub}_{\mathrm{syn}}(x\sigma)$

5   Minimal Solutions

Let σ be a normal ground substitution and $\mathcal{C}$ be a constraint system. We say that σ is *bound* in $\mathcal{C}$ if, for every $s \in \mathrm{Sub}(\mathrm{Var}(\mathcal{C})\sigma)$, if $\mathrm{SIG}(s) = 1$ then s is bound by σ in $\mathrm{Sub}(\mathcal{C})$. The goal of this section is to prove that whenever a constraint system $\mathcal{C}$ is satisfiable, there exists a normal ground substitution σ bound in $\mathcal{C}$ such that $\sigma \models \mathcal{C}$. The last key ingredient to this proof is the notion of quasi well-formed derivations.

Definition 5. *A derivation $E_0 \to^* E_n$ and of goal t is* quasi well-formed *if for every term $u \in \mathrm{Sub}(E_n)$ we have $\mathrm{SIG}(u) = 1$ implies $u \in \mathrm{Sub}(E_0, t) \cup C_{\mathrm{spe}}$.*

Let $\mathcal{I} = \langle \mathcal{F}, \mathcal{S}, \mathcal{E} \rangle$ be a well-moded intruder that satisfies HYPOTHESIS 1 w.r.t. this mode function.

Lemma 6. *Assume* $c_{min} \in E$ *and* E *is in normal form. If* $t \in \overline{E}^S$ *there exists a quasi well-formed derivation starting from* E *of goal* t.

Lemma 7. *Let* E *and* F *be finite sets of normalized terms with* $c_{min} \in E$. *Let* s, t *be two normalized terms not in* C_{spe} *with* $s \in \overline{E} \setminus \mathrm{Sub}(E)$, $\mathrm{SIG}(s) = 1$ *and* $t \in \overline{E \cup F}$. *We have:*

$$(t\delta_s)\downarrow \in \overline{((E \cup F)\delta_s)\downarrow}$$

We can now prove that a satisfiable constraint system is satisfied by a bound solution.

Proposition 1. *Let* $\mathcal{C}$ *be a satisfiable constraint system. There exists a normal bound substitution* σ *such that* $\sigma \models \mathcal{C}$.

If we denote $\mathrm{Sub}_1(T)$ the terms of signature 1 in $\mathrm{Sub}(T)$, this implies the equality: $\mathrm{Sub}_1((\mathrm{Sub}_1(\mathcal{C})\sigma)\downarrow) = (\mathrm{Sub}_1(\mathcal{C})\sigma)\downarrow$

6   Application to Security Protocols

We present now a decision procedure for the exponentiation operator which is used *e.g.* with Diffie-Hellman scheme for the collaborative construction of a secret key by two principals. We define the *union* of two intruder systems as the intruder system having the deduction rules of both intruder systems.

In order to support properties of the exponential operator in cryptographic protocols analysis our goal is to prove the decidability of ordered satisfiability for an intruder able to exploit the properties of exponentiations. Note that the specification of the exponentiation operation is dependent on the specification of the multiplication, and thus Theorem 1 of [8] cannot be applied directly.

Note also that simple extensions of the theory we consider here would lead to undecidability of intruder constraints even when they are reduced to equational unification problems. See [17] for a survey of several exponentiation theories and their unification problems. The axiomatization we consider here was to our knowledge first introduced in [21].

Intruder Deduction System. We consider the union $\mathcal{F}$ of the two signatures $\mathcal{F}_0 = \{\_ \cdot \_, i(\_), 1\}$ and $\mathcal{F}_1 = \{\exp(\_, \_)\}$. We consider terms in $T(\mathcal{F}, \mathcal{X})$ modulo the following equational theory $\mathcal{E}$:

$$x \cdot (y \cdot z) = (x \cdot y) \cdot z \qquad (A)$$
$$x \cdot y = y \cdot x \qquad (C)$$
$$x \cdot 1 = x \qquad (U)$$
$$x \cdot i(x) = 1 \qquad (I)$$
$$\exp(x, 1) = x \qquad (E_0)$$
$$\exp(\exp(x, y), z) = \exp(x, y \cdot z) \qquad (E_1)$$

Modes. One easily checks that for the following mode and signature functions the theory $\mathcal{E}$ is a well-moded theory:

- $\text{M}(\cdot, 1) = \text{M}(\cdot, 2) = \text{M}(i, 1) = 0$;
- $\text{M}(\exp, 1) = 1$ and $\text{M}(\exp, 2) = 0$;
- $\text{SIG}(\cdot) = \text{SIG}(i) = \text{SIG}(1) = 0$
- $\text{SIG}(\exp) = 1$

According to this definition of mode and signature we define $\mathcal{E}$ to be the union of $\mathcal{E}_0 = \{(A), (C), (U), (I)\}$ and $\mathcal{E}_1 = \{(E_0), (E_1)\}$. The set $\mathcal{E}_0$ generates the theory of a free Abelian group whose generators are the atomic symbols in C. We denote by R an o-completion of $\mathcal{E}$ with the same congruence classes as $\mathcal{E}$ and such that for each term $t = \exp(t_1, t_2)$, if t is in normal form for R then t_1 is not an exponential term (*i.e.* $\text{SIG}(t_1) \neq 1$)[3].

Let $T = \{x \cdot y, i(x), 1, \exp(x, y)\}$. We now consider the intruder system $\mathcal{I}_{\exp} = \langle \mathcal{F}, T, \mathcal{E} \rangle$ that represents the modular exponentiation operation as employed for Diffie-Hellman-like construction of secret keys. According to mode and signature functions, this permits to define two intruder systems by taking $\mathcal{S}_0 = \{x \cdot y, i(x), 1\}$ and $\mathcal{S}_1 = \{\exp(x, y)\}$. Let $\mathcal{I}_{ag}$ be the intruder $\langle \{\cdot, i, 1\}, \{x \cdot y, i(x), 1\}, \mathcal{E}_0 \rangle$. In the rest of this section we present and justify an algorithm that runs in NP time and permits to reduce ordered satisfiability for $\mathcal{I}_{\exp}$ deterministic constraint systems to ordered satsifiability for $\mathcal{I}_{ag}$ deterministic constraint systems. Before proceeding further, let us first prove that the intruder $\mathcal{I}_{\exp}$ satisfies HYPOTHESIS 1.

Lemma 8. *Let E be a finite set of terms in normal form, and let r, t be two terms in normal form such that:*

$$E \to_{\mathcal{S}_1} E, r \to_{\mathcal{S}_1} E, r, t$$

If $r \notin \text{Sub}(E, t)$ and $E \nrightarrow E, t$ then there exists a term u such that:

$$E \to_{\mathcal{S}_0} E, u \to_{\mathcal{S}_1} E, u, t$$

Proof. Assume $r \notin \text{Sub}(t)$ and $E \nrightarrow E, t$. Since $r \notin \text{Sub}(E)$ it is necessary an exponential by Lemma 4. Let τ be the substitution with which the second rule $x, y \to \exp(x, y)$ is applied. Since $E \nrightarrow E, t$ one must have either $r = x\tau$ or $r = y\tau$.

First let us prove that w.l.o.g. one can assume $r \neq y\tau$. If $x\tau$ is not an exponential, then since E and r are in normal form, so is $\exp(x\tau, y\tau)$, and thus $r \in \text{Sub}(t)$, which contradicts the hypothesis. If $x\tau$ is an exponential, say $x\tau = \exp(x_1\tau, y_1\tau)$, then:

$$\exp(x\tau, r) =_{\mathcal{E}} t' = \exp(x_1\tau, y_1\tau \times r)$$

By convergence of R we have $(t')\!\downarrow = t$. Since either $x\tau \in E$ or $r = x\tau$, the assumption $r \notin \text{Sub}(E)$ implies that r is not a strict subterm of $x\tau$, and thus

---
[3] such a system R can be obtained by o-completion with a suitable ordering.

$r \notin \mathrm{Sub}(x_1\tau, y_1\tau)$. Since the factors of t' are in normal form and $r \neq t$, we have $(t'\delta_r)\!\downarrow = (t\delta_r)\!\downarrow$, and thus $r \notin \mathrm{Sub}(t)$ implies $(t'\delta_r)\!\downarrow = t$. In turn, this implies that $x\tau, c_{\min} \to t$ is ground instance of a rule in $\mathcal{S}_1$ that can be applied on E, r to deduce t.

The claim and $E \nrightarrow t$ implies $x\tau = r$ and $y\tau \neq r$ and thus $y\tau \in E$. It suffices now to consider the ground instance $s_1, s_2 \to (\exp(s_1, s_2))\!\downarrow = r$ of the rule that permits to deduce r from E. Since $s_1, s_2 \in E$ we have the following derivation:

$$E \to_{\mathcal{S}_0} E, s_2 \times y\tau \to_{\mathcal{S}_1} E, s_2 \times y\tau, (\exp(s_1, s_2 \times y\tau))\!\downarrow$$

The equality E_2 implies that this last term is equal to t.

As a consequence the exponential intruder enjoys quasi well-formed derivations and by Proposition 1, a satisfiable constraint system can be satisfied by a bound substitution. Thus we can bound the number of exponential subterms in quasi well-formed derivations. We can therefore design a correct, complete and terminating algorithm for solving the $\mathcal{I}_{\exp}$-constraints.

Properties of Bound Solutions. Let $\mathcal{C} = ((E_i \rhd v_i)_{1 \leq i \leq n}, \mathcal{S})$ be a constraint system and σ be a solution of $\mathcal{C}$. Given $t \in \mathrm{Sub}(\mathcal{C})$ let us define $I_t = \{j \mid (t\sigma)\!\downarrow \in \mathrm{Sub}((\mathrm{Sub}(E_j)\sigma)\!\downarrow, v_j\sigma)\}$. If $I_t \neq \emptyset$ we say that the term t is *deduction-bound*. In this case we define the *indice* of t, and denote i_t, the minimum indice in I_t. If $t \in \mathrm{Sub}(\mathcal{C})$ is deduction bound, we say it is *past-bound* if $t \in \mathrm{Sub}((\mathrm{Sub}(E_{j_t})\sigma)\!\downarrow)$ and *past-free* otherwise. Finally, given a past-bound term t of indice i_t, we say that a term m is a *complete prefix* of t if:

1. $\mathrm{SIG}(m) = \mathrm{SIG}((t\sigma)\!\downarrow)$ and $(m\sigma)\!\downarrow = (t\sigma)\!\downarrow$;
2. For all factor u of m; either $(u\sigma)\!\downarrow$ is past-free or $\mathrm{SIG}(u) = \mathrm{SIG}((u\sigma)\!\downarrow)$
3. $\mathrm{Var}(m) \subseteq \{v_1, \ldots, v_{i_t}\}$

Lemma 9. *It is possible to compute a complete prefix of $(t\sigma)\!\downarrow$ for all past-bound terms t in $\mathrm{Sub}(\mathcal{C})$.*

Algorithm. We present here a decision procedure for the exponential intruder $\mathcal{I}_{\exp}$ that takes as input a constraint system $\mathcal{C} = ((E_i \rhd v_i)_{1 \leq i \leq n}, \mathcal{S})$ and a linear ordering $<_i$ on variables and constants of $\mathcal{C}$. Let $m = |\mathrm{Sub}(\mathcal{C})|$ be the number of subterms in $\mathcal{C}$.

Step 1: Choose m triples $(e_i, x_i, y_i)_{i \in \{1, \ldots, m\}}$ of new variables and m^2 variables $\{y_{i,j}\}_{i,j \in \{1, \ldots, m\}}$. Add to $\mathcal{S}$ equations $e_i \overset{?}{=} \exp(x_i, y_i)$ for $i \in \{1, \ldots, m\}$ and $y_i \cdot y_{i,j} \overset{?}{=} y_j$ for $i, j \in \{1, \ldots, m\}$. Let $\mathcal{S}_e$ be the obtained unification problem and X_e be the set of these new variables.

Step 2: Choose an equivalence $\equiv_\sigma$ relation among subterms of $\mathcal{C}$ and $\mathcal{S}_e$. Let $Q = \{q_1, \ldots, q_n\}$ be a set of new variables each denoting an equivalence class. Add to $\mathcal{S}_e$ the equation $t \overset{?}{=} q$ for each $t \in q$ for each equivalence class $q \in Q$. Let $\mathcal{S}''$ be the obtained constraint system. Choose a subterm relation on Q.

Step 3: Guess a subset of Q_d of Q, and let $L = Q \cup \{v_1, \ldots, v_n\}$ and let $L = \{l_1, \ldots, l_k\}$. Let $<$ be a total order on L such that $i < j$ implies $v_i < v_j$ and form the constraint system $\mathcal{C}' = ((F_i \triangleright l_i)_{1 \leq i \leq k}, \mathcal{S}'')$ with

$$\begin{cases} F_1 = E_1 \\ F_{i+1} = F_i \cup (E_{j+1} \setminus E_j) & \text{If } l_i = v_j \\ F_{i+1} = F_i, l_i & \text{Otherwise} \end{cases}$$

Step 4: Replace each past-bound term in $\mathcal{C}'$ with a complete prefix and past-free terms with the representative q of their equivalence class. Reduce with equation (E_1) to form the constraint system $\mathcal{C}''$.

Step 5: Guess which constraints $E \triangleright v$ in $\mathcal{C}''$ must be solved by derivations ending with a rule in $\mathcal{S}_1$, and reduce them (if possible) to constraints to solve with $\mathcal{S}_0$

Step 6: Reduce $\mathcal{S}''$ to a system of general unification modulo $\mathcal{E}_0$ according to algorithm employed in [21], p. 7, proof of main theorem and purify the deduction constraints.

Step 7: Solve the resulting $\mathcal{I}_{\mathrm{ag}}$ deterministic intruder system with the linear constant restriction $<_i$.

Comments on the Algorithm. We assume in the following that the ordered satisfiability problem $(\mathcal{C}, <_i)$ is satisfied by a ground substitution σ_0.

Step 1: If $\mathcal{C}$ is satisfiable, it is satisfied by a bound substitution for which there are less than m different exponential terms. The $y_{i,j}$ will denote the exponents that we have to build so that $\exp(e_i, y_{i,j}) = e_j$.

Step 2: The subterm relation and the equivalence classes are needed to compute past-free and past-bound terms.

Step 3: The construction amounts to concatenating all derivations from $(E_i\sigma)\!\downarrow$ of goal $v_i\sigma$ into one derivation that has to deduce the terms $v_i\sigma$ at some point and in which in some steps the set of term is arbitrarily extended (case $l_i = v_j$). From this pseudo-derivation we extract in turn all applications of the $\mathcal{S}_1$ rule and all applications of the $\mathcal{S}_0$ rule that yield a past-free term. The first one permits a complete reduction at next step of the algorithm, while the second one permits to ensure that the resulting constraint system is determinitic once past-free terms are replaced by variables. The rational for this is that by definition the normal form q of a past-free term will be deduced (*i.e.* appear in a constraint $F \triangleright q$) before a term in this equivalence class (that will be replaced by the variable q) appears in any knowledge set.

Step 4: Note here that if a $\mathcal{S}_1$ rule permits to deduce a non-exponential term q, this term is past-bound. Thus the replacement made at previous step permits to ensure that q will never appear again in the deduction part of the constraint system, and thus that erasing this constraint during the reduction will not turn the constraint system into a non-deterministic one. If a $\mathcal{S}_1$ rule permits to deduce an exponential term, it will be seen as a constant when solving the resulting constraint system w.r.t. the $\mathcal{I}_{\mathrm{ag}}$ intruder. It is thus safe to erase the constraint in this case.

Step 5: In [21] unification systems with constants modulo $\mathcal{E}$ are reduced to general unification systems modulo $\mathcal{E}_0$ containing the exp as a free binary symbol. We go one step further and turn this unification system into a unification system with linear constant restrictions but without non-constant free symbols in order to syntactically eliminate the exp symbol. The deduction constraints are purified by replacing all equivalence classes of *exponential* terms by a representative *constant*.

As a consequence of this algorithm we have a decidability result for ordered satisfiability w.r.t. exponential intruder.

Proposition 2. *The ordered satisfiability problem for deterministic constraints and intruder $\mathcal{I}_{\mathrm{exp}}$ is decidable (with complexity NP).*

7   Conclusion

We have introduced a combination scheme for intruder theories that extends disjoint combination. We have shown how it can be used to derive new decidability results for security protocols. The scheme relies on an extension of the notion of locality. Unfortunately it does not apply to homomorphism properties (handled in a specific way in [18]) because they are ill-moded by nature and more investigations are needed to see whether it can be extended in this direction.

References

1. R. Amadio, D. Lugiez, and V. Vanackère. On the symbolic reduction of processes with cryptographic functions. *Theor. Comput. Sci.*, 290(1):695–740, 2003.
2. F. Baader and K. U. Schulz. Unification in the union of disjoint equational theories. combining decision procedures. *J. Symb. Comput.*, 21(2):211–243, 1996.
3. D. Basin, S. Mödersheim, and L. Viganò. An On-The-Fly Model-Checker for Security Protocol Analysis. In Einar Snekkenes and Dieter Gollmann, editors, *Proceedings of ESORICS'03*, LNCS 2808, pages 253–270. Springer-Verlag, 2003.
4. M. Boreale. Symbolic trace analysis of cryptographic protocols. In *Proceedings of the 28th ICALP'01*, LNCS 2076, pages 667–681. Springer-Verlag, Berlin, 2001.
5. N. Borisov, I. Goldberg, and D. Wagner. Intercepting mobile communications: the insecurity of 802.11. In *Proceedings of MOBICOM 2001*, pages 180–189, 2001.
6. Y. Chevalier, R. Kuesters, M. Rusinowitch, and M. Turuani. An NP Decision Procedure for Protocol Insecurity with XOR. In *Proceedings of the Logic In Computer Science Conference, LICS'03*, June 2003.
7. Y. Chevalier, R. Küsters, M. Rusinowitch, and M. Turuani. Deciding the Security of Protocols with Diffie-Hellman Exponentiation and Products in Exponents. In *Proceedings of the Foundations of Software Technology and Theoretical Computer Science, FSTTCS'03*, Lecture Notes in Computer Science. Springer, December 2003.
8. Y. Chevalier and M. Rusinowitch. Combining intruder theories. In Luís Caires, Giuseppe F. Italiano, Luís Monteiro, Catuscia Palamidessi, and Moti Yung, editors, *ICALP*, volume 3580 of *Lecture Notes in Computer Science*, pages 639–651. Springer, 2005.

9. Y. Chevalier and L. Vigneron. A Tool for Lazy Verification of Security Protocols. In *Proceedings of the Automated Software Engineering Conference (ASE'01)*. IEEE Computer Society Press, 2001.

10. H. Comon-Lundh. Intruder theories (ongoing work). In Igor Walukiewicz, editor, *7th International Conference, FOSSACS 2004*, volume 2987 of *Lecture Notes on Computer Science*, pages 1–4, Barcelona, Spain, March 2004. Springer Verlag.

11. H. Comon-Lundh and V. Shmatikov. Intruder Deductions, Constraint Solving and Insecurity Decision in Presence of Exclusive or. In *Proceedings of the Logic In Computer Science Conference, LICS'03*, pages 271–280, 2003.

12. H. Comon-Lundh and R. Treinen. Easy intruder deductions. In *Verification: Theory and Practice*, volume 2772 of *Lecture Notes in Computer Science*, pages 225–242, 2003.

13. S. Delaune and F. Jacquemard. A decision procedure for the verification of security protocols with explicit destructors. In *Proceedings of the 11th ACM Conference on Computer and Communications Security (CCS'04)*, pages 278–287, Washington, D.C., USA, October 2004. ACM Press.

14. N. Dershowitz and J-P. Jouannaud. Rewrite systems. In *Handbook of Theoretical Computer Science, Volume B*, pages 243–320. Elsevier, 1990.

15. R. Givan and D. A. McAllester. New results on local inference relations. In *KR*, pages 403–412, 1992.

16. J. Hsiang and M. Rusinowitch. On word problems in equational theories. In *ICALP*, volume 267 of *LNCS*, pages 54–71. Springer, 1987.

17. D. Kapur, P. Narendran, and L. Wang. An e-unification algorithm for analyzing protocols that use modular exponentiation. In Robert Nieuwenhuis, editor, *RTA*, volume 2706 of *Lecture Notes in Computer Science*, pages 165–179. Springer, 2003.

18. P. Lafourcade, D. Lugiez, and R. Treinen. Intruder deduction for ac-like equational theories with homomorphisms. In *Proceedings of the 16th International Conference on Rewriting Techniques and Applications (RTA'05)*, Lecture Notes in Computer Science, Nara, Japan, April 2005. Springer. To appear.

19. David A. McAllester. Automatic recognition of tractability in inference relations. *J. ACM*, 40(2):284–303, 1993.

20. C. Meadows. The NRL protocol analyzer: an overview. *Journal of Logic Programming*, 26(2):113–131, 1996.

21. C. Meadows and P. Narendran. A unification algorithm for the group Diffie-Hellman protocol. In *Workshop on Issues in the Theory of Security (in conjunction with POPL'02)*, Portland, Oregon, USA, January 14-15, 2002.

22. J. Millen and V. Shmatikov. Constraint solving for bounded-process cryptographic protocol analysis. In *ACM Conference on Computer and Communications Security*, pages 166–175, 2001.

23. J. Millen and V. Shmatikov. Symbolic protocol analysis with an abelian group operator or Diffie-Hellman exponentiation. *Journal of Computer Security*, 2005.

24. M. Rusinowitch and M. Turuani. Protocol insecurity with finite number of sessions is NP-complete. In *Proc.14th IEEE Computer Security Foundations Workshop*, Cape Breton, Nova Scotia, June 2001.

25. V. Shmatikov. Decidable analysis of cryptographic protocols with products and modular exponentiation. In *Proceedings of ESOP'04*, volume 2986 of *Lecture Notes in Computer Science*, pages 355–369,. Springer-Verlag, 2004.

26. T. Wu. The srp authentication and key exchange system. Technical Report RFC 2945, IETF – Network Working Group, september 2000. available at http://www.ietf.org/rfc/rfc2945.txt.

Feasible Trace Reconstruction for Rewriting Approximations

Yohan Boichut[1] and Thomas Genet[2]

[1] LIFC / Université de Franche-Comté
16, route de Gray
F-25030 Besançon cedex
INRIA/CASSIS
boichut@lifc.univ-fcomte.fr
[2] IRISA / Université de Rennes 1
Campus de Beaulieu
F-35042 Rennes Cedex
LANDE
genet@irisa.fr

Abstract. Term Rewriting Systems are now commonly used as a modeling language for programs or systems. On those rewriting based models, reachability analysis, i.e. proving or disproving that a given term is reachable from a set of input terms, provides an efficient verification technique. For disproving reachability (i.e. proving non reachability of a term) on non terminating and non confluent rewriting models, Knuth-Bendix completion and other usual rewriting techniques do not apply. Using the tree automaton completion technique, it has been shown that the non reachability of a term t can be shown by computing an over-approximation of the set of reachable terms and prove that t is not in the approximation. However, when the term t is in the approximation, nothing can be said. In this paper, we refine this approach and propose a method taking advantage of the approximation to compute a rewriting path to the reachable term when it exists, i.e. produce a counter example. The algorithm has been prototyped in the Timbuk tool. We present some experiments with this prototype showing the interest of such an approach w.r.t. verification of rewriting models.

1 Introduction

In the rewriting theory, the reachability problem is the following: given a term rewriting system (TRS) $\mathcal{R}$ and two terms s and t, can we decide whether $s \rightarrow_{\mathcal{R}}^{*} t$ or not? This problem, which can easily be solved on strongly terminating TRS (by rewriting s into all its possible reduced forms and compare them to t), is undecidable on non terminating TRS. There exists several syntactic classes of TRSs for which this problem becomes decidable: some are surveyed in [FGVTT04], more recent ones are [GV98, TKS00]. In general, the decision procedures for those classes compute a finite tree automaton recognizing the possibly infinite set of terms reachable from a set $E \subseteq \mathcal{T}(\mathcal{F})$ of initial term, by $\mathcal{R}$, denoted by

F. Pfenning (Ed.): RTA 2006, LNCS 4098, pp. 123–135, 2006.
© Springer-Verlag Berlin Heidelberg 2006

$\mathcal{R}^*(E)$. Then, provided that $s \in E$, those procedures check whether $t \in \mathcal{R}^*(E)$ or not. On the other hand, outside of those decidable classes, one can prove $s \not\rightarrow_\mathcal{R}^* t$ using over-approximations of $\mathcal{R}^*(E)$ [Jac96, Gen98, FGVTT04] and proving that t does not belong to this approximation.

Recently, reachability analysis turned out to be a very efficient verification technique for proving properties on infinite systems modeled by TRS. Some of the most successful experiments, using proofs of $s \not\rightarrow_\mathcal{R}^* t$, were done on cryptographic protocols [Mon99, GK00, OCKS03, GTTVTT03], [BHK05] where protocols and intruders are described using a TRS $\mathcal{R}$, E represents the set of initial configurations of the protocol and t a possible flaw. Then reachability analysis can detect the flaw (if $s \rightarrow_\mathcal{R}^* t$) or prove its absence (if $\forall s \in E : s \not\rightarrow_\mathcal{R}^* t$). However, the main drawback of those techniques based on tree automata, is that if $t \in \mathcal{R}^*(E)$ then we have the proof but not the rewriting path (also denoted by *trace* in the following). Indeed, from the tree automaton recognizing $\mathcal{R}^*(E)$ it is not possible to reconstruct the rewrite path from a possible s to t (i.e. the attack leading to a flaw in the context of cryptographic protocols). On the other hand, when dealing with an over-approximation $App \supseteq \mathcal{R}^*(E)$ if $t \in App$ then there is no way to check whether $t \in \mathcal{R}^*(E)$ (t is really reachable from s) or if $t \in App \setminus \mathcal{R}^*(E)$ (t is an artefact of the approximation). This problem becomes crucial when using approximations to prove security and safety properties. In that case, producing counter examples on a faulty specification makes the user more confident with the tool when it finally claims that the property is proven on a fixed version of the specification.

This paper tackles those two problems and proposes a solution that automatically gives the rewrite path when $\mathcal{R}^*(E)$ is constructed exactly and helps to discriminate between terms of $\mathcal{R}^*(E)$ and terms of the approximation when $\mathcal{R}^*(E)$ is over-approximated.

This paper is organized as follows. In section 2, we give the basic definitions for TRS and tree automata. In section 3, we recall the tree automata completion technique. In section 4, we define the trace reconstruction method we propose. Finally, in section 5, we present some experimentations done with our prototype implemented within Timbuk [GVTT00] a tree automata completion tool.

2   Preliminaries

Comprehensive surveys for TRSs and tree automata can be found respectively in [BN98] and in [CDG$^+$02].

Let $\mathcal{F}$ be a finite set of symbols, each one with an arity and let $\mathcal{X}$ be a countable set of variables. $\mathcal{T}(\mathcal{F}, \mathcal{X})$ denotes the set of terms, and $\mathcal{T}(\mathcal{F})$ denotes the set of ground terms (terms without variables). The set of variables of a term t is denoted by $\mathcal{V}ar(t)$. A substitution is a function σ from $\mathcal{X}$ into $\mathcal{T}(\mathcal{F}, \mathcal{X})$, which can uniquely be extended to an endomorphism of $\mathcal{T}(\mathcal{F}, \mathcal{X})$. A position p for a term t is a word over $\mathbb{N}$. The empty sequence ϵ denotes the top-most position. The set $\mathcal{P}os(t)$ of positions of a term t is inductively defined by:

- $\mathcal{P}os(t) = \{\epsilon\}$ if $t \in \mathcal{X}$
- $\mathcal{P}os(f(t_1, \ldots, t_n)) = \{\epsilon\} \cup \{i.p \mid 1 \leq i \leq n \text{ and } p \in \mathcal{P}os(t_i)\}$

If $p \in \mathcal{P}os(t)$, then $t|_p$ denotes the subterm of t at position p and $t[s]_p$ denotes the term obtained by replacement of the subterm $t|_p$ at position p by the term s. For any term $s \in \mathcal{T}(\mathcal{F}, \mathcal{X})$, we denote by $\mathcal{P}os_{\mathcal{F}}(s)$ the set of functional positions in s, i.e. $\{p \in \mathcal{P}os(s) \mid \mathcal{R}oot(s|_p) \in \mathcal{F}\}$ where $\mathcal{R}oot(t)$ denotes the symbol at position ϵ in t. Similarly, we denote by $\mathcal{P}os_x(s)$ the set of positions of variable $x \in \mathcal{X}$ occuring in s, i.e. the set $\{p \in \mathcal{P}os(s) \mid s|_p = x\}$.

A TRS $\mathcal{R}$ is a set of *rewrite rules* $l \rightarrow r$, where $l, r \in \mathcal{T}(\mathcal{F}, \mathcal{X})$, $l \notin \mathcal{X}$, and $\mathcal{V}ar(l) \supseteq \mathcal{V}ar(r)$. A rewrite rule $l \rightarrow r$ is *left-linear* (resp. *right-linear*) if each variable of l (resp. r) occurs only once in l (resp. in r). A rule is linear if it is both left and right-linear. A TRS $\mathcal{R}$ is linear (resp. left-linear, right-linear) if every rewrite rule $l \rightarrow r$ of $\mathcal{R}$ is linear (resp. left-linear, right-linear). The TRS $\mathcal{R}$ induces a rewriting relation $\rightarrow_{\mathcal{R}}$ on terms whose reflexive transitive closure is denoted by $\rightarrow_{\mathcal{R}}^*$. The set of $\mathcal{R}$-descendants of a set of ground terms E is $\mathcal{R}^*(E) = \{t \in \mathcal{T}(\mathcal{F}) \mid \exists s \in E \text{ s.t. } s \rightarrow_{\mathcal{R}}^* t\}$.

Let $\mathcal{Q}$ be an infinite set of symbols, with arity 0, called *states* such that $\mathcal{Q} \cap \mathcal{F} = \emptyset$. $\mathcal{T}(\mathcal{F} \cup \mathcal{Q})$ is called the set of *configurations*.

Definition 1 (Transition and normalized transition). *A* transition *is a rewrite rule* $c \rightarrow q$, *where* c *is a configuration i.e.* $c \in \mathcal{T}(\mathcal{F} \cup \mathcal{Q})$ *and* $q \in \mathcal{Q}$. *A* normalized transition *is a transition* $c \rightarrow q$ *where* $c = f(q_1, \ldots, q_n)$, $f \in \mathcal{F}$, $\mathcal{A}rity(f) = n$, *and* $q_1, \ldots, q_n \in \mathcal{Q}$.

An epsilon transition is a transition of the form $q \rightarrow q'$ where q and q' are states. Any set of transition $\Delta \cup \{q \rightarrow q'\}$ can be equivalently replaced by $\Delta \cup \{c \rightarrow q' \mid c \rightarrow q \in \Delta\}$.

Definition 2 (Bottom-up non-deterministic finite tree automaton). *A bottom-up non-deterministic finite tree automaton (tree automaton for short) is a quadruple* $\mathcal{A} = \langle \mathcal{F}, \mathcal{Q}, \mathcal{Q}_f, \Delta \rangle$, *where* $\mathcal{Q}_f \subseteq \mathcal{Q}$ *and* Δ *is a set of normalized transitions.*

The *rewriting relation* on $\mathcal{T}(\mathcal{F} \cup \mathcal{Q})$ induced by the transitions of $\mathcal{A}$ (the set Δ) is denoted by $\rightarrow_{\Delta}$. When Δ is clear from the context, $\rightarrow_{\Delta}$ will also be denoted by $\rightarrow_{\mathcal{A}}$. Similarly, by notation abuse, we will often note $q \in \mathcal{A}$ and $t \rightarrow q \in \mathcal{A}$ respectively for $q \in \mathcal{Q}$ and $t \rightarrow q \in \Delta$.

Definition 3 (Recognized language). *The tree language recognized by* $\mathcal{A}$ *in a state* q *is* $\mathcal{L}(\mathcal{A}, q) = \{t \in \mathcal{T}(\mathcal{F}) \mid t \rightarrow_{\mathcal{A}}^* q\}$. *The language recognized by* $\mathcal{A}$ *is* $\mathcal{L}(\mathcal{A}) = \bigcup_{q \in \mathcal{Q}_f} \mathcal{L}(\mathcal{A}, q)$. *A tree language is* regular *if and only if it can be recognized by a tree automaton. A state* q *is a* dead state *if* $\mathcal{L}(\mathcal{A}, q) = \emptyset$.

Example 1. Let $\mathcal{A}$ be the tree automaton $\langle \mathcal{F}, \mathcal{Q}, \mathcal{Q}_f, \Delta \rangle$ such that $\mathcal{F} = \{f, g, a\}$, $\mathcal{Q} = \{q_0, q_1, q_2\}$, $\mathcal{Q}_f = \{q_0\}$ and $\Delta = \{f(q_0) \rightarrow q_0, g(q_1) \rightarrow q_0, g(q_2) \rightarrow q_2, a \rightarrow q_1\}$. In Δ transitions are normalized. A transition of the form $f(g(q_2)) \rightarrow q_0$ is not normalized. The term $g(a)$ is a term of $\mathcal{T}(\mathcal{F} \cup \mathcal{Q})$ (and of $\mathcal{T}(\mathcal{F})$) and can be rewritten by Δ in the following way: $g(a) \to_\Delta g(q_1) \to_\Delta q_0$. Note that $\mathcal{L}(\mathcal{A}, q_1) = \{a\}$ and $\mathcal{L}(\mathcal{A}, q_0) = \{f(g(a)), f(f(g(a))), \ldots\} = \{f^\star(g(a))\}$. Note also that $\mathcal{L}(\mathcal{A}, q_2) = \emptyset$ since no term of $\mathcal{T}(\mathcal{F})$ rewrites to q_2, hence q_2 is a dead state.

3   Tree Automata Completion

Given a tree automaton $\mathcal{A}$ and a TRS $\mathcal{R}$, the tree automata completion algorithm, proposed in [Gen98, FGVTT04], computes a tree automaton $\mathcal{A}_k$ such that $\mathcal{L}(\mathcal{A}_k) = \mathcal{R}^*(\mathcal{L}(\mathcal{A}))$ when it is possible (for the classes of TRSs covered by this algorithm see [FGVTT04]) and such that $\mathcal{L}(\mathcal{A}_k) \supseteq \mathcal{R}^*(\mathcal{L}(\mathcal{A}))$ otherwise.

The tree automata completion works as follows. From $\mathcal{A} = \mathcal{A}_0$ completion builds a sequence $\mathcal{A}_0.\mathcal{A}_1 \ldots \mathcal{A}_k$ of automata such that if $s \in \mathcal{L}(\mathcal{A}_i)$ and $s \to_\mathcal{R} t$ then $t \in \mathcal{L}(\mathcal{A}_{i+1})$. If we find a fixpoint automaton $\mathcal{A}_k$ such that $\mathcal{R}^*(\mathcal{L}(\mathcal{A}_k)) = \mathcal{L}(\mathcal{A}_k)$, then we have $\mathcal{L}(\mathcal{A}_k) = \mathcal{R}^*(\mathcal{L}(\mathcal{A}_0))$ (or $\mathcal{L}(\mathcal{A}_k) \supseteq \mathcal{R}^*(\mathcal{L}(\mathcal{A}))$ if $\mathcal{R}$ is not in one class of [FGVTT04]). To build $\mathcal{A}_{i+1}$ from $\mathcal{A}_i$, we achieve a *completion step* which consists in finding *critical pairs* between $\to_\mathcal{R}$ and $\to_{\mathcal{A}_i}$. For a substitution $\sigma : \mathcal{X} \mapsto \mathcal{Q}$ and a rule $l \to r \in \mathcal{R}$, a critical pair is an instance $l\sigma$ of l such that there exists $q \in \mathcal{Q}$ satisfying $l\sigma \to^*_{\mathcal{A}_i} q$ and $l\sigma \to_\mathcal{R} r\sigma$. For every critical pair detected between $\mathcal{R}$ and $\mathcal{A}_i$ such that $r\sigma \not\to^*_{\mathcal{A}_i} q$, $\mathcal{A}_{i+1}$ is constructed by adding a new transition $r\sigma \to q$ to $\mathcal{A}_i$ such that $\mathcal{A}_{i+1}$ recognizes $r\sigma$ in q, i.e. $r\sigma \to_{\mathcal{A}_{i+1}} q$.

However, the transition $r\sigma \to q$ is not necessarily a normalized transition of the form $f(q_1, \ldots, q_n) \to q$ and so it has to be normalized first. For example, to normalize a transition of the form $f(g(a), h(q')) \to q$, we need to find some states q_1, q_2, q_3 and replace the previous transition by a set of normalized transitions: $\{a \to q_1, g(q_1) \to q_2, h(q') \to q_3, f(q_2, q_3) \to q\}$.

Assume that q_1, q_2, q_3 are new states, then adding the transition itself or its normalized form does not make any difference. Now, assume that $q_1 = q_2$, the normalized form becomes $\{a \to q_1, g(q_1) \to q_1, h(q') \to q_3, f(q_1, q_3) \to q\}$. This set of normalized transitions represents the regular set of non normalized transitions of the form $f(g^\star(a), h(q')) \to q$ which contains the transition we want to add but also many others. Hence, this is an over-approximation. We could have made an even more drastic approximation by identifying q_1, q_2, q_3 with q, for instance.

For every transition, there exists an equivalent set of normalized transitions. Normalization consists in decomposing a transition $s \to q$, into a set Norm($s \to q$) of normalized transitions. The method consists in abstracting subterms s' of s s.t. $s' \notin \mathcal{Q}$ by states of $\mathcal{Q}$.

Definition 4 (Abstraction function). *Let $\mathcal{F}$ be a set of symbols, and $\mathcal{Q}$ a set of states. An* abstraction *function α maps every normalized configuration into a state:*

$$\alpha : \{f(q_1, \ldots, q_n) \mid f \in \mathcal{F}^n \text{ and } q_1, \ldots q_n \in \mathcal{Q}\} \mapsto \mathcal{Q}$$

Definition 5 (Abstraction state). *Let $\mathcal{F}$ be a set of symbols, and $\mathcal{Q}$ a set of states. For a given abstraction function α and for all configuration $t \in \mathcal{T}(\mathcal{F} \cup \mathcal{Q})$ the abstraction state of t, denoted by $top_\alpha(t)$, is defined by:*

1. *if $t \in \mathcal{Q}$, then $top_\alpha(t) = t$,*
2. *if $t = f(t_1, \ldots, t_n)$ then $top_\alpha(t) = \alpha(f(top_\alpha(t_1), \ldots, top_\alpha(t_n)))$.*

Definition 6 (Normalization function). *Let $\mathcal{F}$ be a set of symbols, $\mathcal{Q}$ a set of states, Δ a set of normalized transitions, $s \to q$ a transition s.t. $s \in \mathcal{T}(\mathcal{F} \cup \mathcal{Q})$ and $q \in \mathcal{Q}$, and α an abstraction function. The set $\mathrm{Norm}_\alpha(s \to q)$ of normalized transitions is inductively defined by:*

1. *if $s = q$, then $\mathrm{Norm}_\alpha(s \to q) = \emptyset$, and*
2. *if $s \in \mathcal{Q}$ and $s \neq q$, then $\mathrm{Norm}_\alpha(s \to q) = \{c \to q \mid c \to s \in \Delta\}$, and*
3. *if $s = f(t_1, \ldots, t_n)$, then $\mathrm{Norm}_\alpha(s \to q) =$*
   $\{f(top_\alpha(t_1), \ldots, top_\alpha(t_n)) \to q\} \cup \bigcup_{i=1}^{n} \mathrm{Norm}_\alpha(t_i \to top_\alpha(t_i))$.

Example 2. Let α an abstraction function such that: $\alpha = \{g(q_1, q_0) \mapsto q, b \mapsto q_1, a \mapsto q_2)\}$. Consequently, $top_\alpha = \{q_0 \mapsto q_0, b \mapsto q_1, a \mapsto q_2, g(b, q_0) \mapsto q\}$. The transition $f(a, g(b, q_0)) \to q$ can be normalized using Norm_α in the following way :

$$\mathrm{Norm}_\alpha(f(a, g(b, q_0)) \to q) = \{f(top_\alpha(a), top_\alpha(g(b, q_0))) \to q\}$$
$$\cup \mathrm{Norm}_\alpha(a \to q_2) \cup \mathrm{Norm}_\alpha(g(b, q_0) \to q)$$

By applying Definition 6 on $\mathrm{Norm}_\alpha(a \to q_2)$ and $\mathrm{Norm}_\alpha(g(b, q_0) \to q)$, we obtain that $\mathrm{Norm}_\alpha(f(a, g(b, q_0)) \to q) = \{f(q_2, q) \to q, a \to q_2, b \to q_1, g(q_1, q_0) \to q\}$.

With different abstraction function, on the same transition, one can obtain different normalizations. Hence, the precision of the fixpoint automaton $\mathcal{A}_k$ depends on the abstraction α.

Definition 7 (Automaton completion). *Let $\mathcal{A}_i = \langle \mathcal{F}, \mathcal{Q}_i, \mathcal{Q}_f, \Delta_i \rangle$ be a tree automaton, $\mathcal{R}$ a TRS and α an abstraction function. The one step completed automaton $\mathcal{A}_{i+1}$ is a tree automaton $\langle \mathcal{F}, \mathcal{Q}_{i+1}, \mathcal{Q}_f, \Delta_{i+1} \rangle$ such that:*

$$\Delta_{i+1} = \Delta_i \cup \bigcup_{l \to r \in \mathcal{R},\, q \in \mathcal{Q},\, \sigma: \mathcal{X} \mapsto \mathcal{Q},\, l\sigma \to^*_{\Delta_i} q} \mathrm{Norm}_\alpha(r\sigma \to q)$$

$$\mathcal{Q}_{i+1} = \{q \mid c \to q \in \Delta_{i+1}\}$$

4   Reconstruction Method

To make the reading of this paper easier, we give once for all the notations used in the remainder of this paper:

- $\mathcal{R}$ is a left-linear TRS;
- α is a given abstraction function (following Definition 4);
- $\mathcal{A}_0, \mathcal{A}_1, \ldots, \mathcal{A}_k$ is a finite sequence of automata obtained by the completion algorithm presented in Definition 7 for a given abstraction function α;
- $\mathcal{A}_k$ is a fixpoint automaton obtained from $\mathcal{A}_0, \mathcal{R}$ and α;
- Δ_i is the set of transitions of the automaton $\mathcal{A}_i$;
- $\mathcal{Q}_f$ is the set of final states of automata $\mathcal{A}_0, \mathcal{A}_1, \ldots, \mathcal{A}_k$.

We suppose that for all $i = 0, \ldots, k$: Δ_i are a sets of normalized transitions. In particular all Δ_i do not contain epsilon transitions (see Definition 1).

Definition 8 (Δ–Unifier). *Let Δ be a set of transitions, $t_1 \in \mathcal{T}(\mathcal{F}, \mathcal{X})$, and $t_2 \in \mathcal{T}(\mathcal{F} \cup \mathcal{Q})$. A substitution $\sigma : \mathcal{X} \mapsto \mathcal{T}(\mathcal{F} \cup \mathcal{Q})$ is a Δ-unifier of t_1 and t_2 if and only if*

1. *$t_1\sigma \rightarrow^*_\Delta t_2$, and*
2. *for all $x \in \mathcal{V}ar(t_1)$:*
   - *if $\mathcal{P}os_x(t_1) \cap \mathcal{P}os_\mathcal{F}(t_2) \neq \emptyset$ then $\sigma(x) = t_2|_{p'}$ for some $p' \in \mathcal{P}os_x(t_1) \cap \mathcal{P}os_\mathcal{F}(t_2)$.*
   - *otherwise, $\sigma(x) = q$ where $q \in \mathcal{Q}$.*

We denote by $\Uparrow_\Delta(t_1, t_2)$ the set of Δ–unifiers of t_1 and t_2.

The example below illustrates that, in general, for two terms t_1 and t_2 there exist several Δ-unifiers. The example also illustrates the case when t_1 is not linear. After the example, in Lemma 1, we show that for two terms t_1 and t_2 the set $\Uparrow_\Delta(t_1, t_2)$ is finite.

Example 3. Let $t_1 = f(g(x), h(y, y))$, $t_2 = f(q_1, h(g(a), g(q_2)))$ and a set of transitions Δ containing at least the following transitions $a \rightarrow q_2$, $g(q) \rightarrow q_1$, $g(q_1) \rightarrow q_1$ and $g(q_2) \rightarrow q_1$. The following two substitutions $\sigma_1 = \{x \mapsto q, y \mapsto g(a)\}$ and $\sigma_2 = \{x \mapsto g(q_1), y \mapsto g(a)\}$ are Δ-unifiers of t_1 and t_2, but $\sigma_3 = \{x \mapsto q_1, y \mapsto g(q_2)\}$ is not since $g(q_2) \not\rightarrow^*_\Delta g(a)$ and thus $t_1\sigma_3 \not\rightarrow^*_\Delta t_2$.

Lemma 1. *For all Δ, t_1 and t_2, the set $\Uparrow_\Delta(t_1, t_2)$ is finite.*

Proof. Given a term $t \in \mathcal{T}(\mathcal{F} \cup \mathcal{Q})$, let us denote by $Sub(t)$ the set of all subterms of t, i.e. $Sub(t) = \{t|_p \mid p \in \mathcal{P}os(t)\}$. By definition of $\Uparrow_\Delta(t_1, t_2)$, any substitution σ of this set maps a variable either to a state or to a subterm of t_1. Hence $\Uparrow_\Delta(t_1, t_2) \subset (\mathcal{V}ar(t_1) \mapsto (\mathcal{Q} \cup Sub(t_1)))$. Since $\mathcal{V}ar(t_1)$, $\mathcal{Q}$ and $Sub(t_1)$ are finite then so is $(\mathcal{V}ar(t_1) \mapsto (\mathcal{Q} \cup Sub(t_1)))$ and thus $\Uparrow_\Delta(t_1, t_2)$ is finite.

During the completion algorithm presented in Definition 7, critical pairs are detected. The role of *Occurency of a critical pair (OCCP)* is to store the information on every critical pair found during completion (the applied rule, the substitution, the position) so as to infer information on feasible traces afterwards.

Definition 9 (Occurency of a Critical Pair($OCCP$)). *Let $\mathcal{A}_k$ be a fixpoint tree automaton obtained from $\mathcal{A}_0$ using the completion algorithm of Definition 7 such that $\mathcal{A}_k = \mathcal{A}_{k+1}$ and $k \geq 0$. An OCCP is a triple $\langle l \rightarrow r, \rho, q \rangle$ where:*

- $l \rightarrow r$ *is a rewriting rule of $\mathcal{R}$;*
- ρ *a substitution of variables in $\mathcal{V}ar(l)$ by states in $\mathcal{Q}$;*
- $q \in \mathcal{Q}$;
- $l\rho \rightarrow^*_{\Delta_k} q$

Let $OCCP_k$ be the set of all OCCP built on $\mathcal{A}_k$.

Our method of trace reconstruction is based on data back-tracking on the automaton $\mathcal{A}_k = \langle \mathcal{F}, \mathcal{Q}_k, \mathcal{Q}_f, \Delta_k \rangle$ obtained by the completion of an automaton $\mathcal{A}_0 = \langle \mathcal{F}, \mathcal{Q}_0, \mathcal{Q}_f, \Delta_0 \rangle$ by a TRS $\mathcal{R}$.

From $\mathcal{A}_k$, $OCCP_k$ and a term $t \in \mathcal{L}(\mathcal{A}_k)$, we need to find $t' \in \mathcal{L}(\mathcal{A}_k)$ a predecessor of t i.e. such that $t' \rightarrow_{\mathcal{R}} t$. Then we search for a predecessor of t' and so on until reaching a term $t_0 \in \mathcal{L}(\mathcal{A}_0)$. Before defining formally the set of predecessors of a term t from $\mathcal{A}_k$ and $OCCP_k$, we define a particular substitution constructor.

Definition 10. $\sigma \bigsqcup \rho = \sigma \cup \{x \mapsto t \mid x \notin dom(\sigma) \wedge x \mapsto t \in \rho\}$

Example 4. Let $\sigma_1, \sigma_2 : \mathcal{X} \mapsto \mathcal{T}(\mathcal{F} \cup \mathcal{Q})$ be two substitutions such that:

- $\sigma_1 = \{x \rightarrow a, y \rightarrow q_1\}$ and
- $\sigma_2 = \{x \rightarrow q_2, y \rightarrow q_3, z \rightarrow b\}$.

Thus, $\sigma_1 \bigsqcup \sigma_2 = \{x \mapsto a, y \mapsto q_1, z \mapsto b\}$ and $\sigma_2 \bigsqcup \sigma_1 = \{x \mapsto q_2, y \mapsto q_3, z \mapsto b\} = \sigma_2$.

Definition 11 (Pred). *Let $cp = \langle l \rightarrow r, \rho, q \rangle$ be an OCCP such that $cp \in OCCP_k$. The set of predecessors of $t \in \mathcal{T}(\mathcal{F} \cup \mathcal{Q})$ w.r.t. cp at position $p \in \mathcal{P}os(t)$ is defined by $Pred(t, cp, p) = \{t[l\sigma \bigsqcup \rho]_p \mid \sigma \in \Uparrow_{\mathcal{A}_k}(r, t|_p)$ and $r\sigma \rightarrow^*_{\Delta_k} r\rho\}$.*

Example 5. Let Δ be a set of transitions containing the transitions $g(q_2) \rightarrow q_3, a \rightarrow q_2, g(q_1) \rightarrow q_1, g(q_4) \rightarrow q_1$ and $l \rightarrow r = f(x, y) \rightarrow f(g(x), h(y, y))$ be a rewriting rule. Let $cp = \langle l \rightarrow r, q, \rho \rangle$ be an OCCP where $\rho = \{x \mapsto q_1, y \mapsto q_3\}$ and $t = f(q_1, h(g(a), g(q_2)))$ be a term over $\mathcal{T}(\mathcal{F} \cup \mathcal{Q})$. For the position ϵ and the term t, $Pred(t, cp, \epsilon) = \{f(q_1, g(a))\}$. Indeed, the only Δ−unifier $\sigma \in \Uparrow_{\Delta}(r, t)$ respecting the condition $r\sigma \rightarrow^*_{\Delta} r\rho$ is $\sigma = \{x \mapsto g(q_1), y \mapsto g(a)\}$. Consequently, by applying the substitution $\sigma \bigsqcup \rho$ on l, we obtain $f(g(q_1), g(a))$.

Thus, by iterating this process on the terms obtained at each step, we are able to build some sequences of terms as shown in Definition 12.

Definition 12 (Sequence). *Let $t_0, \ldots, t_n \in \mathcal{T}(\mathcal{F} \cup \mathcal{Q})$ and $q \in \mathcal{Q}$ such that $\forall i = 1 \ldots n : t_i \rightarrow^*_{\Delta_k} q$. Let $cp_1, \ldots, cp_n \in OCCP_k$ and $p_1, \ldots, p_n \in \mathbb{N}^*$ such that $\forall i = 1 \ldots n : cp_i = \langle l_i \rightarrow r_i, \rho_i, q_i \rangle$. If $\forall i = 1 \ldots n : t_{i-1} \in Pred(t_i, cp_i, p_i)$ then*

$$t_n \overset{cp_n, p_n}{\longleftarrow} t_{n-1} \ldots \overset{cp_1, p_1}{\longleftarrow} t_0.$$

is a sequence from t_n to t_0.

The following theorem relates sequences to rewriting paths. Roughly, if a sequence starts from t and ends up on a term $t_0 \in \mathcal{L}(\mathcal{A}_0)$ then there exist a rewriting path from t_0 to t. Note that if $t_0 \notin \mathcal{T}(\mathcal{F})$ (but $t_0 \in \mathcal{T}(\mathcal{F} \cup \mathcal{Q})$) then there exists a term $t_0' \in \mathcal{T}(\mathcal{F})$ and $t_0' \in \mathcal{L}(\mathcal{A}_0)$ such that $t_0' \rightarrow_{\mathcal{A}_0} t_0$ and $t_0' \rightarrow_{\mathcal{R}}^* t$.

Theorem 1 (Correctness). *Given* $t \in \mathcal{L}(A_k)$ *and a final state* $q_f \in \mathcal{Q}_f$ *such that* $t \rightarrow_{\Delta_k}^* q_f$, *if there exists a sequence* $t_n \overset{cp_n,p_n}{\longleftarrow} t_{n-1} \ldots \overset{cp_1,p_1}{\longleftarrow} t_0$ *such that* $t = t_n$, $\forall i = 0 \ldots n : t_i \rightarrow_\Delta^* q_f$ *and* $t_0 \rightarrow_{\Delta_0}^* q_f$ *then there exists a term*

$$s \in \mathcal{L}(\mathcal{A}_0), \ s \rightarrow_{\Delta_0}^* t_0 \ and \ s \rightarrow_{\mathcal{R}}^* t.$$

Moreover, there exists a rewrite path (or trace)

$$s_n \overset{rl_n,p_n}{\rightarrow} s_{n-1} \ldots s_1 \overset{rl_0,p_0}{\rightarrow} t$$

where $rl_i = l_i \rightarrow r_i$ *is the rule of the OCCP* cp_i, $s_n = s$, *and* $s_{i-1} = s_i[r_i\mu_i]_{p_i}$ *and* $l\mu_i = s_i|_{p_i}$.

Proof (Sketch). The proof can be done by induction on the length of the sequence. See [BG06] for more details.

Not only to be correct, our method is also complete in the sense that if there exists a rewriting path between two terms then our method finds at least one path. However, because of Definition 8, only minimal paths are constructed. Thus, between two terms s and t it is not possible to find all the possible traces but only minimal ones.

Theorem 2 (Completeness). *Let* $t, u \in \mathcal{L}(\mathcal{A}_k)$. *If* $u \rightarrow_{\mathcal{R}}^* t$ *then there exists* $t_0, \ldots, t_{n-1} \in \mathcal{L}(\mathcal{A}_k)$, $cp_1, \ldots, cp_n \in OCCP_k$, $p_1, \ldots p_n \in \mathbb{N}^*$ *and a sequence* $t \overset{cp_n,p_n}{\longleftarrow} t_{n-1} \ldots \overset{cp_1,p_1}{\longleftarrow} t_0$ *such that* $u \rightarrow_\Delta^* t_0$.

Proof (Sketch). The proof is done by induction of the length of the rewrite derivation $u \rightarrow_{\mathcal{R}}^* t$. See [BG06] for details.

Thus, thanks to Theorem 2 and Theorem 1, we can define an algorithm that builds a valid sequence. For constructing a valid sequence, we start from a term t, construct the finite set of predecessors of t for all positions of t and all the computed *OCCP*s. Then, we repeat non deterministically the same operation on all the predecessors of t until finding a term $t_0 \rightarrow_{\Delta_0}^* q_f$ where q_f is a final state of $\mathcal{A}_0$. Of course, since we are dealing with infinite models, it is not always possible to conclude because trace reconstruction may diverge if it starts from a term that is not reachable.

5   Experimental Results

Timbuk[GVTT01, GVTT00] was used to prototype the method presented in Section 4. Timbuk is a collection of tools for achieving proofs of reachability over TRS and for manipulating tree automata (bottom-up non-deterministic finite tree automata).

In particular, Timbuk implements the tree automata completion algorithm presented in Section 3. To reconstruct a trace, the process is the following: Let $\mathcal{A}_0$, $\mathcal{R}$, α and t be respectively a tree automaton, a TRS, an abstraction function and a term to find. Timbuk performs the completion of $\mathcal{A}_0$ by $\mathcal{R}$ using the abstraction function α.

If the term t is not recognized by the completed tree automaton, then t is not reachable. Otherwise, either it is reachable or it is in the over-approximation part. To discriminate between the two solutions, we can use trace reconstruction. First, the set of $OCCP$ is computed. Then, given t and a natural N, we search the tree of possible sequences for a predecessor of t in $\mathcal{L}(\mathcal{A}_0)$, breadth-first and up to a depth N. Our system returns the first found trace which is one of the minimal traces.

Three different results can be obtained:

1. $t_0 \to_{\mathcal{R}} \ldots \to_{\mathcal{R}} t$: a trace is found and returned;
2. **Term of the approximation**: No trace can be provided because t is in the approximation part. This can be shown when the tree of predecessors of t is of depth M such that $M \leq N$ and no term of the tree belongs to the initial set, i.e. $\mathcal{L}(\mathcal{A}_0)$.
3. **Cannot conclude**: The tree of predecessors of t has been explored up to a depth N (and depth of the tree is not bounded by N) without finding a term of the initial set.

Note that in practice, we also use a generalization of this: patterns of forbidden terms, $t_p \in \mathcal{T}(\mathcal{F}, \mathcal{X})$, instead of a single term t. In those cases, the trace reconstruction process is very similar: we look for substitutions $\sigma : \mathcal{X} \mapsto \mathcal{Q}$ such that $t_p\sigma$ is recognized by the over-approximation. Then, we can start reconstruction from $t_p\sigma$. If such a σ exists then we know that there exists at least one substitution $\rho : \mathcal{X} \mapsto \mathcal{T}(\mathcal{F})$ such that $t_p\rho$ is recognized by the over-approximation (no dead states). Then we can start reconstruction from $t_p\rho$.

Now, let us present some experimentations on the verification of a simple two processes counting system. The following TRS describes the behavior of two processes each one equipped with an input list and a FIFO. Each process receives a list of symbols '+' and '−' to count, as an input. One of the processes, say P_+, is counting the '+' symbols and the other one, say P_- is counting the '−' symbols. When P_+ receives a '+', it counts it and when it receives a '−', it adds the symbol to P_-'s FIFO. The behavior of P_- is symmetric. When a process' input list and FIFO is empty then it stops and gives the value of its counter.

Here is a possible rewrite specification of this system, given in the Timbuk language, where $S(\_,\_,\_,\_)$ represents a configuration with a process P_+, a process P_-, P_+'s FIFO and P_-'s FIFO. The term $Proc(\_,\_)$ represents a process with an input list and a counter, $add(\_,\_)$ implements adding of an element in a FIFO, and cons, nil, s, o are the usual constructors for lists and natural numbers.

```
Ops
        S:4 Proc:2 Stop:1 cons:2 nil:0 plus:0 minus:0 s:1 o:0 end:0 add:2
Vars    x y z u c m n
TRS R1
  add(x, nil) -> cons(x, nil)
  add(x, cons(y, z)) -> cons(y, add(x, z))
  S(Proc(cons(plus, y), c), z, m, n) -> S(Proc(y, s(c)), z, m, n)
  S(Proc(cons(minus, y), c), u, m, n) -> S(Proc(y, c), u, m, add(minus, n))
  S(x, Proc(cons(minus, y), c), m, n) -> S(x, Proc(y, s(c)), m, n)
  S(x, Proc(cons(plus, y), c), m, n) -> S(x, Proc(y, c), add(plus, m), n)
  S(Proc(x, c), z, cons(plus,m), n) -> S(Proc(x, s(c)), z, m, n)
  S(x, Proc(z, c), m, cons(minus,n)) -> S(x, Proc(z, s(c)), m ,n)
  S(Proc(nil, c), z, nil, n)  -> S(Stop(c), z, nil, n)
  S(x, Proc(nil, c), m, nil) -> S(x, Stop(c), m, nil)
```

On this specification, we aim at proving that, for any input lists, there is no possible deadlock. In this example, a deadlock is a configuration where a process has stopped but there are still symbols to count in its FIFO, i.e. terms of the form (pattern t_p): S(Stop(x), z, cons(plus, u), c). The set of initial configurations of the system is described by the following tree automaton, where each process has a counter initialized to 0 and has an unbounded input list (with both '+' and '−') and with at least one symbol.

```
Automaton A1
States q0 qinit qzero qnil qlist qsymb
Final States q0
Transitions
  cons(qsymb, qnil) ->qlist        minus -> qsymb        nil -> qnil
  Proc(qlist, qzero) -> qinit      plus -> qsymb           o -> qzero
  S(qinit, qinit, qnil, qnil) -> q0   cons(qsymb, qlist) -> qlist
```

Let α_1 be the (constant) abstraction function normalizing every configuration into a single state q, i.e. $\alpha_1 : \mathcal{T}(\mathcal{F} \cup \mathcal{Q}) \mapsto \{q\}$ and let α_2 be the abstraction function normalizing every configuration into a new state. Roughly, α_1 and α_2 are respectively the worst and the better abstraction functions. Using completion, α_1 (resp. α_2) produces the most approximated (resp. precise) automaton. By running Timbuk on the previous specification, with α_1, we can obtain within a few seconds a tree automaton over-approximating R1*($\mathcal{L}$(A1)). However, we cannot prove that the system is deadlock free. Indeed, when looking for patterns S(Stop(x), z, cons(plus, u), c) in the over-approximation, some solutions are found, i.e. there exists substitutions $\sigma : \mathcal{X} \mapsto \mathcal{T}(\mathcal{F})$ such that S(Stop(x), z, cons(plus, u), c)σ is recognized by the tree automaton. Without trace reconstruction, there was no mean to figure out if it was a real problem or an approximation artefact. Now, using trace reconstruction, we can find automatically a counter-example, i.e. the smallest rewriting path between a particular term of the infinite language $\mathcal{L}$(A1) and a term matching S(Stop(x), z, cons(plus, u), c). The rewriting path obtained by our prototype is given using the following syntax: s -[| *applied rule, position*]-> s' ...

```
Statistics:
 - Number of nodes visited: 23921
 - Computation Time: 11.39 seconds
 - Trace(s):
S(Proc(cons(plus,nil),o),Proc(cons(plus,nil),o),nil,nil)
-[|S(Proc(cons(plus,y),c),z,m,n) -> S(Proc(y,s(c)),z,m,n),epsilon|]->
   S(Proc(nil,s(o)),Proc(cons(plus,nil),o),nil,nil)
   -[|S(Proc(nil,c),z,nil,n) -> S(Stop(c),z,nil,n),epsilon|]->
     S(Stop(s(o)),Proc(cons(plus,nil),o),nil,nil)
     -[|S(x,Proc(cons(plus,y),c),m,n) ->
        S(x,Proc(y,c),add(plus,m),n),epsilon|]->
          S(Stop(s(o)),Proc(nil,o),add(plus,nil),nil)
          -[|add(x,nil) -> cons(x,nil),epsilon.3|]->
            S(Stop(s(o)),Proc(nil,o),cons(plus,nil),nil)
```

Thanks to trace reconstruction, our system has found the two necessary conditions for the problem to occur:

- the P_- process has to have at least one '+' symbol in its input list (initial term), and
- the P_+ process needs to count all the symbols of its list and terminates before the P_- process starts to store '+' symbols in the P_+ FIFO (rewriting sequence)

then P_+ is stopped with a non-empty FIFO. Note that, using α_2 instead of α_1, completion does not terminate but after a finite number of completion steps, a similar trace can be found by visiting fewer nodes (113 nodes in 2.86 seconds) thanks to a more precise approximation.

The problem found here can be fixed by adding an additional symbol: 'end' which has to be added by process P_+ to P_- FIFO when P_+ has reached the end of its list, and symmetrically for P_-. Then, a process can stop if and only if it has reached the end of its list and if it has read the 'end' symbol in its FIFO. On the corrected TRS, it is possible to compute an over-approximation of all reachable terms with Timbuk. In the obtained approximation, no dead-lock situation occurs, proving the property [BG06].

In some other experiments, we used approximations and trace reconstruction, to find attacks in rewrite specifications of cryptographic protocols [BG06]. On those examples with more complex search trees, the obtained results show that computing first an over-approximation of reachable terms and then searching for a particular reachable term provides a very efficient alternative to usual breath-first search in the rewriting tree. In particular, in [BG06] we give an example where trace reconstruction succeeds and Maude [CDE+01] exhausts memory.

6   Conclusion

In this paper, we have presented a trace reconstruction method for over-approximations of sets of reachable terms. The proposed algorithm takes advantage of the completion-based approximation construction to prune the search space in the set of all possible traces. Completeness of the approach ensures that if a trace exists then it can be obtained by trace reconstruction, whatever the approximation may be. However, since we are dealing with infinite models, it is not always possible to conclude because trace reconstruction may diverge if it starts from a term that is not reachable.

With regards to other works in the domain, the first main interest of our technique is that it can find reachable terms on infinite sets (regular tree languages) of initial input terms. As shown on the counting processes example, trace reconstruction permits to find the exact rewriting path to a reachable term from an infinite set of possible input terms. When (1) the set of initial terms is infinite and (2) the term rewriting system is not confluent not terminating, this problem can hardly be tackled by usual rewriting tools such as Elan[BKK+98] or Maude[CDE+01] as well as by completion based tools like Waldmeister[GHLS03]. However, in order to compare with some existing tools, we also achieved some experiments on finite sets of initial terms in the case of cryptographic protocols [BG06]. It comes up that, for reachability analysis and in some particular cases, our prototype can compete with a cutting edge rewrite engine like Maude.

Furthermore, when every problem is corrected in the rewrite specification, the usual tree automata completion algorithm is able to prove that problem/attack/ deadlock are not reachable, hence cannot happen in the system, by over-approximating the set of reachable terms. This is to be used in the TA4SP tool (based on Timbuk) which is part of the AVISPA [ABB+05] protocol verification tool so as to discriminate between reachable terms (real attacks) and terms of the approximation.

References

[ABB+05]   A. Armando, D. Basin, Y. Boichut, Y. Chevalier, L. Compagna, J. Cuellar, P. Hankes Drielsma, P.-C. Héam, O. Kouchnarenko, J. Mantovani, S. Mödersheim, D. von Oheimb, M. Rusinowitch, J. Santos Santiago, M. Turuani, L. Viganò, and L. Vigneron. The AVISPA Tool for the automated validation of internet security protocols and applications. In *CAV'2005*, volume 3576 of *LNCS*, pages 281–285, Edinburgh, Scotland, 2005. Springer.

[BG06]   Y. Boichut and Th. Genet. Trace reconstruction. Research Report RR2006-02, LIFC - Laboratoire d'Informatique de l'Université de Franche Comtéand IRISA / Université de Rennes, 2006. http://lifc.univ-fcomte.fr/publis/papers/Year/2006.html.

[BHK05]   Y. Boichut, P.-C. Héam, and O. Kouchnarenko. Automatic Verification of Security Protocols Using Approximations. Research Report RR-5727, INRIA-Lorraine - CASSIS Project, October 2005.

[BKK+98]   P. Borovanský, C. Kirchner, H. Kirchner, P.-E. Moreau, and C. Ringeissen. An overview of elan. In *Proc. 2nd WRLA*, ENTCS, Pont-à-mousson (France), 1998. Elsevier.

[BN98]   F. Baader and T. Nipkow. *Term Rewriting and All That*. Cambridge University Press, 1998.

[CDE+01]   Manuel Clavel, Francisco Durán, Steven Eker, Patrick Lincoln, Narciso
             Martí-Oliet, José Meseguer, and José F. Quesada. Maude: Specification
             and programming in rewriting logic. *Theoretical Computer Science*,
             2001.
[CDG+02]   H. Comon, M. Dauchet, R. Gilleron, F. Jacquemard, D. Lugiez, S. Ti-
             son, and M. Tommasi. Tree automata techniques and applications.
             http://www.grappa.univ-lille3.fr/tata/, 2002.
[FGVTT04]   G. Feuillade, T. Genet, and V. Viet Triem Tong. Reachability Analysis
             over Term Rewriting Systems. *JAR*, 33 (3-4):341–383, 2004.
[Gen98]   T. Genet. Decidable approximations of sets of descendants and sets of
             normal forms. In *Proc. 9th RTA Conf., Tsukuba (Japan)*, volume 1379
             of *LNCS*, pages 151–165. Springer-Verlag, 1998.
[GHLS03]   J.-M. Gaillourdet, Th. Hillenbrand, B. Löchner, and H. Spies. The new
             WALDMEISTER loop at work. In F. Baader, editor, *Proc. CADE'2003*,
             volume 2741 of *LNAI*, pages 317–321. Springer-Verlag, 2003.
[GK00]   T. Genet and F. Klay. Rewriting for Cryptographic Protocol Verifica-
             tion. In *In Proc. CADE'2000*, volume 1831 of *LNAI*. Springer-Verlag,
             2000.
[GTTVTT03]   T. Genet, Y.-M. Tang-Talpin, and V. Viet Triem Tong. Verification
             of Copy Protection Cryptographic Protocol using Approximations of
             Term Rewriting Systems. In *In Proc. of WITS'2003*, 2003.
[GV98]   P. Gyenizse and S. Vágvölgyi. Linear Generalized Semi-Monadic
             Rewrite Systems Effectively Preserve Recognizability. *TCS*, 194(1-
             2):87–122, 1998.
[GVTT00]   T. Genet and V. Viet Triem Tong. Timbuk 2.0 – a Tree
             Automata Library. IRISA / Université de Rennes 1, 2000.
             http://www.irisa.fr/lande/genet/timbuk/.
[GVTT01]   T. Genet and Valérie Viet Triem Tong. Reachability Analysis of Term
             Rewriting Systems with *timbuk*. In *Proc. 8th LPAR Conf., Havana
             (Cuba)*, volume 2250 of *LNAI*, pages 691–702. Springer-Verlag, 2001.
[Jac96]   F. Jacquemard. Decidable approximations of term rewriting systems.
             In H. Ganzinger, editor, *Proc. 7th RTA Conf., New Brunswick (New
             Jersey, USA)*, pages 362–376. Springer-Verlag, 1996.
[Mon99]   D. Monniaux. Abstracting Cryptographic Protocols with Tree Au-
             tomata. In *Proc. 6th SAS, Venezia (Italy)*, 1999.
[OCKS03]   F. Oehl, G. Cécé, O. Kouchnarenko, and D. Sinclair. Automatic Ap-
             proximation for the Verification of Cryptographic Protocols. In *In
             Proceedings of FASE'03*, volume 2629 of *LNCS*, pages 34–48. Springer-
             Verlag, 2003.
[TKS00]   T. Takai, Y. Kaji, and H. Seki. Right-linear finite-path overlapping
             term rewriting systems effectively preserve recognizability. In *Proc.
             11th RTA Conf., Norwich (UK)*, volume 1833 of *LNCS*. Springer-
             Verlag, 2000.

Rewriting Models of Boolean Programs

Ahmed Bouajjani[1,*] and Javier Esparza[2,**]

[1] LIAFA, University of Paris 7
abou@liafa.jussieu.fr
[2] IFMCS, University of Stuttgart
esparza@informatik.uni-stuttgart.de

Abstract. We show that rewrite systems can be used to give semantics to imperative programs with boolean variables, a class of programs used in software model-checking as over- or underapproximations of real programs. We study the classes of rewrite systems induced by programs with different features like procedures, concurrency, or dynamic thread creation, and survey a number of results on their word problem and their symbolic reachability problem.

1 Introduction

Software Model Checking is an active research area whose goal is the application of model-checking techniques to the analysis and verification of programs. It devotes a lot of attention to *boolean programs*, which are imperative, possibly nondeterministic programs acting on variables of boolean type. The reason is that boolean programs can be used as over- or underapproximations of the real program one wishes to analyze. In order to obtain underapproximations, one restricts the range of the variables to a small, finite domain. Once this has been done, an instruction of the program can be simulated by an instruction acting on a number of boolean variables, one for each bit needed to represent the finite range (for instance, if we restrict the range of an integer variable v to the interval $[0..3]$ we can simulate an assignment to v by a simultaneous assignment to two boolean variables). The executions of the underapproximation correspond to the executions of the program in which the values of the variables stay within the specified range.

Overapproximations are obtained by *predicate abstraction* [22]. In this approach, one defines a set of boolean predicates on the variables of the program (e.g., $x \leq y$ for two integer variables x and y) and defines an abstraction function that assigns to a valuation of the program variables the set of predicates that it satisfies. Using standard abstract interpretation techniques [15], one can then construct a boolean program having the same control structure as the original one, but now acting on a set of boolean variables, one for each predicate. In Software Model Checking, these approximations are progressively refined in an automatic way until the property is proved, refuted, or until the tools run out of memory. This technique is called *counterexample-guided abstraction refinement* [14].

---

* Partially supported by the French Ministry of Research ACI project Persée.
** Partially supported by the DFG project "Algorithms for Software Model Checking".

F. Pfenning (Ed.): RTA 2006, LNCS 4098, pp. 136–150, 2006.

Boolean while programs have a finite state space, and can be analyzed using standard model-checking techniques. However, modern software goes far beyond while programs: Programs can exhibit recursion, parallelism, and thread creation. Each one of these features leads to an infinite state space, and to questions about the decidability and complexity of analysis problems. In order to attack these questions we need to find semantic models linking boolean programs to formal models with a strong theory and powerful analysis algorithms. This has been the subject of intensive research since the late 90s.

This paper shows that semantic models for boolean programs can be elegantly formulated as rewrite systems. In this approach, program states are formalized as terms, and program instructions as rewrite rules. A step of the program is matched by a rewrite step in its corresponding rewrite system. The nature of the program determines the class of terms we use. In particular, we use string-rewriting and multiset-rewriting as special cases of term rewriting.

Once we have a rewrite model, we wish to analyze it. From the model-checking or program analysis point of view questions like termination and confluence play a minor rôle. One is far more interested in the word problem, and actually on a generalization of it: Given a rewriting system and two (possibly infinite!) sets of terms T and T', can some element of T be rewritten into an element of T'? The software model checking community has attacked this question by studying *symbolic reachability* techniques. In this approach, one tries to find data structures providing finite representations of a sufficiently interesting class of infinite sets of terms, and satisfying at least one of the two following properties: (1) if a set T is representable, then the set $post^*(T)$ of terms reachable from T by means of an arbitrary number of rewriting steps is also representable; moreover, its representation can be effectively computed from the representation of T, and (2) same property with the set $pre^*(T)$ of terms that can be rewritten into terms of T instead of $post^*(T)$.

We survey a number of results on symbolic reachability algorithms for different classes of programs. We start with sequential programs, move to concurrent programs without recursion and, finally, consider the difficult case of concurrent programs with recursive procedures. For each class we give a small example of a program and its semantics, and then present analysis results.

2   Sequential Programs

Consider the program of Figure 1. It consists of two procedures, $main()$ and $p()$, and has no variables. The intended semantics of **if ? then** c_1 **else** c_2 **fi** is a nondeterministic choice between c_1 and c_2. The program state is not determined by the current value of the program counter only; we also need information about the procedure calls that have not terminated yet. This suggests to represent a state of the program as a *string* $p_0 p_1 \ldots p_n$ where p_0 is the current value of the program counter and $p_1 \ldots p_n$ is the stack of return addresses of the procedure calls whose execution has not terminated yet. For instance, the initial state of the program of Figure 1 is m_0, but the state reached after the execution of $m_1 : \textbf{call} p()$ is not p_0, it is the string $p_0 m_2$.

procedure $p()$;

p_0:	**if** (?) **then**	$p_0 \rightarrow p_1$
p_1:	**call** *main*();	$p_0 \rightarrow p_3$
p_2:	**if** ? **then call** $p()$ **fi**	$p_1 \rightarrow m_0\, p_2$
	else	$p_2 \rightarrow p_0\, p_4$
p_3:	**call** $p()$	$p_2 \rightarrow p_4$
	fi;	$p_3 \rightarrow p_0\, p_4$
p_4:	**return**	$p_4 \rightarrow \varepsilon$
		$m_0 \rightarrow \varepsilon$
procedure *main*();		$m_0 \rightarrow m_1$
m_0:	**if** ? **then return fi**;	$m_1 \rightarrow p_0\, m_2$
m_1:	**call** p;	$m_2 \rightarrow \varepsilon$
m_2:	**return**	

Fig. 1. A sequential program and its semantics

We can capture the behaviour of the program by the set of *string-rewriting* rules on the right of Figure 1. A procedure call is modelled by a rule of the form $X \rightarrow YZ$, where X is the current program point, Y the initial program point of the callee, and Y the return address of the caller. A return instruction is modelled by a rule $X \rightarrow \varepsilon$, where ε denotes the empty string. However, with the ordinary rewriting policy of string-rewriting systems

$$\frac{X \rightarrow w}{uXv \xrightarrow{r} uwv}$$

where $\xrightarrow{r}$ denotes a rewrite step, we have $m_0\, p_2\, m_2 \xrightarrow{r} m_0\, p_0\, p_4\, m_2$ (rule $p_2 \rightarrow p_0\, p_4$), which is not allowed by the intuitive semantics. We need to use the *prefix-rewriting policy*

$$\frac{X \rightarrow w}{Xv \xrightarrow{r} wv}$$

instead. We also need to interpret ε as the empty string. With these changes we have for instance the rewriting chain

$$m_0 \xrightarrow{r} m_1 \xrightarrow{r} p_0\, m_2 \xrightarrow{r} p_1\, m_2 \xrightarrow{r} m_0\, p_2\, m_2 \xrightarrow{r} p_2\, m_2 \xrightarrow{r} p_4\, m_2 \xrightarrow{r} m_2 \xrightarrow{r} \varepsilon$$

Notice that the string-rewriting system of Figure 1 is *monadic*, i.e., the left-hand-side of the rewrite rules consists of one single symbol.

2.1 Adding Variables

Consider the program of Figure 2, where b is a global variable and l is a local variable of the function *foo*(). In the presence of variables, a state of a sequential program can be modelled as a string over the alphabet containing

- a symbol for every valuation of the global variables; and
- a symbol $\langle v, p \rangle$ for every program point p and for every valuation v of the local variables of the procedure p belongs to.

```
bool function foo(l);
f₀:   if l then
f₁:       return false
      else
f₂:       return true
      fi

procedure main();
m₀:   while b do
m₁:       b := foo(b)
      od
m₂:   return
```

$$b\langle t, f_0\rangle \to b\langle t, f_1\rangle$$
$$b\langle f, f_0\rangle \to b\langle f, f_2\rangle$$
$$b\langle l, f_1\rangle \to f$$
$$b\langle l, f_2\rangle \to t$$
$$t\, m_0 \to t\, m_1$$
$$f\, m_0 \to \varepsilon$$
$$b\, m_1 \to b, \langle b, f_0\rangle\, m_0$$
$$b\, m_2 \to \varepsilon$$

Fig. 2. A sequential program with global and local variables and its semantics

States are modelled by strings of the form $g\langle v_1, p_1\rangle \ldots \langle v_n, p_n\rangle$, where g encodes the current values of the global variables, and each pair $\langle v_i, p_i\rangle$ corresponds to a procedure call that has not terminated yet. The symbol v_i encodes the values of the local variables of the caller right before the call takes place, while p_i encodes the return address at which execution must be resumed once the callee terminates. It is straightforward to assign rewrite rules to the program instructions. For instance, the call to $foo(b)$ in $main()$ is modelled by the rules

$$t\, m_1 \to t\langle t, f_0\rangle\, m_0 \qquad \text{and} \qquad f\, m_1 \to f\langle f, f_0\rangle\, m_0$$

indicating that control is transferred to f_0, that the local variable l gets assigned the current value of the global variable b, and that the return address is m_0. The complete set of rules is shown on the right of Figure 2. The symbols b and l stand in the rules for either **true** or **false**.

Notice that, due to the presence of global variables, the rewrite system is no longer monadic, although the left-hand-sides of the rules are strings of length at most 2.

String-rewriting systems using the prefix-rewriting policy are called *pushdown systems*, due to their similarity with pushdown automata: Given a string $g\langle v_1, p_1\rangle \ldots \langle v_n, p_n\rangle$ modelling a program state, the valuation g of the global variables corresponds to the current control state of the automaton, while the rest of the string corresponds to the current stack content.

2.2 Analysis

String-rewriting systems with prefix-rewriting have an interesting story. They seem to have been studied for the first time by Büchi [9], who called them *canonical systems* (see also Chapter 5 of his unfinished book [10]). Büchi proved the fundamental result that given a regular set S of strings, the sets $pre^*(S)$ and $post^*(S)$ are also regular. The result was rediscovered by Caucal [12]. Book and Otto (who were also unaware of Büchi's work) proved that $pre^*(S)$ is regular for *monadic* string-rewriting systems with *ordinary* rewriting and presented a very simple algorithm that transform a finite automaton accepting S into another one accepting $pre^*(S)$. This algorithm was adapted to pushdown systems in [2,21]), and their performance was improved in [18].

Theorem 1. *[2,21,18] Given a pushdown system R and a finite-state automaton A, the sets post* (L(A)) and pre* (L(A)) are regular and effectively constructible. in polynomial time in the sizes of R and A.*

More precisely, let P be the set of control states of R, and let Q and δ be the sets of states and transitions of the automaton A, respectively. Let $n = \max\{|Q|, |P|\}$. The automaton representing post* (L(A)) can be constructed in $O(|P||R|(n+|R|) + |P||\delta|)$ time and space, and the automaton representing pre* (L(A)) can be constructed in $O(n^2|R|)$ time and $O(n|R| + |\delta|)$ space.

The theory of pushdown systems and related models (canonical systems, monadic string-rewriting systems, recursive state machines, context-free processes, Basic Process Algebra, etc.) is very rich, and even a succinct summary would exceed the scope of this paper. A good summary of the results up to the year 2000 can be found in [11].

The algorithms of Theorem 1 have found interesting applications. They constitute the core of the Moped tool, Schwoon's back-end for model-checking software, and of its Java front-end jMoped [34]. They are also at the basis of the MOPS tool [13].

3   Concurrent Programs Without Procedures

Programming languages deal with concurrency in many different ways. In scientific computing *cobegin-coend* sections are a popular primitive, while object-oriented languages usually employ *threads*. We consider both variants. Languages also differ in their synchronization and communication mechanisms: shared variables, rendezvous, asynchronous message passing. This point is less relevant for this paper, and we only consider the shared variables paradigm. In this section we consider programs without procedures. the combination of concurrency and procedures is harder to analyze, and we consider it in the next section.

3.1   Threads

The program on the left of Figure 3 spawns a new thread $p()$ each time the while loop of *main()* is executed. This thread runs concurrently with *main()* and with the other instances of $p()$ spawned earlier. Threads communicate with each other through shared variables, in this case the global variable b. Since $p()$ nondeterministically decides whether b should be set to **true** or **false**, *main()* can create an unbounded number of instances of $p()$.

The state of the program can be modelled as a *multiset* containing the following elements:

- the current value of the global variable b,
- the current value of the program counter for the *main()* thread, and
- the current value of the program counter for each thread $p()$.

For instance, the multiset $\{0, m_1, p_1, p_2, p_2\}$ is a possible (and in fact reachable) state of the program with four threads. In order to model the program by means of rewrite rules we introduce a parallel composition operator $\parallel$ and model the state as $(0 \parallel m_1 \parallel p_1 \parallel p_2 \parallel p_2)$. Intuitively, we consider a global variable as a process running in parallel with

```
thread p();
p0:    if ? then
p1:        b := true;
       else
p2:        b := false
       fi;
p3:    end

thread main();
m0:    while b do
m1:        fork p()
       od;
m2:    end
```

$$b \parallel p_0 \rightarrow b \parallel p_1$$
$$b \parallel p_0 \rightarrow b \parallel p_2$$
$$b \parallel p_1 \rightarrow t \parallel p_3$$
$$b \parallel p_2 \rightarrow f \parallel p_3$$
$$b \parallel p_3 \rightarrow b \parallel \varepsilon$$
$$t \parallel m_0 \rightarrow t \parallel m_1$$
$$f \parallel m_0 \rightarrow f \parallel m_2$$
$$b \parallel m_1 \rightarrow b \parallel m_0 \parallel p_0$$
$$b \parallel m_2 \rightarrow b \parallel \varepsilon$$

Fig. 3. A program with dynamic thread generation and its semantics

the program and communicating with it. We rewrite modulo the equational theory of $\parallel$, which states that $\parallel$ is associative, commutative, and has the empty multiset (denoted again by ε) as neutral element:

$$u \parallel (v \parallel w) = (u \parallel v) \parallel w \qquad u \parallel v = v \parallel u \qquad u \parallel \varepsilon = u.$$

Observe that, since we rewrite modulo the equational theory, it does not matter which rewriting policy we use (ordinary or prefix-rewriting). The complete set of rewrite rules for the program of Figure 3 is shown on the right of the figure. As in the non-concurrent case, if the program has no global variables then the rewrite system is monadic. Observe that without global variables *no communication between threads is possible*.

Notice that instructions like $p : \mathbf{wait}(b); p' : \ldots$ forcing a thread to wait until the global variable b becomes true can be modelled by the rule $t \parallel p \rightarrow t \parallel p'$.

Analysis. While the word problem for pushdown systems can be solved in polynomial time (Theorem 1), it becomes harder for multiset rewriting.

Theorem 2. *[24,17] The word problem for monadic multiset-rewriting systems is NP-complete.*

NP-hardness can be proved by a straightforward reduction to SAT, while membership in NP requires a little argument. We can also prove a result similar to Theorem 1. In order to formulate the result, observe first that a multiset M over an alphabet $A = \{a_1, \ldots, a_n\}$ can be represented by the vector $\langle x_1, \ldots, x_n \rangle \in \mathbf{N}^n$, where x_i, $i \in \{1, \ldots, n\}$, is the number of occurrences of a_i in M. This encoding allows to represent sets of multisets by means of arithmetical constraints on integer vectors. The sets of vectors definable by formulas of Presburger arithmetic are called *semi-linear sets*. This name is due to the fact that every semi-linear set is a finite union of *linear sets*, defined as follows. A set $V \subseteq \mathbf{N}^n$ is linear if there is a *root vector* $v_0 \in \mathbf{N}^n$ and a finite number of *periods* $v_1, \ldots, v_k \in \mathbf{N}^n$ such that

$$V = \{v_0 + n_1 v_1 + \ldots, n_k v_k \mid n_1, \ldots, n_k \in \mathbf{N}\}.$$

Semi-linear sets share many properties with regular sets. They are closed under boolean operations. Moreover, if we associate to each word w of a regular language its *Parikh image* (the multiset containing as many copies of each symbol a as there are occurrences of a in w) we get a semi-linear set of multisets[1]. Conversely, every semi-linear set is the Parikh image of some regular language.

Intuitively, the following theorem states that semi-linear sets are to monadic multiset-rewriting what regular sets are to prefix-rewriting (see Theorem 1).

Theorem 3. *[17] Given a monadic multiset-rewriting system and a semi-linear set of states S, the sets $post^*(S)$ and $pre^*(S)$ are semi-linear and effectively constructible.*

Unfortunately, Theorem 3 does not hold for non-monadic multiset-rewriting systems. It is easy to see that these systems are equivalent to (place/transition) Petri nets. In a nutshell, a rewrite rule

$$(X_1 \parallel \ldots \parallel X_n) \rightarrow (Y_1 \parallel \ldots \parallel Y_m)$$

corresponds to a Petri net transition that takes a token from the places $X_1, \ldots, X_n$ and puts a token on the places $Y_1, \ldots, Y_m$. It is well-known that for Petri nets $post^*(S)$ can be a non semi-linear set of states even when S is a singleton [23].

The word problem for multiset-rewriting systems is equivalent to the reachability problem for Petri nets, and so, using well-known results of net theory we obtain:

Theorem 4. *[29,25,26] The word-problem for multiset-rewriting systems is decidable and EXPSPACE-hard.*

The known algorithms for the reachability problem of Petri nets are too complicated for practical use (not to speak of their complexity, which exceeds any primitive-recursive function). However, many program analysis problems can be stated as *control point reachability* problems in which we wish to know if a program point can be reached by a thread, independently of which or how many other threads run in parallel with it. In multiset-rewriting terms, the question is if the rewrite system associated to the program can reach a state of the form $X \parallel t$ for some multiset t. This target set of states is *upward-closed*: if some term t belongs to the set, then $t \parallel t'$ also belongs to the set for every multiset t'. Moreover, multiset-rewriting systems have the following important property: if $t \xrightarrow{r} t'$, then $t \parallel t'' \xrightarrow{r} t' \parallel t''$ for every multiset t''. This makes them *well-structured* systems in the sense of [1,20], and allows to apply a generic backward reachability algorithm to the control-reachability problem. More precisely, one can show that (1) every upward-closed set admits a finite representation (its set of minimal multisets), (2) if U is upward-closed then $U \cup pre(U)$ is upward-closed, where $pre(U) = \{t \mid \exists u \in U : t \xrightarrow{r} u\}$, and (3) every sequence $U_1 \subseteq U_2 \subseteq U_3 \ldots$ of upward-closed sets reaches a fixpoint after finitely many steps. The generic backwards reachability algorithm iteratively computes (the finite representations of) $U, U \cup pre(U), U \cup pre(U) \cup pre^2(U) \ldots$ until the fixpoint is reached. So we have:

Theorem 5. *Given a multiset-rewriting system and an upward-closed set of states S, the set $pre^*(S)$ is upward-closed and effectively constructible.*

The approach we described above has been adopted for instance in [16] for the verification of multithreaded Java programs.

---

[1] Parikh's theorem states the same result for context-free languages.

3.2   Cobegin-Coend Sections

Another popular way of introducing concurrency is by means of cobegin-coend sections. Intuitively, in the program $(\textbf{cobegin}\, c_1 \parallel c_2\, \textbf{coend})\,;c_3$ the code c_1 and c_2 is executed in parallel, and execution continues with c_3 after *both* c_1 and c_2 have terminated. The fundamental difference with threads is the existence of an implicit synchronization point at the end of the execution of c_1 and c_2.

Modelling the semantics requires to use term rewriting with two operators, one for sequential and another for parallel composition, which we denote by $\cdot$ and $\parallel$, respectively. For instance, if p_1,p_2,p_3 are the control locations associated to c_1,c_2,c_3 in the expression above, then we model the expression by the term $(v \parallel p_1 \parallel p_2) \cdot p_3$, where v is the current valuation of the global variables. Rewriting takes place modulo the equational theory of the $\cdot$ and $\parallel$ operators:

$$u \cdot (v \cdot w) = (u \cdot v) \cdot w \qquad u \parallel (v \parallel w) = (u \parallel v) \parallel w$$
$$\varepsilon \cdot u = u \qquad\qquad \varepsilon \parallel u = u$$
$$u \parallel v = v \parallel u$$

We also have to make the rewriting policy precise. Intuitively, it says that we can only rewrite the leftmost part of the syntax tree of a term. Formally,

$$\frac{X \to w}{X \cdot v \xrightarrow{r} w \cdot v} \quad \text{and} \quad \frac{X \to w}{X \parallel v \xrightarrow{r} w \parallel v}.$$

This model was introduced by Mayr [30,31] under the name of *Process Rewrite Systems* (PRS). Figure 4 shows a program and its rewriting semantics as PRS. Notice the rule $b \cdot m_0 \to b \parallel m_0$, which allows to make progress after the execution of the instruction $(b := \neg b \parallel b := \textbf{true})$.

m_0:	**while** b **do**	$t \parallel m_0 \to t \parallel m_1$
m_1:	**cobegin**	$f \parallel m_0 \to \varepsilon$
	m_2: $b := \neg b \parallel m_3$: $b := \textbf{true}$	$b \parallel m_1 \to (b \parallel m_2 \parallel m_3) \cdot m_0$
	coend	$b \parallel m_2 \to \neg b$
	od	$b \parallel m_3 \to t$
		$b \cdot m_0 \to b \parallel m_0$

Fig. 4. A program with a cobegin-coend section and its semantics

PRSs can also model at least part of the interaction between procedures and concurrency, and therefore we delay their analysis until the next section.

4   Putting Procedures and Concurrency Together

The analysis of programs containing both procedures and concurrency is notoriously difficult. It is easy to show that a two-counter machine can be simulated by a boolean program consisting of two recursive procedures running in parallel and accessing one single global boolean variable. Intuitively, the two recursion stacks associated to the two procedures are used to simulate the two counters; the depth of the stack corresponds to the current value of the counter. Increments and decrements can be simulated by calls and returns. The global variable is used as a semaphore indicating which counter has to be accessed next. Since two-counter machines are Turing powerful, all interesting analysis problems about these programs are bound to be undecidable.

In programs with procedures and concurrency the same code unit can be called following different policies: procedural call (caller waits until callee terminates), thread call (caller runs concurrently with callee), cobegin-coend call (caller waits, may call several callees). We use the keyword **process** to denote such a unit.

4.1 Procedural Programs with Cobegin-Coend Sections

In a while program with cobegin-coend sections the maximum number of processes that can be executed concurrently is syntactically bounded. This is no longer the case in the presence of recursion. For instance, a process may contain a cobegin-coend section one of whose branches calls the process itself. The program of Figure 5 is an example. In the absence of global variables, we can model the program as a monadic PRS (the rules are shown on the right of the figure).

process *main*();
m_0: **if** ? **then**
    cobegin
        m_1: **call** *main*() $\|$ m_2: **skip**
    coend
  fi;
m_3: **return**

$$m_0 \rightarrow (m_1 \| m_2) \cdot m_3$$
$$m_0 \rightarrow m_3$$
$$m_1 \rightarrow m_0$$
$$m_2 \rightarrow \varepsilon$$
$$m_3 \rightarrow \varepsilon$$

Fig. 5. A procedural program with global variables and its semantics

Unfortunately, the addition of global variables leads to complications. In order to understand why, consider the program of Figure 6. It is very similar to the program of Figure 5, but has a global variable b. The right side of the figure shows an attempt at a semantics following the ideas we have used so far. However, the derivations of the rewrite system do not match the intuitive semantics, in which *main*() should be able to call itself, and then execute $b = \textbf{true}$ immediately thereafter. No derivation of the rewrite systems allows to do so. The only derivation having a chance would be

$$b\,m_0 \xrightarrow{r} (b \| m_1 \| m_2) \cdot m_3 \xrightarrow{r} (b \| m_0 \| m_2) \cdot m_3 \xrightarrow{r} (((b \| m_1 \| m_2) \cdot m_3) \| m_2) \cdot m_3$$

but now the rule $b \| m_2 \rightarrow b \| \varepsilon$ can only be applied to the *innermost* m_2, which corresponds to the incarnation of *main*() as callee, not to its incarnation as caller.

So we conclude that, while monadic PRS are a suitable formalism for modelling programs without global variables, PRS do not match the interplay between recursion and concurrency in conventional programming languages.

process *main*();
m_0: **if** ? **then**
        cobegin
           m_1: **call** *main*() $\|$ m_2: $b :=$ **true**
        coend
    fi;
m_3: **return**

$$b \| m_0 \rightarrow (b \| m_1 \| m_2) \cdot m_3$$
$$b \| m_0 \rightarrow b \| m_3$$
$$b \| m_1 \rightarrow b \| m_0$$
$$b \| m_2 \rightarrow t \| \varepsilon$$
$$b \| m_3 \rightarrow b \| \varepsilon$$

Fig. 6. A program with global variables and an incorrect PRS semantics

Analysis. Mayr has shown that the word problem for PRS is decidable [30,31] but, since PRS contain Petri nets as a subclass, the problem is EXPSPACE-hard. Fortunately, in the case of monadic PRS (which, as we have seen, seems to be more useful for modelling programs), there exist far more efficient approaches based on symbolic reachability analysis.

The design of symbolic reachability analysis procedures for PRS (or even for its monadic fragment) is not easy. First, we need to represent sets of PRS terms, and a natural idea is to use finite-state tree automata for that. However, we have the problem that the commutative closure of a regular set of terms is not regular. To see this, consider the regular set of terms of the form $a\|b\|a\|b\| \ldots \|a\|b$. Its commutative closure is the set of all parallel terms with the same number of a's and b's, which is clearly not regular. As a consequence, neither the $post^*$ nor the pre^* operation preserves regularity of a set of terms. Moreover, since PRS subsume Petri nets, they do not preserve semi-linearity either.

Nevertheless, in some classes of multiset-rewriting systems, including the monadic class, the $post^*$ operation does preserve semi-linearity, and there exists an algorithm that computes $post^*$ image of any given semi-linear set (see Theorem 3). Let us call a class of multiset-rewriting systems satisfying this property a *semi-linear class*. Given a semi-linear class C, let PRS$[C]$ be the class of PRS whose sets of rules can be partitioned into two sets M and P, where M is a multiset-rewriting system belonging to the class C, and the rules of P only contain occurrences of the sequential composition operator. Notice that the sets P and M may share constants. We can ask whether the semi-linearity of C can be exploited to define an algorithm for symbolic reachability analysis of PRS$[C]$. This question was addressed in [8] from an automata-theoretic point of view. An important issue is the representation of sets of PRS terms which are closed under the equational theories of $\cdot$ and $\|$. Since these operators are associative, PRS terms can be seen as trees with unbounded width. Each node labelled with $\|$ (resp. $\cdot$) may have an arbitrary number of children labelled with simple symbols (process constants) and an arbitrary number of children labelled with $\cdot$ (resp. $\|$). Therefore, a natural idea is to use *unranked tree automata* (also called *hedge automata*) as symbolic representations for sets of PRS terms. Furthermore, since parallel composition is commutative, we should use *commutative hedge automata (CHA)*. CHA are closed under boolean operations and have a decidable emptiness problem (see also [27,33]). Then, we have the following generic result:

Theorem 6. *[8] Let C be a semi-linear class of multiset rewrite systems. For every system in $PRS[C]$ and for every CHA-definable set of terms T, the sets $post^*(T)$ and $pre^*(T)$ are CHA-definable and effectively constructible.*

By Theorem 3 we know that the class of monadic multiset rewrite systems is semi-linear. Therefore, Theorem 6 gives a procedure for symbolic reachability analysis for monadic PRS:

Theorem 7. *[8] Given a monadic PRS, for every CHA-definable set of terms T, the sets $post^*(T)$ and $pre^*(T)$ are CHA-definable and effectively constructible.*

Actually, using the same approach, it is possible to extend Theorem 7 to a larger subclass of PRS whose rules contain no occurrence of the parallel operator on the left-hand-side (but possibly occurrences of $\cdot$) [8]. This class is called PAD in the literature.

 Another approach to the symbolic reachability problem for monadic PRS constructs not the sets $post^*(T)$ or $pre^*(T)$ themselves, but a set of *representatives* w.r.t. the equational theories of sequential and parallel composition. This is sufficient to solve reachability problems where the origin and target sets of terms are closed modulo these equational theories. In particular, the approach is powerful enough to solve control point reachability problems. The approach was first studied in [28] and later in [19]. We rephrase here the result of [19]. Notice that, by essentially the same procedure used to put a context-free grammar in Chomsky normal form, we can transform a monadic PRS into a normal form in which the right-hand-side of all rules has the shape $X \cdot Y$ or $X \parallel Y$.

Theorem 8. *[28,19] Let R be a monadic PRS in normal form, and let A be a bottom-up tree automaton recognizing a set $L(A)$ of PRS terms. One can construct in $O(|R| \cdot |A|)$ time two bottom-up tree automata recognizing for every term $t \in post^*(L(A))$ ($t \in pre^*(L(A))$) a term t' such that $t = t'$ in the equational theory of the $\parallel$ and $\cdot$ operators.*

This approach was extended to the case of PAD systems in [7].

4.2   Multithreaded Procedural Programs

Process Rewrite Systems are also inadequate for modelling the combination of multi-threading and procedures, even in the absence of variables. Consider the program of Figure 7. If we model the **fork** operation by a rule like $m_1 \to m_3 \parallel p_0$, we get the derivation

$$m_0 \xrightarrow{r} m_2 \xrightarrow{r} m_0 \cdot m_3 \xrightarrow{r} m_1 \cdot m_3 \xrightarrow{r} (m_3 \parallel p_0) \cdot m_3 \xrightarrow{r} p_0 \cdot m_3$$

But this is not the intended semantics. The main thread (corresponding to m_3 in the term $p_0 \cdot m_3$) can only terminate *after* the new thread (corresponding to p_0 has terminated.

 A new approach has been proposed by the first author, Müller-Olm and Touili in [6]. The idea is to represent a state at which n-threads are active by a string $\#w_n\#w_{n-1}\ldots\#w_1$. Here, $w_1,\ldots w_n$ are the strings modelling the states of the threads, and they are ordered according to the following criterion: for every $1 \leq i < j \leq n$, the i-th thread (i.e., the thread in state w_i) must have been created no later than the j-th thread. The reason for putting younger threads *to the left* of older ones will be clear in a moment.

process $p()$;

p_0:   **if** (?) **then**

p_1:   **call** $p()$

      else

p_2:   **skip**

      fi;

p_3:   **return**

process $main()$;

m_0:   **if** (?) **then**

m_1:   **fork** $p()$

      else

m_2:   **call** $main()$

      fi;

m_3:   **return**

$$\# p_0 \rightarrow \# p_1$$
$$\# p_0 \rightarrow \# p_2$$
$$\# p_1 \rightarrow \# p_0\, p_3$$
$$\# p_2 \rightarrow \# p_3$$
$$\# p_3 \rightarrow \#\varepsilon$$
$$\# m_0 \rightarrow \# m_1$$
$$\# m_0 \rightarrow \# m_2$$
$$\# m_1 \rightarrow \# p_0 \# m_3$$
$$\# m_2 \rightarrow \# m_0\, m_3$$
$$\# m_3 \rightarrow \#\varepsilon$$
$$\#\# \rightarrow \#$$

Fig. 7. A program with dynamic thread generation and its semantics

We can now try to capture the semantics of the program by string-rewriting rules. Notice however that we cannot use the prefix-rewriting policy. Loosely speaking, a thread in the middle of the string should also be able to make a move, and this amounts to rewriting "in the middle", and not only "on the left"". So we must go back to the ordinary rewriting policy

$$\frac{X \rightarrow w}{u X v \xrightarrow{r} u w v}$$

Instructions not involving thread creation are modelled as in the non-concurrent case, with one difference: Since we can only rewrite on the left of a w_i substring, we "anchor" the rewrite rules, and use for instance $\# p_1 \rightarrow \# p_0\, p_3$ instead of $p_1 \rightarrow p_0\, p_3$. The thread creation at program point m_1 is modelled by the rule $\# m_1 \rightarrow \# p_0 \# m_3$. Notice that we would not be able to give a rule if we wanted to place the new thread to the right of its creator, because the stack length of the creator at the point of creating the new thread can be arbitrarily large. This class of string-rewriting systems is called *dynamic networks of pushdown systems* (DPN) in [6]. The complete set of rewrite rules for the program of Figure 7 is shown on the right of the same figure.

Analysis. Notice that DPNs are neither prefix-rewriting nor monadic. However, we still have good analizability results. First of all, it can be proved that the *pre** operation preserves regularity:

Theorem 9. *[6] For every regular set S of states of a DPN, the set $pre^*(S)$ is regular and a finite-state automaton recognizing it can be effectively constructed in polynomial time.*

The *post** operation, however, does not preserve regularity. To see this, consider a program which repeatedly creates new threads and counts (using its stack) the number of threads it has created. The set of reachable states is not regular, because in each of them the number of spawned threads must be equal to the length of the stack. Nevertheless, the *post** operation preserves context-freeness.

Theorem 10. *[6] For every context-free (pushdown automata definable) set S of states of a DPN, the set post*(S) is context-free and a pushdown automaton recognizing it can be effectively constructed in polynomial time.*

Since intersection of a regular language with a context-free language is always context-free, and since the emptiness problem of context-free languages is decidable, this result allows to solve the reachability problem between a context-free initial set of configurations and a regular set of target configurations.

So far we have only considered the variable-free case. The results above can be extended to the case in which processes have local variables, but global variables make the model Turing powerful. In this case over/underapproximate analysis approaches can be adopted, which are outside the scope of this paper (see, e.g., [4,5,32,3]).

5 Summary

We have studied rewriting models for sequential and concurrent boolean programs where concurrent processes communicate through shared variables.

Sequential boolean programs with procedure calls can be modelled by prefix-rewriting systems (pushdown systems). The word problem and the symbolic reachability problem for regular sets of states can be solved in polynomial time. The algorithms have been implemented in the Moped and MOPS tools and applied to the analysis of large programs.

Concurrent programs with dynamic thread creation but without procedures can be modelled by multiset-rewriting systems (Petri nets). The word problem is decidable, but the algorithm is not useful in practice. The control reachability problem can be solved by a simple algorithm based on the theory of well-structured systems. The symbolic reachability problem can be solved for the monadic fragment and semi-linear sets. The monadic fragment corresponds to programs without global variables, and so to absence of communication between threads.

Process Rewrite Systems (PRS) combine prefix-rewriting and multiset-rewriting. PRS have a decidable word problem, but do not match the interplay between procedures and concurrency in conventional programming languages. Monadic PRS model parallel programs with cobegin-coend sections and procedure calls, but without global variables. The word problem is NP-complete. The control reachability problem can be solved very efficiently using bottom-up tree automata. The symbolic reachability problem can be solved for sets of states recognizable by commutative hedge automata.

Concurrent programs with thread creation and procedures, but without communication between threads, can be model by dynamic networks of pushdown systems [6], a class of string-rewriting systems. The word problem can be solved in polynomial time. The *pre** operation preserves regularity (and can be computed in polynomial time), while the *post** operation preserves context-freeness.

Concurrent programs with procedures and one single global variable are already Turing powerful, and so very difficult to analyze. Several approximate analysis have been proposed based on the automata techniques presented in this paper (see e.g. [4,5,32,3]). The constrained dynamic networks of [6] replace global variables by a more restricted form of communication in which a process can wait for a condition on the threads it created, or for a result computed by a procedure it called.

References

1. P. A. Abdulla, K. Cerans, B. Jonsson, and Y.-K. Tsay. General decidability theorems for infinite-state systems. In *LICS*, pages 313–321, 1996.
2. A. Bouajjani, J. Esparza, and O. Maler. Reachability analysis of pushdown automata: Application to model-checking. In *CONCUR*, volume 1243 of *Lecture Notes in Computer Science*, pages 135–150. Springer, 1997.
3. A. Bouajjani, J. Esparza, S. Schwoon, and J. Strejcek. Reachability analysis of multithreaded software with asynchronous communication. In *FSTTCS*, volume 3821 of *Lecture Notes in Computer Science*. Springer, 2005.
4. A. Bouajjani, J. Esparza, and T. Touili. A generic approach to the static analysis of concurrent programs with procedures. *Int. J. Found. Comput. Sci.*, 14(4):551–, 2003.
5. A. Bouajjani, J. Esparza, and T. Touili. Reachability analysis of synchronized pa systems. *Electr. Notes Theor. Comput. Sci.*, 138(3):153–178, 2005.
6. A. Bouajjani, M. Müller-Olm, and T. Touili. Regular symbolic analysis of dynamic networks of pushdown systems. In *CONCUR*, volume 3653 of *Lecture Notes in Computer Science*. Springer, 2005.
7. A. Bouajjani and T. Touili. Reachability analysis of process rewrite systems. In *FSTTCS*, volume 2914 of *Lecture Notes in Computer Science*, pages 74–87. Springer, 2003.
8. A. Bouajjani and T. Touili. On computing reachability sets of process rewrite systems. In *RTA*, volume 3467 of *Lecture Notes in Computer Science*. Springer, 2005.
9. J. R. Büchi. Regular canonical systems. *Arch. Math. Logik Grundlag.*, 6:91–111, 1964.
10. J. R. Büchi. *The collected works of J. Richard Büchi*. Springer-Verlag, New-York, 1990.
11. O. Burkart, D. Caucal, F. Moller, and B. Steffen. Verification on Infinite Structures. In *Handbook of Process Algebra*. North-Holland Elsevier, 2001.
12. D. Caucal. On the regular structure of prefix rewriting. *Theor. Comput. Sci.*, 106(1):61–86, 1992.
13. H. Chen and D. Wagner. MOPS: an infrastructure for examining security properties of software. In *ACM Conference on Computer and Communications Security*, pages 235–244, 2002.
14. E.M. Clarke, O. Grumberg, S. Jha, Y. Lu, and H. Veith. Counterexample-guided abstraction refinement for symbolic model checking. *J. ACM*, 50(5):752–794, 2003.
15. P. Cousot and R. Cousot. Abstract interpretation: A unified lattice model for static analysis of programs by construction or approximation of fixpoints. In *POPL*, pages 238–252, 1977.
16. G. Delzanno, J.-F. Raskin, and L. Van Begin. Towards the automated verification of multithreaded java programs. In *TACAS*, volume 2280 of *Lecture Notes in Computer Science*, pages 173–187. Springer, 2002.
17. J. Esparza. Petri nets, commutative context-free grammars, and basic parallel processes. *Fundam. Inform.*, 31(1):13–25, 1997.
18. J. Esparza, D. Hansel, P. Rossmanith, and S. Schwoon. Efficient algorithms for model checking pushdown systems. In *CAV*, volume 1855 of *Lecture Notes in Computer Science*, pages 232–247. Springer, 2000.
19. J. Esparza and A. Podelski. Efficient algorithms for pre* and post* on interprocedural parallel flow graphs. In *POPL*, pages 1–11, 2000.
20. A. Finkel and Ph. Schnoebelen. Well-structured transition systems everywhere! *Theor. Comput. Sci.*, 256(1-2):63–92, 2001.
21. A. Finkel, B. Willems, and P. Wolper. A direct symbolic approach to model checking pushdown systems. *Electr. Notes Theor. Comput. Sci.*, 9, 1997.
22. S. Graf and H. Saïdi. Construction of abstract state graphs with PVS. In *CAV*, volume 1254 of *Lecture Notes in Computer Science*, pages 72–83. Springer, 1997.

23. J.E. Hopcroft and J.-J. Pansiot. On the reachability problem for 5-dimensional vector addition systems. *Theor. Comput. Sci.*, 8:135–159, 1979.
24. D.T. Huynh. Commutative grammars: The complexity of uniform word problems. *Information and Control*, 57(1):21–39, 1983.
25. S. R. Kosaraju. Decidability of reachability in vector addition systems (preliminary version). In *STOC*, pages 267–281. ACM, 1982.
26. R. Lipton. The Reachability Problem Requires Exponential Space. Technical Report 62, Yale University, 1976.
27. D. Lugiez. Counting and equality constraints for multitree automata. In *FoSSaCS*, volume 2620 of *Lecture Notes in Computer Science*, pages 328–342. Springer, 2003.
28. D. Lugiez and Ph. Schnoebelen. The regular viewpoint on PA-processes. In *CONCUR*, volume 1466 of *Lecture Notes in Computer Science*, pages 50–66. Springer, 1998.
29. E.W. Mayr. An algorithm for the general Petri net reachability problem. In *STOC*, pages 238–246. ACM, 1981.
30. R. Mayr. *Decidability and Complexity of Model Checking Problems for Infinite-State Systems*. PhD thesis, Technische Universität München, 1998.
31. R. Mayr. Process rewrite systems. *Inf. Comput.*, 156(1-2):264–286, 2000.
32. S. Qadeer and J. Rehof. Context-bounded model checking of concurrent software. In *TACAS*, volume 3440 of *Lecture Notes in Computer Science*, pages 93–107. Springer, 2005.
33. H. Seidl, Th. Schwentick, and A. Muscholl. Numerical Document Queries. In *PODS'03*. ACM press, 2003.
34. D. Suwimonteerabuth, S. Schwoon, and J. Esparza. jMoped: A Java bytecode checker based on Moped. In *TACAS*, volume 3440 of *Lecture Notes in Computer Science*, pages 541–545, Edinburgh, UK, 2005. Springer.

Syntactic Descriptions: A Type System for Solving Matching Equations in the Linear λ-Calculus

Sylvain Salvati

National Institute of Informatics
National Center of Sciences
Tokyo, Japan
salvati@nii.ac.jp

Abstract. We introduce syntactic descriptions, an extended type system for the linear λ-calculus. With this type system checking that a linear λ-term normalizes to another one reduces to type-checking. As a consequence this type system can be seen as a formal tool to design matching algorithms. In that respect, solving matching equations becomes a combination of type inference and proof search. We present such an algorithm for linear matching equations.In the case of second order equations, this algorithm stresses the similarities between linear matching in the linear λ-calculus and linear context matching. It uses tabular techniques and is a practical alternative to Huet's algorithm for those equations.

1 Introduction

Matching is the problem of the existence of an instantiation of a term so that it equates, modulo some equivalence relation, another term. This problem can be formulated in many term algebra and with many equivalence relations. The most well known matching problem is the higher order matching in the simply typed λ-calculus modulo βη-equivalence. This problem has been defined by Huet [5] as a restriction of the problem of unification. Still the decidability of higher order matching is not known. In some particular cases, second [5], third [4] and fourth order [9] this problem has been proved to be decidable. While choosing β-equivalence as equivalence relation, Loader has proved that sixth order matching is undecidable [8]. Other classes of matching problems involve some simpler algebraic languages and relations as for example context matching. Context matching uses trees (unknowns represent tree contexts) and syntactic equality and is known to be decidable. Its complexity has been studied in details by Schmidt-Schauß and Stuber [12].

The class of matching problem we study in this paper is the matching in the linear λ-calculus, the equivalence relation being βη-equivalence. Motivated by the central role it plays in parsing Abstract Categorial Grammars, a grammatical formalism defined by Philippe de Groote [3], this problem has recently focused many attention. This class of matching problem has been shown decidable and

F. Pfenning (Ed.): RTA 2006, LNCS 4098, pp. 151–165, 2006.

NP-complete by de Groote [2] and its complexity has been extensively studied [11], [13].

To cope with this kind of matching, we introduced [10] a new formal tool: syntactic descriptions. This formalism is actually an extended type system for linear λ-calculus and allows one to statically check whether a term t $\beta\eta$-normalizes to a term u. With this type system we design a new algorithm for solving linear matching in the linear λ-calculus which avoids many drawbacks of the adaptation of uet's algorithm proposed by de Groote [2]. Furthermore, when applied to solve second order linear equations, this algorithm shows many similarities with the algorithm proposed by [12] to solve linear context matching. This has to be contrasted with our previous result [11] where we showed that second order linear matching in the linear λ-calculus was NP-complete which stressed the differences between this problem and linear context matching. Notice that when dealing with linear matching in the linear λ-calculus, we face two kinds of linearities, the linearity of the matching equation and the linearity of the λ-calculus. One must clearly identify those two different definitions of linearity not to get confused while reading this paper.

The four remaining sections of this paper are organized as follows. Section 2 will give the necessary formal definitions. Syntactic descriptions will be defined in section 3. This section will also show their basic properties and how they are related to linear matching in the linear λ-calculus. Section 4 will describe the algorithm that we propose to solve linear matching in the linear λ-calculus. Finally section 5 will conclude, discuss related issues and explain future work.

2   Definitions

In this section we define the notion of linear λ-terms. For some technical convenience, we adopt a Church-style type system. So before defining linear λ-terms, considering we are given a finite set of atomic types $\mathcal{A}$, we define linear types.

Definition 1. *The set $\mathcal{LT}$ is the set of linear implicative types and is given by the following grammar:*

$$\mathcal{LT} ::= \mathcal{A} \mid (\mathcal{LT} \multimap \mathcal{LT})$$

For short, we will·write $\alpha_1 \multimap \ldots \multimap \alpha_n \multimap \beta$ instead of $(\alpha_1 \multimap (\ldots \multimap (\alpha_n \multimap \beta)\ldots))$.

Definition 2. *The size of a type α, noted $|\alpha|$ is defined by:*

1. $|\alpha| = 1$ if α is atomic
2. $|\alpha \multimap \beta| = |\alpha| + |\beta|$

We then suppose we are given for any $\alpha \in \mathcal{LT}$ an infinite enumerable set of variables x^α, y^α, $z^\alpha \ldots$ an infinite set of unknowns $\mathbf{X}^\alpha$, $\mathbf{Y}^\alpha$, $\mathbf{Z}^\alpha \ldots$ Finally we suppose we are given a finite set of constants $a_1^{\alpha_1}, \ldots, a_n^{\alpha_n}$. Variables, unknowns and constants are all decorated with a type and are called *atomic terms*. Often, when it is not relevant, we will omit the type annotation. The sets $\mathcal{V}$, $\mathcal{X}$ and $\mathcal{C}$ are respectively the set of variables, the set of unknowns and the set of constants.

We can now define raw λ-terms:

Definition 3. *The set Λ of raw λ-terms is defined with the following grammar:*

$$\Lambda ::= \mathcal{V} \mid \mathcal{X} \mid \mathcal{C} \mid \lambda \mathcal{V}.\Lambda \mid (\Lambda \Lambda)$$

We will write $\lambda x_1 \ldots x_n.t$ for the term $\lambda x_1.\lambda x_2.\ldots.\lambda x_n.t$ and $t_0 t_1 \ldots t_n$ for the term $((\ldots(t_0 t_1)\ldots)t_n)$. The notion of free variables, β-reduction, η-reduction, α-conversion *etc.*...are defined as usual [1]. We write $FV(t)$ for the set of free variables in t and $U(t)$ for the set of unknowns which occur in t. The infix symbols $\to_{\beta\eta}, \overset{*}{\to}_{\beta\eta}, =_{\beta\eta}$ denote respectively, $\beta\eta$-contraction, $\beta\eta$-reduction and $\beta\eta$-conversion.

Definition 4. *The size of a λ-term t, noted $|t|$ is given by:*

1. $|h| = 1$ *where h is an atomic term*
2. $|\lambda x.t| = |t|$
3. $|t_1 t_2| = |t_1| + |t_2|$

Among raw λ-terms we distinguish linear λ-terms.

Definition 5. *The set $\mathcal{L}\Lambda^\alpha$ of linear λ-terms of type α is the smallest set such that:*

1. $x^\alpha \in \mathcal{L}\Lambda^\alpha$, $\mathbf{X}^\alpha \in \mathcal{L}\Lambda^\alpha$ *and* $a^\alpha \in \mathcal{L}\Lambda^\alpha$
2. $t_1 \in \mathcal{L}\Lambda^{\beta \multimap \alpha}$, $t_2 \in \mathcal{L}\Lambda^\beta$ *and* $FV(t_1) \cap FV(t_2) = \emptyset$ *imply* $(t_1 t_2) \in \mathcal{L}\Lambda^\alpha$
3. $t \in \mathcal{L}\Lambda^\alpha$ *and* $x^\beta \in FV(t)$ *imply* $\lambda x^\beta.t \in \mathcal{L}\Lambda^{\beta \multimap \alpha}$

The set $\mathcal{L}\Lambda$ of linear λ-terms is defined to be $\mathcal{L}\Lambda = \bigcup_{\alpha \in \mathcal{L}T} \mathcal{L}\Lambda^\alpha$.

Those terms are said linear because variables may have at most one free occurrence and bound variables have exactly one free occurrence below their binding abstraction.

The definition of syntactic descriptions requires a precise notion of occurrence of subterms. Because what is elegant for programming is also convenient for mathematics, we have chosen Huet's zippers [6] to denote subterms. To this end, we need contexts and some of their technical properties.

Definition 6. *A context is a λ-term with a hole. Contexts are defined by the following grammar:*

$$\mathbf{C} = [] \mid \Lambda \mathbf{C} \mid \mathbf{C}\Lambda \mid \lambda \mathcal{V}.\mathbf{C}$$

Contexts will be written $C[]$ with indices or superscripts. Given a term t, we will write $C[t]$ the term obtained by inserting t at the place of the hole in $C[]$. Note that $C[]$ can bind variables which are free in t.

We are now in position to define subterms by means of zippers.

Definition 7. *Given a term u, the set of subterms of u, $\mathcal{S}_u$, is defined as:*

$$\mathcal{S}_u = \{(C[], v) \mid C[v] = u\}$$

We may refer to the subterms of u which have type α (i.e. $\mathcal{S}_u \cap (\mathbf{C} \times \mathcal{L}\Lambda^\alpha)$) as $\mathcal{S}_u^\alpha$. Subterms are partially ordered; $(C_1[], v_1) \prec (C_2[], v_2)$ means that there is $C_3[]$ such that $C_1[] = C_2[C_3[]]$. Whenever neither $(C_1[], v_1) \prec (C_2[], v_2)$ nor $(C_2[], v_2) \prec (C_1[], v_1)$ hold then we write $(C_1[], v_1) \prec\succ (C_2[], v_2)$.

Linear λ-terms have some remarkable properties. These properties are rather well known and easy to prove so that we don't sketch their proofs.

Proposition 1. *Given $t \in \mathcal{L}\Lambda$ such that $x \in FV(t)$, there is a unique context $C_{t,x}[]$ such that $C_{t,x}[x] = t$ and $x \notin FV(C_{t,x}[])$.*

The context $C_{t,x}[]$ denotes the position of the free occurrence of x in t.

Lemma 1 (Type uniqueness). *If $t \in \mathcal{L}\Lambda^\alpha$ and $t \in \mathcal{L}\Lambda^\beta$ then $\alpha = \beta$.*

Lemma 2 (Substitution). *Given $t, u \in \mathcal{L}\Lambda$, if $u \in \mathcal{L}\Lambda^\alpha$, $t \in \mathcal{L}\Lambda^\beta$, $FV(u) \cap FV(t) = \emptyset$ and $x^\alpha \in FV(t)$, then $t[x^\alpha := u] \in \mathcal{L}\Lambda^\beta$.*

Lemma 3 (Extraction). *Given $t \in \mathcal{L}\Lambda$ and $u \in \mathcal{L}\Lambda^\alpha$ such that $x^\alpha \in FV(t)$ and $t[x^\alpha := u] \in \mathcal{L}\Lambda^\beta$ then $t \in \mathcal{L}\Lambda^\beta$.*

Proposition 2 (Subject Reduction). *If $t \in \mathcal{L}\Lambda^\alpha$ and $t \xrightarrow{*}_{\beta\eta} t'$ then $t' \in \mathcal{L}\Lambda^\alpha$.*

Proposition 3 (Subject Expansion). *If $t \in \mathcal{L}\Lambda$, $t' \in \mathcal{L}\Lambda^\alpha$ and $t \xrightarrow{*}_{\beta\eta} t'$ then $t \in \mathcal{L}\Lambda^\alpha$.*

For the subject expansion property to be proved, it is mandatory that t is a linear λ-term, otherwise, of course, it does not hold.

Definition 8. *Given $t \in \mathcal{L}\Lambda$, we say that t is in η-long form if for all $(C[], v) \in \mathcal{S}_t^{\alpha\multimap\beta}$ one of the following properties holds:*

1. *$v = \lambda x.w$ for some w.*
2. *$C[] = C'[[]w]$ for some $C'[]$ and some w.*

For every $t \in \mathcal{L}\Lambda$, there is t' in η-long form such that $t' \xrightarrow{*}_\eta t$. Furthermore, Huet [5] has proved that the terms in η-long form are closed under β-reduction.

The elements of $\mathcal{L}\Lambda^\alpha$ (resp. $\mathcal{L}\Lambda$) which are both in η-long form and in normal form are denoted by lnf^α (resp. lnf). Remark that given $t \in \mathrm{lnf}$, for all $(C[], v) \in \mathcal{S}_t^\alpha$ if $v \in \mathrm{lnf}$ then either α is atomic or $v = \lambda x.v'$.

Definition 9. *A matching equation is a pair (t, u), written $t \overset{?}{=} u$, such that $U(u) = \emptyset$ and there is α such that $t, u \in \mathrm{lnf}^\alpha$. The matching equation $t \overset{?}{=} u$ has a solution if there is a substitution σ with the following properties:*

1. *the domain of σ is equal to $U(t)$ and $t.\sigma \in \mathcal{L}\Lambda$*
2. *$t.\sigma =_{\beta\eta} u$*

We say that a matching equation is linear whenever each unknown has at most one occurrence in t.

3   Syntactic Descriptions

In this section we introduce the main conceptual tool of this paper, syntactic descriptions. Syntactic descriptions make the problem of checking whether a certain linear λ-term has a certain normal form become a problem of type-checking. Consequently syntactic descriptions formalize how to build a linear λ-term which must have a certain normal form. As solving a matching equation $t \stackrel{?}{=} u$ consists in finding a substitution σ such that $t.\sigma$ normalizes to u, syntactic descriptions transform that problem in finding a substitution σ such that $t.\sigma$ respects the construction rules to make $t.\sigma$ be $\beta\eta$-convertible to u.

Technically syntactic descriptions may be seen as a type system for linear λ-calculus. Thus, in a certain way, syntactic descriptions transform a matching problem in a proof search problem. We now present the system and the properties which relate it to matching.

Definition 10. *Given $u \in \mathrm{lnf}$, we define the set of u-descriptions of type α, $\mathcal{D}_u^\alpha$ as follows:*

1. *if α is atomic then $\mathcal{D}_u^\alpha = \mathcal{S}_u^\alpha$*
2. *if $d \in \mathcal{D}_u^\alpha$ and $e \in \mathcal{D}_u^\beta$ then $d \multimap e \in \mathcal{D}_u^{\alpha \multimap \beta}$*

The set of all descriptions built on u is denoted $\mathcal{D}_u$.

Remark that given a type α the size of $\mathcal{D}_u^\alpha$ is in $\mathcal{O}(|u|^{|\alpha|})$.

For typing terms with syntactic descriptions, we will use the following typing contexts.

Definition 11. *Given $u \in \mathrm{lnf}$, then a u-context is a finite subset, Γ, of $\mathcal{V} \times \mathcal{D}_u$ such that:*

1. *if $(x^\alpha, d) \in \Gamma$ then $d \in \mathcal{D}_u^\alpha$*
2. *if $(x^\alpha, d) \in \Gamma$ and $(x^\alpha, e) \in \Gamma$ then $d = e$.*

Given a u-context $\Gamma = \{(x_1^{\alpha_1}, d_1); \ldots ; (x_n^{\alpha_n}, d_n)\}$ we will denote Γ by $x_1^{\alpha_1} : d_1, \ldots, x_n^{\alpha_n} : d_n$. If Γ and Δ are two u-contexts such that $\Gamma \cap \Delta = \emptyset$ and $\Gamma \cup \Delta$ is a u-context, then we write Γ, Δ for the context $\Gamma \cup \Delta$. We may also associate descriptions to unknowns in u-context when they occurre linearly in a term.

Some particular descriptions are needed both to type constants with descriptions and to characterize the terms which are $\beta\eta$-convertible to subterms of u.

Definition 12. *Given $u \in \mathrm{lnf}$ and $(C[], t) \in \mathcal{S}_u$, we define $\theta(C[], t)$ as follows:*

1. *if $C[] = C'[[]t']$ then $\theta(C[], t) = \theta(C'[t[]], t') \multimap \theta(C'[], tt')$*
2. *if $t = \lambda x.t'$ then $\theta(C[], t) = \theta(C[\lambda x.C_{t',x}[]], x) \multimap \theta(C[\lambda x.[]], t')$*
3. *$\theta(C[], t) = (C[], t)$ otherwise*

This definition is not ambiguous because if $(C[], t) \in \mathcal{S}_u$, $C[t]$ is a linear λ-term in normal form. Firstly if $C[] = C'[[]t']$, it is not possible to have $t = \lambda x.t''$ otherwise we would have that $u = C'[(\lambda x.t'')t']$ is not in normal form. Secondly when

$t = \lambda x.t'$, having $C[] = C'[[]t'']$ will also contradict that $u \in \mathrm{lnf}$. Furthermore, it is easy to check that, by induction on the type of t that the definition of $\theta(C[],t)$ does not loop. Moreover, the last case can only arise when t has an atomic type and therefore this last case eventually produces a syntactic description which obeys the definition we have given. Thus, a trivial induction proves that if $(C[],t) \in \mathcal{S}_u$, then $\theta(C[],t) \in \mathcal{D}_u$.

In the following, given $u \in \mathrm{lnf}$, $(C[],t) \in \mathcal{S}_u$ such that $FV(t) = \{x_1, \ldots, x_n\}$ we write $\varPhi(C[],t)$ for the context:

$$x_1 : \theta(C[C_{t,x_1}[]], x_1), \ldots, x_n : \theta(C[C_{t,x_n}[]], x_n).$$

As it will be shown by lemma 8, the context $\varPhi(C[],t)$ must be seen as the context in which t can be typed by the description $\theta(C[],t)$. If $S = \{x_1, \ldots, x_n\}$ is a subset of $FV(u)$, we then write $\varPhi(u,S)$ for the u-context

$$x_1 : \theta(C_{u,x_1}[], x_1), \ldots, x_n : \theta(C_{u,x_n}[], x_n).$$

The typing rules of syntactic descriptions are given by the following rules:

$$\frac{d \in \mathcal{D}_u^\alpha}{u; x^\alpha : d \vdash x^\alpha : d} \text{ Axiom} \qquad \frac{(C[],a) \in \mathcal{S}_u}{u; \ \vdash a : \theta(C[],a)} \text{ Constant}$$

$$\frac{u; \Gamma, x : d \vdash t : e}{u; \Gamma \vdash \lambda x.t : d \multimap e} \lambda\text{-abst.} \qquad \frac{u; \Gamma_1 \vdash t_1 : d \multimap e \quad u; \Gamma_2 \vdash t_2 : d}{u; \Gamma_1, \Gamma_2 \vdash t_1 t_2 : e} \text{ App.}$$

Implicitely, we assume that the rule App can only be applied when Γ_1, Γ_2 is an actual u-context. Thus, it can easily established that all derived contexts are u-contexts, that all derived types are u-descriptions.

As shown by the next two lemmas, this system can only type linear λ-terms.

Lemma 4. *The sequent* $u; \Gamma, x^\alpha : d \vdash t : e$ *is derivable iff* $x^\alpha \in FV(t)$.

Proof. Trivial induction on the derivation of $u; \Gamma \vdash t : e$.

Lemma 5. *If* $d \in \mathcal{D}_u^\alpha$ *and* $u; \Gamma \vdash t : d$ *is derivable then* $t \in \mathcal{L}\Lambda^\alpha$.

Proof. This lemma can be proved by induction on the derivation of $u; \Gamma \vdash t : d$, but one has to use the previous lemma for the case of abstraction.

We associate a formal semantics to syntactic descriptions and we prove that the derivation system is sound with this semantics.

Definition 13. *If* $u \in \mathrm{lnf}$, *then the semantics of the elements of* $\mathcal{D}_u$ *is:*

- *if* α *is atomic and* $(C[], v) \in \mathcal{S}_u^\alpha$, $[\![(C[], v)]\!] = \{w \in \mathcal{L}\Lambda^\alpha \mid C[w] =_{\beta\eta} C[v]\}$
- *if* $d \multimap e \in \mathcal{D}_u^{\alpha \multimap \beta}$ *then* $[\![d \multimap e]\!] = \{w \in \mathcal{L}\Lambda^{\alpha \multimap \beta} \mid \forall v \in [\![d]\!].wv \in [\![e]\!]\}$

Before we prove that this semantics is correct for the system, we need show technical properties which are necessary to initiate the induction.

Lemma 6. *Given* $u \in \mathrm{lnf}$ *and* $(C[], t) \in \mathcal{S}_u^{\alpha}$, *if* $t \in \mathrm{lnf}$ *then*

$$\{w \in \mathcal{LA}^{\alpha} \mid C[w] =_{\beta\eta} C[t]\} = \{w \in \mathcal{LA}^{\alpha} \mid w =_{\beta\eta} t\}.$$

Proof. When $t \in \mathrm{lnf}$, the context $C[]$ cannot be of the form $C'[[]t']$, thus for any w the term $C[w]$ contains only the redices of w. This entails that the normal form of $C[w]$ is $C[v]$ where v is the normal form of w, and we finally get the result.

We now prove that the semantics of $\theta(C[], t)$ is the set of terms v such that $C[v] =_{\beta\eta} C[t]$. The description $\theta(C[], t)$ characterizes the terms which can replace t in the context $C[]$ without changing the normal form. This, somehow, explains the definition of $\theta(C[], t)$. For example, $\theta(C[[]t'], t) = \theta(C[t[]], t') \multimap \theta(C[]tt')$ expresses that the terms that can replace t in the context $C[[]t']$ are the one which, when applied to terms that can replace t' in $C[t[]]$, can replace tt' in the context $C[]$. The other equalities of the definition come from a similar idea.

Lemma 7. *If* $u \in \mathrm{lnf}$ *and* $(C[], t) \in \mathcal{S}_u$, *then* $[\![\theta(C[], t)]\!] = \{v \mid C[v] =_{\beta\eta} C[t]\}$.

Proof. We proceed by induction on the type of t:

Case 1: in case $C[] = C'[[]t']$ we have that $t \in \mathcal{LA}^{\alpha \multimap \beta}$ and:

$$[\![\theta(C[], t)]\!] = [\![\theta(C'[t[]], t') \multimap \theta(C'[], tt')]\!]$$
$$= \{v \in \mathcal{LA}^{\alpha \multimap \beta} \mid \forall w \in [\![\theta(C'[t[]], t')]\!].vw \in [\![(\theta C'[], tt')]\!]\}$$

By induction hypothesis and lemma 6 (because t' is in long form), we have:

$$[\![\theta(C'[t[]], t')]\!] = \{w \mid C'[tw] =_{\beta\eta} C'[tt']\} = \{w \mid w =_{\beta\eta} t'\} \tag{1}$$

$$\text{and } [\![\theta(C'[], tt')]\!] = \{w \mid C'[w] =_{\beta\eta} C'[tt']\} \tag{2}$$

If we take $w \in [\![\theta(C[], t)]\!]$ and $v \in [\![\theta(C'[t[]], t')]\!]$ we have that $wv \in [\![\theta(C'[], tt')]\!]$. So from (2) we have $C'[wv] =_{\beta\eta} C'[tt'] = C[t]$ but, from (1), $v =_{\beta\eta} t'$. Thus, we obtain $C[t] =_{\beta\eta} C'[wt'] = C[w]$. This shows that if $w \in [\![\theta(C[], t)]\!]$ then $C[w] =_{\beta\eta} C[t]$.

Conversely, given w such that $C[w] =_{\beta\eta} C[t]$ and $v \in [\![\theta(C'[t[]], t')]\!]$, from (1) we have that $v =_{\beta\eta} t'$. To prove that $w \in [\![\theta(C[], t)]\!]$, we have to show that $wv \in [\![\theta(C'[], tt')]\!]$, which is, by (2), equivalent to prove that $C'[wv] =_{\beta\eta} C'[tt']$. As $v =_{\beta\eta} t'$, $C'[wv] =_{\beta\eta} C'[wt'] = C[w]$ and we already know that $C[w] =_{\beta\eta} C[t]$. This finally entails that $[\![\theta(C[], t)]\!] = \{v \mid C[v] =_{\beta\eta} C[t]\}$.

Case 2: in case $t = \lambda x.t'$ we have that $t \in \mathcal{LA}^{\alpha \multimap \beta}$, and:

$$[\![\theta(C[], t)]\!] = [\![\theta(C[\lambda x.C_{t,x}[]], x) \multimap \theta(C[\lambda x.[]], t')]\!]$$
$$= \{w \in \mathcal{LA}^{\alpha \multimap \beta} \mid \forall v \in [\![\theta(C[\lambda x.C_{t,x}[]], x)]\!].wv \in [\![\theta(C[\lambda x.[]], t')]\!]\}$$

By induction hypothesis, the fact that t is in lnf because u is in lnf and lemma 6, we have:

$$[\![\theta(C[\lambda x.C_{t,x}[]], x)]\!] = \{w \mid C[\lambda x.C_{t,x}[w]] =_{\beta\eta} C[\lambda x.C_{t,x}[x]]\}, \tag{3}$$

$$[\![\theta(C[\lambda x.[]], t')]\!] = \{w \mid C[\lambda x.[w]] =_{\beta\eta} C[\lambda x.[t']]\} \tag{4}$$

$$\text{and } \{w \mid C[w] =_{\beta\eta} C[t]\} = \{w \mid w =_{\beta\eta} t\} \tag{5}$$

If $v \in [\![\theta(C[], t)]\!]$, then, as $x \in [\![\theta(C[\lambda x.C_{t,x}[]], x)]\!]$, we have $vx \in [\![\theta(C[\lambda x.[]], t')]\!]$ i.e, from (4), $C[\lambda x.vx] =_{\beta\eta} C[\lambda x.t'] = C[t]$. But $C[v] =_{\eta} C[\lambda x.vx]$, we then get that $C[v] =_{\beta\eta} C[t]$ and as a conclusion $[\![\theta(C[], t)]\!] \subseteq \{v \mid C[v] =_{\beta\eta} C[t]\}$.

Now given $w \in [\![\theta(C[\lambda x.C_{t,x}[]], x)]\!]$ and v such that $C[v] =_{\beta\eta} C[t]$, from (5), we have $v =_{\beta\eta} t$, we have to prove that $vw \in [\![\theta(C[\lambda x.[]], t')]\!]$. This amounts to show that $C[\lambda x.vw] =_{\beta\eta} C[t]$, but

$$C[\lambda x.(vw)] =_{\beta\eta} C[\lambda x.(tw)] = C[\lambda x.((\lambda x.C_{t,x}[x])w)] \to_{\beta} C[\lambda x.C_{t,x}[w]]$$

by (3) $C[\lambda x.C_{t,x}[w]] =_{\beta\eta} C[\lambda x.C_{t,x}[x]] = C[t]$, which finally completes the proof of $[\![\theta(C[], t)]\!] = \{v \mid C[v] =_{\beta\eta} C[t]\}$.

Case 3: in case $t \in \mathcal{L}\Lambda^{\alpha}$ with α atomic, the result is a direct consequence of the definitions.

We are now in position to prove that the derivation system we propose is sound for that semantics.

Proposition 4. (Soundness) *Given $u \in \mathsf{lnf}$, if $u; x_1 : d_1, \ldots, x_n : d_n \vdash t : d$ is derivable then for all families $(t_k)_{k \in [1,n]}$ such that $t_k \in [\![d_k]\!]$ we have $t[x_1 := t_1, \ldots, x_n := t_n] \in [\![d]\!]$.*

Proof. This lemma can easily be proved by induction on the derivation of $u; x_1 : d_1, \ldots, x_n : d_n \vdash t : d$. The only case which is not trivial is the case where one derives that $u; \ \vdash a : \theta(C[], a)$ from $(C[], a) \in \mathcal{S}_u$. This case can be solved with the previous lemma which implies $a \in [\![\theta(C[], a)]\!]$.

The derivation system for syntactic descriptions is not complete for the semantics. Indeed proving that $t \in [\![d]\!]$ may necessitate some classical reasoning and the derivation system is intrinsically intuitionnistic. This is not an issue; as we will see later, this system is powerful enough to solve matching equations. The following lemma shows that it can establish that $t \in [\![\theta(C[], t)]\!]$.

Lemma 8 (Reflexivity). *Given $u \in \mathsf{lnf}$ and $(C[], t) \in \mathcal{S}_u$, $u; \Phi(C[], t) \vdash t : \theta(C[], t)$ is derivable.*

Proof. This lemma is proved by a trivial induction on the structure of t.

We can now prove the derivation system is able to establish $v \in [\![\theta(C[], t)]\!]$ whenever $v =_{\beta\eta} t$. This further result relies on the usual properties of the linear λ-calculus we emphasized in the previous section.

Lemma 9 (Substitution). *If $u; \Gamma, x^{\alpha} : d \vdash t : e$ and $u; \Delta \vdash t' : d$ are derivable, $\Gamma \cap \Delta = \emptyset$ and $\Gamma \cup \Delta$ is a u-context, then $u; \Gamma, \Delta \vdash t[x^{\alpha} := t'] : e$ is derivable.*

Proof. Induction on the derivation of $u; \Gamma, x^\alpha : d \vdash t : e$.

Lemma 10 (Extraction). *Given* $t, t' \in \mathcal{L}\Lambda$, *if* $u; \Gamma \vdash t[x^\alpha := t'] : e$ *is derivable and* $x^\alpha \in FV(t)$ *then there is a description* d *and two* u-*contexts* Γ_1 *and* Γ_2 *such that,* $\Gamma = \Gamma_1, \Gamma_2$ *and* $u; \Gamma_1, x^\alpha : d \vdash t : e$ *and* $u; \Gamma_2 \vdash t' : d$ *are derivable.*

Proof. Induction on the derivation of $u; \Gamma \vdash t[x^\alpha := t'] : e$.

Proposition 5 (Subject Reduction). *Given* $t, t' \in \mathcal{L}\Lambda$, *if* $u; \Gamma \vdash t : d$ *is derivable and* $t \twoheadrightarrow_{\beta\eta} t'$, *then* $u; \Gamma \vdash t' : d$ *is derivable.*

Proof. This proposition can be proved with the usual techniques.

Proposition 6 (Subject Expansion). *Given* $t, t' \in \mathcal{L}\Lambda$, *if* $u; \Gamma \vdash t' : d$ *is derivable and* $t \twoheadrightarrow_{\beta\eta} t'$, *then* $u; \Gamma \vdash t : d$ *is derivable.*

Proof. The η-expansion is trivial, concerning β-expansion we need two steps: 1. one shows by induction on the structure of $C[]$ that if $u; \Gamma \vdash C[t[x := t']] : d$ is derivable then $u; \Gamma \vdash C[(\lambda x.t)t'] : d$ is derivable. The basic case can be proved with the extraction lemma. 2. it suffices to iterate the previous result to obtain the proposition.

The subject reduction property, the subject expansion property and the reflexivity lemma are the key properties for reducing the normalization problem to the type-checking problem.

Proposition 7. *Given* $u \in \mathsf{lnf}$ *and* $(C[], t) \in \mathcal{S}_u$, *if* $t \in \mathsf{lnf}$, *then* $v =_{\beta\eta} t$ *if and only if* $u; \Phi(C[], t) \vdash v : \theta(C[], t)$ *is derivable.*

Proof. The *only if* part of the equivalence can be obtained by using the reflexivity lemma, the subject expansion property and the subject reduction property. The *if* part uses the soundness property and the lemma 6.

We can now express in terms of syntactic descriptions what it means to solve a matching equation.

Proposition 8. *Given a linear matching equation* $t \overset{?}{=} u$, $(S_k)_{k \in [1,n]}$, $(t_k)_{k \in [1,n]}$ *with* $U(t) = \{\mathbf{X}_1^{\alpha_1}; \ldots; \mathbf{X}_n^{\alpha_n}\}$, $FV(t)$, $S_1, \ldots, S_n$ *is a partition of* $FV(u)$ *then:*

1. *for all* $k \in [1,n]$, $FV(t_k) = S_k$ *and* $t_k \in \mathcal{L}\Lambda^{\alpha_k}$
2. $t[\mathbf{X}_1^{\alpha_1} := t_1, \ldots, \mathbf{X}_n^{\alpha_n} := t_n] =_{\beta\eta} u$

if and only if there is a family $(d_k)_{k \in [1,n]}$ *such that*

1. *for all* $k \in [1,n]$, $d_k \in \mathcal{D}_u^{\alpha_k}$ *and* $u; \Phi(u, S_k) \vdash t_k : d_k$ *is derivable*
2. $u; \mathbf{X}_1^{\alpha_1} : d_1, \ldots, \mathbf{X}_n^{\alpha_n} : d_n, \Phi(u, FV(t)) \vdash t : \theta([], u)$ *is derivable*

Proof. The first part of the equivalence is proved using the previous proposition and the extraction lemma. The second part is proved using the substitution lemma and the previous proposition.

As a consequence of the previous proposition, solving a linear matching equation $t \overset{?}{=} u$ where $U(t) = \{\mathbf{X}_1^{\alpha_1}, \ldots, \mathbf{X}_n^{\alpha_n}\}$ can be seen as:

1. partionning $FV(u) \backslash FV(t)$ within $S_1, \ldots, S_n$
2. choosing descriptions $d_1 \in \mathcal{D}_u^{\alpha_1}, \ldots, d_n \in \mathcal{D}_u^{\alpha_n}$
3. verifying that $u; \mathbf{X}_1^{\alpha_1} : d_1, \ldots, \mathbf{X}_n^{\alpha_n} : d_n, \Phi(u, FV(t)) \vdash t : \theta([], u)$ is derivable
4. proving that for all k there is t_k such that $u; \Phi(u, S_k) \vdash t_k : d_k$ is derivable

The first step of the procedure gives rise to finitely many possibilities. Since we must have $d_k \in \mathcal{D}_u^{\alpha_k}$, the second step also gives rise to finitely many possibilities. The third step only amounts to type-checking and can be performed in linear time in the size of t. The fourth step amounts to proof-search. With an analogy with implicative linear logic one can easily see that this step is decidable[1]. This entails the decidability of the problem of solving linear matching in the linear λ-calculus. This can easily be extended to the case of matching in the linear λ-calculus by remarking that such a matching equation can be considered as a linear matching equation together with some equality constraints between the terms substituted to the unknowns. This gives an alternative proof to the decidability of matching in the linear λ-calculus and if one looks more carefully at each step, then it is easy to see the problem is in NP.

4   Using Syntactic Descriptions to Solve Linear Matching Equations

The algorithm we propose uses tabular techniques. Given an equation $t \overset{?}{=} u$ it memorises triplets of the form $(\Gamma; (C[], v); d)$ where Γ is a u-context, $(C[], v) \in \mathcal{S}_t^\alpha$ and $d \in \mathcal{D}_u^\alpha$. Such triplets are called *partial solutions* of $t \overset{?}{=} u$. Given a partial solution $(\Gamma; (C[], v); d)$ with $U(v) = \{\mathbf{X}_1^{\alpha_1}; \ldots; \mathbf{X}_n^{\alpha_n}\}$, it represents the existence of descriptions $d_1, \ldots, d_n$ such that $d_i \in \mathcal{D}_u^{\alpha_i}$, $u; \Gamma, \mathbf{X}_1^{\alpha_1} : d_1, \ldots, \mathbf{X}_n^{\alpha_n} : d_n \vdash v : d$ is derivable, and there is $\Delta_i \subseteq \Phi([], u)$ and t_i such that $u; \Delta_i \vdash t_i : d_i$ is derivable.

In what follows, we suppose we are given a linear equation $t \overset{?}{=} u$ we are trying to solve. The algorithm and its properties are expressed with respect to that equation.

As a subroutine, this algorithm uses a proof-search procedure that we don't describe here. It acts on a set M of partial solutions with the five following rules:

1. if $(C[], \mathbf{X}v_1 \ldots v_n) \in \mathcal{S}_t^\alpha$ and the following hypothesis hold:
   (a) α is an atomic type,
   (b) for all $i \in [1, n]$, $(\Gamma_i; (C[\mathbf{X}v_1 \ldots v_{i-1}[] \ldots v_n], v_i); d_i) \in M$
   (c) there is $(C'[], v) \in \mathcal{S}_u^\alpha$ and there is $\Delta \subseteq \Phi([], u)$ and w such that $u; \Delta \vdash w : d_1 \multimap \ldots \multimap d_n \multimap (C'[], v)$ is derivable

---

[1] The exponential character of the constant is harmless since their type are somehow monotonous, *i.e.* the conclusions of the arguments are strict subterms of the conclusion of the type itself.

then:
$$M \to_{\mathcal{M}} M \cup \{(\Gamma_1, \ldots, \Gamma_n; (C[], \mathbf{X}^\beta v_1 \ldots v_n); (C'[], v))\}$$

2. if $(C[], x v_1 \ldots v_n) \in \mathcal{S}_t^\alpha$, and the following hypothesis hold:
   (a) α is an atomic type,
   (b) $C[]$ binds x
   (c) for all $i \in [1, n]$, $(\Gamma_i; (C[x v_1 \ldots v_{i-1}[] \ldots v_n], v_i); d_i) \in M$
   (d) $(C'[], v) \in \mathcal{S}_u^\alpha$ and $d = d_1 \multimap \cdots \multimap d_n \multimap (C'[], v)$
   then:

$$M \to_{\mathcal{M}} M \cup \{(\Gamma_1, \ldots, \Gamma_n, x : d; (C[], x v_1 \ldots v_n); (C'[], v))\}$$

3. if $(C[\lambda x.[]], v) \in \mathcal{S}_t$ and $(\Gamma, x : d; (C[\lambda x.[]], v); e) \in M$ then:

$$M \to_{\mathcal{M}} M \cup \{(\Gamma; (C[], \lambda x.v); d \multimap e)\}$$

4. if $(C[], x v_1 \ldots v_n) \in \mathcal{S}_t^\alpha$, and the following hypothesis hold:
   (a) α is an atomic type,
   (b) $C[]$ does not bind x
   (c) for all $i \in [1, n]$, $(\Gamma_i; (C[x v_1 \ldots v_{i-1}[] \ldots v_n], v_i); d_i) \in M$
   (d) $x \in FV(u) \cap FV(t)$ and $\theta(C_{u,x}[], x) = d_1 \multimap \cdots \multimap d_n \multimap (C'[], v)$
   then:

$$M \to_{\mathcal{M}} M \cup \{(\Gamma_1, \ldots, \Gamma_n, x : \theta(C_{u,x}[], x); (C[], x v_1 \ldots v_n); (C'[], v))\}$$

5. if $(C[], a v_1 \ldots v_n) \in \mathcal{S}_t^\alpha$, and the following hypothesis hold:
   (a) α is an atomic type,
   (b) for all $i \in [1, n]$, $(\Gamma_i; (C[a v_1 \ldots v_{i-1}[] \ldots v_n], v_i); d_i) \in M$
   (c) there is $(C_a[], a) \in \mathcal{S}_u$ such that $\theta(C_a[], a) = d_1 \multimap \cdots \multimap d_n \multimap (C'[], v)$
   then:
$$M \to_{\mathcal{M}} M \cup \{(\Gamma_1, \ldots, \Gamma_n; (C[], a v_1 \ldots v_n); (C'[], v))\}$$

The proof-search procedure is only needed for the first rule. In a first time, such a procedure can be implemented as a slight modification of a proof-search procedure for intuitionnistic implicative linear logic (**IILL**).

Lemma 11. *If $M \overset{*}{\to}_{\mathcal{M}} M_1$ and $M \overset{*}{\to}_{\mathcal{M}} M_2$ then $M \overset{*}{\to}_{\mathcal{M}} M_1 \cup M_2$.*

Proof. This is a simple consequence the memorisation technique.

Lemma 12 (Correctness). *If $\emptyset \overset{p}{\to}_{\mathcal{M}} M$ then for all $(\Gamma; (C[], v); d) \in M$ with $U(v) = \{\mathbf{X}_1^{\alpha_1}; \ldots; \mathbf{X}_n^{\alpha_n}\}$, there is $(\Delta_i)_{i \in [1,n]}$, $(t_i)_{i \in [1,n]}$ and $(d_i)_{i \in [1,n]}$ such that:*

1. *$u; \Gamma, \mathbf{X}_1^{\alpha_1} : d_1, \ldots, \mathbf{X}_n^{\alpha_n} : d_n \vdash v : d$ is derivable*
2. *$u; \Delta_i \vdash t_i : d_i$ is derivable for all $i \in [1, n]$*

Proof. This proof is a simple induction on p where one only has to check that the five different cases of the algorithm respect the derivation rules of syntactic descriptions.

Lemma 13 (Completeness). *Given* $(C[], v) \in \mathcal{S}_t^\alpha$ *such that* $v \in \ln f$, *if* $U(v) = \{\mathbf{X}_1^{\alpha_1}; \ldots; \mathbf{X}_n^{\alpha_n}\}$ *and if there is* $d \in \mathcal{D}_u^\alpha$, *a u-context* Γ, $(\Delta_i)_{i \in [1,n]}$, $(t_i)_{i \in [1,n]}$ *and* $(d_i)_{i \in [1,n]}$ *such that:*

1. *if* $x : e \in \Gamma$ *and* x *is not bound by* $C[]$ *then* $x \in FV(u)$ *and* $e = \theta(C_{u,x}[], x)$
2. $u; \Gamma, \mathbf{X}_1^{\alpha_1} : d_1, \ldots, \mathbf{X}_n^{\alpha_n} : d_n \vdash v : d$ *is derivable*
3. $\Delta_i \subseteq \Phi([], u)$ *and* $u; \Delta_i \vdash t_i : d_i$ *is derivable for all* $i \in [1,n]$

then $\emptyset \xrightarrow{*}_\mathcal{M} M$ *with* $(\Gamma; (C[], v); d) \in M$.

Proof. We proceed by induction on the structure of v. There are five cases:

Case 1: in case $v = \mathbf{X}^\beta v_1 \ldots v_n$, then, because $v \in \ln f$, v has an atomic type and d must be equal to some $(C'[], w) \in \mathcal{S}_u^\alpha$. By hypothesis, for all $\mathbf{Y}^\gamma \in U(v)$ we have $\Delta_{\mathbf{Y}^\gamma} \subseteq \Phi([], u)$, $t_{\mathbf{Y}^\gamma} \in \mathcal{L}\Lambda^\gamma$ and $d_{\mathbf{Y}^\gamma} \in \mathcal{D}_u^\gamma$ such that $u; \Delta_{\mathbf{Y}^\gamma} \vdash t_{\mathbf{Y}^\gamma} : d_{\mathbf{Y}^\gamma}$ is derivable. Now let $\Gamma_i' = \{\mathbf{Y}^\gamma : d_{\mathbf{Y}^\gamma} \mid \mathbf{Y}^\gamma \in U(v_i)\}$, by hypothesis, we have:

$$u; \Gamma, \Gamma_1', \ldots, \Gamma_n', \mathbf{X}^\beta : d_{\mathbf{X}^\beta} \vdash \mathbf{X}^\beta v_1 \ldots v_n : (C'[], w)$$

We must have $d_{\mathbf{X}^\beta} = d_1 \multimap \cdots \multimap d_n \multimap (C'[], v)$ for some $d_1, \ldots, d_n$. Therefore there must be $(\Gamma_i)_{i \in [1,n]}$ such that $\Gamma = \Gamma_1, \ldots, \Gamma_n$ and for all $i \in [1,n]$ the sequent $u; \Gamma_i, \Gamma_i' \vdash v_i : d_i$ is derivable. Then by induction hypothesis, for all $i \in [1,n]$, $\emptyset \xrightarrow{*}_\mathcal{M} M_i$ such that $(\Gamma_i; (C_i[], v_i); d_i) \in M_i$. By lemma 11, this entails that

$$\emptyset \xrightarrow{*}_\mathcal{M} M' \text{ where } M' = \bigcup_{i=1}^{n} M_i$$

and for all $i \in [1,n]$, $(\Gamma_i; (C_i[], v_i); d_i) \in M'$. We then have $M' \rightarrow_\mathcal{M} M' \cup \{(\Gamma; (C[], v); (C'[], w))\}$ by using the first rule.

Case 2: $v = x v_1 \ldots v_n$ and x is bound by $C[]$, then Γ must be of the form $\Gamma', x : d_1 \multimap \cdots \multimap d_n \multimap (C'[], w)$ and d must be equal to $(C'[], w)$. We can use the induction hypothesis in a similar way as in the previous case and then conclude by using the second rule.

Case 3: $v = \lambda x.v'$, in that case, $d = d_1 \multimap d_2$ and $u; \Gamma \vdash \lambda x.v' : d_1 \multimap d_2$ is derivable if and only if $u; \Gamma, x : d_1 \vdash v' : d_2$ is derivable. We can then use the induction hypothesis and conclude with the third rule.

Case 4: $v = x v_1 \ldots v_n$ and x is not bound by $C[]$, in that case Γ must be of the form $\Gamma', x : \theta(C_{u,x}[], x)$ there and if $\theta(C_{u,x}[], x) = d_1 \multimap \cdots \multimap d_n \multimap (C'[], w)$ then $d = (C'[], w)$. We can use the induction hypothesis as in the first case and then conclude with the fourth rule.

Case 5: $v = a v_1 \ldots v_n$, in that case we must have $d = (C'[], a w_1 \ldots w_n)$ and therefore to derive the sequent $u; \Gamma, \mathbf{X}_1^{\alpha_1} : d_1, \ldots, \mathbf{X}_n^{\alpha_n} : d_n \vdash v : (C'[], a w_1 \ldots w_n)$ the assumption $u; \vdash a : \theta(C'[[]w_1 \ldots w_n], a)$ must have been used. This allows us to conclude by induction as for case 1 but using the fifth rule.

Given $(C[], v) \in \mathcal{S}_t$, if $v \in \mathcal{L}\Lambda^{\alpha_0}$ and $\{x_1^{\alpha_1}, \ldots, x_n^{\alpha_n}\}$ is the set of free variables in v which are bound by $C[]$, then there can be $\prod_{i=0}^{n} |u|^{|\alpha_i|}$ partial solutions of the form $(\Gamma; (C[\,], v); d)$ stored in M. Furthermore, if $(C[\,], \mathbf{X}v_1 \ldots v_n) \in \mathcal{S}_t$ (*resp.* $(C[\,], xv_1 \ldots v_n) \in \mathcal{S}_t$) and if v_i can be in at most n_{v_i} partial solution then the rule concerning the unknown (*resp.* variable) can be used $|u| \times \prod_{i=1}^{n} n_{v_i}$ times.

Therefore if for all $(C[\,], v) \in \mathcal{S}_t$ such that $v \in \mathcal{L}\Lambda^{\alpha_0}$ and $\{x_1^{\alpha_1}, \ldots, x_n^{\alpha_n}\}$ is the set of free variables in v which are bound by $C[\,]$, we write $\rho(C[\,], v) = \sum_{i=0}^{n} |\alpha_i|$. Let $p = \max_{(C[\,], v) \in \mathcal{S}_t} (\rho(C[\,], v))$ and q be the maximal arity of the types of variables or unknowns in t, then the algorithm runs in $\mathcal{O}(|u|^{p \times q + 1})$ in time (without taking proof-search for descriptions into account) and $\mathcal{O}(|u|^p)$ in space.

Of course as soon as p and q become big, this algorithm becomes intractable. Nevertheless, in practice, for natural language processing, p and q are rather small, and this algorithm can give very quickly refutation when an equation has no solution whereas the adaptation of huet's algorithm proposed by de Groote will have to try all the possible substitutions before ending. It is also possible to add heuristic to its behavior so that it can find solutions faster.

In particular, it is possible to show that only particular descriptions can be used in partial solutions in order to obtain a solution of the equation. This constraint gives a way of saving space and time. Furthermore, given such a description d, one can prove there is a unique context $\Delta \subseteq \Phi([\,], u)$ so that if there is t_d and $\Delta' \subseteq \Phi([\,], u)$ such that $u; \Delta' \vdash t_d : d$ then $\Delta = \Delta'$. This context can easily be inferred from d. Finally because of the particular form of descriptions involved in matching, the search for t_d is greatly eased by very simple heuristics (we don't know yet if this proof-search problem polynomial or NP-hard). Another way of improving the algorithm is also to determine in t which are the variable which will be substituted during reduction (active variables) and the one which won't (passive variables). Passive variables are the variables introduced because of the presence of constants and free variables common in u and t. Therefore this restricts the possible descriptions that can be associated to them. Because of the lack of space, we cannot go into all those details. We will instead emphasize how this algorithm relates linear matching in the linear λ-calculus to the linear context matching. To this end, we show how the algorithm behaves when it attempts to solve second order linear equations.

Let's consider a linear equation $t \overset{?}{=} u$ where the unknowns of t are at most second order and the constants and the free variables of u are at most second order. When applying the algorithm to this kind of equations, partial solutions can be represented as pairs $((C'[\,], v); (C[\,], w))$ because the restrictions on the equation entail that the context will always be empty and that at any time the description will be atomic. The rules that will be used are only the first, the fourth and the fifth one. For the first one, we will need to find whether there is a $\Delta \subseteq \Phi([\,], u)$ and a term t such that $u; \Delta \vdash t : (C_1[\,], v_1) \multimap \cdots \multimap (C_n[\,], v_n) \multimap (C_0[\,], v_0)$ is derivable. One can easily see that this is the case if and only if:

1. for all $i \in [1, n]$ $(C_i[\,], v_i) \prec (C_0[\,], v_0)$
2. for all $i, j \in [1, n], i \neq j$ implies $(C_i[\,], v_i) \prec\succ (C_j[\,], v_j)$

Therefore rephrasing the rules of the algorithm for that particular case gives the three following rules:

1. if $(C[], \mathbf{X}v_1 \ldots v_n) \in \mathcal{S}_t^\alpha$, and the following hold:
   (a) α is an atomic type
   (b) for all $i \in [1, n]$, there is $((C[\mathbf{X}v_1 \ldots v_{i-1}[] \ldots v_n], v_i), (C_i'[], w_i)) \in M$
       such that if $i \neq j$ $C_i'[] \prec\!\!\succ C_j'[]$.
   (c) $(C'[], w) \in \mathcal{S}_u^\alpha$ is such that for all $i \in [1, n]$ $C_i'[] \prec C'[]$
       then $M \to_{\mathcal{M}} M \cup \{(C[], \mathbf{X}v_1 \ldots v_n), (C'[], w))\}$
2. if $(C[], xv_1 \ldots v_n) \in \mathcal{S}_t^\alpha$ and $(C'[], xw_1 \ldots w_n) \in \mathcal{S}_u^\alpha$ and for all $i \in [1, n]$,

$$((C[xv_1 \ldots v_{i-1}[] \ldots v_n], v_i), (C'[xw_1 \ldots w_{i-1}[] \ldots w_n], w_i)) \in M$$

then $M \to_{\mathcal{M}} M \cup \{((C[], xv_1 \ldots v_n), (C'[], xw_1 \ldots w_n))\}$
3. if $(C[], av_1 \ldots v_n) \in \mathcal{S}_t^\alpha$ and $(C'[], aw_1 \ldots w_n) \in \mathcal{S}_u^\alpha$ and for all $i \in [1, n]$,

$$((C[av_1 \ldots v_{i-1}[] \ldots v_n], v_i), (C'[aw_1 \ldots w_{i-1}[] \ldots w_n], w_i)) \in M$$

then $M \to_{\mathcal{M}} M \cup \{((C[], av_1 \ldots v_n), (C'[], aw_1 \ldots w_n))\}$

In the second order case, the algorithm associates subterms of t to subterms of u in M as soon as there is a substitution which unifies those subterms. This is the principle used by the algorithm proposed by Schmidt-Schauß and Stuber [12] in order to prove the polynomiality of linear context matching. What makes linear matching in the linear λ-calculus be NP-complete at second order while linear context matching is polynomial is that the condition (b) of the first rule is more restrictive in the case of linear context matching. Together with the fact that $i \neq j$ implies $C_i'[] \prec\!\!\succ C_j'[]$, there is a further constraint which imposes that $i < j$ implies that $C_i'[]$ is on the left of $C_j'[]$. This last ordering condition makes condition (b) be checkable in linear time for linear context matching whereas its complexity has n as exponent in the case of linear second order matching in the linear λ-calculus.

5   Conclusion

In this paper, we have presented a new approach for coping with higher order matching in the linear λ-calculus. This approach gives some precise theoretical understanding of matching. The underlying ideas can be generalized in order to study many matching problems. Firstly, the fact that we have focused the linear matching problem can be easily overcome by seeing non-linear equations as linear equations with some further equality constraints on the solutions. Secondly the restriction to the linear λ-calculus can also be overcome. For example, this approach can be used for matching in the calculus which represents proofs in the Lambek system [7] or in other substructural logics. It can also be adapted to higher order $\beta\eta$ matching in the simply typed λ-calculus by using some intersection types. Thus this approach may find some natural application for pattern matching in XML documents.

We used the syntactic descriptions to design an algorithm for linear matching equations in the linear λ-calculus. This algorithm generalizes an efficient approach for linear context matching equations. This algorithm can be really fine tuned by including some simple heuristics and by proving some basic properties on syntactic descriptions. The proof-search algorithm which is needed, we could not present within the space of this article, is also simple and can solve the problem of the emptiness of descriptions efficiently. A further use of syntactic descriptions [10] we made is for parsing second order Abstract Categorial Gammars [3]. The algorithm we obtain is very efficient and general. It is proved to be as efficient as the best known algorithms for parsing.

References

1. Henk P. Barendregt. *The Lambda Calculus: Its Syntax and Semantics*, volume 103. Studies in Logic and the Foundations of Mathematics, North-Holland Amsterdam, 1984. revised edition.
2. Philippe de Groote. Higher-order linear matching is np-complete. In *proc of the 11th International Conference on Rewriting Techniques and Applications*, Lecture Notes in Computer Science, pages 127–140. RTA, 2000.
3. Philippe de Groote. Towards abstract categorial grammars. In Association for Computational Linguistic, editor, *Proceedings 39th Annual Meeting and 10th Conference of the European Chapter*, pages 148–155. Morgan Kaufmann Publishers, 2001.
4. Gilles Dowek. Third order matching is decidable. *Annals of Pure and Applied Logic*, 69:135–155, 1994.
5. Gérard Huet. *Résolution d'équations dans des langages d'ordre 1,2,...,ω*. Thèse de doctorat es sciences mathématiques, Université Paris VII, 1976.
6. Gérard Huet. The zipper. *J. Funct. Program.*, 7(5):549–554, 1997.
7. Joachim Lambek. The mathematics of sentence structure. *American Mathematical Monthly*, 65:154–170, 1958.
8. Ralph Loader. Higher order β matching is undecidable. *Logic Journal of the IGPL*, 11(1):51–68, 2002.
9. Vincent Padovani. *Filtrage d'odre supérieur*. Thèse de doctorat, Université de Paris 7, 1994.
10. Sylvain Salvati. *Problèmes de filtrage et problèmes d'analyse pour les grammaires catégorielles abstraites*. PhD thesis, Institut National Polytechnique de Lorraine, 2005.
11. Sylvain Salvati and Philippe de Groote. On the complexity of higher-order matching in the linear lambda-calculus. In R. Nieuwenhuis, editor, *Rewriting Techniques and Applications, 14th International Conference, RTA-03*, LNCS 2706, pages 234–245, 2003.
12. Manfred Schmidt-Schauß and Jürgen Stuber. On the complexity of linear and stratified context matching problems. *Theory of Computing Systems*, 37:717–740, 2004.
13. Ryo Yoshinaka. Higher-order matching in the linear lambda calculus in the absence of constants is np-complete. In *proc of the 16th International Conference on Rewriting Techniques and Applications*, Lecture Notes in Computer Science. RTA, 2005.

A Terminating and Confluent Linear Lambda Calculus

Yo Ohta and Masahito Hasegawa

Research Institute for Mathematical Sciences, Kyoto University
Kyoto 606-8502, Japan

Abstract. We present a rewriting system for the linear lambda calculus corresponding to the $\{!, \multimap\}$-fragment of intuitionistic linear logic. This rewriting system is shown to be strongly normalizing, and Church-Rosser modulo the trivial commuting conversion. Thus it provides a simple decision method for the equational theory of the linear lambda calculus. As an application we prove the strong normalization of the simply typed computational lambda calculus by giving a reduction-preserving translation into the linear lambda calculus.

1 Introduction

In the literature, there exist many proposals of linearly typed lambda calculi which correspond to Girard's linear logic [7] via the Curry-Howard correspondence. However, only a few of them have studied the equality between terms (or proofs) seriously. Just like the simply typed lambda calculus with the $\beta\eta$-equality is sound and complete for semantic models given by cartesian closed categories [13,5], it is desirable for a linear lambda calculus to be equipped with an equational theory which is sound and complete for the now well-established categorical models of linear logic [19,3,4,16].

Barber and Plotkin's *Dual Intuitionistic Linear Logic (DILL)* [1,2] is one of such calculi: its equational theory, determined by the standard $\beta\eta$-axioms and a few axioms for commuting conversions (for identifying the terms representing the same proof modulo trivial proof permutations), has been shown to be sound and complete for the categorical models of the multiplicative exponential fragment of the intuitionistic linear logic. Together with its natural-deduction style simple term expressions, DILL can be considered as one of the canonical calculi for this fragment of linear logic.

However, DILL is not equipped with a rewriting system. There is a symmetric un-orientable axiom for commuting conversions, thus it is not clear if the equational theory of DILL has a simple decision procedure based on a rewriting system, while it is the case for many of the standard typed lambda calculi.

Regarding decidability, the answer is actually known: Barber [1] in his PhD thesis, and independently Ghani [6] in an unpublished manuscript, have shown that the equational theory of DILL is decidable. However, their proofs are long and complicated, using some new notations and/or advanced techniques which

F. Pfenning (Ed.): RTA 2006, LNCS 4098, pp. 166–180, 2006.

are not always easy to follow. Barber's approach involves a translation into a net-like system and rewriting on equivalence classes of expressions. Ghani have used the η-expansion technique which again is a rather heavy machinery. At least, they do not present a simple and intuitively understandable rewriting system in the traditional sense.

Here we propose a simpler solution for the $\{-\circ, !\}$-fragment (which is enough to mimic the simply typed lambda calculus via Girard's translation as $\sigma \to \tau = !\sigma -\circ \tau$) by a classical rewriting-theoretic method. Specifically, we appeal to the seminal result by Huet on *reduction modulo equivalence* [12]. We provide a rewriting system $\succ$ together with a (trivially) decidable equational theory $\sim^*$ generated by the symmetric commuting conversion $\sim$ on linear lambda terms such that (following the terminology of Terese [20])

1. The equivalence relation generated from $\succ$ and $\sim$ agrees with the equational theory of the linear lambda calculus,
2. $\succ$ is strongly normalizing,
3. $\succ$ is locally confluent modulo $\sim^*$, and
4. $\succ$ is locally coherent modulo $\sim^*$.

Then Huet's theorem implies that $\succ$ is Church-Rosser modulo $\sim^*$, and deciding the equality in this linear lambda calculus is reduced to comparing the $\succ$-normal forms up to the easily decidable equality $\sim^*$.

From rewriting-theoretical point of view, this work does not present much new idea. However, it does give an interesting case motivated by the study on the semantic and logical foundations of functional programming languages. Recent work [8,9,10] suggest that there exist many interesting translations of various calculi into this linear lambda calculus, including monadic and CPS translations. As an interesting example, we prove the strong normalization of the simply typed version of Moggi's computational lambda calculus by giving a reduction-preserving translation into the linear lambda calculus. Together with this result, our work can be considered as a follow-up of the work by Maraist et al. [15] and Sabry and Wadler [18].

The rest of this paper is organized as follows. We introduce the linear lambda calculus in Section 2, and our rewriting system in Section 3. Section 4 is a quick reminder of the classical definitions and result from the theory of reduction modulo equivalence. Section 5, 6 and 7 are devoted to show the strong normalization, local confluence modulo $\sim^*$, and local coherence modulo $\sim^*$, which jointly imply the Church-Rosser property modulo equivalence. Section 8 gives a reduction-preserving translation from the simply typed computational lambda calculus to the linear lambda calculus. Some concluding remarks are given in Section 9.

2   The Linear Lambda Calculus with $-\circ$ and !

The calculus to be considered below is a dual-context natural deduction system for the $\{!, -\circ\}$-fragment of IMELL, based on DILL of Barber and Plotkin [1,2].

The identical calculus appears in [8]. In this formulation of the linear lambda calculus, a typing judgement takes the form Γ ; $\Delta \vdash M : \tau$ in which Γ represents an intuitionistic (or additive) context whereas Δ is a linear (multiplicative) context. We assume that all variables in Γ and Δ are distinct. While the variables in Γ can be used in the term M as many times as we like, those in Δ must be used exactly once. A typing judgement $x_1 : \sigma_1, \ldots, x_m : \sigma_m$; $y_1 : \tau_1, \ldots, y_n : \tau_n \vdash M : \sigma$ can be considered as the proof of the sequent $!\sigma_1, \ldots, !\sigma_m, \tau_1, \ldots, \tau_n \vdash \sigma$, or the proposition $!\sigma_1 \otimes \ldots \otimes !\sigma_m \otimes \tau_1 \otimes \ldots \otimes \tau_n \multimap \sigma$.

Types and Terms

$$\sigma ::= b \mid \sigma \multimap \sigma \mid !\sigma$$
$$M ::= x \mid \lambda x^\sigma.M \mid M\,M \mid !M \mid \text{let } !x^\sigma \text{ be } M \text{ in } M$$

where b ranges over a set of base types. We may omit the type subscripts for ease of presentation.

Typing

$$\frac{}{\Gamma \; ; \; x : \tau \vdash x : \tau} \; \text{LinAx} \qquad\qquad \frac{}{\Gamma_1, x : \tau, \Gamma_2 \; ; \; \emptyset \vdash x : \tau} \; \text{IntAx}$$

$$\frac{\Gamma \; ; \; \Delta, x : \tau_1 \vdash M : \tau_2}{\Gamma \; ; \; \Delta \vdash \lambda x^{\tau_1}.M : \tau_1 \multimap \tau_2} \; \multimap \text{Intro} \qquad \frac{\Gamma ; \Delta_1 \vdash M : \tau_1 \multimap \tau_2 \quad \Gamma ; \Delta_2 \vdash N : \tau_1}{\Gamma \; ; \; \Delta_1 \sharp \Delta_2 \vdash MN : \tau_2} \; \multimap \text{Elim}$$

$$\frac{\Gamma \; ; \; \emptyset \vdash M : \tau}{\Gamma \; ; \; \emptyset \vdash !M : !\tau} \; !\text{Intro} \qquad \frac{\Gamma \; ; \; \Delta_1 \vdash M : !\tau_1 \quad \Gamma, x : \tau_1 \; ; \; \Delta_2 \vdash N : \tau_2}{\Gamma \; ; \; \Delta_1 \sharp \Delta_2 \vdash \text{let } !x^{\tau_1} \text{ be } M \text{ in } N : \tau_2} \; !\text{Elim}$$

where $\emptyset$ is the empty context, and $\Delta_1 \sharp \Delta_2$ is a merge of Δ_1 and Δ_2 [1,2]. Thus, $\Delta_1 \sharp \Delta_2$ represents one of possible merges of Δ_1 and Δ_2 as finite lists. More explicitly, we can define the relation "Δ is a merge of Δ_1 and Δ_2" inductively as follows [1]:

- Δ is a merge of $\emptyset$ and Δ
- Δ is a merge of Δ and $\emptyset$
- if Δ is a merge of Δ_1 and Δ_2, then $x : \sigma, \Delta$ is a merge of $x : \sigma, \Delta_1$ and Δ_2
- if Δ is a merge of Δ_1 and Δ_2, then $x : \sigma, \Delta$ is a merge of Δ_1 and $x : \sigma, \Delta_2$

We assume that, when we introduce $\Delta_1 \sharp \Delta_2$, there is no variable occurring both in Δ_1 and in Δ_2. We note that any typing judgement has a unique derivation (hence a typing judgement can be identified with its derivation).

Axioms

$$\begin{array}{llll} \beta_\multimap & (\lambda x.M)\,N & = M[N/x] \\ \eta_\multimap & \lambda x.M\,x & = M \\ \beta_! & \text{let } !x \text{ be } !M \text{ in } N & = N[M/x] \\ \eta_! & \text{let } !x \text{ be } M \text{ in } !x & = M \\ com & C[\text{let } !x \text{ be } M \text{ in } N] & = \text{let } !x \text{ be } M \text{ in } C[N] \end{array}$$

where $M[N/x]$ denotes the capture-free substitution, while $C[-]$ is a linear context (no ! binds $[-]$):

$$C ::= [-] \mid \lambda x.C \mid C\,M \mid M\,C \mid \text{let } !x \text{ be } C \text{ in } M \mid \text{let } !x \text{ be } M \text{ in } C$$

The use of linear contexts is crucial: *com* is not allowed for non-linear contexts, e.g. the "idempotency equation" [2] $!(\text{let } !x \text{ be } M \text{ in } x) = \text{let } !x \text{ be } M \text{ in } !x$ (which implies the idempotency of !, i.e., $!!\sigma \simeq !\sigma$) is not derivable. The equality judgement $\Gamma ; \Delta \vdash M = N : \sigma$, where $\Gamma ; \Delta \vdash M : \sigma$ and $\Gamma ; \Delta \vdash N : \sigma$, is defined as the congruence relation on the well-typed terms of the same type under the same typing context, generated from these axioms.

In the sequel, we work on terms up to the α-congruence. We may write $M = N$ as a shorthand for the equality judgement $\Gamma ; \Delta \vdash M = N : \sigma$, while we will use $M \equiv N$ for expressing that M and N are the same modulo α-congruence.

The axiom *com* expresses the commuting conversions. By induction on the construction of linear contexts, *com* can be expressed by five explicit instances:

Proposition 1. *The axiom com can be replaced by the following five axioms.*

com_1 (let $!x$ be M in N) L $\qquad = \text{let } !x \text{ be } M \text{ in } N L$
com_2 let $!y$ be (let $!x$ be M in N) in $L = \text{let } !x \text{ be } M \text{ in let } !y \text{ be } N \text{ in } L$
com_3 $\lambda y.(\text{let } !x \text{ be } M \text{ in } N)$ $\qquad = \text{let } !x \text{ be } M \text{ in } \lambda y.N$
com_4 L (let $!x$ be M in N) $\qquad = \text{let } !x \text{ be } M \text{ in } L N$
com_5 let $!x$ be L in let $!y$ be M in N $\quad = \text{let } !y \text{ be } M \text{ in let } !x \text{ be } L \text{ in } N$

□

Remark 1. As stated above, we only consider the equality on the well-typed terms under the same typing contexts. Thus, for example, in com_3, y cannot be free in M; and in com_5, x and y cannot be free in L and M.

Remark 2. As noted in [11], this linear lambda calculus allows a yet simpler axiomatization:

$\beta_{\multimap}$ $(\lambda x.M) N$ $\qquad = M[N/x]$
$\eta_{\multimap}$ $\lambda x.M x$ $\qquad = M$
$\beta_!$ let $!x$ be $!M$ in N $\quad = N[M/x]$
$\eta'_!$ let $!x$ be M in $L (!x) = L M$

While this is very compact, it does not immediately hint a terminating confluent rewriting system. Nevertheless, we will see later that a rewrite rule similar to this $\eta'_!$ is needed for obtaining such a rewriting system.

3   A Rewriting System for the Linear Lambda Calculus

3.1   Motivating the Rewriting Rules

Now let us derive a rewriting system for the linear lambda calculus from its axioms. As a natural starting point, we orient the $\beta\eta$-axioms from left to right, as the case of the standard $\beta\eta$ lambda calculus. The commuting conversions are tricky, however. First of all, it is not possible to orient the symmetric axiom com_5, so it needs to be treated separately. Here we follow the tradition of *reduction modulo equivalence*: we design our system so that com_5-reasoning can be postponed after all other rewriting steps are done. For $com_{1\sim4}$, it seems natural to orient the axioms so that the let-bindings are pulled outside the contexts, i.e.,

com_1 (let $!x$ be M in N) L $\succ$ let $!x$ be M in $N\,L$
com_2 let $!y$ be (let $!x$ be M in N) in $L \succ$ let $!x$ be M in let $!y$ be N in L
com_3 $\lambda y.$(let $!x$ be M in N) $\succ$ let $!x$ be M in $\lambda y.N$
com_4 L (let $!x$ be M in N) $\succ$ let $!x$ be M in $L\,N$

thus flattening the let-expressions as possible as we can. Alas, there is a problem on these rules and $\eta_!$:

- $\eta_!$ and $com_{1\sim4}$ give a non-joinable critical pair, e.g.

$$\text{let } !x \text{ be } M \text{ in } L\,(!x) \xleftarrow{com_4} L\,(\text{let } !x \text{ be } M \text{ in } !x) \xrightarrow{\eta_!} L\,M$$

- The same problem happens with com_5:

let $!x$ be M in let $!y$ be N in $!x \xleftarrow{com_5}$ let $!y$ be N in let $!x$ be M in $!x \xrightarrow{\eta_!}$ let $!y$ be N in M

To overcome this difficulty, we introduce a refined version of $\eta_!$

$$\eta'_! \quad \text{let } !x \text{ be } M \text{ in } C[!x] \succ C[M]$$

(where C ranges over the linear contexts as before) for which this problem disappears.

3.2 Rewriting System

Our rewriting system features the following rules.

$\beta_{\multimap}$	$(\lambda x.M)\,N$	$\succ M[N/x]$
$\eta_{\multimap}$	$\lambda x.M\,x$	$\succ M$
$\beta_!$	let $!x$ be $!M$ in N	$\succ N[M/x]$
$\eta'_!$	let $!x$ be M in $C[!x]$	$\succ C[M]$
com_1	(let $!x$ be M in N) L	$\succ$ let $!x$ be M in $N\,L$
com_2	let $!y$ be (let $!x$ be M in N) in L	$\succ$ let $!x$ be M in let $!y$ be N in L
com_3	$\lambda y.$(let $!x$ be M in N)	$\succ$ let $!x$ be M in $\lambda y.N$
com_4	L (let $!x$ be M in N)	$\succ$ let $!x$ be M in $L\,N$

We may use $\succ$ for the compatible relation on the well-typed terms generated by these rules (one-step rewriting), and $\succ^*$ will denote its reflexive transitive closure (many-step rewriting). We note that the com-rewriting rules can be summarized as

$$D[\text{let } !x \text{ be } M \text{ in } N] \;\succ\; \text{let } !x \text{ be } M \text{ in } D[N]$$

where $D ::= [-]\,L \mid \text{let } !y \text{ be } [-] \text{ in } L \mid \lambda y.[-] \mid L\,[-]$.

We also have to consider the symmetric rule com_5:

com_5	let $!x$ be L in let $!y$ be M in $N \;\sim\;$ let $!y$ be M in let $!x$ be L in N

We write $\sim$ for the compatible relation generated by com_5 (one-step reasoning via com_5), and $\sim^*$ for its reflexive transitive closure. A few easy facts:

Proposition 2. *The reflexive symmetric transitive closure of* $\succ \cup \sim$ *coincides with the equality of the linear lambda calculus.*   □

Proposition 3. *Each equivalence class of* $\sim^*$ *is finite, and thus* $\sim^*$ *is decidable.*
                                                                                       □

The following result, easily shown by induction, will be useful in proving the local confluence:

Lemma 1. $C[\text{let } !x \text{ be } M \text{ in } N] \succ^* \cdot \sim^* \text{let } !x \text{ be } M \text{ in } C[N]$.   □

Remark 3. In passing, we shall note that our rewriting system $\succ$ can simulate the $\beta\eta$-reduction in the simply typed lambda calculus via Girard translation [7]: types are translated as $b^\circ = b$ and $(\sigma \to \tau)^\circ =\, !\sigma^\circ \multimap \tau^\circ$, and for terms we have

$$x^\circ \equiv x$$
$$(\lambda x.M)^\circ \equiv \lambda y.\text{let } !x \text{ be } y \text{ in } M^\circ$$
$$(MN)^\circ \equiv M^\circ(!N^\circ)$$

For further details, see e.g. [8]. It is immediate to see that each $\beta\eta$-reduction in the simply typed lambda calculus is sent to non-trivial reduction in $\succ$:

$$
\begin{aligned}
((\lambda x.M)\,N)^\circ &\equiv (\lambda y.\text{let } !x \text{ be } y \text{ in } M^\circ)\,(!N^\circ) \\
&\succ \text{let } !x \text{ be } !N^\circ \text{ in } M^\circ && (\beta_\multimap) \\
&\succ M^\circ[N^\circ/x] && (\beta_!) \\
&\equiv (M[N/x])^\circ
\end{aligned}
$$

$$
\begin{aligned}
(\lambda x.M\,x)^\circ &\equiv \lambda y.\text{let } !x \text{ be } y \text{ in } M^\circ\,(!x) \\
&\succ \lambda y.M^\circ\,y && (\eta'_!) \\
&\succ M^\circ && (\eta_\multimap)
\end{aligned}
$$

4   Rewriting Modulo Equivalence

In the following sections, we will show that our rewriting system together with $\sim$ gives a decision procedure of the equality on the linear lambda terms. Fortunately, it turns out that a classical result due to Huet is directly applicable to our case. Below we recall basic definitions on reduction modulo equivalence for abstract rewriting systems (ARS's) and state Huet's theorem. We follow the treatment in Terese (Chapter 14.3) [20].

Definition 1. *Let* $(A, \to)$ *be an ARS, and* $\sim$ *be an equivalence relation on* A. *We say:*

1. a, b *are* joinable modulo $\sim$ *if there exist* c, d *such that* $a \to^* c$, $b \to^* d$ *and* $c \sim d$.
2. $\to$ *is* locally confluent modulo $\sim$ *if, for any* a, b, c, $a \to b$ *and* $a \to c$ *imply* b *and* c *are joinable modulo* $\sim$.

3. $\to$ is locally coherent modulo $\sim$ if, for any a, b, c, $a \to b$ and $a \sim c$ imply that b and c are joinable modulo $\sim$.
4. $\to$ is Church-Rosser modulo $\sim$ if $a \approx b$ implies a and b are joinable modulo $\sim$, where $\approx$ is $(\sim \cup \to \cup \leftarrow)^*$.

What we wish to establish for our system on linear lambda terms is the strong normalization of $\succ$ and the Church-Rosser property of $\succ$ modulo $\sim^*$. The following result provides a sufficient condition for this.

Theorem 1 (Huet [12]). *Let $(A, \succ)$ be an ARS, and $\sim$ be an equivalence relation on A. If $\succ$ is strongly normalizing, locally confluent modulo $\sim$, and locally coherent with $\sim$, then $\succ$ is Church-Rosser modulo $\sim$.* $\qquad\square$

In the following three sections, we show that $\succ$ is (i) strongly normalizing, (ii) locally confluent modulo $\sim^*$, and (iii) locally coherent with $\sim^*$.

5   Strong Normalization

Theorem 2 (strong normalization). $\succ$ *is strongly normalizing.*

For proving this, we proceed as follows. First, by showing that a translation into the simply typed lambda calculus weakly preserves the reduction, we reduce the problem to that of the smaller rewriting system. We then show the termination of this subsystem by assigning natural numbers to expressions which are strictly decreasing with respect to the reduction steps.

5.1   Translation into the Simply Typed Lambda Calculus

There is an obvious translation from the linear lambda calculus into the simply typed $\beta\eta$-lambda calculus (an inverse to Girard's translation [8]) which weakly preserves the reductions.

$$
\begin{aligned}
b^\bullet &= b \\
(!\tau)^\bullet &= \tau^\bullet \\
(\tau_1 \multimap \tau_2)^\bullet &= \tau_1^\bullet \to \tau_2^\bullet \\[1em]
x^\bullet &\equiv x \\
(\lambda x^\tau.M)^\bullet &\equiv \lambda x^{\tau^\bullet}.M^\bullet \\
(M\,N)^\bullet &\equiv M^\bullet\,N^\bullet \\
(!M)^\bullet &\equiv M^\bullet \\
(\text{let } !x^\tau \text{ be } M \text{ in } N)^\bullet &\equiv N^\bullet[M^\bullet/x]
\end{aligned}
$$

Straightforward inductions show the following facts:

Lemma 2 (type soundness). $\Gamma\,;\,\Delta \vdash M : \tau$ *implies* $\Gamma^\bullet, \Delta^\bullet \vdash M^\bullet : \tau^\bullet$. $\qquad\square$

Lemma 3 (substitution lemma). $(M[N/x])^\bullet \equiv M^\bullet[N^\bullet/x]$. $\qquad\square$

Now we see how reductions in the linear lambda calculus are related to those on the simply typed lambda calculus.

Proposition 4. *If $M \succ N$ in the linear lambda calculus, then $M^\bullet \succ_{\beta\eta} N^\bullet$ or $M^\bullet \equiv N^\bullet$ in the simply typed lambda calculus.*

Proof. It suffices to look at the reduction rules.

$$((\lambda x.M)\,N)^\bullet$$
$$\equiv (\lambda x.M^\bullet)\,N^\bullet$$
$$\succ_\beta M^\bullet[N^\bullet/x]$$
$$\equiv (M[N/x])^\bullet \quad \text{by Lemma 3}$$

$$(\lambda x.M\,x)^\bullet$$
$$\equiv \lambda x.M^\bullet\,x$$
$$\succ_\eta M$$

$$(\text{let } !x \text{ be } !M \text{ in } N)^\bullet$$
$$\equiv N^\bullet[M^\bullet/x]$$
$$\equiv (N[M/x])^\bullet$$

$$(\text{let } !x \text{ be } M \text{ in } C[!x])^\bullet$$
$$\equiv (C[!x])^\bullet[M^\bullet/x]$$
$$\equiv (C[M])^\bullet$$

$$((\text{let } !x \text{ be } M \text{ in } N)\,L)^\bullet$$
$$\equiv N^\bullet[M^\bullet/x]\,L^\bullet$$
$$\equiv (\text{let } !x \text{ be } M \text{ in } N\,L)^\bullet$$

$$(\text{let } !y \text{ be } (\text{let } !x \text{ be } M \text{ in } N) \text{ in } L)^\bullet \equiv L^\bullet[N^\bullet[M^\bullet/x]/y]$$
$$\equiv L^\bullet[N^\bullet/y][M^\bullet/x]$$
$$\equiv (\text{let } !x \text{ be } M \text{ in let } !y \text{ be } N \text{ in } L)^\bullet$$

$$(L\,(\text{let } !x \text{ be } M \text{ in } N))^\bullet$$
$$\equiv L^\bullet\,(N^\bullet[M^\bullet/x])$$
$$\equiv (\text{let } !x \text{ be } M \text{ in } L\,N)^\bullet$$

$$(\lambda y.\text{let } !x \text{ be } M \text{ in } N)^\bullet$$
$$\equiv \lambda y.N^\bullet[M^\bullet/x]$$
$$\equiv (\text{let } !x \text{ be } M \text{ in } \lambda y.N)^\bullet$$

□

Corollary 1. *Strong normalization of $\beta_!, \eta_!', com_1, com_2, com_3$ and com_4 implies that of $\succ$.*

Proof. Suppose that $\succ$ is not strongly normalizing, thus there exists an infinite strict reduction sequence $M_0 \succ M_1 \succ \ldots$ in the linear lambda calculus. We then have an infinite sequence $M_0^\bullet, M_1^\bullet, \ldots$ in the simply typed lambda calculus, where $M_i^\bullet \succ_{\beta\eta} M_{i+1}^\bullet$ or $M_i^\bullet \equiv M_{i+1}^\bullet$ holds by the last proposition. Since the $\beta\eta$-reduction of the simply typed lambda calculus is strongly normalizing, there exists some n such that $M_m^\bullet \equiv M_n^\bullet$ holds for any $m \geq n$. This means that the infinite reduction sequence $M_n \succ M_{n+1} \succ \ldots$ consists just of the non-$\beta_{\multimap}\eta_{\multimap}$ reductions. □

5.2 Termination of the Subsystem

We now complete our proof of strong normalization by showing that $\beta_!, \eta_!', com_1, com_2, com_3, com_4$ is indeed strongly normalizing.

Proposition 5. *The following set of rewriting rules is strongly normalizing.*

$\beta_!$	let $!x$ be $!M$ in N	$\succ N[M/x]$
$\eta_!'$	let $!x$ be M in $C[!x]$	$\succ C[M]$
com_1	(let $!x$ be M in N) L	$\succ$ let $!x$ be M in $N\,L$
com_2	let $!y$ be (let $!x$ be M in N) in L	$\succ$ let $!x$ be M in let $!y$ be N in L
com_3	$\lambda y.$(let $!x$ be M in N)	$\succ$ let $!x$ be M in $\lambda y.N$
com_4	L (let $!x$ be M in N)	$\succ$ let $!x$ be M in $L\,N$

Proof. We assign a positive natural number $|M|$ to each (possibly non-well-typed) term M by

$$
\begin{aligned}
|x| &= 1 \\
|\lambda x^\tau.M| &= 2|M| \\
|M\,N| &= 2|M| + 2|N| \\
|!M| &= |M| \\
|\text{let } !x^\tau \text{ be } M \text{ in } N| &= 2|M| + |N[M/x]|
\end{aligned}
$$

(note that the last line is well-defined — compare the depths of let-bindings) and show that $M \succ N$ implies $|M| > |N|$. Note that this assignment is monotone with respect to each argument. Therefore $|L[M/x]| \geq |L[N/x]|$ holds if we know $|M| \geq |N|$.

- $\beta_!$: $|\text{let } !x \text{ be } !M \text{ in } N| = 2|!M| + |N[!M/x]| = 2|M| + |N[M/x]| > |N[M/x]|$.
- $\eta_!'$: $|\text{let } !x \text{ be } M \text{ in } C[!x]| = 2|M| + |C[!M]| = 2|M| + |C[M]| > |C[M]|$.
- com_1: $|(\text{let } !x \text{ be } M \text{ in } N)\,L| = 4|M| + 2|N[M/x]| + 2|L|$, while
  $|\text{let } !x \text{ be } M \text{ in } N\,L| = 2|M| + 2|N[M/x]| + 2|L|$.
- com_2:

$$
\begin{aligned}
&|\text{let } !y \text{ be (let } !x \text{ be } M \text{ in } N) \text{ in } L| \\
&= 2|\text{let } !x \text{ be } M \text{ in } N| + |L[\text{let } !x \text{ be } M \text{ in } N/y]| \\
&= 4|M| + 2|N[M/x]| + |L[\text{let } !x \text{ be } M \text{ in } N/y]| \\
&\geq 4|M| + 2|N[M/x]| + |L[N[M/x]/y]|
\end{aligned}
$$

while

$$
\begin{aligned}
&|\text{let } !x \text{ be } M \text{ in let } !y \text{ be } N \text{ in } L| \\
&= 2|M| + |\text{let } !y \text{ be } N[M/x] \text{ in } L| \\
&= 2|M| + 2|N[M/x]| + |L[N[M/x]/y]|
\end{aligned}
$$

- com_3, com_4: similar to the case of com_1.   □

6   Local Confluence Modulo Equivalence

Theorem 3 (local confluence modulo $\sim^*$). $\succ$ *is locally confluent modulo* $\sim^*$: *if $L \succ M_1$ and $L \succ M_2$ then there exist N_1 and N_2 such that $M_1 \succ^* N_1$, $M_2 \succ^* N_2$ and $N_1 \sim^* N_2$.*

Proof (sketch). There are 16 cases to be considered. Many of them are joinable without $\sim^*$, except the following three cases.

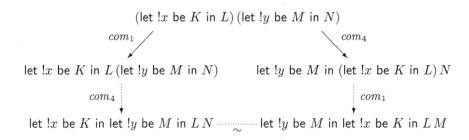

Other two cases involve $\sim^*$ via Lemma 1 (Section 3.2).

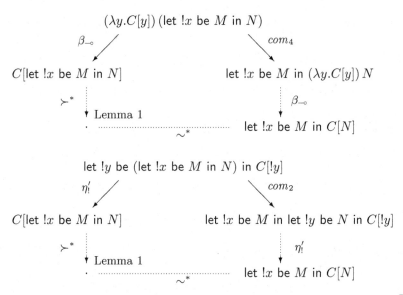

7   Local Coherence Modulo Equivalence

Theorem 4 (local coherence modulo $\sim^*$). $\succ$ *is locally coherent modulo* $\sim^*$, *i.e., if* $L \sim^* M \succ N$ *then there exists some* L', N' *such that* $L \succ^* L'$, $N \succ^* N'$ *and* $L' \sim^* N'$.

Proof (sketch). Note that it suffices to show: if $L \sim M \succ N$ then there exists some L', N' such that $L \succ^* L'$, $N \succ^* N'$ and $L' \sim^* N'$. There are six cases to be considered. The first four are rather obvious:

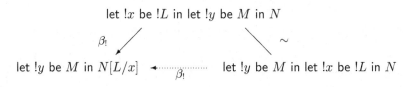

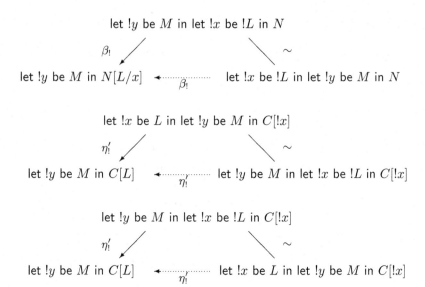

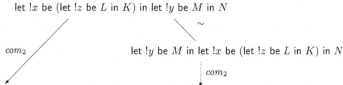

The remaining two cases are less trivial:

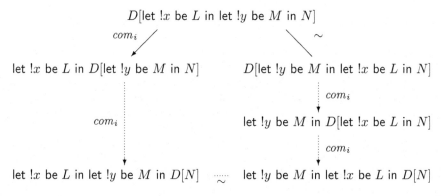

let !y be M in let !x be L in D[N] ⋯⋯ let !y be M in let !x be L in D[N]

□

Now we can state the fruit of the last three sections, thanks to Theorem 1.

Theorem 5 (Church-Rosser modulo ~*). $\succ$ *is Church-Rosser modulo* $\sim^*$*;*
if $M = N$*, there exist* M', N' *such that* $M \succ^* M'$*,* $N \succ^* N'$ *and* $M' \sim^* N'$*.*   □

8   Translation from the Computational Lambda Calculus

8.1   The Simply Typed Computational Lambda Calculus λ_c

The simply typed computational lambda calculus λ_c (its untyped version was introduced by Moggi [17]) has the same syntax as the simply typed lambda calculus plus the let-binding

$$\frac{\Gamma \vdash M : \sigma \quad \Gamma, x : \sigma \vdash N : \tau}{\Gamma \vdash \mathsf{let}\ x^\sigma\ \mathsf{be}\ M\ \mathsf{in}\ N : \tau}$$

It is a *call-by-value* calculus however, and its rewriting / equational theory is valid for reasoning about programs in the call-by-value programming languages like ML and Scheme. λ_c features the following reduction rules:

$(\beta.v)$	$(\lambda x^\sigma.M)\,V$	$\succ M[V/x]$	
$(\eta.v)$	$\lambda x^\sigma.V\,x$	$\succ V$	$(x \notin \mathrm{FV}(V))$
$(\beta.let)$	$\mathsf{let}\ x^\sigma\ \mathsf{be}\ V\ \mathsf{in}\ M$	$\succ M[V/x]$	
$(\eta.let)$	$\mathsf{let}\ x^\sigma\ \mathsf{be}\ M\ \mathsf{in}\ x$	$\succ M$	
$(assoc)$	$\mathsf{let}\ y^\tau\ \mathsf{be}\ (\mathsf{let}\ x^\sigma\ \mathsf{be}\ L\ \mathsf{in}\ M)\ \mathsf{in}\ N$	$\succ \mathsf{let}\ x^\sigma\ \mathsf{be}\ L\ \mathsf{in}\ \mathsf{let}\ y^\tau\ \mathsf{be}\ M\ \mathsf{in}\ N$	
$(let.1)$	$P\,M$	$\succ \mathsf{let}\ x^\sigma\ \mathsf{be}\ P\ \mathsf{in}\ x\,M$	$(P : \sigma)$
$(let.2)$	$V\,Q$	$\succ \mathsf{let}\ y^\sigma\ \mathsf{be}\ Q\ \mathsf{in}\ V\,y$	$(Q : \sigma)$

where V, W range over values (variables and lambda abstractions) while P, Q over non-values (applications and let expressions).

8.2   The Kernel Computational Lambda Calculus λ_{c*}

Interestingly, the reductions in the λ_c-calculus can be simulated within a smaller sublanguage λ_{c*} called *kernel computational lambda calculus* [18]. In λ_{c*}, applications $M\,N$ are restricted to those of values $V\,W$, and we no longer have reduction rules $(let.1)$ and $(let.2)$. Its reduction rules are given as follows.

$(\beta.v)$	$(\lambda x^\sigma.M)\,V$	$\succ M[V/x]$	
$(\eta.v)$	$\lambda x^\sigma.V\,x$	$\succ V$	$(x \notin \mathrm{FV}(V))$
$(\beta.let)$	$\mathsf{let}\ x^\sigma\ \mathsf{be}\ V\ \mathsf{in}\ M$	$\succ M[V/x]$	
$(\eta.let)$	$\mathsf{let}\ x^\sigma\ \mathsf{be}\ M\ \mathsf{in}\ x$	$\succ M$	
$(assoc)$	$\mathsf{let}\ y^\tau\ \mathsf{be}\ (\mathsf{let}\ x^\sigma\ \mathsf{be}\ L\ \mathsf{in}\ M)\ \mathsf{in}\ N$	$\succ \mathsf{let}\ x^\sigma\ \mathsf{be}\ L\ \mathsf{in}\ \mathsf{let}\ y^\tau\ \mathsf{be}\ M\ \mathsf{in}\ N$	

Here is a reduction-preserving inclusion $(-)^*$ from λ_c into λ_{c*}:

$$
\begin{aligned}
x^* &\equiv x \\
(\lambda x^\sigma.M)^* &\equiv \lambda x^\sigma.M^* \\
(P\,M)^* &\equiv \mathsf{let}\ x\ \mathsf{be}\ P^*\ \mathsf{in}\ (\lambda y.(y\,M)^*)\,x \\
(V\,Q)^* &\equiv \mathsf{let}\ y\ \mathsf{be}\ Q^*\ \mathsf{in}\ (\lambda x.V^*\,x)\,y \\
(V\,W)^* &\equiv V^*\,W^* \\
(\mathsf{let}\ x^\sigma\ \mathsf{be}\ M\ \mathsf{in}\ N)^* &\equiv \mathsf{let}\ x^\sigma\ \mathsf{be}\ M^*\ \mathsf{in}\ N^*
\end{aligned}
$$

Lemma 4. *If $\Gamma \vdash M : \sigma$ is derivable in λ_c, so is $\Gamma \vdash M^* : \sigma$ in λ_{c*}.*   □

Lemma 5. $M^*[V^*/x] \equiv (M[V/x])^*$.   □

Proposition 6. *If $M \succ_1 N$ in λ_c, then $M^* \succ_1 N^*$ in λ_{c*}.*

Proof (sketch). The key cases are

$$
\begin{aligned}
(PM)^* &\equiv \text{let } x \text{ be } P^* \text{ in } (\lambda y.(y\,M)^*)\,x \\
&\overset{\beta.v}{\succ} \text{let } x \text{ be } P^* \text{ in } (x\,M)^* \\
&\equiv (\text{let } x \text{ be } P \text{ in } x\,M)^*
\end{aligned}
$$

$$
\begin{aligned}
(VQ)^* &\equiv \text{let } y \text{ be } Q^* \text{ in } (\lambda x.V^*\,x)\,y \\
&\overset{\beta.v \text{ or } \eta.v}{\succ} \text{let } y \text{ be } Q^* \text{ in } V^*\,y \\
&\equiv (\text{let } y \text{ be } Q \text{ in } V\,y)^*
\end{aligned}
$$

□

Corollary 2. λ_{c*} *is strongly normalizing if and only if λ_c is strongly normalizing.*   □

Remark 4. This embedding $(-)^*$ is inspired from the translation $*_1 : \lambda_c \to \lambda_{c*}$ given by Sabry and Wadler [18], but not quite the same. For $*_1$, the translations of $P\,M$ and $V\,Q$ are simply

$$(PM)^* \equiv \text{let } x \text{ be } P^* \text{ in } (x\,M)^* \qquad (VQ)^* \equiv \text{let } y \text{ be } Q^* \text{ in } V^*\,y$$

while our embedding introduces additional redices so that the reduction steps are strictly preserved.

8.3   Embedding λ_{c*} into the Linear Lambda Calculus

Now it is fairly easy to give a reduction-preserving translation $(-)^\circ$ from λ_{c*} into the linear lambda calculus (the "call-by-value Girard translation"): let $b^\circ = b$, $(\sigma_1 \to \sigma_2)^\circ = !\sigma_1^\circ \multimap !\sigma_2^\circ$ and

$$
\begin{aligned}
x^\dagger &\equiv x \\
(\lambda x^\sigma.M)^\dagger &\equiv \lambda y^{!\sigma^\circ}.\text{let } !x^{\sigma^\circ} \text{ be } y \text{ in } M^\circ
\end{aligned}
$$

$$
\begin{aligned}
V^\circ &\equiv !V^\dagger \\
(V\,W)^\circ &\equiv V^\dagger\,W^\circ \\
(\text{let } x^\sigma \text{ be } M \text{ in } N)^\circ &\equiv \text{let } !x^{\sigma^\circ} \text{ be } M^\circ \text{ in } N^\circ
\end{aligned}
$$

Lemma 6 (type soundness). *If $\Gamma \vdash M : \sigma$ is derivable in λ_{c*}, so is $\Gamma^\circ ; \emptyset \vdash M^\circ :!\sigma^\circ$ in the linear lambda calculus [18,9].*   □

Lemma 7 (substitution lemma). $M^\circ[V^\dagger/x] \equiv (M[V/x])^\circ$.   □

Proposition 7 (preservation of reduction). *If $M \succ N$ in λ_{c*}, then $M^\diamond \succ^+$ $N^\diamond$ in the linear lambda calculus.*   □

Corollary 3 (strong normalization). λ_c *is strongly normalizing.*   □

We note that a different proof of this result via the reducibility argument has been given by Lindley and Stark [14].

Remark 5. For reasoning about *commutative effects* like non-termination and non-determinism, it makes sense to add the *commutativity axiom*

com let x be L in let y be M in N = let y be M in let x be L in N

We conjecture that our translation also preserves reduction modulo the equivalence relation generated by this *com*.

9 Concluding Remarks

We have given a rather simple-minded rewriting system on the linear lambda calculus which enjoys strong normalization and Church-Rosser property modulo trivial commuting conversion. We hope that this gives a reasonably understandable and feasible tool for reasoning about equivalence of terms in the linear lambda calculus. We shall conclude this paper by a few additional remarks.

9.1 Call-by-Name, Call-by-Value, and the Linear Lambda Calculus

This work can be considered as a refinement of some of the results in [15] where reduction-preserving translations between the (simply typed) call-by-name, call-by-value, call-by-need and linear lambda calculi were discussed. In *ibid.*, weaker non-extensional theories without η-rules were considered. In contrast, here we have studied the semantically complete theories (DILL-based linear lambda calculus and the computational lambda calculus, as well as the simply typed $\beta\eta$-lambda calculus) and the translations into the linear lambda calculus. We conjecture that the CPS translation from the computational lambda calculus into the linear lambda calculus [9] also enjoys good property with respect to the reduction theories.

9.2 Other Connectives

It is natural to ask if this approach would work well for other logical connectives in DILL, i.e., tensor $\otimes$ and unit I. While the tensor does not seem to cause any significant trouble, the unit is really problematic. For example we have let $*$ be M in N = let $*$ be N in M and $M \otimes N = N \otimes M$ for any $M, N : I$. For overcoming this problem with unit, perhaps we need to use the η-expansions as considered by Ghani [6].

Acknowledgements. This work was partially supported by the Japanese Ministry of Education, Culture, Sports, Science and Technology, Grant-in-Aid for Young Scientists (B) 17700013.

References

1. Barber, A. (1997) *Linear Type Theories, Semantics and Action Calculi.* PhD Thesis ECS-LFCS-97-371, University of Edinburgh.
2. Barber, A. and Plotkin, G. (1997) Dual intuitionistic linear logic. Manuscript. An earlier version available as Technical Report ECS-LFCS-96-347, LFCS, University of Edinburgh.
3. Barr, M. (1991) *-autonomous categories and linear logic. *Math. Struct. Comp. Sci.* **1**, 159–178.
4. Bierman, G.M. (1995) What is a categorical model of intuitionistic linear logic? In *Proc. Typed Lambda Calculi and Applications*, Springer Lecture Notes in Comput. Sci. **902**, pp. 78–93.
5. Crole, R. (1993) *Categories for Types.* Cambridge University Press.
6. Ghani, N. (1996) Adjoint rewriting and the !-type constructor. Manuscript.
7. Girard, J.-Y. (1987) Linear logic. *Theoret. Comp. Sci.* **50**, 1–102.
8. Hasegawa, M. (2000) Girard translation and logical predicates. *J. Funct. Programming* **10**(1), 77–89.
9. Hasegawa, M. (2002) Linearly used effects: monadic and CPS transformations into the linear lambda calculus. In *Proc. 6th Functional and Logic Programming*, Springer Lecture Notes in Comput. Sci. **2441**, pp. 67–182.
10. Hasegawa, M. (2004) Semantics of linear continuation-passing in call-by-name. In *Proc. 7th Functional and Logic Programming*, Springer Lecture Notes in Comput. Sci. **2998**, pp. 229–243.
11. Hasegawa, M. (2005) Classical linear logic of implications. *Math. Struct. Comput. Sci.* **15**(2), 323–342.
12. Huet, G. (1980) Confluent reductions: abstract properties and applications to term rewriting systems. *J. ACM* **27**(4), 797–821.
13. Lambek, J. and Scott, P. (1986) *Introduction to Higher-order Categorical Logic.* Cambridge University Press.
14. Lindley, S. and Stark, I. (2005) Reducibility and $\top\top$-lifting for computation types. In *Proc. Typed Lambda Calculi and Applications*, Springer Lecture Notes in Comput. Sci. **3461**, pp. 262–277.
15. Maraist, J., Odersky, M., Turner, D.N. and Wadler, P. (1995) Call-by-name, call-by-value, call-by-need and the linear lambda calculus. In *Proc. 11th Mathematical Foundations of Programming Semantics*, Electr. Notes Theor. Comput. Sci. **1**, pp. 370–392.
16. Mellies, P.-A. (2003) Categorical models of linear logic revisited. To appear in *Theoret. Comp. Sci.*
17. Moggi, E. (1989) Computational lambda-calculus and monads. In *Proc. 4th Annual Symposium on Logic in Computer Science*, pp. 14–23; a different version available as Technical Report ECS-LFCS-88-86, University of Edinburgh, 1988.
18. Sabry, A. and Wadler, P. (1997) A reflection on call-by-value. *ACM Transactions on Programming Languages and Systems*, **19**(6), 916–941.
19. Seely, R.A.G. (1989) Linear logic, *-autonomous categories and cofree coalgebras. In *Categories in Computer Science, AMS Contemp. Math.* **92**, pp. 371–389.
20. Terese (2003) *Term Rewriting Systems.* Cambridge University Press.

A Lambda-Calculus with Constructors

Ariel Arbiser[1], Alexandre Miquel[2], and Alejandro Ríos[1]

[1] Departamento de Computación – Facultad de Ciencias Exactas y Naturales
Universidad de Buenos Aires, Argentina
{arbiser, rios}@dc.uba.ar
[2] PPS & Université Paris 7 – Case 7014, 2 Place Jussieu
75251 PARIS Cedex 05 – France
alexandre.miquel@pps.jussieu.fr

Abstract. We present an extension of the $\lambda(\eta)$-calculus with a case construct that propagates through functions like a head linear substitution, and show that this construction permits to recover the expressiveness of ML-style pattern matching. We then prove that this system enjoys the Church-Rosser property using a semi-automatic 'divide and conquer' technique by which we determine all the pairs of commuting subsystems of the formalism (considering all the possible combinations of the nine primitive reduction rules). Finally, we prove a separation theorem similar to Böhm's theorem for the whole formalism.

1 Introduction

Lambda-calculus has been introduced by Church in the 30's [6] as a universal language to express computations of functions. Despite its remarkable simplicity, λ-calculus is rich enough to express all recursive functions. Since the rise of computers, λ-calculus has been used fruitfully as the basis of all functional programming languages, from LISP to the languages of the ML family. From the theoretical point of view, untyped λ-calculus enjoys many good properties [3], such as Church and Rosser's property expressing determinism of computations. In Logic, λ-calculus is also a fundamental tool to describe the computational contents of proofs via the Curry-Howard correspondence.

Although arbitrarily complex data structures can be encoded in the pure λ-calculus, modern functional programming languages provide primitive constructs for most data structures, for which a purely functional encoding would be inefficient. One of the most popular extensions of λ-calculus is pattern-matching on constructed values (a.k.a. variants), a problem that has been widely investigated in functional programming [12,10,13] and in rewriting [14,7,5,11,9].

However, introducing objects of different kinds—functions and constructed values—in the same formalism raises the problem of their interaction. What does it mean to apply a constructed value to an argument? Should the constructed value accumulate the extra argument? Or should it produce an error? Similarly, what does it mean to perform case analysis on a function?

Unfortunately, these problems are usually not addressed in the literature because they are irrelevant in a typed setting—applications go with functions, case

F. Pfenning (Ed.): RTA 2006, LNCS 4098, pp. 181–196, 2006.

analyses with variants. However, one should not forget that one of the reasons of the success of the λ-calculus in computer science and in logic lies in its excellent operational semantics in the untyped case. The best example is given by Böhm's separation theorem [4] that expresses that two observationally equivalent $\beta\eta$-normal λ-terms are intentionally equal. In the pure λ-calculus, a $\beta\eta$-normal term is a canonical form not only because it cannot be further reduced, but also because its computational behaviour cannot be expressed by another $\beta\eta$-normal term.

The situation is far from being as clear when we add pattern-matching to the untyped λ-calculus. As far as we know, there is no generalisation of Böhm's theorem for this kind of extension. One reason for that is that the notion of normal form is not as clear as in the pure λ-calculus, precisely because the traditional operational semantics says nothing about the computational behaviour of ill-typed constructions, such as a case analysis over an abstraction.

An extended operational semantics of case analysis. In this paper, we propose an extension of the untyped λ-calculus with constructors and case analysis that fills the holes of the traditional operational semantics. Technically, the main novelty is that we let application and case analysis (written $\{\!\{\theta\}\!\}.M$) commute via the (ill-typed[1]) reduction rule

(CASEAPP) $\{\!\{\theta\}\!\}.(MN) \quad\rightarrow\quad (\{\!\{\theta\}\!\}.M)N$.

(Here, θ denotes a *case binding*, that is a finite map from constructors to terms.) Symmetrically, we introduce a reduction rule

(CASELAM) $\{\!\{\theta\}\!\}.(\lambda x.M) \quad\rightarrow\quad \lambda x.(\{\!\{\theta\}\!\}.M) \qquad (x \notin FV(\theta))$

to let case analysis go through abstractions. In this way, case analysis can be understood as a form of head linear explicit substitution... of constructors.

Surprisingly, the system we obtain is not only computationally sound—we will show (section 3) that it is confluent and conservative over the untyped $\lambda\eta$-calculus—but it also permits to decompose ML-style pattern matching (with patterns of any arity) from the construction $\{\!\{\theta\}\!\}.M$ that only performs case analysis on constant constructors (section 2).

Finally, we will show (section 4) a theorem of weak separation for the whole calculus, using a separation technique inspired by Böhm's [4,3]. For this reason, the formalism provides a special constant written ✚ and called the *daimon* (following the terminology and notation of [8]) that requests the termination of the program—something like an `exit` system call—and which will be used as the main technical device to observe normal forms and separate them.

Proofs and technical details are omitted from this extended abstract, but are available in the long version of the paper [1].

---

[1] Observe that M is treated as a function in the l.h.s. of the rule whereas it is treated as a constructed value in the r.h.s. This rule should not be confused with the rule of *commutative conversion* $(\{\!\{\theta\}\!\}.M)N = \{\!\{\theta N\}\!\}.M$ that comes from logic, a rule which is well-typed... but incompatible with the reduction rules of our calculus!

2 Syntax and Reduction Rules

2.1 Syntax

The λ-calculus with constructors distinguishes two kinds of names: *variables* (written x, y, z, etc.) and *constructors* (written c, c', etc.) The set of variables and the set of constructors are written $\mathcal{V}$ and $\mathcal{C}$, respectively. In what follows, we assume that both sets $\mathcal{V}$ and $\mathcal{C}$ are denumerable and disjoint.

The terms (written M, N, etc.) and the case bindings (written θ, ϕ, etc.) of the λ-calculus with constructors are inductively defined as follows:

Terms	$M, N ::= x$	(Variable)
	$\mid c$	(Constructor)
	$\mid \maltese$	(Daimon)
	$\mid MN$	(Application)
	$\mid \lambda x \,.\, M$	(Abstraction)
	$\mid \{\!|\theta|\!\} \,.\, M$	(Case construct)
Case bindings	$\theta, \phi ::= c_1 \mapsto M_1; \ldots; c_n \mapsto M_n$	$(c_i \neq c_j$ for $i \neq j)$

The sets of terms and case bindings are denoted by $\Lambda_{\mathcal{C}}$ and $\mathcal{B}$, respectively, and their disjoint union by $\Lambda_{\mathcal{C}} + \mathcal{B}$.

Constructor Binding. Each case binding θ is formed as a finite unordered list of constructor bindings of the form $(c \mapsto M)$ whose l.h.s. are pairwise distinct. We say that a constructor c is *bound* to a term M in a case binding θ if the binding $(c \mapsto M)$ belongs to the list θ. From the definition of case bindings, it is clear that a constructor c is bound to at most one term in a given case binding θ. When there is no such binding, we say that the constructor c is *unbound* in θ.

The *size* of a case binding $\theta = (c_1 \mapsto M_1; \ldots; c_n \mapsto M_n)$ is written $|\theta|$ and defined by $|\theta| = n$.

We also introduce an (external) operation of *composition* between two case bindings θ and ϕ, which is written $\theta \circ \phi$ and defined by:

$$\theta \circ (c_1 \mapsto M_1; \ldots; c_n \mapsto M_n) \quad \equiv \quad c_1 \mapsto \{\!|\theta|\!\}. M_1; \ldots; c_n \mapsto \{\!|\theta|\!\}. M_n$$

(where $\phi \equiv (c_1 \mapsto M_1; \ldots; c_n \mapsto M_n)$). Notice that this operation is not syntactically associative, since:

$$(\theta \circ \phi) \circ (c_i \mapsto M_i)_{i=1..n} \quad \equiv \quad (c_i \mapsto \{\!|\theta \circ \phi|\!\}. M_i)_{i=1..n}$$

whereas

$$\theta \circ (\phi \circ (c_i \mapsto M_i)_{i=1..n} \quad \equiv \quad (c_i \mapsto \{\!|\theta|\!\}. \{\!|\phi|\!\}. M_i)_{i=1..n}$$

However, composition of case bindings only makes sense in the presence of the case conversion reduction rule $\{\!|\theta|\!\}. \{\!|\phi|\!\}. M \rightarrow \{\!|\theta \circ \phi|\!\}. M$ (see 2.2), for which both right hand sides above are convertible.

Free Variables and Substitution. The notions of bound and free occurrences of a variable are defined as expected. The set of free variables of a term M (resp. a case binding θ) is written $FV(M)$ (resp. $FV(\theta)$).

As in the (ordinary) λ-calculus, terms are considered up to α-conversion (i.e. up to a renaming of bound variables). Notice that the renaming policy of the λ-calculus with constructors is strictly the same as in the λ-calculus: it only affects (bound) *variable names*, but leaves *constructor names* unchanged.

The external substitution operation of the λ-calculus, written $M\{x := N\}$, is extended to the λ-calculus with constructors as expected. The same operation is also defined for case bindings (notation: $\theta\{x := N\}$).

2.2   Reduction Rules

The λ-calculus with constructors has 9 primitive reduction rules that are depicted in Fig. 1.

Beta-reduction

| AppLam | (AL) | $(\lambda x . M)N \rightarrow M\{x := N\}$ | |
| AppDai | (AD) | $✠ N \rightarrow ✠$ | |

Eta-reduction

| LamApp | (LA) | $\lambda x . Mx \rightarrow M$ | $(x \notin FV(M))$ |
| LamDai | (LD) | $\lambda x . ✠ \rightarrow ✠$ | |

Case propagation

CaseCons	(CO)	$\{\!	\theta	\!\} . c \rightarrow M$	$((c \mapsto M) \in \theta)$
CaseDai	(CD)	$\{\!	\theta	\!\} . ✠ \rightarrow ✠$	
CaseApp	(CA)	$\{\!|\theta|\!\} . (MN) \rightarrow (\{\!|\theta|\!\} . M)N$	
CaseLam	(CL)	$\{\!|\theta|\!\} . \lambda x . M \rightarrow \lambda x . \{\!|\theta|\!\} . M$	$(x \notin FV(\theta))$

Case conversion

| CaseCase | (CC) | $\{\!|\theta|\!\} . (\{\!|\phi|\!\} . M) \rightarrow \{\!|\theta \circ \phi|\!\} . M$ | |

Fig. 1. Reduction rules of the λ-calculus with constructors

In what follows, we will be interested not only in the system induced by the 9 reduction rules taken together, but more generally in the subsystems formed by all subsets of these 9 rules. We write $\lambda \mathcal{B}_{\mathcal{C}}$ the calculus generated by all rules of Fig. 1, and $\mathcal{B}_{\mathcal{C}}$ the calculus generated by all rules but AppLam (a.k.a. β).

Notice that AppLam (a.k.a β) and LamApp (a.k.a. η) are the only reduction rules that may apply to an ordinary λ-term in $\lambda \mathcal{B}_{\mathcal{C}}$.

2.3   An Example

In $\lambda \mathcal{B}_\mathcal{C}$, the predecessor function (over unary integers) is implemented as

$$\text{pred} \equiv \lambda n \,.\, \{\!| 0 \mapsto 0;\; \mathsf{s} \mapsto \lambda z \,.\, z |\!\} \,.\, n$$

(where 0 and s are two distinct constructors). From the rules APPLAM $(=\beta)$ and CASECONS it is obvious that

$$\text{pred } 0 \;\rightarrow\; \{\!| 0 \mapsto 0;\; \mathsf{s} \mapsto \lambda z \,.\, z |\!\} . 0 \;\rightarrow\; 0 \,.$$

More interesting is the case of pred $(\mathsf{s}\ N)$ (where N is an arbitrary term)

$$\begin{aligned}
\text{pred } (\mathsf{s}\ N) &\rightarrow \{\!| 0 \mapsto 0;\; \mathsf{s} \mapsto \lambda z \,.\, z |\!\} \,.\, (\mathsf{s}\ N) \\
&\rightarrow (\{\!| 0 \mapsto 0;\; \mathsf{s} \mapsto \lambda z \,.\, z |\!\} . \mathsf{s})\ N \;\rightarrow\; (\lambda z \,.\, z)\ N \;\rightarrow\; N
\end{aligned}$$

which shows how the case construct captures the head occurrence of the constructor s via the reduction rule CASEAPP. More generally, ML-style pattern-matching (on disjoint patterns) is translated in $\lambda \mathcal{B}_\mathcal{C}$ as follows:

match N with		
| $c_1(x_1, \ldots, x_{n_1}) \;\mapsto\; M_1$	becomes	$\{\!| c_1 \mapsto \lambda x_1 \cdots x_{n_1} \,.\, M_1 \,;$
| $c_2(x_1, \ldots, x_{n_2}) \;\mapsto\; M_2$		$c_2 \mapsto \lambda x_1 \cdots x_{n_2} \,.\, M_2 \,;$
| $\cdots$		$\cdots$
		$|\!\} \cdot N$

3   The Church-Rosser Property

In this section, we aim to prove that $\lambda \mathcal{B}_\mathcal{C}$ is confluent. For that, we will prove a much more general result by characterising among the $2^9 = 512$ possible subsets of the 9 primitive reduction rules which subsets induce a subsystem of $\lambda \mathcal{B}_\mathcal{C}$ which is confluent, and which ones do not.

3.1   Preliminary Definitions

Let us first recall some classic definitions.

Definition 1. — *An* Abstract Rewriting System (ARS) *is a pair $A = (|A|, \rightarrow_A)$ formed by an arbitrary set $|A|$ (called the carrier of A) equipped with a binary relation $\rightarrow_A$ on $|A|$. We denote by $\rightarrow_A^*$ the reflexive-transitive closure of $\rightarrow_A$, and by $\rightarrow_A^=$ the reflexive closure of $\rightarrow_A$.*

Given an ARS A, the set $\text{SN}(A)$ of *strongly normalising* elements of A is defined as the least subset $\text{SN}(A) \subset |A|$ which is closed under the rule

If for all $y \in |A|$, $x \rightarrow_A y$ entails $y \in \text{SN}(A)$, then $x \in \text{SN}(A)$.

Intuitively, an element x of A is strongly normalising if there is no infinite reduction sequence of the form $x = x_0 \rightarrow_A x_1 \rightarrow_A x_2 \rightarrow_A \cdots$

Definition 2. — *An ARS A is strongly normalising (SN) if all the elements of A are strongly normalising, i.e.* $\mathrm{SN}(A) = |A|$.

Definition 3. — *Let $A = (S, \to_A)$ and $B = (S, \to_B)$ be two ARSs defined on the same carrier set S. We say that:*

- *A weakly commutes with B, written $A \mathrel{/\!/_w} B$, if for all M, M_1, M_2 s.t. $M \to_A M_1$ and $M \to_B M_2$ there exists M_3 s.t. $M_1 \to_B^* M_3$ and $M_2 \to_A^* M_3$.*
- *A commutes with B, written $A \mathrel{/\!/} B$, if for all M, M_1, M_2 s.t. $M \to_A^* M_1$ and $M \to_B^* M_2$ there exists M_3 s.t. $M_1 \to_B^* M_3$ and $M_2 \to_A^* M_3$.*

An ARS A is said to be weakly confluent or weakly Church-Rosser (WCR) (resp. confluent, or Church-Rosser (CR)) if $A \mathrel{/\!/_w} A$ (resp. if $A \mathrel{/\!/} A$).

Given two ARSs A and B defined on the same carrier set, we write $A + B$ the (set-theoretic) union of both relations. The confluence proof of $\lambda \mathcal{B}_C$ relies on standard results of rewriting [2], and in particular in the following two lemmas:

Lemma 1. — *If $A \mathrel{/\!/_w} B$ and $A + B$ is SN, then $A \mathrel{/\!/} B$.*

PROOF: Same proof-technique as for Newman's lemma [2].   □

Lemma 2. — *If $A \mathrel{/\!/} B$ and $A \mathrel{/\!/} C$ then $A \mathrel{/\!/} (B + C)$.*

3.2 Critical Pairs and Closure Conditions

Each of the 9 primitive reduction rules of $\lambda \mathcal{B}_C$ describes the interaction between two syntactic constructs of the language, which is reflected by the name of the rule: APPLAM for 'APplication over a LAMbda', etc. These reduction rules induce 13 different critical pairs, that are summarised in Fig. 2 and 3.

Critical pairs occur for all pairs of rules of the form FOOBAR/BARBAZ. A quick examination of Fig. 2 and 3 reveals that each time we have to close such a critical pair, we need to use the third rule FOOBAZ when this rule exists. This occurs for the 6 critical pairs (2), (4), (5), (6), (7) and (8) of Fig. 2; in the other cases, the critical pair is closed by the only rules FOOBAR and BARBAZ.

This remark naturally suggests the following definition:

Definition 4 (Closure conditions). — *We say that a subset s of the 9 rules given in Fig. 1 fulfils the closure conditions and write $s \models \mathrm{CC}$ if:*

(CC1)	APPLAM $\in s$	$\wedge$	LAMDAI $\in s$	$\Rightarrow$	APPDAI	$\in s$
(CC2)	LAMAPP $\in s$	$\wedge$	APPDAI $\in s$	$\Rightarrow$	LAMDAI	$\in s$
(CC3)	CASEAPP $\in s$	$\wedge$	APPLAM $\in s$	$\Rightarrow$	CASELAM $\in s$
(CC4)	CASEAPP $\in s$	$\wedge$	APPDAI $\in s$	$\Rightarrow$	CASEDAI $\in s$
(CC5)	CASELAM $\in s$	$\wedge$	LAMAPP $\in s$	$\Rightarrow$	CASEAPP $\in s$
(CC6)	CASELAM $\in s$	$\wedge$	LAMDAI $\in s$	$\Rightarrow$	CASEDAI $\in s$

Intuitively, a subset that fulfils the 6 closure conditions defines a system in which all critical pairs can be closed, and thus constitutes a good candidate for Church-Rosser. The aim of this section is to turn this intuition into the

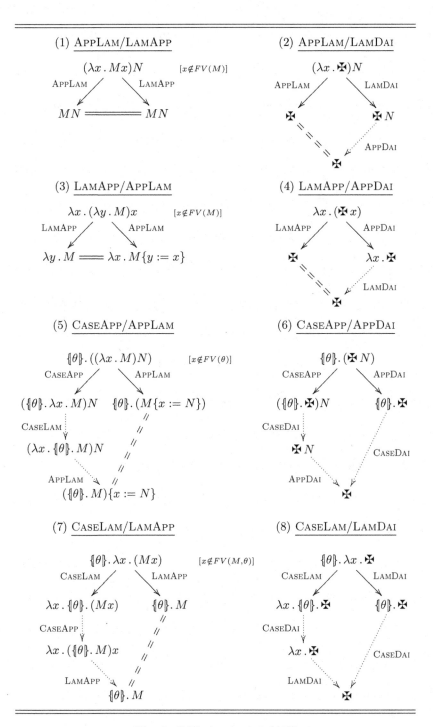

Fig. 2. Critical pairs 1–8 (/13)

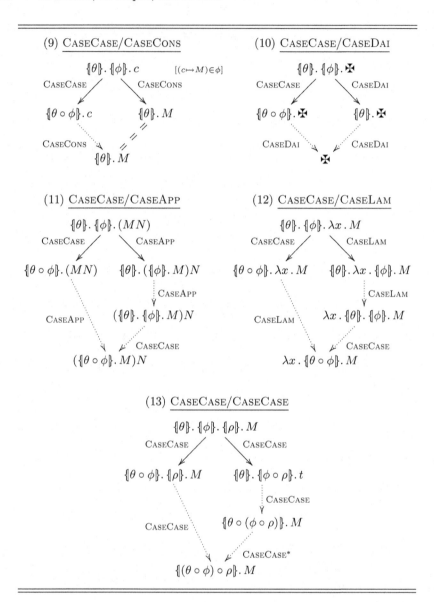

Fig. 3. Critical pairs 9–13 (/13)

Theorem 1 (Church-Rosser). — *For each of the 512 subsystems s of $\lambda\mathcal{B}_C$ the following propositions are equivalent:*

1. *s fulfils the closure conditions (CC1)–(CC6);*
2. *s is weakly confluent;*
3. *s is confluent.*

Since the full system (i.e. $\lambda\mathcal{B}_\mathcal{C}$) obviously fulfils all closure conditions, we will get as an immediate corollary:

Corollary 1 (Church-Rosser). — $\lambda\mathcal{B}_\mathcal{C}$ *is confluent.*

The proof of theorem 1 relies on a systematic analysis of the commutation properties of all pairs of subsystems (s_1, s_2) of $\lambda\mathcal{B}_\mathcal{C}$. For that, we first have to generalise the notion of closure condition to any pair (s_1, s_2) of subsystems. This leads us to adopt the following definition:

Definition 5 (Binary closure conditions). — *We say that a pair (s_1, s_2) of subsystems fulfils the* binary closure conditions *and write $(s_1, s_2) \models \mathrm{BCC}$ if*

(BCC1)	AppLam	$\in s_1$	$\wedge$	LamDai	$\in s_2$	$\Rightarrow$	AppDai $\in s_1$
(BCC2)	LamApp	$\in s_1$	$\wedge$	AppDai	$\in s_2$	$\Rightarrow$	LamDai $\in s_1$
(BCC3)	CaseApp	$\in s_1$	$\wedge$	AppLam	$\in s_2$	$\Rightarrow$	CaseLam $\in s_2$
(BCC4)	CaseApp	$\in s_1$	$\wedge$	AppDai	$\in s_2$	$\Rightarrow$	CaseDai $\in (s_1 \cap s_2)$
(BCC5)	CaseLam	$\in s_1$	$\wedge$	LamApp	$\in s_2$	$\Rightarrow$	CaseApp $\in s_2$
(BCC6)	CaseLam	$\in s_1$	$\wedge$	LamDai	$\in s_2$	$\Rightarrow$	CaseDai $\in (s_1 \cap s_2)$
(BCC7)	CaseCase	$\in s_1$	$\wedge$	CaseDai	$\in s_2$	$\Rightarrow$	CaseDai $\in s_1$
(BCC8)	CaseCase	$\in s_1$	$\wedge$	CaseApp	$\in s_2$	$\Rightarrow$	CaseApp $\in s_1$
(BCC9)	CaseCase	$\in s_1$	$\wedge$	CaseLam	$\in s_2$	$\Rightarrow$	CaseLam $\in s_1$

as well as the 9 symmetric conditions (obtained by exchanging s_1 with s_2).

Again, the 9 binary closure conditions come from an analysis of critical pairs. For example (BCC1) comes from the observation that critical pair (2) of Fig. 2 can be formed as soon as s_1 contains AppLam and s_2 contains LamDai, and that it can be closed only if s_1 contains AppDai.

We can also remark that when we take $s_1 = s_2 = s$, the binary closure conditions (BCC1)–(BCC6) degenerate to the (simple) closure conditions (CC1)–(CC6) whereas (BBC7)–(BCC9) become tautologies, so that:

Fact 1. — *For all subsystems s of $\lambda\mathcal{B}_\mathcal{C}$: $s \models \mathrm{CC}$ iff $(s, s) \models \mathrm{BCC}$.*

We first show that:

Proposition 1. — *For all pairs (s_1, s_2) of subsystems of $\lambda\mathcal{B}_\mathcal{C}$ the following propositions are equivalent:*

1. *$(s_1, s_2) \models \mathrm{BCC}$ (binary closure conditions);*
2. *$s_1 /\!/_w s_2$ (weak commutation).*

PROOF: $(1 \Rightarrow 2)$ By structural induction on the reduced term, closing critical pairs using BCCs. $(2 \Rightarrow 1)$ By contraposition, exhibiting a suitable counter-example for each BCC that does not hold. □

Now it remains to be shown that all weakly commuting pairs commute.

3.3 Strong Normalisation of the $\mathcal{B}_C$-Calculus

The first step is to check that the subsystem $\mathcal{B}_C = (\lambda\mathcal{B}_C \setminus \text{AppLam})$ is SN.

Proposition 2 (SN of $\mathcal{B}_C$-calculus). — *The $\mathcal{B}_C$-calculus is SN.*

PROOF: Consider the function $h : \Lambda_C + \mathcal{B} \to \mathbb{N}$ recursively defined by

$$
\begin{aligned}
h(x) &= h(c) = h(\maltese) = 1 & h(\{\!\{\theta\}\!\}.\,M) &= h(\theta) + (|\theta| + 2)h(M) \\
h(\lambda x.\,M) &= h(M) + 1 \\
h(MN) &= h(M) + h(N) & h((c_i \mapsto M_i)_{i=1..n}) &= \textstyle\sum_{i=1}^n h(M_i)
\end{aligned}
$$

It is routine to check that h decreases at each $\mathcal{B}_C$-reduction step.   □

From Lemma 1 and Prop. 1 we get:

Proposition 3. — *If $(s_1, s_2) \models \text{BCC}$ and $\text{AppLam} \notin (s_1 + s_2)$, then $s_1 /\!/ s_2$.*

3.4 Propagation of Commutation Lemmas

Let us now consider the 512×512 matrix formed by all $\binom{512}{1} + \binom{512}{2} = 131,328$ (unordered) pairs of subsystems of $\lambda\mathcal{B}_C{}^2$. With the help of a small computer program[3], we easily check that $13,396$ of the $131,328$ pairs of systems fulfil BCCs—and thus weakly commute. Moreover $5,612$ of these $13,396$ weakly commuting pairs do not involve AppLam—and thus we know that they commute.

The situation is summarised in the following table:

	Pairs (s_1, s_2)	$s_1 = s_2$
SN + commuting $(= \neg\text{AppLam} + \text{BCC})$	5,612	160
Weakly commuting $(= \text{BCC})$	13,396	248
Total	131,328	512

The problem is now to check that the $13,396 - 5,612 = 7,784$ remaining weakly commuting pairs commute too. For that, we notice that:

Fact 2. — *If the 12 pairs of subsystems of Table 1 commute, then all $13,396$ weakly commuting pairs of systems commute.*

Again, this fact can be mechanically checked by considering the set formed by all $5,612$ SN-commuting pairs extended with the 12 pairs of Table 1, and by checking that the closure of this set of $5,624$ pairs under Lemma 2 yields the set of all $13,396$ pairs that fulfil BCCs. To conclude, it suffices to prove:

Proposition 4. — *The 12 pairs of Table 1 commute.*

The details of the 12 commutation proofs can be found in [1].

From that we deduce that all pairs of subsystems that fulfil BCCs commute, and the proof of Theorem 1 is now complete.

---

[2] In what follows, we count (s_1, s_2) and (s_2, s_1) as a single pair of systems.

[3] This program can be downloaded from the web pages of the authors.

Table 1. The 12 initial commutation lemmas

(1)	APPLAM // APPLAM
(2)	APPLAM // APPDAI
(3)	APPLAM // LAMAPP
(4)	APPLAM // CASECONS
(5)	APPLAM // CASEDAI
(6)	APPLAM // CASELAM
(7)	APPLAM // CASECASE
(8)	APPLAM + APPDAI // LAMDAI
(9)	APPLAM + APPDAI // LAMAPP + LAMDAI
(10)	APPLAM + CASELAM // CASEAPP
(11)	APPLAM + CASELAM // LAMAPP + CASEAPP
(12)	APPLAM + APPDAI + CASEDAI + CASELAM //
	LAMAPP + LAMDAI + CASEDAI + CASEAPP

Corollary 2. — $\lambda\mathcal{B}_C$ *is conservative over $\lambda\eta$-calculus, in the sense that:*

$$\forall M_1, M_2 \in \Lambda \quad (\lambda\mathcal{B}_C \models M_1 = M_2 \quad \Rightarrow \quad \lambda\eta \models M_1 = M_2).$$

PROOF: Follows from Cor. 1 using the concluding remark of subsection 2.2.   □

4   Separation

The aim of this section is to establish the theorem of (weak) separation, expressing that observationally equivalent normal terms are syntactically equal. For that, we will show that for all normal terms[4] $M_1 \not\equiv M_2$ of $\lambda\mathcal{B}_C$ there exists a context $C[]$ such that $C[M_1]$ converges whereas $C[M_2]$ diverges—or vice-versa—using notions of convergence and divergence that will be precised.

Separation [4] can be understood as some kind of completeness of the formalism. Intuitively, it expresses that the calculus provides sufficiently many reduction rules to identify observationally equivalent terms, or—which is the same dually—that it provides sufficiently many syntactic constructs (i.e. observers) to discriminate different normal forms.

4.1   Quasi-normal Forms

Let us first analyse the shape of normal forms in the calculus.

Definition 6 (Head term). — *We call a* head term *(and write H, H_1, H', etc.) any term that has one of the following four forms:*

Head term $H \quad ::= \quad x \quad | \quad c \quad | \quad \{\!|\theta|\!\}.x \quad | \quad \{\!|\theta|\!\}.c \quad (c \notin \mathrm{dom}(\theta))$

---

[4] Actually, we will prove our separation theorem only for *completely defined* normal terms (cf subsection 4.2).

When a head term H is of one of the first three forms (variable, constructor, case binding on a variable), we say that H is defined. *When H is of the last form (case binding on an unbound constructor), we say that H is* undefined.

Definition 7 (Quasi-head normal form). — *A term M is said to be in* quasi-head normal form *(quasi-hnf) if it has one of the following two forms*

Quasi-hnf M $::=$ ✠ $\mid$ $\lambda x_1 \cdots x_n . H N_1 \cdots N_k$ $(n, k \geq 0)$

where H is an arbitrary head term, called the head *of M, and where $N_1, \ldots, N_k$ are arbitrary terms.*

Here, the prefix 'quasi-' expresses that such terms are in head normal form w.r.t. all reduction rules, but (possibly) the rule LAMAPP $(= \eta)$. In what follows, 'quasi-' systematically refers to 'all reduction rules but LAMAPP'.

As for head terms, we distinguish *defined* quasi-hnfs from *undefined* ones. We say that a quasi-hnf M is *defined* when either $M \equiv \lambda x_1 \cdots x_n . H N_1 \cdots N_k$ with H defined, or when $M \equiv$ ✠; and we say that M is *undefined* when $M \equiv \lambda x_1 \cdots x_n . (\{\!\{\theta\}\!\}. c) N_1 \cdots N_k$ with $c \notin \mathrm{dom}(\theta)$.

More generally, we call a *defined* term (resp. an *undefined* term) any term that reduces to a defined (resp. undefined) quasi-hnf. The class of defined terms is closed under arbitrary reduction, as for the class of undefined terms. Moreover, the class of undefined terms is closed under arbitrary substitution.

Definition 8 (Quasi-normal form). — *A term (resp. a case binding) is said to be in* quasi-normal form *when it is in normal form w.r.t. all the reduction rules but LAMAPP $(= \eta)$.*

Terms (resp. case bindings) that are in quasi-normal form are simply called *quasi-normal terms* (resp. *quasi-normal case bindings*). In particular, we call a *quasi-normal head term* any head term H which is in quasi-normal form. These notions have the following syntactic characterisation:

Proposition 5. — *Quasi-normal terms, quasi-normal head terms, and quasi-normal case bindings are (mutually) characterised by the following BNF:*

Q.n.-terms N $::=$ ✠ $\mid$ $\lambda x_1 \cdots x_n . H N_1 \cdots N_k$

Q.n.-head-terms H $::=$ x $\mid$ c $\mid$ $\{\!\{\theta\}\!\}. x$ $\mid$ $\{\!\{\theta\}\!\}. c$ $(c \notin \mathrm{dom}(\theta))$

Q.n.-case bind. θ $::=$ $c_1 \mapsto N_1; \ldots; c_p \mapsto N_p$

4.2   Separation Contexts

The notion of *context with one hole* is defined in $\lambda \mathcal{B}_{\mathcal{C}}$ as expected. The term obtained by filling the hole of a context $C[]$ with a term M is written $C[M]$, and the composition of two contexts $C[]$ and $C'[]$ is written $C'[C[]]$. In what follows, we will use contexts of a particular form, namely, *evaluation contexts*:

Evaluation contexts $E[]$ $::=$ $[] N_1 \cdots N_n$ $\mid$ $(\{\!\{\theta\}\!\}. []) N_1 \cdots N_n$

Notice that the composition $E'[E[]]$ of two evaluation contexts $E[]$ and $E'[]$ is not always an evaluation context, but that it always reduces to an evaluation context using zero, one or several steps of the CASEAPP rule, possibly followed by a single step of the CASECASE rule.

The daimon ✠ which represents immediate termination naturally absorbs all evaluation contexts:

Lemma 3. — *In any evaluation context $E[]$ one has* $E[✠] \to^* ✠$.

Symmetrically, each sub-term of the form $\{\theta\}.c$ (with $c \notin \mathrm{dom}(\theta)$) blocks the computation process at head position so that undefined terms "absorb" all evaluation contexts as well:

Lemma 4. — *Given an undefined term U, the term $E[U]$ is undefined for all evaluation contexts $E[]$.*

The daimon ✠ and undefined terms are thus natural candidates to define the notion of separability:

Definition 9 (Separability). — *We say that two terms M_1 and M_2 are:*

- weakly separable *if there exists a context with one hole $C[]$ such that either:*
  - $C[M_1] \to^* ✠$ *and $C[M_2]$ is undefined, or*
  - $C[M_2] \to^* ✠$ *and $C[M_1]$ is undefined;*
- strongly separable *if there exists two contexts $C_1[]$ and $C_2[]$ such that*
  - $C_1[M_1] \to^* ✠$ *and $C_1[M_2]$ is undefined, and*
  - $C_2[M_2] \to^* ✠$ *and $C_2[M_1]$ is undefined.*

Since undefined terms cannot be separated from each other (because undefined heads block all computations), we have to exclude them[5] from our study:

Definition 10 (Completely defined quasi-normal term). — *A term M in quasi-normal form is said to be* completely defined *if it contains no sub-term of the form $\{\theta\}.c$, where $c \notin \mathrm{dom}(\theta)$.*

4.3   Disagreement

The separation theorem is proved in two steps:

1. First we define a syntactic relation between terms, called *disagreement at depth $d \in \mathbb{N}$*, and we show that any pair of distinct normal forms have η-expansions that disagree at some depth (this subsection).
2. Then we show (by induction on the depth of disagreement) that any pair of disagreeing quasi-normal terms are weakly separable (subsection 4.5).

Definition 11 (Skeleton equivalence). — *We say that two defined head terms H_1 and H_2 have the same* skeleton *and write $H_1 \approx H_2$ if either:*

---

[5] Semantically, this means that we identify undefined terms with non weakly normalising terms, and thus interpret them as Ω (Scott's bottom).

- $H_1 \equiv H_2 \equiv x$ *for some variable* x; *or*
- $H_1 \equiv H_2 \equiv c$ *for some constructor* c; *or*
- $H_1 \equiv \{\!|\theta_1|\!\}.x$ *and* $H_2 \equiv \{\!|\theta_2|\!\}.x$ *for some variable* x *and for some* θ_1, θ_2 *such that* $\mathrm{dom}(\theta_1) = \mathrm{dom}(\theta_2)$.

Definition 12 (Disagreement at depth d). — *For each* $d \in \mathbb{N}$, *we define a binary relation on the class of completely defined quasi-normal terms, called the disagreement relation at depth d. This relation, written* $\mathrm{dis}_d(M_1, M_2)$ *('M_1 and M_2 disagree at depth d'), is defined by induction on $d \in \mathbb{N}$ as follows:*

- *(Base case)*   We write $\mathrm{dis}_0(M_1, M_2)$ *if either:*
  - $M_1 = \maltese$ *and* $M_2 = \lambda x_1 \cdots x_n . H N_1 \cdots N_k$; *or*
  - $M_1 = \lambda x_1 \cdots x_n . H N_1 \cdots N_k$ *and* $M_2 = \maltese$; *or*
  - $M_1 = \lambda x_1 \cdots x_n . H_1 N_{1,1} \cdots N_{1,k_1}$ *and*
    $M_2 = \lambda x_1 \cdots x_n . H_2 N_{2,1} \cdots N_{2,k_2}$ *and*
    $H_1 \not\approx H_2$.
- *(Inductive case)*   *For all* $d \in \mathbb{N}$, *we write* $\mathrm{dis}_{d+1}(M_1, M_2)$ *if*
  $M_1 = \lambda x_1 \cdots x_n . H_1 N_{1,1} \cdots N_{1,k_1}$ *and*
  $M_2 = \lambda x_1 \cdots x_n . H_2 N_{2,1} \cdots N_{2,k_2}$ *and*
  $H_1 \approx H_2$, *and if either*
  - $H_1 = \{\!|\theta_1|\!\}.y$ *and* $H_2 = \{\!|\theta_2|\!\}.y$ *for some case bindings* θ_1, θ_2 *and for some variable* y, *and there is a constructor* $c \in \mathrm{dom}(\theta_1) = \mathrm{dom}(\theta_2)$ *such that* $\mathrm{dis}_d(\theta_1(c), \theta_2(c))$; *or*
  - *There is a position* $1 \le k \le \min(k_1, k_2)$ *such that* $\mathrm{dis}_d(N_{1,k}, N_{2,k})$.

Lemma 5 (Cooking lemma). — *If M_1 and M_2 are completely defined normal terms (w.r.t. all reduction rules including* LamApp *$= \eta$) such that $M_1 \not\equiv M_2$, then one can find two completely defined quasi-normal terms M_1' and M_2' such that $M_1' \to_\eta^* M_1$, $M_2' \to_\eta^* M_2$, and $\mathrm{dis}_d(M_1', M_2')$ for some $d \in \mathbb{N}$.*

4.4   Ingredients for Separation

Separating disagreeing quasi-normal terms relies on definitions and techniques that are fully described in [1]. Here we briefly present some of them.

Tuples. In order to retrieve arbitrary sub-terms of a given normal form (the so called 'Böhm-out' technique), we need tuples that are encoded as in the pure λ-calculus as $\langle M_1; \ldots; M_n \rangle \equiv \lambda e . e M_1 \cdots M_n$. In what follows, we use a more general notation to represent partial application of the n-tuple constructor to its first k arguments and waiting the remaining $n - k$ arguments:

$$\langle M_1; \ldots; M_k; *_{n-k} \rangle \equiv \lambda x_{k+1} \cdots x_n e . e M_1 \cdots M_k x_{k+1} \cdots x_n \qquad (0 \le k \le n)$$

With these notations, the n-tuple constructor is written $\langle *_n \rangle$.

Encoding names. Separation of distinct free variables is achieved by substituting them by easily separable closed terms. For that, we associate to each variable name x a unique Church numeral written $\mathtt{x}$ (using the same name written in typewriter face), which we call the *symbol* of x.

Substitutions. A *substitution* is a finite association list which maps pairwise distinct variables to terms. A substitution σ can be applied to a term M, and the result (which is defined as expected) is written $M[\sigma]$.

Separation is achieved (Prop. 6) using a particular substitution σ_X^K parameterised by an integer $K \geq 0$ and a finite set of variables X, namely, the substitution that maps each variable $x \in X$ to the term $\langle \mathbf{x}; *_K \rangle$ representing the partial application of the $(K+1)$-tuple constructor to the symbol of x.

4.5 The Separation Theorem

Let M be a term in quasi-normal form. We call the *application strength* of M the largest integer $k \geq 0$ such that M has a sub-term of the form $HN_1 \cdots N_k$.

Proposition 6 (Separation of disagreeing terms). — *Let $K \geq 0$ be a natural number, and M_1 and M_2 two completely defined quasi-normal terms whose application strength is less than or equal to K and such that M_1 and M_2 disagree at some depth $d \in \mathbb{N}$. Then there exists an evaluation context $E[]$ such that either*

- *$E\big[M_1[\sigma_X^K]\big] \to^* \maltese$ and $E\big[M_2[\sigma_X^K]\big]$ is undefined, or*
- *$E\big[M_2[\sigma_X^K]\big] \to^* \maltese$ and $E\big[M_1[\sigma_X^K]\big]$ is undefined;*

where X is any finite set of variables that contains at least the free variables of M_1 and M_2, and where σ_X^K is the substitution defined in subsection 4.4.

From this proposition and lemma 5 we easily conclude:

Theorem 2 (Separation). — *Let M_1 and M_2 be completely defined terms in normal form. If $M_1 \not\equiv M_2$, then M_1 and M_2 are weakly separable.*

5   Conclusion

We have introduced an extension of λ-calculus, $\lambda\mathcal{B}_\mathcal{C}$, in which pattern matching is implemented via a mechanism of case analysis that behaves like a head linear substitution over constructors. We have shown that the reduction relation of $\lambda\mathcal{B}_\mathcal{C}$ is confluent and conservative over the $\lambda\eta$-calculus, but also that it is complete in the sense that it provides sufficiently many reduction rules to identify all observationally equivalent normalising terms.

Using the divide-and-conquer method for other proofs of confluence. An original aspect of this work is the way we proved confluence by systematically studying the commutation properties of all pairs of subsystems of $\lambda\mathcal{B}_\mathcal{C}$. Surprisingly, the mechanical propagation rule "if $A // B$ and $A // C$ then $A // (B+C)$" (combined with the primitive knowledge of all commutation properties between subsystems that do not involve APPLAM) is sufficient to reduce the proof of the expected 7,784 non-trivial commutation lemmas to only 12 primitive lemmas, that are established by hand. It would be interesting to investigate further to see whether the same method can be used to prove the confluence of other rewrite systems with many reduction rules—typically, systems with explicit substitutions.

A notion of Böhm tree for $\lambda\mathcal{B}_C$. The separation theorem we proved suggests that head normal forms of $\lambda\mathcal{B}_C$ could be the adequate brick to define a notion of Böhm-tree [4,3] for $\lambda\mathcal{B}_C$—and more generally, for ML-style pattern-matching. However, the fact that it is a *weak* separation theorem also suggests that the observational ordering is non-trivial on the set of normal forms. Characterising observational ordering on normal forms could be the next step to deepen our understanding of both operational and denotational semantics of $\lambda\mathcal{B}_C$.

Which type system for $\lambda\mathcal{B}_C$? The reduction rules CASEAPP and CASELAM which are the starting point of this work deeply challenge the traditional intuition of the notion of type, for which functions and constructed values live in different worlds. However, the good operational semantics of the calculus naturally raises the exciting question of finding a suitable type system for $\lambda\mathcal{B}_C$.

References

1. A. Arbiser, A. Miquel, and A. Ríos. A λ-calculus with constructors. Manuscript, available from the web pages of the authors, 2006.
2. F. Baader and T. Nipkow. *Rewriting and All That.* Addison-Wesley, 1999.
3. H. Barendregt. *The Lambda Calculus: Its Syntax and Semantics*, volume 103 of *Studies in Logic and The Foundations of Mathematics*. North-Holland, 1984.
4. C. Böhm, M. Dezani-Ciancaglini, P. Peretti, and S. Ronchi Della Rocha. A discrimination algorithm inside lambda-beta-calculus. *Theoretical Computer Science*, 8(3):265–291, 1979.
5. S. Cerrito and D. Kesner. Pattern matching as cut elimination. In *Logics In Computer Science (LICS'99)*, pages 98–108, 1999.
6. A. Church. *The calculi of lambda-conversion*, volume 6 of *Annals of Mathematical Studies*. Princeton, 1941.
7. H. Cirstea and C. Kirchner. Rho-calculus, the rewriting calculus. In *5th International Workshop on Constraints in Computational Logics*, 1998.
8. J.-Y. Girard. Locus solum: From the rules of logic to the logic of rules. *Mathematical Structures in Computer Science*, 11(3):301–506, 2001.
9. C. Barry Jay. The pattern calculus. *ACM Transactions on Programming Languages and Systems*, 26(6):911–937, 2004.
10. S. Peyton Jones et al. *The Revised Haskell 98 Report.* Cambridge Univ. Press, 2003. Also on http://haskell.org/.
11. W. Kahl. Basic Pattern Matching Calculi: A Fresh View on Matching Failure. In Y. Kameyama and P. Stuckey, editors, *Functional and Logic Programming, Proceedings of FLOPS 2004*, volume 2998 of *LNCS*, pages 276–290. Springer, 2004.
12. R. Milner, M. Tofte, and R. Harper. *The definition of Standard ML.* MIT Press, 1990.
13. The Objective Caml language. http://caml.inria.fr/.
14. V. van Oostrom. Lambda calculus with patterns. Technical Report IR-228, Vrije Universiteit, Amsterdam, 1990.

Structural Proof Theory as Rewriting

J. Espírito Santo[1], M.J. Frade[2], and L. Pinto[1,*]

[1] Departamento de Matemática, Universidade do Minho, Braga, Portugal
[2] Departamento de Informática, Universidade do Minho, Braga, Portugal
{jes, luis}@math.uminho.pt, mjf@di.uminho.pt

Abstract. The multiary version of the λ-calculus with generalized applications integrates smoothly both a fragment of sequent calculus and the system of natural deduction of von Plato. It is equipped with reduction rules (corresponding to cut-elimination/normalisation rules) and permutation rules, typical of sequent calculus and of natural deduction with generalised elimination rules. We argue that this system is a suitable tool for doing structural proof theory as rewriting. As an illustration, we investigate combinations of reduction and permutation rules and whether these combinations induce rewriting systems which are confluent and terminating. In some cases, the combination allows the simulation of non-terminating reduction sequences known from explicit substitution calculi. In other cases, we succeed in capturing interesting classes of derivations as the normal forms w.r.t. well-behaved combinations of rules. We identify six of these "combined" normal forms, among which are two classes, due to Herbelin and Mints, in bijection with normal, ordinary natural deductions. A computational explanation for the variety of "combined" normal forms is the existence of three ways of expressing multiple application in the calculus.

1 Introduction

The study of proof systems by means of associated term calculi increases the efficiency of the study and offers a computational perspective over logical phenomena. This applies to the study of a proof system in isolation, and to the study of the relationship between proof systems, typical in structural proof theory [10].

The multiary version of the λ-calculus with generalized applications (named λ**Jm**-calculus [5,6]) integrates smoothly both a fragment of sequent calculus and the system of natural deduction of von Plato, for intuitionistic implication. Its unary fragment corresponds to the ΛJ-calculus of [8], whereas its cut-free fragment captures the multiary cut-free sequent terms of [12]. The system λ**Jm** is equipped with reduction rules (corresponding to cut-elimination/normalisation rules) and permutation rules, typical of sequent calculus and of natural deduction with generalised elimination rules. This calculus offers the possibility of

---

* All authors are supported by FCT through the Centro de Matemática da Universidade do Minho (first and last authors) and through the Centro de Ciências e Tecnologias da Computação da Universidade do Minho (second author); all authors are also supported by the european thematic networks APPSEM II and TYPES.

F. Pfenning (Ed.): RTA 2006, LNCS 4098, pp. 197–211, 2006.

an integrated study of the relationship between sequent calculus and natural deduction and is a suitable tool for doing structural proof theory as rewriting.

As an illustration, we investigate combinations of reduction and permutation rules, in order to study the interaction between cut-elimination / normalisation and permutative conversions. The relationship between sequent calculus and natural deduction has very much to do with permutative conversions. Typically, the fragments of sequent calculus closer to natural deduction are those whose derivations are permutation-free [7,9,1], or, even better, those whose derivations are the normal forms w.r.t. permutation rules of bigger fragments [2,12]. On the other hand, systems of natural deduction closer to sequent calculus contain general elimination rules and, therefore, "hidden convertibilities" [13]. However, in the literature, the interaction between normalisation / cut-elimination and permutative conversions is usually avoided. In [7] the cut-free derivations are also permutation-free but the system does not include permutation rules. In [2,12] permutation rules are studied in a cut-free system. In [13] the "hidden convertibilities" are seen as belonging to the normalisation process.

We investigate whether the combinations of reduction and permutation rules of $\lambda \mathbf{Jm}$ induce rewriting systems which are confluent and terminating. In some cases, the combination allows the simulation of non-terminating reduction sequences known from explicit substitution calculi. In other cases, we succeed in capturing interesting classes of derivations as the normal forms w.r.t. well-behaved combinations of rules. We identify six "combined" normal forms, among which are two classes, due to Herbelin and Mints, in bijection with normal natural deductions. In order to achieve this, we proceed the study, initiated in [6], of the "overlaps" between the constructors of the calculus and the permutation rules they generate. In particular, the "overlap" between the features of multi-arity and generality is explained as a manifestation of the existence of various ways of expressing multiple application in the system.

The paper is organised as follows. Section 2 recalls system $\lambda \mathbf{Jm}$. Section 3 considers combined normal forms resulting from (slight modifications of) rules introduced in [5]. These suffice to capture Herbelin normal forms. Section 4 offers a deeper study of $\lambda \mathbf{Jm}$ in order to capture Mints normal forms. Section 5 concludes, giving some computational interpretation of these results.

2   The System λJm

Expressions and typing rules: We assume a denumerable set of variables and x, y, w, z to range over it. In the generalised multiary λ-calculus $\lambda \mathbf{Jm}$ there are two kinds of expressions, *terms* and *lists*, described in the following grammar:

$$(terms\ of\ \lambda\mathbf{Jm}) \quad t, u, v ::= x \mid \lambda x.t \mid t(u, l, (x)v)$$
$$(lists\ of\ \lambda\mathbf{Jm}) \quad l ::= t{::}l \mid []$$

A term of the form $t(u, l, (x)v)$ is called a *generalised multiary application* (gm-application for short) and t is called the *head* of such term. In terms $\lambda x.v$ and $t(u, l, (x)v)$, occurrences of x in v are bound.

Informally, a generalised multiary application $t(u, l, (x)v)$ can be thought of as the application of a function t to a list of arguments, whose head is u and tail is l, explicitly substituted for x in term v. Multiarity is the capability of applying a function t to more than one argument and generality is the capability of specifying the term v where the result of applying t to its arguments is going to be used.

Formulas (= *types*) A, B, C, ... are built up from propositional variables using just $\supset$ (for implication) and *contexts* Γ are finite sets of *variable : formula* pairs, associating at most one formula to each variable. *Sequents* of $\lambda\mathbf{Jm}$ are of one of two forms: $\Gamma \vdash t : A$ and $\Gamma; B \vdash l : C$. The typing rules of $\lambda\mathbf{Jm}$ are as follows:

$$\frac{}{x:A, \Gamma \vdash x:A} \; Axiom \qquad \frac{x:A, \Gamma \vdash t:B}{\Gamma \vdash \lambda x.t : A \supset B} \; Right$$

$$\frac{\Gamma \vdash t:A \supset B \quad \Gamma \vdash u:A \quad \Gamma; B \vdash l:C \quad x:C, \Gamma \vdash v:D}{\Gamma \vdash t(u, l, (x)v):D} \; gm - Elim$$

$$\frac{\Gamma \vdash u:A \quad \Gamma; B \vdash l:C}{\Gamma; A \supset B \vdash u::l:C} \; Lft \qquad \frac{}{\Gamma; C \vdash []:C} \; Ax$$

with the proviso that $x \notin \Gamma$ in Right and in gm-Elim. An instance of rule gm-Elim is called a *generalised multiary elimination* (or gm-elimination, for short).

$\lambda\mathbf{Jm}$ corresponds to an extension, with cuts of a certain form, of Schwichtenberg's cut-free, multiary, sequent calculus of [12]. This view splits gm-applications $t(u, l, (x)v)$ into those where the head term t is a variable, called *multiary-Left introductions*, and those where t is not a variable, called *cuts*. Thus cut-elimination in $\lambda\mathbf{Jm}$ is about the elimination of cuts in this sense. The rules to perform cut-elimination are called *reduction rules*.

Reduction rules: The *reduction rules* for $\lambda\mathbf{Jm}$ are as follows:

(β_1) $(\lambda x.t)(u, [], (y)v) \rightarrow \mathbf{s}(\mathbf{s}(u, x, t), y, v)$
(β_2) $(\lambda x.t)(u, v::l, (y)v') \rightarrow \mathbf{s}(u, x, t)(v, l, (y)v')$
(π) $t(u, l, (x)v)(u', l', (y)v') \rightarrow t(u, l, (x)v(u', l', (y)v'))$
(μ) $t(u, l, (x)x(u', l', (y)v)) \rightarrow t(u, \mathbf{a}(l, u' :: l'), (y)v), \quad x \notin u', l', v$

The auxiliary operators of substitution $\mathbf{s}(t, x, v)$, called *generalised multiary substitution* (*gm-substitution* for short) and of appending $\mathbf{a}(l, u :: l')$ are as follows:

$$\mathbf{s}(t, x, x) = t \qquad\qquad\qquad \mathbf{a}([], u :: l) = u :: l$$
$$\mathbf{s}(t, x, y) = y, \; y \neq x \qquad\qquad \mathbf{a}(u' :: l', u :: l) = u' :: \mathbf{a}(l', u :: l)$$
$$\mathbf{s}(t, x, \lambda y.u) = \lambda y.\mathbf{s}(t, x, u)$$
$$\mathbf{s}(t, x, u(v, l, (y)v')) = \mathbf{s}(t, x, u)(\mathbf{s}(t, x, v), \mathbf{s}'(t, x, l), (y)\mathbf{s}(t, x, v'))$$
$$\mathbf{s}'(t, x, []) = []$$
$$\mathbf{s}'(t, x, v :: l) = \mathbf{s}(t, x, v) :: \mathbf{s}'(t, x, l)$$

At the typing level these two operations are associated to the admissibility in $\lambda\mathbf{Jm}$ of certain cut rules [5]. Let $\beta = \beta_1 \cup \beta_2$. The notation $\to_{\beta,\pi,\mu}$ stands for the compatible closure of $\beta \cup \pi \cup \mu$ and the notations $\to^+_{\beta,\pi,\mu}$ and $\to^*_{\beta,\pi,\mu}$ stand for the transitive and the reflexive-transitive closure of $\to_{\beta,\pi,\mu}$ respectively. In the sequel we use similar conventions and notations for reduction relations. Normal forms w.r.t. $\to_{\beta,\pi}$ ($\beta\pi$-nfs for short) are the terms whose occurrences of gm-applications as sub-terms are of the form $x(u, l, (y)v)$, i.e. the head is a variable; they correspond exactly to Schwichtenberg's multiary cut-free sequent terms. $\beta\pi\mu$-nfs in turn correspond to Schwichtenberg's "multiary normal forms".

[6] shows that $\to_{\beta,\pi,\mu}$ enjoys properties of confluence, strong normalisation of typable terms and subject reduction.

Permutative conversion rules: Permutative conversions correspond to certain oriented permutations in the order of inferences in derivations. They aim at reducing gm-eliminations to a particular form that corresponds to the elimination rule of natural deduction.

In $\lambda\mathbf{Jm}$ we have two forms of permutative conversion (*permutation* for short): **p**-permutation and **q**-permutation. **p**-permutation aims at converting every gm-application to an application of the form $t(u, l, (x)x)$, that is a form that makes no real use of the generality feature. The **p**-permutation rules are:

$$(p_1) \qquad\qquad t(u, l, (x)y) \to y, \ x \neq y$$
$$(p_2) \qquad\qquad t(u, l, (x)\lambda y.v) \to \lambda y.t(u, l, (x)v)$$
$$(p_3) \ t_1(u_1, l_1, (x)t_2(u_2, l_2, (y)v)) \to$$
$$t_1(u_1, l_1, (x)t_2)(t_1(u_1, l_1, (x)u_2), \mathbf{p'_3}(t_1, u_1, l_1, x, l_2), (y)v) \ \text{if} \ x \notin v,$$

where
$$\mathbf{p'_3}(t, u, l, x, []) = []$$
$$\mathbf{p'_3}(t, u, l, x, u'::l') = t(u, l, (x)u') :: \mathbf{p'_3}(t, u, l, x, l') \ .$$

$p = p_1 \cup p_2 \cup p_3$. **q**-permutation aims at converting every gm-application to an application of the form $t(u, [], (x)v))$, that is a form that makes no use of the multiarity feature. The unique **q**-permutation rule is

$$(q) \ t(u, v::l, (x)v') \to t(u, [], (y)y)(v, l, (x)v') \ .$$

Permutations preserve typing. $\to_p$, $\to_q$ and $\to_{pq}$ are confluent and terminating. The p-nf (resp. q-nf, pq-nf) of a $\lambda\mathbf{Jm}$-term t is denoted $\mathbf{p}(t)$ (resp. $\mathbf{q}(t)$, $\phi(t)$). These properties of permutations are proved in [5].

Subsystems of $\lambda\mathbf{Jm}$: We present several subsystems of $\lambda\mathbf{Jm}$ obtained by constraining the construction $t(u, l, (x)v)$ either by forcing $l = []$ or $v = x$ or both. The systems thus obtained correspond to previously known systems. They are identified in the following commutative diagram, alongside with mappings to interpret amongst them, defined in [5].

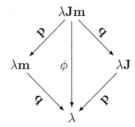

The terms of $\lambda\mathbf{J}$ are obtained by constraining l in $t(u, l, (x)v)$ to be $[]$. A gm-application of the form $t(u, [], (x)v)$ is called a *generalised application* (or *g-application*, for short) and is abbreviated to $t(u, (x)v)$. The reduction rules (resp. permutative conversion rules) for $\lambda\mathbf{J}$ are β_1 and π (resp. p_1, p_2 and p_3). The β_1, π-nfs are the terms whose g-application sub-terms have the form $x(u, (x)v)$, i.e. the head term is a variable. Let ΛJ be the Curry-Howard counterpart to von Plato's system of natural deduction with generalised elimination [13]. This system was studied by Joachimski and Matthes in [8]. The system $\lambda\mathbf{J}$ is isomorphic to ΛJ, if one disregards permutative conversion rules.

The terms of $\lambda\mathbf{m}$ are obtained by constraining v in $t(u, l, (x)v)$ to be x. A gm-application of the form $t(u, l, (x)x)$ is called a *multiary application* (or *m-application*, for short) and is written as $t(u, l)$.

In order to define the reduction rules of $\lambda\mathbf{m}$, we introduce the following auxiliary reduction rule in $\lambda\mathbf{Jm}$, corresponding to a combination of π and μ:

$$(h) \quad t(u, l, (x)x)(u', l', (y)v) \rightarrow t(u, \mathbf{a}(l, u' :: l'), (y)v) \tag{1}$$

The reduction rules for $\lambda\mathbf{m}$ are β_1, β_2 and h. The unique permutative conversion rule for $\lambda\mathbf{m}$ is q. The β, h-nfs are the terms where all m-applications occurring as subterms have the form $x(u, l)$, i.e. the head is a variable. If we disregard the permutative conversion rule, the system thus obtained is isomorphic to the λPh-calculus defined in [3,4].

A gm-application of the form $t(u, [], (x)x)$ is called a *(simple) application* and is written as $t(u)$. A λ-term is a term t such that every application occurring in t is simple. The set of λ-terms is closed for rule β_1, and λ-terms are exactly the pq-nfs. We obtain thus an isomorphic copy of the λ-calculus inside $\lambda\mathbf{Jm}$.

3 Combining Reduction and Permutation Rules I

We study the interaction between normalisation / cut-elimination and permutative conversions, by combining reduction and permutation rules of $\lambda\mathbf{Jm}$, and by analysing the resulting normal forms, which we call *combined normal forms*.

Figure 1 shows how the system captures important classes that show up in structural proof theory. $\boxed{\mathrm{Am}}$ represents the set of $\beta\pi$-nfs of $\lambda\mathbf{Jm}$, which are precisely the multiary-cut-free forms of [12]. $\boxed{\mathrm{A}}$ represents the set of "usual" (or unary) cut-free forms, which is the same as von Plato's "fully-normal" forms, and correspond to the $\beta\pi$-normal λJ-terms. The permutation-free multiary-cut-free forms of [12] are precisely the βh-normal λm-terms, which in turn capture Herbelin's cut-free $\overline{\lambda}$-terms. We call these *Herbelin-nfs*. Mint's "normal" cut-free derivation (or *Mints-nfs*, for short) are formalized, in the style of [2], as a subset of the $\beta\pi$-normal λJ-terms, as follows.

Definition 1. *A term* $v \in \lambda$Jm *is* x-normal *if* $v = x$ *or* $v = x(u, l, (y)v')$, *with* $x \notin u, l, v'$ *and* v' y-normal. *A* λJm-*term is* normal *if, for every* gm-*application* $t(u, l, (x)v)$ *occurring in it,* v *is* x-normal. *A* λJ-*term is a* Mints-normal form *if it is normal and a* $\beta\pi$-normal form.

The dotted arrows in the figure indicate the "place" of the permutation systems studied in [2,12]. It is well-known that the sets of Herbelin-nfs, Mints-nfs, and β-normal λ-terms (represented by $\boxed{B}$) are in bijective correspondence.

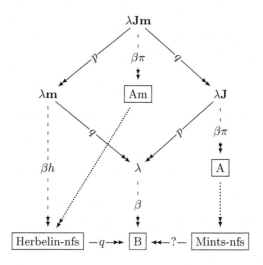

Fig. 1. Combining reduction and permutation in λJm

Combined nfs already available: Herbelin-nfs and β-normal λ-terms have immediate characterisations in terms of combinations of reduction and permutation rules of λJm.

Proposition 1. *1. t is a Herbelin-nf iff t is a βph-nf.*
    2. t is a β-normal λ-term iff t is a βpq-nf.

Some variants of these characterisations are possible. For instance, we may adjoin the μ rule to each of the above combinations, without changing the set of normal forms, since a μ-redex is also a p-redex.

    Unfortunately, $\to_{\beta pq}$ and $\to_{\beta ph}$ are non-terminating. Indeed, $\to_{\beta p}$ is already non-terminating. In order to prove non-termination of $\to_{\beta p}$, we need to recall the λx-calculus [11], a λ-calculus with explicit substitution. Its terms are given by

$$M, N ::= x \mid \lambda x.M \mid MN \mid \langle N/x \rangle M \ ,$$

and this set of terms is equipped with six reduction rules:

(B) $(\lambda x.M)N \to \langle N/x \rangle M$ (x_2) $\langle N/x \rangle (\lambda y.M) \to \lambda y.\langle N/x \rangle M$

(x_0) $\langle N/x \rangle x \to N$ (x_3) $\langle N/x \rangle (MM') \to (\langle N/x \rangle M)\langle N/x \rangle M'$

(x_1) $\langle N/x \rangle y \to y$ (x_4) $\langle N/x \rangle \langle N'/y \rangle M \to \langle \langle N/x \rangle N'/y \rangle M, \ x \notin M$

Theorem 1. *There is a typed* $t \in \lambda\mathbf{J}$ *such that* t *is not* βp-*SN.*

Proof: Let $I = \lambda x.x$ and $A = \lambda mn.I(n, (z)m(z))$. Define $(\_)^\star : \lambda\mathbf{x} \to \lambda\mathbf{J}$ as follows:

$$x^\star = x \qquad\qquad (MN)^\star = A(M^\star)(N^\star)$$
$$(\lambda x.M)^\star = \lambda x.M^\star \qquad (\langle N/x \rangle M)^\star = I(N^\star, (x)M^\star)$$

This mapping has the following property: If $R \in \{B, \mathsf{x}_0, \mathsf{x}_1, \mathsf{x}_2, \mathsf{x}_3, \mathsf{x}_4\}$ and $M \to_R N$ in $\lambda\mathbf{x}$, then $M^\star \to_{\beta p}^+ N^\star$. Let M be a typed λ-term such that M is not $B\mathsf{x}_0\mathsf{x}_1\mathsf{x}_2\mathsf{x}_3\mathsf{x}_4$-SN (one such term exists - see for instance [11]). Then $M^\star$ is a typed $\lambda\mathbf{J}$-term which is not βp-SN.   ∎

Another permutation rule: In order to overcome non-termination, we replace permutation rule p by a new permutation rule called s:

$$(s) \qquad t(u, l, (x)v) \to \mathsf{s}(t(u, l), x, v), \ v \neq x$$

Naturally, if one replaces p by s in Proposition 1, one gets another characterisation of Herbelin-nfs and β-normal λ-terms. This time, the characterisations are in terms of combinations of rules that are both confluent and terminating on typed terms.

Proposition 2 (Confluence). *Any of the following kinds of reduction is confluent:* s, βs, βsq *and* βsh.

Proof: By confluence of β in $\lambda\mathbf{m}$ or λ and βh in $\lambda\mathbf{m}$, together with the following properties of $\mathbf{p}$ and ϕ: (1) $\mathbf{p}$ maps a β (resp. h) step in $\lambda\mathbf{Jm}$ to zero or more β (resp. h) steps in $\lambda\mathbf{m}$, and collapses s steps. (2) For all $t \in \lambda\mathbf{Jm}$, $t \to_s^* \mathbf{p}(t)$. (3) ϕ maps a β step in $\lambda\mathbf{Jm}$ to zero or more β steps in λ, and collapses s and q steps. (4) For all $t \in \lambda\mathbf{Jm}$, $t \to_{sq}^* \phi(t)$.   ∎

The mapping $(\_)^\bullet : \lambda\mathbf{Jm} \to \lambda\mathbf{m}$ is given by

$$x^\bullet = x \qquad\qquad\qquad\qquad\qquad\qquad []^\bullet = []$$
$$(\lambda x.t)^\bullet = \lambda x.t^\bullet \qquad\qquad\qquad\qquad (u :: l)^\bullet = u^\bullet :: l^\bullet$$
$$(t(u, l, (x)v))^\bullet = \begin{cases} (\lambda x.v^\bullet)(t^\bullet(u^\bullet, l^\bullet)) & \text{if } v \neq x \\ t^\bullet(u^\bullet, l^\bullet) & \text{if } v = x \end{cases}$$

Proposition 3.   1. *If* $t \to_\beta u$ *in* $\lambda\mathbf{Jm}$, *then* $t^\bullet \to_\beta^+ u^\bullet$ *in* $\lambda\mathbf{m}$.

                   2. *If* $t \to_s u$ *in* $\lambda\mathbf{Jm}$, *then* $t^\bullet \to_\beta^+ u^\bullet$ *in* $\lambda\mathbf{m}$.

                   3. *For all* $t \in \lambda\mathbf{Jm}$, *if* $t^\bullet$ *is* β-*SN, then* t *is* βs-*SN.*

Proof: 1. and 2. are straightforward inductions and use $s(t, x, v)^{\bullet} = s(t^{\bullet}, x, v^{\bullet})$. 3. is immediate from 1. and 2.   ∎

Corollary 1 (SN). *If $t \in \lambda\mathbf{Jm}$ is typable, then t is βs-SN.*

Proof: If $t \in \lambda\mathbf{Jm}$ is typable, then so is $t^{\bullet}$. Hence $t^{\bullet}$ is β-SN and, by the previous proposition, t is βs-SN.   ∎

Let $R \in \{q, h\}$. The next Proposition, together with termination of $\to_R$, reduce termination of $\to_{\beta sR}$ to termination of $\to_{\beta s}$.

Proposition 4 (Postponement). *Let $R \in \{q, h\}$ and $S \in \{\beta, s\}$. If $t_1 \to_R t_2 \to_S t_3$, then there is t_4 such that $t_1 \to_S t_4 \to_R^* t_3$.*

Corollary 2 (SN). *If $t \in \lambda\mathbf{Jm}$ is typable, then t is βsq-SN and βsh-SN.*

A consequence of Proposition 4 is that βsq-reduction or βsh-reduction can always be split into two stages: first, a βs stage; next, a q or h stage . An illustration of this fact is in the following diagram.

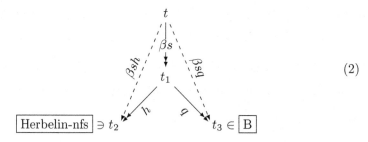

$$(2)$$

A βs-nf (*i.e.* a $\lambda\mathbf{m}$-term in β-nf) is a term whose applications are of the form

$$x(u_1, l_1)...(u_n, l_n) \tag{3}$$

for some $n \geq 1$. In addition, a βsh-nf requires that $n = 1$, whereas a βsq requires each l_i to be $[]$. For instance, the βs-nf $x(u_1, [v_{11}, v_{12}])(u_2, [v_{21}])$ has a h-nf of the form $x(u_1', [v_{11}', v_{12}', u_2', v_{21}'])$ and a q-nf of the form $x(u_1'')(v_{11}'')(v_{12}'')(u_2'')(v_{21}'')$. A βsq or βsh reduction splits into a βs-stage, followed by a q- or h-stage. The later stage simply organizes in a certain way the arguments of applications of the form (3).

4   Combining Reduction and Permutation Rules II

Now we study Mints-nfs and obtain a characterisation for them in terms of a well-behaved combination of reduction and permutation rules. It turns out that this result requires a deeper understanding of the constructors of $\lambda\mathbf{Jm}$ and their "overlaps", together with the rules that manifest such "overlaps". This leads to a systematic study of combined normal forms and, in particular, to a clarification of the relationship between Mints-nfs, Herbelin-nfs and β-nfs of the λ-calculus.

The overlap between multiarity and generality: In [6] one can find a study of the "overlap" between the multiarity and generality features of $\lambda\mathbf{Jm}$. Consider the following particular case of μ^{-1}, which we call ν: $t(u, u' :: l, (y)v) \to t(u, (x)x(u', l, (y)v))$ (x fresh). Repeated application of this rule eliminates uses of multiarity (*i.e.* occurrences of cons) at the expense of uses of generality. Conversely, rule μ shows that we may use cons as a shorthand for specific uses of generality.

The following mapping calculates the μ-normal form of each $\lambda\mathbf{Jm}$-term:

$$\mu(x) = x$$
$$\mu(\lambda x.t) = \lambda x.\mu(t)$$
$$\mu(t(u, l, (x)v)) = \begin{cases} \mu(t)(\mu(u), \mathbf{a}(\mu'(l), u' :: l'), (y)v'), \\ \qquad \text{if } \mu(v) = x(u', l', (y)v') \text{ and } x \notin u', l', v' \\ \mu(t)(\mu(u), \mu'(l), (x)\mu(v)), \quad \text{otherwise} \end{cases}$$

$$\mu'([]) = []$$
$$\mu'(u :: l) = \mu(u) :: \mu'(l)$$

In *op. cit.* it is proved that this mapping is a bijection between the set of $\lambda\mathbf{J}$-terms and the set of μ-normal forms (which is another manifestation of overlap). The inverse of μ is called ν and is given by:

$$\nu(x) = x$$
$$\nu(\lambda x.t) = \lambda x.\nu(t)$$
$$\nu(t(u, l, (x)v)) = \nu(t)(\nu(u), (z)\nu'(z, l, x, \nu(v))), \quad z \text{ fresh}$$
$$\nu'(z, [], x, v) = \mathbf{s}(z, x, v)$$
$$\nu'(z, u :: l, x, v) = z(\nu(u), (w)\nu'(w, l, x, v)), \quad w \text{ fresh}$$

Actually, still according to [6], this bijection can be turned into an isomorphism. First consider the variant π' of rule π, given by

$$t(u, l, (x)v)(u', l', (y)v') \to t(u, l, (x)v@_x(u', l', (y)v')) \ , \tag{4}$$

where $v@_x(u', l', (y)v') = x(u, l, (z)v@_z(u', l', (y)v'))$, if $v = x(u, l, (z)v)$ and $x \notin u, l, v$; and $v@_x(u', l', (y)v') = v(u', l', (y)v')$, otherwise.

Notice that t is a π-nf iff is a π'-nf. From now on we consider $\lambda\mathbf{J}$ equipped with π' instead of π. Second, for $R \in \{\beta, \pi\}$, equip the set of μ-normal forms with relation $\to_{R_\mu}$, defined as $\to_R$ followed by reduction to μ-normal form. Then, μ, ν establish an isomorphism between $\to_{\beta_\mu}$ (resp. $\to_{\pi_\mu}$), in the set of μ-nfs, and $\to_\beta$ (resp. $\to_{\pi'}$), in $\lambda\mathbf{J}$.

A refined analysis of the overlap between multiarity and generality: We now aim at refining this isomorphism. It may be helpful to have Figure 2 in mind.

Consider the set of *normal* $\lambda\mathbf{Jm}$-terms (recall Definition 1). This set is closed for $\to_\beta$ and $\to_{\pi'}$, hence naturally equipped with these rules. We obtain the system $\lambda\mathbf{nm}$. Also the set of normal $\lambda\mathbf{J}$-terms is naturally equipped with $\to_\beta$ and $\to_{\pi'}$. The latter system is denoted $\lambda\mathbf{n}$. It is, simultaneously, the unary (=cons-free) fragment of $\lambda\mathbf{nm}$ and the normal fragment of $\lambda\mathbf{J}$. On the other hand:

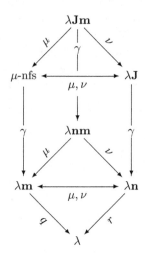

Fig. 2. Another view of the internal structure of $\lambda\mathbf{Jm}$

Lemma 1. 1. For all $t \in \lambda\mathbf{m}$: $t \to_{\beta_\mu} t'$ iff $t' \in \lambda\mathbf{m}$ and $t \to_\beta t'$ in $\lambda\mathbf{m}$.
  2. For all $t \in \lambda\mathbf{m}$: $t \to_{\pi_\mu} t'$ iff $t' \in \lambda\mathbf{m}$ and $t \to_h t'$ in $\lambda\mathbf{m}$.

Notice that in $\lambda\mathbf{m}$ there is no distiction between $\to_{\pi_\mu}$ and $\to_{\pi'_\mu}$.

Now, the restriction of μ to $\lambda\mathbf{n}$-terms and the restriction of ν to $\lambda\mathbf{m}$-terms are mutually inverse. From the previous lemma follows that these restrictions of μ and ν establish an isomorphism between $\to_\beta$, $\to_h$ in $\lambda\mathbf{m}$ and $\to_\beta$, $\to_{\pi'}$ in $\lambda\mathbf{n}$, respectively.

Theorem 2 (Isomorphism). Let R be β (resp. h) and let S be β (resp. π').

  1. $t \to_R t'$ in $\lambda\mathbf{m}$ iff $\nu(t) \to_S \nu(t')$ in $\lambda\mathbf{n}$.
  2. $t \to_S t'$ in $\lambda\mathbf{n}$ iff $\mu(t) \to_R \mu(t')$ in $\lambda\mathbf{m}$.

In particular, μ, ν establish a bijection between the set of normal $\lambda\mathbf{J}$-terms that are $\beta\pi$-nfs and the set of $\lambda\mathbf{m}$-terms that are βh-nfs. That is:

Corollary 3. The appropriate restriction of mapping μ is a bijection between the set of Mints-nfs and the set of Herbelin-nfs, whose inverse is the appropriate restriction of mapping ν.

A more systematic analysis of overlaps in $\lambda\mathbf{Jm}$: Consider the following diagram:

$$t(u,\mathbf{a}(l,u' :: l'),(y)v) \xleftrightarrows[\nu]{\mu} t(u,l,(x)x(u',l',(y)v))$$

$$\underbrace{t(u,l)\ (u',l',(y)v)}_{t(u,l,(x)x)}$$

proviso:
$x \notin u',l',v$ (5)

Any of the terms in this diagram consists of a function t, a first argument u, at least another argument u' and a "continuation" $(y)v$. The diagram shows three alternative ways of accommodating the extra argument u': either by using the list facility (top left corner), or by using a restricted form of the generality feature, sometimes called *normal* generality (top right corner), or by iterated application. So the diagram illustrates three ways of expressing *multiple* application in $\lambda\mathbf{Jm}$.

We adopt the extensions to rules q and ν suggested in the diagram, *e.g.*:

$$(q) \quad t(u, \mathbf{a}(l, u' :: l'), (y)v) \rightarrow t(u, l, (x)x(u', l', (y)v)) \ . \tag{6}$$

The versions of these rules considered so far correspond to the case $l = []$. Also a new rule r is defined in $\lambda\mathbf{Jm}$:

$$(r) \quad t(u, l, (x)v@_x(u', l', (y)v')) \rightarrow t(u, l, (x)v)(u', l', (y)v') \ , \tag{7}$$

where v is x-normal, v' is y-normal and $x \notin u', l', v'$. The particular case $v = x$ gives the version of the rule in diagram (5).

The example $t(u, (x)x(u', (y)y(u'', (z)v))) \rightarrow_r t(u, (x)(x(u')(u'', (z)v)))$ shows that neither $\lambda\mathbf{nm}$ nor $\underline{\lambda\mathbf{n}}$ is closed for $\rightarrow_r$. The problem is that the contracted r-redex (the underlined term) is x-normal, but the reduct is not. Similar observations apply to rule q. In order to overcome this fact, we define, for $\mathcal{R} \in \{r, q\}$, a new relation $\leadsto_{\mathcal{R}} \subseteq \rightarrow_{\mathcal{R}}$ that "respects" the normal fragment: in $\leadsto_{\mathcal{R}}$, reduction is allowed in the sub-expressions of an application $t(u, l, (x)v)$ (that is in t, u, l, v) only if v is x-normal; moreover, if v is the sub-expression where the reduction happens, the redex contracted is not v itself.

Not all occurrences of cons are eliminated by $\leadsto_q$. This is why rule q has to be supplemented with

$$t(u, l, (x)t'@_x(u', \mathbf{a}(l', u'' :: l''), (y)v)) \rightarrow t(u, l, (x)t'@_x(u', l', (z)z))(u'', l'', (y)v) \ , \tag{8}$$

where t' is x-normal and x does not occur outside t'. We consider this rule in reverse to belong to rule h.

$\lambda\mathbf{nm}$ and $\lambda\mathbf{n}$ are closed for $\leadsto_r$ and $\lambda\mathbf{nm}$ and $\lambda\mathbf{m}$ are closed for $\leadsto_q$. In $\lambda\mathbf{m}$, $\leadsto_q = \rightarrow_q$. In $\lambda\mathbf{n}$ $\leadsto_r$ has also a simple, alternative characterisation: $t \leadsto_r t'$ in $\lambda\mathbf{n}$ iff $\mu(t) \rightarrow_q \mu(t')$. So μ and ν establish an isomorphism between $\rightarrow_q$ in $\lambda\mathbf{m}$ and $\leadsto_r$ in $\lambda\mathbf{n}$.

Since $\rightarrow_q$ in $\lambda\mathbf{m}$ is terminating and confluent, so is $\leadsto_r$ in $\lambda\mathbf{n}$. For each $t \in \lambda\mathbf{n}$, let $r(t)$ denote the normal form of t w.r.t. $\leadsto_r$. Mappings μ, ν establish a bijection between q-normal $\lambda\mathbf{m}$-terms (*i.e* λ-terms) and $\lambda\mathbf{n}$-terms normal w.r.t $\leadsto_r$. Since ν leaves λ-terms invariant, the $\lambda\mathbf{n}$-terms normal w.r.t $\leadsto_r$ are exactly the λ-terms. Hence, we have the commutation of the lower triangle in Figure 2.

Proposition 5. $r \circ \nu = q$ *and* $q \circ \mu = r$.

Proof: If $t \in \lambda\mathbf{m}$, then $\nu(q(t)) = r(\nu(t))$, by the isomorphism between the $\rightarrow_q$ reduction of t and the $\leadsto_r$ reduction of $\nu(t)$. But $\nu(q(t)) = q(t)$. Hence $q(t) = r(\nu(t))$. The other statement follows from $\nu \circ \mu = id$. ∎

Corollary 4. *1. If $t \rightarrow_\beta t'$ in $\lambda\mathbf{n}$ then $r(t) \rightarrow_\beta r(t')$ in λ.*
         2. If $t \rightarrow_{\pi'} t'$ in $\lambda\mathbf{n}$ then $r(t) = r(t')$.

Proof: From Theorem 2, the previous proposition and the facts: (i) if $t \rightarrow_{\beta_i} t'$ in $\lambda\mathbf{m}$ then $\mathbf{q}(t) \rightarrow_{\beta_1} \mathbf{q}(t')$ in λ; (ii) if $t \rightarrow_h t'$ in $\lambda\mathbf{m}$ then $\mathbf{q}(t) = \mathbf{q}(t')$. ∎

Six combined nfs: We can now converge towards our ultimate goal, which is the diagram in Figure 3. From now on, $\mathcal{R}$-reduction refers to $\leadsto_\mathcal{R}$ and not to $\rightarrow_\mathcal{R}$, when $\mathcal{R} \in \{q, r, h, \pi'\}$. For instance, t is a βr-nf if t is irreducible for both $\rightarrow_\beta$ and $\leadsto_r$.

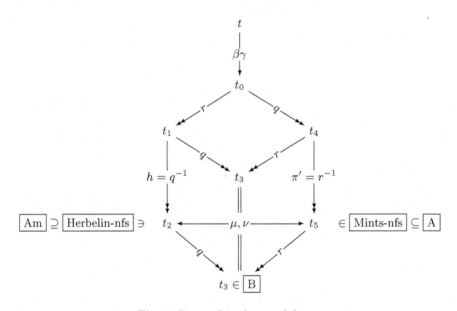

Fig. 3. Six combined normal forms

The results obtained so far give us another view of the internal structure of $\lambda\mathbf{Jm}$ and contribute to the diagram in Figure 3. Corollary 4 guarantees the commutation of the triangles with vertices t_1, t_2, t_3 and t_3, t_4, t_5, whereas Proposition 5 gives the commutation of triangle t_2, t_3, t_5 (by the way, the latter commutation extends Corollary 3). The last ingredient required by the diagram is the following rule:

$$(\gamma) \quad t(u, l, (x)v) \rightarrow s(t(u, l), x, v), \quad v \text{ is not } x\text{-normal.}$$

Notice that a $\lambda\mathbf{Jm}$-term is γ-normal iff is a $\lambda\mathbf{nm}$-term. Also observe that $\gamma \subset s$. On the other hand, $s \subset \gamma \cup r$, and t is s-normal iff t is γr-normal. So we do not need s anymore.

In order to guarantee Proposition 6 below, we will have to restrict several of the rules considered so far. But, since the aim is to combine those rules with γ, the restrictions will be harmless (the normal forms do not change). First, rule β is from now on restricted to the case where a redex $(\lambda x.t)(u, l, (y)v)$ satisfies:

t is normal and v is y-normal. Notice that, in the context of λ**nm** (hence of λ-calculus), this restriction is empty. Moreover, a λ**Jm**-term is $\beta\gamma$-normal in the old sense iff is $\beta\gamma$-normal in the new sense. Second, we impose v y-normal in rule h (see (1) and (8) in reverse) and in rule q (see (6) and (8)); and impose v x-normal and v' y-normal in rule π' (see (4)). In this way, $h = q^{-1}$ and $\pi' = r^{-1}$.

Theorem 3.   *1. t is a Herbelin-nf iff t is a $\beta\gamma q^{-1}r$-nf.*
          2. t is a Mints-nf iff t is a $\beta\gamma qr^{-1}$-nf.
          3. t is a β-normal λ-term iff t is a $\beta\gamma qr$-nf.

Proposition 6 (Postponement). *In λ**Jm***:*

1. Let $\mathcal{R} \in \{r, q, h, \pi'\}$. If $t_1 \leadsto_{\mathcal{R}} t_2 \to_{\beta} t_3$ then there is t_4 s.t. $t_1 \to_{\beta} t_4 \leadsto_{\mathcal{R}}^* t_3$.
2. Let $\mathcal{R} \in \{r, q, h, \pi'\}$. If $t_1 \leadsto_{\mathcal{R}} t_2 \to_{\gamma} t_3$ then there is t_4 s.t. $t_1 \to_{\gamma} t_4 \leadsto_{\mathcal{R}}^* t_3$.
3. Let $\mathcal{R} \in \{r, \pi'\}$. If $t_1 \leadsto_{\mathcal{R}} t_2 \leadsto_q t_3$ then there is t_4 s.t. $t_1 \leadsto_q t_4 \leadsto_{\mathcal{R}} t_3$.
4. Let $\mathcal{R} \in \{q, h\}$. If $t_1 \leadsto_{\mathcal{R}} t_2 \leadsto_r t_3$ then there is t_4 s.t. $t_1 \leadsto_r t_4 \leadsto_{\mathcal{R}} t_3$.

Corollary 5 (SN). *Every typable λ**Jm**-term is $\beta\gamma q^{-1}r$-SN, $\beta\gamma qr^{-1}$-SN and $\beta\gamma qr$-SN.*

Proof: It is easy to prove that $\leadsto_{\mathcal{R}}$ is terminating, when $\mathcal{R} \in \{q, r, h, \pi'\}$. These four termination results, together with Proposition 6 reduce strong normalisation for $\beta\gamma q^{-1}r$, $\beta\gamma qr^{-1}$ and $\beta\gamma qr$ to strong normalisation for $\beta\gamma$. Now, every typable λ**Jm**-term is $\beta\gamma$-SN, by Corollary 1. ∎

Proposition 7 (Confluence). *Any of the following kinds of reduction is confluent: $\beta\gamma q^{-1}r$, $\beta\gamma qr$ and $\beta\gamma qr^{-1}$.*

Proof: For $\beta\gamma q^{-1}r$ and $\beta\gamma qr$, the proof is very similar to the proof of Proposition 2. Observe that $\gamma \subset s$, $\mathbf{p}$ and ϕ collapse r-steps, and every λ**Jm**-term t can be γrq-reduced to $\phi(t)$.

As to confluence of $\beta\gamma qr^{-1}$, we could use the properties of mapping $\nu \circ \mathbf{p}$, but we prefer to offer a proof of a different style. Suppose u_0 $\beta\gamma qr^{-1}$-reduces to u_1 and u_2. By Proposition 6, there are v_1, v_2 such that u_0 $\beta\gamma q$-reduces to v_i and v_i r^{-1}-reduces to u_i, $(i = 1, 2)$. u_0, v_1 and v_2 have the same $\beta\gamma qr$-nf, say t. Again by Proposition 6, there are v_1' and v_2' such that v_i $\beta\gamma q$-reduces to v_i' and v_i' r-reduces to t. Now, by the same proposition, r-reduction postpones over $\beta\gamma q$-reduction. As such, there is u_i' such that u_i $\beta\gamma q$-reduces to u_i' and v_i' r^{-1}-reduces to u_i'. Let t_i be a r^{-1}-nf of u_i'. By Corollary 4, $r(t_i) = r(v_i')$, hence $r(t_i) = t$. Since t_1 and t_2 are Mints-nfs, $r(t_1) = r(t_2)$ entails $t_1 = t_2$. But u_i $\beta\gamma qr^{-1}$-reduces to t_i. ∎

Proposition 6 gives, in particular, for Herbelin-nfs *and* Mints-nfs, results analogous to those illustrated in diagram (2), saying that reduction to normal form splits into two stages:

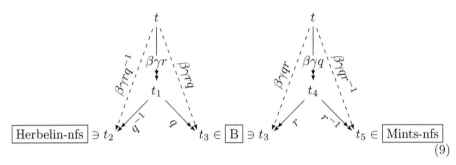

$$(9)$$

A $\lambda\mathbf{n}$-term in β-nf (like t_4) is a term whose applications are of the form

$$x(u_1, (y_1)v_1)...(u_n, (y_n)v_n) \qquad (10)$$

for some $n \geq 1$, and each v_i y_i-normal. The additional requirement of π'-normality imposes $n = 1$, whereas the additional requirement of normality w.r.t. $\leadsto_r$ imposes each v_i to be y_i. For instance, if t_4 is

$$x(u_1, (y_1)y_{11}(v_{11}, (y_{12})y_{12}(v_{12}, (z)z)))(u_2, (y_2)y_2(v_2, (w)w)) \ ,$$

then t_5 is of the form

$$x(u'_1, (y_1)y_{11}(v'_{11}, (y_{12})y_{12}(v_{12'}, (z)z(u'_2, (y_2)y_2(v'_2, (w)w))))))$$

and t_3 is of the form $x(u''_1)(v''_{11})(v''_{12})(u''_2)(v''_{21})$. So t_3, t_4 and t_5 differ only in the organization of the multiple arguments of applications (10).

The consequences of Proposition 6 are illustrated in a fuller way in the diagram of Figure 3. In this diagram, t is an arbitrary $\lambda\mathbf{Jm}$-term, t_0 is a $\lambda\mathbf{nm}$-term, t_1, t_2 and t_3 are $\lambda\mathbf{m}$-terms and t_3, t_4 and t_5 are $\lambda\mathbf{n}$-terms. Hence t_3 is a λ-term. All of them (except t) are in β-nf. Each term t_i ($0 \leq i \leq 5$) is a representative of one among six classes of combined normal forms: $\beta\gamma$, $\beta\gamma r$, $\beta\gamma q$, $\beta\gamma q^{-1}r$, $\beta\gamma qr^{-1}$ and $\beta\gamma qr$. For instance, t_4 is a $\beta\gamma q$-nf. The most inclusive of these classes is the class of $\beta\gamma$-nfs. A $\beta\gamma$-nf (like t_0) is a β-normal $\lambda\mathbf{nm}$-term, that is, a term whose applications are of the form

$$x(u_1, l_1, (y_1)v_1)...(u_n, l_n, (y_n)v_n)$$

for some $n \geq 1$, and each v_i y_i-normal. The remaining five combined normal forms are characterized by restrictions placed on n, l_i or v_i, as explained above when diagrams (2) and (9) were analyzed.

5   Conclusions

This study shows the level of systematization one achieves by doing proof-theoretical studies by means of term calculi. At the computational level, the insight one gains is this: all the interesting classes we identify relate to different ways of organizing the arguments of multiple application. Some ways of doing this organization are homogeneous, in the sense of making use of a single feature (multiarity, "normal" generality, or iterated application) in order to construct a multiple application. The classes determined by homogeneous organization are exactly the classes previously known, that is, the classes due to Herbelin and Mints, as well as the class of normal, ordinary natural deductions. Moreover, the computation to normal form can be organised in two stages, so that the second stage consists of choosing the way of representing multiple application. This suggests a new classification of rules, according to the stage they are involved in, which does not fit the division into reduction and permutation rules.

Acknowledgments. We thank the detailed comments provided by the anonymous referees. Diagrams in this paper were produced with Paul Taylor's macros.

References

1. R. Dyckhoff and L. Pinto. Cut-elimination and a permutation-free sequent calculus for intuitionistic logic. *Studia Logica*, 60:107–118, 1998.
2. R. Dyckhoff and L. Pinto. Permutability of proofs in intuitionistic sequent calculi. *Theoretical Computer Science*, 212:141–155, 1999.
3. J. Espírito Santo. *Conservative extensions of the λ-calculus for the computational interpretation of sequent calculus.* PhD thesis, University of Edinburgh, 2002. Available at http://www.lfcs.informatics.ed.ac.uk/reports/.
4. J. Espírito Santo. An isomorphism between a fragment of sequent calculus and an extension of natural deduction. In M. Baaz and A. Voronkov, editors, *Proceedings of LPAR'02*, volume 2514 of *Lecture Notes in Artificial Intelligence*, pages 354–366. Springer-Verlag, 2002.
5. J. Espírito Santo and Luís Pinto. Permutative conversions in intuitionistic multiary sequent calculus with cuts. In M. Hoffman, editor, *Proc. of TLCA'03*, volume 2701 of *Lecture Notes in Computer Science*, pages 286–300. Springer-Verlag, 2003.
6. J. Espírito Santo and Luís Pinto. Confluence and strong normalisation of the generalised multiary λ-calculus. In Ferruccio Damiani Stefano Berardi, Mario Coppo, editor, *Revised selected papers from the International Workshop TYPES 2003*, volume 3085 of *Lecture Notes in Computer Science*. Springer-Verlag, 2004.
7. H. Herbelin. A λ-calculus structure isomorphic to a Gentzen-style sequent calculus structure. In L. Pacholski and J. Tiuryn, editors, *Proceedings of CSL'94*, volume 933 of *Lecture Notes in Computer Science*, pages 61–75. Springer-Verlag, 1995.
8. F. Joachimski and R. Matthes. Short proofs of normalization for the simply-typed lambda-calculus, permutative conversions and Gödel's T. *Archive for Mathematical Logic*, 42:59–87, 2003.
9. G. Mints. Normal forms for sequent derivations. In P. Odifreddi, editor, *Kreiseliana*, pages 469–492. A. K. Peters, Wellesley, Massachusetts, 1996.
10. S. Negri and J. von Plato. *Structural Proof Theory*. Cambridge, 2001.
11. K. Rose. Explicit substitutions: Tutorial & survey. Technical Report LS-96-3, BRICS, 1996.
12. H. Schwichtenberg. Termination of permutative conversions in intuitionistic gentzen calculi. *Theoretical Computer Science*, 212, 1999.
13. J. von Plato. Natural deduction with general elimination rules. *Annals of Mathematical Logic*, 40(7):541–567, 2001.

Checking Conservativity of Overloaded Definitions in Higher-Order Logic

Steven Obua*

Technische Universität München
D-85748 Garching, Boltzmannstr. 3, Germany
obua@in.tum.de
http://www4.in.tum.de/~obua

Abstract. Overloading in the context of higher-order logic has been used for some time now. We define what we mean by *Higher-Order Logic with Conservative Overloading (HOLCO)*. HOLCO captures how overloading is actually applied by the users of Isabelle.

We show that checking whether definitions obey the rules of HOLCO is not even semi-decidable.

The undecidability proof reveals strong ties between our problem and the *dependency pair method* by Arts and Giesl for proving termination of TRSs via the notion *overloading TRS*. The *dependency graph* of overloading TRSs can be computed exactly. We exploit this by providing an algorithm that checks the conservativity of definitions based on the dependency pair method and a simple form of linear polynomial interpretation; the algorithm also uses the strategy of Hirokawa and Middeldorp of recursively calculating the strongly connected components of the dependency graph. The algorithm is powerful enough to deal with all overloaded definitions that the author has encountered so far in practice.

An implementation of this algorithm is available as part of a package that adds conservative overloading to Isabelle. This package also allows to delegate the conservativity check to external tools like the *Tyrolean Termination Tool* or the *Automated Program Verification Environment*.

1 Introduction

Higher-order logic (HOL) is widely used in the mechanical theorem proving community. There are several implementations of it available, among them HOL 4 [10], HOL-light [11] and Isabelle/HOL [4].

HOL has two theory extension mechanisms:

- It allows to introduce new constants that are either defined in terms of already known ones or uninterpreted. The name of the new constant must be different from the names of all already defined constants.
- New types can be introduced either by defining them as corresponding to a non-empty subset of an already existing type or by leaving them uninterpreted.

---

* Supported by the Ph.D. program "Logik in der Informatik" of the "Deutsche Forschungsgemeinschaft."

F. Pfenning (Ed.): RTA 2006, LNCS 4098, pp. 212–226, 2006.

HOL theories created using only these two mechanisms possess a set theoretic standard model [10]; this property implies consistency.

Isabelle/HOL inherits from its meta logic [2] two additional features, *axiomatic type classes* and *overloaded constant definitions*, both of which have been introduced in [3]. Overloading is most useful together with axiomatic type classes or a similar mechanism but it really is an orthogonal feature of the logic that can be studied separately. We therefore will focus on overloading alone.

In order to provide overloading as it is currently used in the proof-assistant Isabelle we cannot make the assumption that when defining a new constant the name of the constant has never been used before, especially not on the right hand side of the definition. Without this assumption, we are in trouble as the following example shows:

consts
   A :: $\alpha \to$ bool
defs
   A $(x :: \alpha \, \text{list} \times \alpha) \equiv$ A $(\text{snd}\, x \,\#\, \text{fst}\, x)$
   A $(x :: \alpha \, \text{list}) \equiv \neg$ A $(\text{tl}\, x, \text{hd}\, x)$.

There are two definitions for A. The two definitions do not *overlap*, that is A :: $\alpha \, \text{list} \times \alpha \to$ bool and A :: $\beta \, \text{list} \to$ bool do not unify. Nevertheless they lead to an inconsistent theory via

$$A\,[x] \;=\; \neg\,A\,([],x) \;=\; \neg\,A\,[x].$$

Note that this example is accepted by Isabelle 2005! Isabelle treats overloaded definitions just as axioms, merely ensuring that different definitions of the same constant do not overlap.

The trouble with the above example is that while we want to allow a limited *recursion over types*, the constant definition mechanism is not meant to implement *recursion over values*. The latter form of recursion is provided in HOL systems by safe fix point constructions from which the recursive equations are *proven*, not defined.

Thus we need a criterion for legal definitions. The one we will introduce in the next section is composed of two restrictions:

1. Two definitions may not overlap, that is the defined constants (including their respective types) do not unify (after renaming).
2. The process of unfolding definitions always terminates.

Under these two conditions overloading is safe. This will be made precise in the next section under the notion *conservative overloading*. The safety is then stated in Theorem 2.

While our main interest is the extension of theories by defining constants, we also have to take into account the definition of *types* in order to get a proper statement about the safety of overloading. Because a new constant could first be declared and partially defined, then a new type could be defined depending on this constant, then the definition of the constant is augmented by new overloaded definitions, and so on. So Theorem 2 tells us that it is OK to mix theory extensions in any order. This is a major improvement over the results in [3].

2 Higher-Order Logic with Conservative Overloading

The terms and types of higher-order logic (HOL) are those of simply-typed, polymorphic λ-calculus.

A *type* is either a type variable α, or a type constructor c of arity n applied to n type arguments $\tau_1, \ldots, \tau_n$, written $c(\tau_1, \ldots, \tau_n)$. The type variables occurring in a type τ are denoted by $tvars(\tau)$. Type substitutions σ are defined as usual, and for two types τ_1 and τ_2 we write $\tau_2 \leq \tau_1$ if there is a substitution σ such that $\tau_2 = \sigma\,\tau_1$. Two types τ_1 and τ_2 are called *overlapping* if there is a type τ such that $\tau \leq \tau_1$ and $\tau \leq \tau_2$.

A *term* is either a term variable $x :: \tau$ of type τ and with name x, or a constant $C :: \tau$, or an application $f\,g$ of two terms f and g, or an abstraction $\lambda x :: \tau.\,b$ where $x :: \tau$ is a variable and b is a term. Of course we consider only well-typed terms. The type of a term t is denoted by $ty(t)$, the type variables occurring in t by $tvars(t)$. Note that the inclusion $tvars(ty(t)) \subseteq tvars(t)$ can be strict! The set of all constants in t is denoted by $consts(t)$. We call two constants $C_1 :: \tau_1$ and $C_2 :: \tau_2$ *overlapping* if τ_1 and τ_2 are overlapping and $C_1 = C_2$.

A *signature* is a finite pool consisting of type constructors c together with their arities $\iota(c)$, and constant names C together with their most general type $ctype(C)$.

A type τ is *well-formed* for a signature S if all type constructors c in τ are from S and applied to $\iota(c)$ arguments. A term t is *well-formed* for S if it contains only well-formed types and if for all constants $C :: \tau$ in t the name C is from S and also $\tau \leq ctype(C)$. We denote the well-formed terms of S by $Term(S)$.

The *initial signature* S_I contains at least the nullary type prop of propositions, the binary type $\tau \to \tau'$ of functions, the constant $\equiv :: \alpha \to (\alpha \to prop)$ denoting equality, and the two constants rep $:: \alpha \to \beta$, abs $:: \beta \to \alpha$ used for converting between an abstract type and its representation.

We call a term t a *proposition* iff $ty(t) = prop$; therefore every equation $u \equiv v$ is a proposition.

A *type definition* is a triple $(c, [\alpha_1, \ldots, \alpha_n], p)$ such that c is a type constructor name (not necessarily in the signature!), $\alpha_i \neq \alpha_j$ for $i \neq j$, p is a well-formed term of type $ty(p) = \tau \to prop$ for some type τ, p has no free variables and $tvars(p) \subseteq \{\alpha_1, \ldots, \alpha_n\}$. This defines a new type such that the universal set of all elements of this new type is in bijection with the set $\{x :: \tau \mid p\,x\}$.

A *constant definition* is a well-formed proposition $C :: \tau \equiv u$ such that u has no free variables, $tvars(u) = tvars(\tau)$ and $ty(u) = \tau$. Two constant definitions $C_1 :: \tau_1 \equiv u_1$ and $C_2 :: \tau_2 \equiv u_2$ are said to *overlap* iff $C_1 :: \tau_1$ and $C_2 :: \tau_2$ overlap.

A *theory* $\mathfrak{T}$ is a quadruple $(S, Axioms_O, Axioms_T, Axioms_D)$ consisting of a signature S, a set of well-formed propositions $Axioms_O$, a set of type definitions $Axioms_T$ and a set of constant definitions $Axioms_D$.

A *theorem* is a triple written $\Gamma \vdash_{\mathfrak{T}} t$ such that the well-formed proposition t can be deduced in the theory $\mathfrak{T}$ from the *base theorems of* $\mathfrak{T}$ and the finitely many assumptions (well-formed propositions) Γ by the inference rules of higher-order logic [2,10,11]. We will use the shorter notation $\vdash_{\mathfrak{T}} t$ for $\emptyset \vdash_{\mathfrak{T}} t$.

The *base theorems of* $\mathfrak{T}$ arise from the theory $\mathfrak{T}$ in the following way:

- for each $p \in Axioms_0 \cup Axioms_D$ we have the theorem $\vdash_{\mathfrak{T}} p$,
- for each type definition $(c, [\alpha_1, \ldots, \alpha_n], p) \in Axioms_T$ and any variable name x we have the two theorems

$$\{p(x :: \tau)\} \vdash_{\mathfrak{T}} (\mathbf{rep} :: \tau_r)((\mathbf{abs} :: \tau_a)(x :: \tau)) \equiv x :: \tau,$$

$$\vdash_{\mathfrak{T}} (\mathbf{abs} :: \tau_a)((\mathbf{rep} :: \tau_r)(x :: \tau')) \equiv x :: \tau',$$

where $ty(p) = \tau \to \mathbf{prop}$, $\tau' = c(\alpha_1, \ldots, \alpha_n)$, $\tau_r = \tau' \to \tau$, and $\tau_a = \tau \to \tau'$.

2.1 Theory Extensions

Starting from an initial theory, new axioms, type and constant definitions are added to the theory. This is where it gets interesting for us: how do we for example ensure that the definition of possibly overloaded constants does not endanger the consistency of our theory?

A theory $\mathfrak{T}'$ is called an *extension* of a theory $\mathfrak{T}$ if one can arrive at $\mathfrak{T}'$ from $\mathfrak{T}$ by a series of theory extensions.

The *initial theory* is given by $(S_I, Axioms_I, \emptyset, \emptyset)$ where S_I is the initial signature, and $Axioms_I$ the set of axioms specific for HOL (or a weaker logic like Isabelle/Pure). The initial signature must also include all constants that appear in any of the inference rules of the logic.

A *well-formed theory* is a theory that is an extension of the initial theory.

In the following, $\mathfrak{T} = (S, Axioms_0, Axioms_T, Axioms_D)$ denotes the theory that gets extended, and $\mathfrak{T}' = (S', Axioms'_0, Axioms'_T, Axioms'_D)$ denotes the resulting extension. We only write down those resulting theory components which change.

(E1) Extending a theory by declaring a new constant

DECLARE $(\mathfrak{T}, C :: \tau) \leadsto S' = S \cup \{C \text{ with } ctype(C) = \tau\}$ succeeds iff $C :: \tau$ is a constant, τ a well-formed type and C not from S.

(E2) Extending a theory by defining a constant

DEFINE $(\mathfrak{T}, C :: \tau \equiv u) \leadsto Axioms'_D = Axioms_D \cup \{C :: \tau \equiv u\}$ succeeds iff

1. $C :: \tau \equiv u$ is a constant definition (as defined earlier) with respect to $\mathfrak{T}$ such that no τ' exists with τ and τ' overlapping and $C :: \tau' \in Crit_{\mathfrak{T}}$, where

$$Crit_{\mathfrak{T}} = \{C :: ctype(C) \mid C \text{ is from } S_I\} \cup$$

$$\bigcup \{consts(p) \mid p \in Axioms_0\} \cup \{C :: \tau \mid \exists u. C :: \tau \equiv u \in Axioms_D\}.$$

2. the reduction system (see Section 2.2) induced by the set of constant definitions $Axioms'_D$ is *terminating*.

(E3) Extending a theory by declaring a type

TYPEDECL $(\mathfrak{T}, c, n) \leadsto S' = S \cup \{c \text{ with } \iota(c) = n\}$ succeeds iff c is not from S.

(E4) Extending a theory by declaring and defining a type

TYPEDEF $(\mathfrak{T}, d, \vdash_{\mathfrak{T}} pt) \rightsquigarrow S' = S \cup \{c \text{ with } \iota(c) = n\}$, $Axioms'_T := Axioms_T \cup \{d\}$ succeeds iff $d = (c, [\alpha_1, \ldots, \alpha_n], p)$ is a type definition with respect to $\mathfrak{T}$ and also c is not from S. Note that you have to provide a theorem $\vdash_{\mathfrak{T}} pt$ which states that the defined type is non-empty.

(E5) Extending a theory by asserting an axiom

ASSERTAXIOM $(\mathfrak{T}, p) \rightsquigarrow Axioms'_0 = Axioms_0 \cup \{p\}$ succeeds iff p is a well-formed proposition and if for all $c \in consts(p)$ and all $c' \in Crit_D$ we have that c and c' do not overlap, where

$$Crit_D = \{C :: \tau \mid \exists u. C :: \tau \equiv u \in Axioms_D\}.$$

2.2   The Reduction System Induced by a Set of Definitions

For a signature S let D be a non-overlapping set of constant definitions; that is, no two different constant definitions $d_1, d_2 \in D$ overlap.

We define the abstract reduction system [5] $RS(D) = (Term(S), \to_D)$ by

$$t \xrightarrow{(p,d)}_D t' \quad \text{iff} \quad \begin{array}{l} \text{there is } d = C :: \tau \equiv u \in D \text{ and a type substitution} \\ \sigma \text{ such that } C :: \sigma\tau \text{ occurs in } t \text{ at position } p \text{ and } t' \\ \text{can be obtained by replacing this specific occurrence} \\ \text{in } t \text{ with } \sigma u. \end{array}$$

We just write $t \to_D t'$ if (p, d) is of no importance to us.

The reduction system is *terminating* iff there is no infinite chain

$$t \to_D t' \to_D t'' \to_D \ldots.$$

The reduction system is *confluent* iff for $t \to_D^* t'$ and $t \to_D^* t''$ we can always find s with $t' \to_D^* s$ and $t'' \to_D^* s$. As usual, $\to_D^*$ denotes the reflexive and transitive closure of $\to_D$.

Theorem 1. *$RS(D)$ is confluent.*

Proof. See [1, Theorem 1]. □

Note that confluence depends critically on the restriction on constant definitions $C :: \tau \equiv u$ that there may be no type variables in u that do not also occur in τ.

Therefore every terminating $RS(D)$ is *convergent* and hence

$$\mathcal{N}_D(t) = s \quad \text{iff } t \to_D^* s \text{ and there is no } s' \text{ such that } s \to_D s'$$

is a total and uniquely defined function on $Term(S)$. Note that $\mathcal{N}_D$ is *type preserving*, that is $ty(\mathcal{N}_D(t)) = ty(t)$ holds for all $t \in Term(S)$.

For a given well-formed theory $\mathfrak{T} = (S, Axioms_0, Axioms_T, Axioms_D)$ we know by the definition of well-formedness that $Axioms_D$ is non-overlapping and $RS(Axioms_D)$ terminating. Therefore $\mathcal{N}_{Axioms_D}$ is well-defined; we use the shorter notation $\mathcal{N}_{\mathfrak{T}}$ for this function, as well as the notation $RS(\mathfrak{T})$ instead of $RS(Axioms_D)$.

2.3   Overloading Is Conservative

We are now in a position to state that adding overloading to higher-order logic as described in the previous sections is *conservative* in the sense that every well-formed theory $\mathfrak{T} = (S, Axioms_0, Axioms_T, Axioms_D)$ can be reduced to a well-formed theory $\mathfrak{T}^- = (S, Axioms_0, Axioms_T^-, \emptyset)$. We achieve this reduction by setting

$$Axioms_T^- = \{(\mathsf{c}, [\alpha_1, \ldots, \alpha_n], \mathcal{N}_{\mathfrak{T}}(p)) \mid (\mathsf{c}, [\alpha_1, \ldots, \alpha_n], p) \in Axioms_T\}.$$

Theorem 2. *Assume that $\mathfrak{T}$ is a well-formed theory. Then so is $\mathfrak{T}^-$ and if $\{h_1, \ldots, h_n\} \vdash_{\mathfrak{T}} t$ then also $\{\mathcal{N}_{\mathfrak{T}}(h_1), \ldots, \mathcal{N}_{\mathfrak{T}}(h_n)\} \vdash_{\mathfrak{T}^-} \mathcal{N}_{\mathfrak{T}}(t)$. On the other hand, if $\{h_1, \ldots, h_n\} \vdash_{\mathfrak{T}^-} t$, then also $\{h_1, \ldots, h_n\} \vdash_{\mathfrak{T}} t$.*

Proof. The theorem can be split in two parts that can be proved separately, the first part by induction over the inference rules, the second one by induction over the theory extensions; for the proof we refer the interested reader to [1, Corollary 1]. □

The above Theorem 2 states that *overloading is conservative*. What are the implications of this? Let us look for example at *consistency preservation*: Assume that a well-formed theory $\mathfrak{T}$ is inconsistent, that is for all well-formed t the theorem $\vdash_{\mathfrak{T}} t$ is provable. Then Theorem 2 tells us that $\mathfrak{T}^-$ is well-formed and inconsistent, too!

If we now fix $Axioms_0 = Axioms_I$ (that is, we do not use (E5)) and set $Axioms_I$ to the ordinary axioms of HOL, then this is not possible, because then well-formedness of $\mathfrak{T}^-$ implies its consistency: $\mathfrak{T}^-$ is an ordinary HOL theory!

Note that conservativity in the sense of Theorem 2 is different from syntactical conservativity which has been proposed in [3] as a mandatory property of theory extensions. Our theory extensions (only considering (E1)-(E4), of course, for (E5) this violation is clear anyway) actually violate this property which demands that if $\mathfrak{S}$ is an extension of $\mathfrak{T} = (S, \ldots)$ and p a well-formed proposition from $Term(S)$, then $\vdash_{\mathfrak{S}} p$ implies also $\vdash_{\mathfrak{T}} p$. This violation is a natural consequence of the fact that declaration and definition of constants are done separately and that later definitions can give previously only declared constants additional properties.

3   Proving Termination

3.1   Overloading Term Rewriting Systems

A *first-order term rewriting system* (TRS) is a pair (F, R) where F is the signature of the term rewriting system and R is the set of *rules* [5,6]. The signature of a term rewriting system is a set of function symbols with an *arity* associated with each symbol. A rule is a pair of first-order terms over the signature F, written $t \Longrightarrow t'$, such that t is no variable and all variables of t' appear also in t.

Given a theory $\mathfrak{T} = (S, Axioms_0, Axioms_T, Axioms_D)$ we define the term rewriting system $TRS(\mathfrak{T})$ induced by $\mathfrak{T}$ by setting

$$F = \{\text{type constructor names of } S\} \cup \{\text{constant names of } S\}$$

$$arity(f) = \begin{cases} \iota(f) & \text{if } f \text{ is a type constructor name} \\ 1 & \text{if } f \text{ is a constant name} \end{cases}$$

$$R = \{\mathsf{C}(\tau) \Longrightarrow \mathsf{C}'(\tau') \mid \exists u.\, \mathsf{C}' :: \tau' \in consts(u) \wedge \mathsf{C} :: \tau \equiv u \in Axioms_D\}.$$

In the above we assume that the set of type constructor names of S and the set of constant names of S are disjoint.

$TRS(\mathfrak{T})$ is of special shape. Let us be more precise: we say that a TRS is an *overloading TRS* if all of its rules r_i, $i = 1, \ldots, n$, have the form

$$r_i : f_i(a_i) \Longrightarrow g_i(b_i)$$

such that

1. neither f_i nor g_i appear in any of the terms $a_1, b_1, \ldots, a_n, b_n$,
2. if there are substitutions σ, ϑ such that $\sigma(f_i(a_i)) = \vartheta(f_j(a_j))$, then $a_i = a_j$ (and of course also $f_i = f_j$); that is, unifiable left-hand sides are identical.

We call the f_i's and g_i's the *head symbols* of the overloading TRS.

Every TRS can be considered a reduction system on the first-order terms over its signature. For this reduction relation we also write $\Longrightarrow$, for its reflexive and transitive closure we write $\Longrightarrow^*$.

It is not difficult to show directly that $RS(\mathfrak{T})$ terminates iff $TRS(\mathfrak{T})$ terminates.But we gain more insight by using a result that has been established in [7] about the connection between *dependency pairs* and termination of term rewriting systems. This insight is later on put to further good use when we tackle the question of how to practically prove termination of an overloading TRS.

So assume $T = (F, R)$ is an overloading TRS. We call $f \in F$ *defined* if there is a rule $f(\ldots) \Longrightarrow \ldots \in R$. Let us further assume that there is an injective map from the set of defined symbols to the set of not defined symbols which preserves the arity, so that a defined $\mathsf{C} \in F$ is assigned a not defined $\tilde{\mathsf{C}}$ which appears in none of the rules. This is no restriction since we could just add new symbols to F.

We call $\langle \tilde{\mathsf{C}}_1(\tau_1), \tilde{\mathsf{C}}_2(\tau_2) \rangle$ a *dependency pair* if $\mathsf{C}_1(\tau_1) \Longrightarrow \mathsf{C}_2(\tau_2) \in R$ and C_2 is defined. For an overloading TRS this notion of dependency pair actually coincides with the notion found in [7,6].

A *chain* is a (finite or infinite) sequence $\langle s_1, t_1 \rangle, \langle s_2, t_2 \rangle, \ldots$ of dependency pairs such that there are substitutions (with respect to F) σ_i with $\sigma_i t_i \Longrightarrow^* \sigma_{i+1} s_{i+1}$ for all $i = 1, 2, \ldots$. The following theorem has been stated and proven in [7, Theorem 6]:

Theorem 3. *A TRS is terminating iff no infinite chain of dependency pairs exists.*

In general it is not decidable to check whether even only two dependency pairs form a chain; the next theorem teaches that in the case of an overloading TRS this check is just mere syntactic unification. We call a substitution σ *simple* iff the range of σ consists only of terms that contain no head symbols.

Theorem 4. *A simple* chain *is a sequence* $\langle s_1, t_1 \rangle$, $\langle s_2, t_2 \rangle$, ... *of dependency pairs such that there are simple substitutions* σ_i *with* $\sigma_i t_i = \sigma_{i+1} s_{i+1}$.
  In an overloading TRS, the notions chain and simple chain coincide.

Proof. We refer the reader to [1, Theorem 8].   □

One of the anonymous referees pointed out that an overloading TRS belongs to the class of TRSs for which *innermost termination* implies termination, and that for overloading TRSs, simple chains are exactly the *innermost chains*. Instead of using the above theorem, Theorem 16 of [20] could then be used.

Theorem 5. $RS(\mathfrak{T})$ *is terminating iff* $TRS(\mathfrak{T})$ *is terminating.*

Proof. The existence of an infinite reduction sequence in $RS(\mathfrak{T})$ is equivalent to the existence of an infinite simple chain in $TRS(\mathfrak{T})$ [1, Theorem 4]. Theorems 4 and 3 then prove our claim.

Theorem 6. *We can implement a checker for HOLCO iff we can decide termination of any overloading TRS.*

Proof. The difficult part of checking if a theory development conforms to the rules of HOLCO is to decide if $RS(\mathfrak{T})$ for a given theory $\mathfrak{T}$ is terminating. Theorem 5 shows that this can be rephrased as deciding if $TRS(\mathfrak{T})$ is terminating.
  On the other hand, one can encode the termination problem for any overloading TRS as the termination problem of $RS(\mathfrak{T})$ for some theory $\mathfrak{T}$ (see [1, Theorem 6]).   □

3.2   Undecidability of Proving Termination

Because of Theorem 6 we can show the impossibility of checking HOLCO by reducing *Post's Correspondence Problem for Prefix Morphisms* (PCPP) to the termination problem for overloading TRSs. An instance of the PCPP is the following problem: Given pairs $(a_1, b_1), \ldots, (a_n, b_n)$ of non-empty finite words over the alphabet $\{0, 1\}$ such that for $i \neq j$ neither a_i is a prefix of a_j nor b_i is a prefix of b_j, decide if there is a solution, that is a finite sequence $i_1, \ldots, i_m$ with

$$a_{i_1} a_{i_2} \ldots a_{i_m} = b_{i_1} b_{i_2} \ldots b_{i_m}.$$

The PCPP is undecidable [9].
  For the above instance of the PCPP, we construct an overloading TRS. The construction is basically the one that is used in [6] for reducing *Post's Correspondence Problem*, which is the PCPP without demanding that the a_i's and b_j's cannot be prefixes of each other, to the termination problem of a general TRS.

Our TRS has five function symbols: a nullary symbol $\square$, three unary symbols 0, 1, C, and one ternary symbol c. For a binary word $p = q_1 q_2 \ldots q_n$ (that is $q_1, \ldots, q_n \in \{0, 1\}$) we define the term

$$p(t) = q_1(q_2(\ldots(q_n(t))\ldots)).$$

For each pair (a_i, b_i) of the PCPP instance the TRS has a rule

$$\mathsf{C}(\mathsf{c}(a_i(\alpha), b_i(\beta), \gamma)) \Longrightarrow \mathsf{C}(\mathsf{c}(\alpha, \beta, \gamma)),$$

furthermore it has the rules

$$r_0: \quad \mathsf{C}(\mathsf{c}(\square, \square, 0(\alpha))) \Longrightarrow \mathsf{C}(\mathsf{c}(0(\alpha), 0(\alpha), 0(\alpha))),$$
$$r_1: \quad \mathsf{C}(\mathsf{c}(\square, \square, 1(\alpha))) \Longrightarrow \mathsf{C}(\mathsf{c}(1(\alpha), 1(\alpha), 1(\alpha))).$$

Theorem 7. *The TRS we just constructed is an overloading TRS, and it is terminating iff the corresponding PCPP instance has no solution.*

Proof. The left hand sides of no two rules unify, because for $a_i(\alpha)$ and $a_j(\alpha')$ to unify, a_i would need to be a prefix of a_j or the other way around. Therefore the TRS is overloading.

If the PCPP instance has a solution $i_1, \ldots, i_m$, then for $s = a_{i_1} \ldots a_{i_m} = b_{i_1} \ldots b_{i_m}$

$$\mathsf{C}(\mathsf{c}(s(\square), s(\square), s(\square)))$$

starts a cyclic and therefore infinite reduction sequence.

If the TRS is not terminating, then there exists an infinite simple chain. Because all the other rules reduce the size of the term, there must be two dependency pairs in this chain that correspond to r_0 or r_1. The dependency pairs between these two pairs form a solution of the PCPP instance, after throwing away those pairs that correspond to r_0 or r_1. That the solution is not empty is ensured because directly after r_0 or r_1 have been applied, neither of them can be applied again. $\square$

Corollary 1. *It is undecidable if an overloading TRS is terminating. Actually, it is not even semi-decidable.*

Proof. It is semi-decidable if any given PCPP instance has a solution. Considering that the PCPP is undecidable [9], Theorem 7 tells us that the termination of an overloading TRS cannot be semi-decidable. $\square$

3.3   Practically Proving Termination

Corollary 1 shows that any check a proof-assistant might employ for deciding if a set of definitions is associated with a reduction system that terminates will be incomplete: there will always be definitions that *should* pass but *won't* pass.

On the other hand overloading has proven to be a very useful technique which is regularly taken advantage of by users of the proof-assistant Isabelle. Until now the use of overloading in Isabelle was an act of faith; experienced users "knew" how to use overloading only in a sound way. Therefore it is worthwhile to examine how to devise a check that can give definitive and reliable answers on almost all overloaded definitions an experienced user would issue.

We have developed an add-on to the proof-assistant Isabelle that provides functionality for checking if $RS(D)$ terminates for the set D of definitions of an Isabelle theory. The check can either be done by using external TRS termination provers, or by using a built-in termination prover [1].

External Provers and the Dependency Pair Method. There are now several tools available that can prove the termination of a wide range of term rewriting systems, among them *AProVE* [12] and the *Tyrolean Termination Tool* [13]. We have shown that checking definitions can be done by checking if an overloading TRS is terminating. Therefore we can just give this overloading TRS to one of these tools.

We have tested the viability of this approach by checking five formalizations in Isabelle/HOL, all of which make use of overloading:

Main the starting point for Isabelle/HOL users,
Bali which has been concerned with the formalization of various aspects of the programming language Java [16],
Mat a formalization of checking the bounds of real linear programs employing matrices [14],
Nom the implementation of nominal techniques for Isabelle [15],
Ocl an embedding of OCL into HOL [19], example `royals-and-loyals`.

Our add-on extracts the overloading TRS that corresponds to the definitions of the corresponding Isabelle theories. Note that *all* definitions of a theory are collected for this purpose; this also recursively includes the definitions of the parent theories of the theory.

Giving the generated TRS directly to either AProVE or TTT fails with a time-out after several minutes. Can we somehow preprocess the TRS to provide the checkers with easier input?

Both checkers use the dependency pair method invented in [7]. The basic idea there is to calculate a *dependency graph*. This graph has the dependency pairs of the TRS as its nodes and there is an edge from $\langle a, b \rangle$ to $\langle c, d \rangle$ if $\langle a, b \rangle \langle c, d \rangle$ is a chain. Therefore in general the dependency graph is not computable and it is necessary to use an approximation of the dependency graph. We can do better for overloading TRSs. Chains are simple chains and therefore we can compute the exact dependency graph which has an edge from $\langle a, b \rangle$ to $\langle c, d \rangle$ iff there are substitutions σ, ϑ such that $\sigma b = \vartheta c$.

An overloading TRS is not terminating iff it admits an infinite simple chain of dependency pairs. Mapping this infinite chain to the exact dependency graph yields an infinite path p in the graph. We call a maximal set N of nodes such that for $n, m \in N$, there are non-empty paths from n to m and from m to n a *cyclic component* of the graph. Each cyclic component is a *strongly connected component*. If there is an infinite path p then there also must be an infinite path p' with all of its nodes belonging to the same cyclic component, because the dependency graph is finite.

Each cyclic component of the dependency graph is a set of dependency pairs, and can therefore be considered a subset of the rules of the overloading TRS. The overloading TRS is terminating iff all TRSs that correspond to a cyclic component are terminating.

Therefore we have found a method to break up our big overloading TRS into a couple of (much) smaller overloading TRSs whose dependency graphs consist of exactly one cyclic component. APROVE can automatically solve all of these systems; TTT time-outs for several of them in automatic mode but manages to solve all of them in semi-automatic mode with *polynomial interpretations* switched on. Both checkers only need a fraction of a second for all of the smaller term rewriting systems combined.

An Algorithm For Proving Termination. For a proof-assistant it is somewhat unsatisfying and clumsy to have to rely on external TRS termination provers. Therefore we present an algorithm for proving termination of an overloading TRS that is easy to implement and powerful: it can handle all of our examples, that is Bali, Mat, Nom and Ocl (Main is dealt with trivially). It is a variant of the dependency pair method [7] and uses the recursive calculation of cyclic components from [8].

Let us look at a cyclic component of the DG and assume that the component consists of the dependency pairs $d_1, \ldots, d_N$ where $d_i = \langle s_i, t_i \rangle$. If we can find relations $\succ, \succeq \subseteq T \times T$, where T is the set of first-order terms of the overloading TRS, such that for all $x, y, z \in T$ and substitutions σ

- $\succ$ is well-founded, that is there is no infinite chain $u_0 \succ u_1 \succ u_2 \succ \ldots$,
- $x \succ y$ and $y \succeq z$ implies $x \succ z$, and from $x \succeq y$ and $y \succ z$ follows $x \succ z$,
- $x \succ y$ implies $\sigma x \succ \sigma y$, and $x \succeq y$ implies $\sigma x \succeq \sigma y$,

and such that $s_i \succ t_i$ for $i = 1, \ldots, M$, where $1 \leq M \leq N$ and $s_i \succeq t_i$ for $i = M+1, \ldots, N$, then we can drop the nodes $d_1, \ldots, d_M$ and start all over with the cyclic components of this reduced graph.

Theorem 8. *If we can continue this process until we have arrived at the empty graph then we have successfully shown termination.*

Proof. The whole overloading TRS terminates if each of its cyclic components corresponds to a terminating overloading TRS. The TRS that corresponds to the cyclic component we picked is not terminating iff there exists an infinite simple chain of dependency pairs $\langle u_1, v_1 \rangle, \langle u_2, v_2 \rangle, \ldots$ where $\langle u_i, v_i \rangle \in \{d_1, \ldots, d_N\}$ for all i, which means that there are substitutions σ_i with

$$\sigma_1 u_1 \vartriangleright_1 \sigma_1 v_1 = \sigma_2 u_2 \vartriangleright_2 \sigma_2 v_2 = \cdots.$$

In the above $\vartriangleright_i$ is either $\succ$ or $\succeq$, depending on if $j \leq M$ or $j > M$, respectively, where $\langle u_i, v_i \rangle = d_j$. Here we have used the property of $\succ$ and $\succeq$ to be closed under substitutions. If any of the d_j's for $j \leq M$ is equal to $\langle u_i, v_i \rangle$ for infinitely many i then $\vartriangleright_i$ is equal to $\succ$ for infinitely many i. This allows us to construct an infinite descending chain with respect to $\succ$ by using the transitivity-like properties of $\succ$ and $\succeq$. But this contradicts the property of $\succ$ to be well-founded. Therefore there exists an infinite simple chain of dependency pairs in $\{d_1, \ldots, d_N\}$ iff there exists an infinite simple chain of dependency pairs in $\{d_{M+1}, \ldots, d_N\}$.   □

Which relations $\succ$ and $\succeq$ should we choose? This depends on the cyclic component we are currently examining. We use a simple, efficient form of linear polynomial interpretation to find these relations that has been inspired by the method described in [18], but does not lead to combinatorial explosion.

Terms are interpreted as those linear polynomials that can be viewed as functions from $\mathbb{R}_+ \times \cdots \times \mathbb{R}_+$ to $\mathbb{R}_+$. The set $\mathbb{R}_+$ denotes the set of non-negative real numbers.

To each symbol f of the overloading TRS we assign a real constant $c_f \geq 0$, and to each variable α of the TRS we assign a real variable $x_{i(\alpha)}$. The function i converts variables into indices and is injective. A term t is interpreted as $\Theta(t)$:

$$\Theta(\alpha) = x_{i(\alpha)}, \qquad \Theta(f(t_1, \ldots, t_n)) = c_f + \Theta(t_1) + \ldots + \Theta(t_n) .$$

For $\varepsilon \geq 0$ we define the relation $>_\varepsilon$ on functions from $\mathbb{R}_+^n$ to $\mathbb{R}_+$ by

$$p >_\varepsilon q \quad \Longleftrightarrow \quad p(x_1, \ldots, x_n) \geq \varepsilon + q(x_1, \ldots, x_n) \quad \text{for all } x_1, \ldots, x_n \in \mathbb{R}_+.$$

This relation is well-founded for $\varepsilon > 0$. We pull back this relation to the set of terms and define

$$u \succ_\varepsilon v \quad \Longleftrightarrow \quad \Theta(u) >_\varepsilon \Theta(v).$$

We write $\succeq$ instead of $\succ_0$. Can we choose $\varepsilon > 0$ such that by setting $\succ$ to $\succ_\varepsilon$ we obtain the pair of relations we are searching for?

Obviously $\succ_\varepsilon$ and $\succeq$ would have almost all of the desired properties we need in order to apply Theorem 8. In order to get rid of the "almost" we have to show that (after possibly reordering the dependency pairs) we have $s_i \succ_\varepsilon t_i$ for $i = 1, \ldots, M$ and $M \geq 1$ and also $s_i \succeq t_i$ for $i = M+1, \ldots, N$.

For two terms u and v the relation $u \succ_\varepsilon v$ holds iff

1. $\text{diff}(u, v) = \Theta(u)(0, \ldots, 0) - \Theta(v)(0, \ldots, 0) \geq \varepsilon$,
2. and for all $x_1, \ldots, x_n \in \mathbb{R}_+$ and all $i = 1, \ldots, n$ the inequality $\frac{\partial}{\partial x_i}(\Theta(u)(x_1, \ldots, x_n) - \Theta(v)(x_1, \ldots, x_n)) \geq 0$ holds.

The second condition just means that for all α the inequality

$$\text{varcount}(u, \alpha) \geq \text{varcount}(v, \alpha)$$

holds; $\text{varcount}(t, \alpha)$ denotes the number of occurrences of the variable α in the term t. The first condition is also easy to check; the expression $\text{diff}(u, v)$ does not contain any variables any more.

If assignments $f \mapsto c_f$ and $\varepsilon > 0$ exist such that $\succ_\varepsilon$ and $\succeq$ fulfill all the requirements of Theorem 8 then we can actually calculate them automatically. For this we view the c_f's not any longer as fixed constants but treat them as variables. The expressions $D_i = \text{diff}(s_i, t_i)$ are then linear homogeneous polynomials in these variables. We restate our problem as a linear program:

$$
\begin{array}{ll}
\textbf{maximize} & D = D_1 + \cdots + D_N \\
\textbf{subject to} & D_1 \geq 0 \\
& D_2 \geq 0 \\
& \quad \vdots \\
& D_N \geq 0 \; .
\end{array}
$$

It is understood that all variables of the linear program carry non-negativity constraints. See [17] for background information on linear programming.

Theorem 9. *There exist assignments $f \mapsto c_f$ and $\varepsilon > 0$ such that $\succ_\varepsilon$ and $\succeq$ fulfill the conditions of Theorem 8 (modulo a possible reordering of the dependency pairs) iff*

1. *for all α and $i = 1, \ldots, N$, $\text{varcount}(s_i, \alpha) \geq \text{varcount}(t_i, \alpha)$,*
2. *the above real linear program is unbounded.*

In that case those dependency pairs d_i can be dropped from the dependency graph for which $\text{diff}(s_i, t_i) > 0$ holds.

Proof. The linear program is either feasible with D assuming a maximum $D_{\max} = 0$, or feasible with D being unbounded: it cannot be infeasible because the zero vector is a feasible solution, and it cannot be feasible and bounded with D assuming a maximum $\infty > D_{\max} > 0$ because then one could just multiply the solution vector with a constant $k > 1$ which would lead to an objective value $k\,D_{\max} > D_{\max}$.

Assume $D_{\max} = 0$. Then also $D_i = 0$ for all feasible solutions and all $i = 1, \ldots, N$, so our method cannot be applied.

Assume $D_{\max} = \infty$. Then there must be a feasible solution with $D_i > 0$ for at least one i. For this solution define $\varepsilon = \min\{D_i \mid D_i > 0, \, i = 1, \ldots, N\}$. Together with our previous explanations this concludes the proof. $\qquad\square$

Note that the simplex method together with a non-cycling pivoting rule like Bland's rule is well-suited to deal with the above linear program. The dictionary can be obtained directly from our formulation by introducing N slack variables. The normally necessary first phase for discovering a feasible solution can be dropped because the zero vector is an obvious feasible solution. Confidence in the result can be obtained by calculating with exact fractions instead of floating point numbers. Important to note is that in all five examples no pivoting is necessary at all; the starting dictionary already gives away the feasible ray. That means that an even more easily implementable method could be used that works by simply counting variables and constants.

The result of applying the above algorithm to our five example formalizations is shown in Table 1. The algorithm is fast enough for our purposes; most time is

Table 1. Termination Check Statistics (Isabelle 2005 on a 3GHz Pentium 4)

	Main	Bali	Mat	Nom	Ocl
Number of nodes of dependency graph	2082	7078	2822	2710	11016
Number of cyclic components	0	5	6	1	16
Maximum number of nodes per component	0	4	1	10	4
Total number of nodes in components	0	12	6	10	20
Runtime for constructing initial components	10ms	60ms	20ms	20ms	180ms
Number of reduction iterations	0	10	6	7	19
Runtime of reduction	0ms	10ms	10ms	10ms	40ms
Total Runtime (in milli seconds)		10ms	70ms	30ms	30ms	220ms

spent in constructing the cyclic components of the dependency graph, less time for the iterative reductions of these components by polynomial interpretation.

The check actually discovered a bug in the prerelease HOL-OCL [19] distribution; the checked theory files were non-conservative, and this was due to a bug in the generator of these files; this bug has been fixed now.

4   Conclusion

We have presented the first theory extension mechanism that can cope with overloading in higher-order logic as it is actually used in the proof-assistant Isabelle and shown that this mechanism is safe.

Checking if overloaded definitions conform to HOLCO is not even semi-decidable. We have proven this by revealing the ties of this with the problem of checking the termination of a certain kind of term rewriting system.

Fortunately, most practical uses of overloaded definitions can be checked by a simple algorithm. This algorithm incorporates several ideas of recent research on termination of first-order term rewriting systems and has been implemented as an add-on [1] to the proof-assistant Isabelle. Furthermore this add-on is able to export the check in form of several TRSs to external termination provers as TTT or AProVE.

It is now possible to use overloading as just another tool when working in higher-order logic, without (too m)any doubts about its safety.

Acknowledgments. Special thanks to Markus Wenzel for pointing out to me the delicate points of overloading in connection with the HOL type system and in general, and to both Tobias Nipkow and Markus Wenzel for adding axiomatic type classes and overloading to the Isabelle proof assistant. Furthermore I thank Tjark Weber for several heated discussions at the opera and elsewhere concerning the dependency graph that enabled me to see the connection of overloading to the method of Arts and Giesl. Jürgen Giesl helped me understand several points about the dependency pair method. Stefan Berghofer referred me to [6] for background information about TRSs. Thanks to Alexander Krauss, Norbert Schirmer, Tjark Weber, Markus Wenzel and the anonymous referees for reading (an earlier version of) this paper and suggesting several important improvements and corrections, and also to Clemens Ballarin for reminding me not to skip too many proofs (which had to be thrown out of this version due to space considerations). Finally I would like to express my gratitude to Tobias Nipkow for giving me the time to work on this topic and for providing excellent working conditions.

References

1. Steven Obua. *Conservative Overloading in Higher-Order Logic.* Technical Report, Institut für Informatik, Technische Universität München 2006.
   `http://www4.in.tum.de/~obua/checkdefs`
2. Lawrence C. Paulson. The Foundation of a Generic Theorem Prover. *Journal of Automated Reasoning*, Vol. 5, No. 3, 1989, pages 363-397.
3. Markus Wenzel. Type Classes and Overloading in Higher-Order Logic. *TPHOLs '97*, LNCS 1275, Springer 1997, pages 307-322.
4. Tobias Nipkow, Lawrence C. Paulson, Markus Wenzel. *Isabelle/HOL: A Proof Assistant for Higher-Order Logic*, Springer 2002
5. F. Baader, T. Nipkow. *Term Rewriting and All That*, Cambridge U.P. 1998.
6. Terese. *Term Rewriting Systems*, Cambridge U.P. 2003.
7. Thomas Arts, Jürgen Giesl. Termination of term rewriting using dependency pairs. *Theoretical Computer Science*, 2000, Vol. 236, pages 133-178.
8. Nao Hirokawa and Aart Middeldorp. Automating the dependency pair method. *Information and Computation*, vol. 199, 2005, pages 172-199.
9. Keijo Ruohonen. Reversible Machines and Post's Correspondence Problem for Biprefix Morphisms. *Journal of Information Processing and Cybernetics*, 1985, Vol. 21 (12), pages 579-595.
10. *The HOL System Description.* `http://hol.sourceforge.net/`
11. John Harrison. *The HOL Light theorem prover.*
    `http://www.cl.cam.ac.uk/~jrh/hol-light/`
12. J. Giesl, R. Thiemann, P. Schneider-Kamp, and S. Falke. Automated Termination Proofs with AProVE. *RTA-2004*, LNCS 3091, Springer 2004, pages 210-220.
    `http://www-i2.informatik.rwth-aachen.de/AProVE/`
13. Nao Hirokawa and Aart Middeldorp. Tyrolean Termination Tool. *RTA-2005*, LNCS 3467, Springer 2005, pages 175-182.
    `http://cl2-informatik.uibk.ac.at/ttt/`
14. Steven Obua. Proving Bounds for Real Linear Programs in Isabelle/HOL. *TPHOLs '05*, LNCS 3603, Springer 2005, pages 227-244.
15. Christian Urban. Nominal Techniques in Isabelle/HOL. *CADE-20*, LNAI 3632, Springer 2005, pages 38-53.
16. Project Bali. `http://isabelle.in.tum.de/Bali`
17. Robert J. Vanderbei. *Linear Programming*, 2nd ed., Springer 2001.
18. Jürgen Giesl. Generating Polynomial Orderings for Termination Proofs. *RTA-95*, LNCS 914, Springer 1995, pages 426-431.
19. A. D. Brucker and Burkhart Wolff. *A Proposal for a FORMAL OCL Semantics in Isabelle/HOL.* TPHOLs '02, LNCS 2410, Springer 2002, pages 99-114.
20. Jürgen Giesl, Thomas Arts. Verification of Erlang Processes by Dependency Pairs. *Applicable Algebra in Engineering, Communication and Computing*, vol. 12, 2001, pages 39-72.

Certified Higher-Order Recursive Path Ordering

Adam Koprowski

Eindhoven University of Technology
Department of Computer Science
P.O. Box 513, 5600 MB, Eindhoven, The Netherlands
A.Koprowski@tue.nl

Abstract. The paper reports on a formalization of a proof of well-foundedness of the higher-order recursive path ordering (HORPO) in the proof checker Coq. The development is axiom-free and fully constructive. Three substantive parts that could be used also in other developments are the formalizations of the simply-typed lambda calculus, of finite multisets and of the multiset ordering. The Coq code consists of more than 1000 lemmas and 300 definitions.

1 Introduction

A term rewriting system is terminating if all rewrite sequences are finite. Termination of first-order term rewriting, although in general undecidable, is considered to be an important problem in term rewriting. Several techniques have been developed for dealing with this problem and also a number of tools that attempt at proving termination automatically. One of the well-known techniques is the recursive path ordering (RPO) introduced by Dershowitz [6]. It is a well-founded reduction ordering and hence is suitable for proving termination.

In case of higher-order rewriting, a natural extension of first-order rewriting where bound variables may be present, significantly less results are available. Jouannaud and Rubio generalized RPO to higher-order case thus giving rise to higher-order recursive path ordering (HORPO) [8]. Using the notion of computability, introduced by Tait and Girard to prove termination of simply typed lambda calculus ($\lambda^{\rightarrow}$), they succeeded in proving well-foundedness of the union of HORPO and β-reduction of $\lambda^{\rightarrow}$. This is the essential part of the justification why HORPO can be used for proving termination of higher-order rewriting. A corollary of this result is termination of $\lambda^{\rightarrow}$ and well-foundedness of RPO which is embedded in HORPO. Later in [9] they extended and improved the ordering.

Based on those developments we made a formalization of HORPO in the theorem prover Coq, which is the subject of this paper. The formalization is complete (i.e., it does not contain any axioms) and fully constructive. The definition of HORPO is taken from [8] without extensions and improvements from [9], however parts of the theory presented in the latter was formalized as well. The formalization contains all the proofs required to justify the use of HORPO for proving termination of higher-order rewriting. More information along with Coq proof scripts can be found at:

F. Pfenning (Ed.): RTA 2006, LNCS 4098, pp. 227–241, 2006.
© Springer-Verlag Berlin Heidelberg 2006

`http://www.win.tue.nl/~akoprows/coq-horpo.`

To give the reader an impression of the size of this formalization we would like to mention that the Coq scripts consist of 29 files with $> 25,000$ lines of code and with $> 700,000$ total characters. They contain $> 1,100$ lemmas and > 300 definitions.

This formalization has become part of the CoLoR project[1]. CoLoR – Coq library on rewriting and termination – is an initiative to formalize the theory of term rewriting in Coq and ultimately to certify termination proof candidates produced by existing termination proving tools. This work can be seen as a contribution to the CoLoR project.

The structure of this paper is as follows. In Section 2 we shortly discuss motivation of this development, its short history and related work. Then in Section 3 we give a broad overview of the formalization. In Section 4 we introduce some preliminaries and we continue with the definition of higher-order terms and higher-order rewriting in Section 5. Section 6 is devoted to introduction of computability predicate proof method and contains proofs of all the computability properties required in Section 7 where we introduce HORPO ordering and prove some of its properties the main one being well-foundedness. We conclude in Section 8.

2   Motivation, History and Related Work

Formal theorem proving is rather time consuming and often requires enormous amount of work to be completed. Thus one may wonder what is the motivation behind this effort. We will mention three main motivational factors for this development.

- Verification of the proof. Sometimes the goal may be simply to verify the correctness of the proof. This is especially true for complicated proofs that are not very well known such as the work in [8].
  The results from [8,9] are impressive and complicated and as such are inevitably subject to some small slips. This justifies the effort of verification of such results. Indeed in the course of formalization we were able to detect a small flaw, concerning the use of multiset extension of an arbitrary relation, that could be easily repaired (we will discuss it shortly in Sections 4 and 7.1). In general [8,9] turned out to be a very favorable subject for formalization and the structure of the proofs could be followed to the letter in the formalization process.[2]
- Theorem proving is still a rather laborious task. But with constantly improving proof assistants this may not necessarily be so in the future and one

---

[1] `http://color.loria.fr`

[2] Obviously providing formal proofs requires to be more explicit and to include all the results that in normal presentation would be omitted as considered to be straightforward or irrelevant.

of the main stimulus for improvement and growth of theorem provers are big developments accomplished with their use.
- The most pragmatic motivation is the CoLoR initiative. As described in introduction, CoLoR is a project aiming at proving theoretical results from term rewriting in the theorem prover Coq. The ultimate goal is to (automatically) transform termination proof candidates produced by termination tools into formal Coq proofs certifying termination. This requires formalization of the term rewriting theory and this development has become part of the CoLoR library and can be seen as a contribution to this project.

This development started in January 2004 as the author's Master's Thesis [10] at the Free University Amsterdam, supervised by Femke van Raamsdonk. After half a year it was completed only with computability properties left as axioms. The eagerness of having axiom-free development resulted in another one and half year of work at the Eindhoven University of Technology, was finished in February 2006 and is the subject of this paper.

The ideas of formalizing $\lambda^{\rightarrow}$ and RPO are not new and there are some existing formalizations. Persson [14] in his PhD thesis presents a constructive proof of well-foundedness of a general form of recursive path relations. Leclerc [12] presents a formalization in Coq of well-foundedness of RPO with the multiset ordering. Murthy [13] formalizes a classical proof (due to Nash-Williams) of Higman's lemma in Nuprl 3. Berghofer [3] presents a constructive proof (due to Coquand and Fridlender) of Higman's lemma in Isabelle. Coupet-Grimal and Delobel [5] recently formalized RPO within the CoLoR project using parts of our development (for finite multisets and multiset ordering). Berger et al. [2] proved termination of $\lambda^{\rightarrow}$ in three different theorem provers including Coq. They also used the computability proof method but their work was completely independent of this formalization and focuses on extraction of normalization algorithm. However we would like to stress that we are not aware of any attempt at formalizing HORPO so, to the best of our knowledge, the main part of this work was never before a subject of formalization.

3   Overview of the Formalization

The main result of this formalization is well-foundedness of the union of HORPO and β-reduction relation of $\lambda^{\rightarrow}$. The formalization is complete and hence contains development of all the dependant results, most notably formalization of $\lambda^{\rightarrow}$ and of finite multisets and multiset extension of a relation. In this section we give overview of those results.

The development can be divided into 4 main parts. Their very brief description follows.

- **Auxiliary results.** A number of rather simple definitions and results that were not present in the Coq standard library.
- **Multisets.** Since HORPO uses multiset extension to compare arguments of functions some results about finite multisets and multiset ordering were required.

- Finite multisets have been defined using abstract data type paradigm. That is a number of primitive operations for multisets has been specified along with their specifications.
- To show that this axiomatic specification can be fulfilled an implementation of multisets using lists has been provided.
- A number of abstract properties about multisets has been proven. Abstract in the sense that the proofs rely only on abstract specification of multisets. This ensures that given another (say more efficient) implementation of multisets all those results carry on automatically.
- Multiset extension of arbitrary relation has been defined and it has been proven to preserve main properties of the relation, including well-foundedness.

- **Simply typed lambda calculus ($\lambda^\rightarrow$).** Higher-order rewriting uses some form of higher-order metalanguage (we will elaborate more on this subject in Section 5). In this development we decided to use $\lambda^\rightarrow$ in its pure form due to its universality and in the hope that this part of the development can be useful not only for the purpose of this project.

  - $\lambda^\rightarrow$ terms over arbitrary possibly many-sorted signature have been defined using de Bruin indices [4].
  - A number of properties concerning $\lambda^\rightarrow$ terms has been proven. This includes some results concerning typing, including a constructive proof of the decidability of typing and of the decidability of β-reduction from which a certified code for normalization could be extracted.
  - Typed substitution has been defined. Note that this is far from trivial as the substitution operates on typed terms so essentially the entities being substituted are typing judgements and one has to ensure that types and environments do not clash.
  - An equivalence relation on terms has been defined as an extension of α-convertibility to free variables. Convertibility of terms that are equal up to the names of free variables or irrelevant differences in environments and of lifted terms is captured by this relation.
  - Encoding of algebraic terms. As we will see in Section 5 the variant of higher-order rewriting that is of interest for us uses algebraic terms which have been encoded as $\lambda^\rightarrow$ terms.
  - A corollary of the main result of this development is the termination of $\lambda^\rightarrow$.

- **HORPO.** The definition of HORPO and proofs of its properties that constitute the main part of this paper.

  - Definition of HORPO as a slight variant of HORPO from [8].
  - Proofs of computability properties required for the main proof using computability predicate proof method due to Tait and Girard.
  - Main result: well-foundedness of the union of HORPO and β-reduction of $\lambda^\rightarrow$.
  - A consequence of this fact is that HORPO is a higher-order reduction ordering and hence is suitable for proving termination of higher-order term rewriting systems.

The part concerning HORPO will be treated in more details in the following sections. The reader interested in other parts of the development is encouraged to consult [11]. For introduction to $\lambda^{\rightarrow}$ we refer to [1].

4   Preliminaries

For a set A and a relation $>$ we will say that $a \in A$ is *accessible*, $a \in \mathcal{A}cc_>$ if a does not start any infinite reduction $a > a' > \ldots$. If the relation $>$ is clear from the context we will omit the subscript and write $a \in \mathcal{A}cc$. Note that the relation $>$ is well-founded if $\forall a \in A \, . \, a \in \mathcal{A}cc_>$.

Given non-empty set A we define finite multiset over A in the usual way. Now given an arbitrary relation $>$ on A we define its extension to the relation $>_{mul}$ on multisets over A as follows:

$$M >_{mul} N \iff \exists X, Y, Z \, . \, \begin{cases} Y \neq \emptyset \\ M = X \cup Y \\ N = X \cup Z \\ \forall z \in Z \, . \, \exists y \in Y \, . \, y > z \end{cases}$$

We will use following properties of this multiset ordering:

(M_1) If $M >_{mul} N$ then $\forall n \in N \, . \, \exists m \in M \, . \, m \geq n$.

(M_2) If $\forall m \in M \, . \, m \in \mathcal{A}cc_>$ then $M \in \mathcal{A}cc_{>_{mul}}$ (so $>_{mul}$ preserves well-foundedness).

(M_3) If for every $m \in M$, $n \in N$ problem whether $m > n$ is decidable then the problem whether $M >_{mul} N$ is decidable.

Note that in [8] alternative definition of multisets has been used. For orders those definitions are equivalent (which we verified in Coq) however for arbitrary relations with this alternative definition only a weaker variant of (M_2) holds where the conclusion is $m >^* n$. This led to some difficulties as remarked in Section 7.1. See [11] for details.

Given a set A and two relations $>_1$ and $>_2$ on A we define their lexicographic extension $(>_1, >_2)_{lex}$ as:

$$(m_1, m_2)(>_1, >_2)_{lex}(n_1, n_2) \iff m_1 >_1 n_1 \lor (m_1 = n_1 \land m_2 >_2 n_2)$$

We state the well-known property:

(L_1) If $>_1$ and $>_2$ are well-founded then so is $(>_1, >_2)_{lex}$.

5   Higher-Order Rewriting

There are several variants of higher-order rewriting. Here we use the algebraic-functional systems (AFSs) introduced by Jouannaud and Okada [7]. The main difference between AFSs and another popular format of higher-order rewriting systems (HRSs) is that in the second we work modulo beta-eta (using pure $\lambda^{\rightarrow}$

terms) whereas in AFSs we do not (and function symbols have fixed arity). As a consequence rewriting for AFSs is defined using plain pattern matching compared to rewriting modulo $\beta\eta$ of $\lambda^{\rightarrow}$ in HRSs framework. For a broader discussion on this subject we refer the reader to, for instance, [15].

Given a set of *sorts* $\mathcal{S}$ we inductively define a set of simple types $\mathcal{T}$ as:

- $\alpha \in \mathcal{T}$ if $\alpha \in \mathcal{S}$ (*base type*),
- $\alpha \rightarrow \beta \in \mathcal{T}$ if $\alpha, \beta \in \mathcal{T}$ (*arrow type*).

We define a *signature* Σ as a set of function symbols with a fixed arity. For declaration of f expecting n arguments of types $\alpha_1, \dots, \alpha_n$ and an output type β we will write $f : \alpha_1 \times \dots \times \alpha_n \rightarrow \beta$. *Environment* is defined as a set of variable declarations, that is: $\Gamma = \{x_1 : \alpha_1, \dots, x_n : \alpha_n\}$ with $x_i \in \mathcal{V}$ and $\alpha_i \in \mathcal{T}$ for every i and with $x_i \neq x_j$ for $i \neq j$.

The set of preterms over given signature Σ and a set of variables $\mathcal{V}$ is generated according to the following grammar:

$$\mathcal{P}t := \mathcal{V} \mid @(\mathcal{P}t, \mathcal{P}t) \mid \lambda\mathcal{V}{:}\mathcal{T}.\mathcal{P}t \mid \Sigma(\mathcal{P}t, \dots, \mathcal{P}t)$$

denoting variable, application, abstraction and function application. As usually application is left-associative.

Typed terms are identified with *typing judgements* of the form $\Gamma \vdash t : \alpha$ stating that in the environment Γ preterm t has type α. They conform to the following type inference system:

$$\frac{x : \alpha \in \Gamma}{\Gamma \vdash x : \alpha} \qquad \frac{f : \alpha_1 \times \dots \times \alpha_n \rightarrow \beta \in \Sigma \quad \Gamma \vdash t_1 : \alpha_1, \dots, \Gamma \vdash t_n : \alpha_n}{\Gamma \vdash f(t_1, \dots, t_n) : \beta}$$

$$\frac{\Gamma \vdash t : \alpha \rightarrow \beta \quad \Gamma \vdash u : \alpha}{\Gamma \vdash @(t, u) : \beta} \qquad \frac{\Gamma \cup \{x : \alpha\} \vdash t : \beta}{\Gamma \vdash \lambda x{:}\alpha.t : \alpha \rightarrow \beta}$$

From here onwards we assume terms to be typed and often we will omit the environments writing $t : \alpha$ or even only t instead of $\Gamma \vdash t : \alpha$.

Free variables occurring in term t are denoted as $\mathsf{Vars}(t)$. We define the *replacement* of term u in term t at position p in the usual way and denote it by $t[u]_p$. We define the *strict subterm relation* in the standard way and denote it by $\sqsubset$. It is well-founded and will be used for the induction on the structure of terms. List of terms $@(t, u_1, \dots, u_i), u_{i+1}, \dots u_n$ is called *partial left-flattening* of term $@(t, u_1, \dots, u_n)$ for $1 \leq i \leq n$. We will say that a term t is *neutral* if it is not an abstraction. By $\sim$ we denote the equivalence relation on terms that, roughly speaking, extends α-convertibility to free variables. See [11] for details.

We will write substitution as $\gamma = [x_1/u_1, \dots, x_n/u_n]$ with its domain denoted as $Dom(\gamma) = \{x_1, \dots, x_n\}$ and the application of substitution γ to term t as $t\gamma$ resulting in a term in which all free occurrences of variables x_i are replaced with term u_i for $1 \leq i \leq n$. Note that substitution is defined on typing judgements.

The format of *higher-order term rewriting systems* and their *rewrite relation* are of no direct interest in this paper and will be omitted here – they are defined as in [9] (also see [11]).

Note that in the formalization as a metalanguage the pure $\lambda^{\rightarrow}$ terms were used. To avoid dealing with arities an assumption has been made that output types of functions are base types and hence a functional $f : \alpha_1 \times \ldots \times \alpha_n \rightarrow \beta$ is encoded using $\lambda^{\rightarrow}$ constant f of type $\alpha_1 \rightarrow \ldots \rightarrow \alpha_n \rightarrow \beta$ with its application $f(t_1, \ldots, t_n)$ encoded as $@(f, t_1, \ldots, t_n)$.

Again for more throughout introduction to notions presented in this section we refer the reader to [11].

6   Computability

In this section we present the computability predicate proof method due to Tait and Girard. In Section 7 we will use computability with respect to a particular relation (being union of HORPO and β-reduction) but we present computability for an arbitrary relation satisfying given properties.

We begin by defining computability in 6.1, in 6.2 we prove some computability properties and finally in 6.3 we make some remarks on the formalization of computability predicate in Coq.

6.1   Definition of Computability

Definition 1 (Computability). *A term $t : \delta$ is computable with respect to a relation on terms $\rhd$, denoted as $t \in \mathbb{C}_\delta$ (or simply $t \in \mathbb{C}$), if:*

- *δ is a base type and t is strongly normalizable ($t \in Acc_\rhd$) or*
- *$\delta = \alpha \rightarrow \beta$ and $@(t, u) \in \mathbb{C}_\beta$ for all $u \in \mathbb{C}_\alpha$.*

Note that it is usual to assume that variables are computable. We do not do that, following the presentation in [8]. Computability of variables will follow from computability properties.

6.2   Computability Properties

In the formalization we made an attempt at proving computability in an abstract way, that is for an arbitrary relation $\rhd$. Below we present the list of required properties of $\rhd$ that we needed to complete all the computability proofs.

(P_1) Subject reduction: $t : \alpha \rhd u : \beta \implies \alpha = \beta$,
(P_2) Preservation of environments: $\Gamma_t \vdash t : \delta \rhd \Gamma_u \vdash u : \eta \implies \Gamma_t = \Gamma_u$.
(P_3) Free variables consistency: $t \rhd u \implies \mathsf{Vars}(u) \subseteq \mathsf{Vars}(t)$.
(P_4) Normal form of variables: $\neg(x \rhd u)$.
(P_5) Compatibility with $\sim$.[3]
(P_6) Stability under substitution: $t \rhd u \implies t\gamma \rhd u\gamma$.
(P_7) Monotonicity: $u \rhd u' \implies t[u]_p \rhd t[u']_p$.
(P_8) Reductions of abstraction: $\lambda x {:} \alpha.t_b \rhd u \implies \exists u_b \,.\, u = \lambda x {:} \alpha.u_b \wedge t_b \rhd u_b$.

---

[3] Precise formulation of this property requires more detailed introduction of $\sim$ relation; consult [11] for details.

(P_9) Reductions of application: $t = @(t_l, t_r) \rhd u$ implies

- $u = @(u_l, u_r)$, $t_l \unrhd u_l$ and $t_r \unrhd u_r$ and $t_l \rhd u_l \vee t_r \rhd u_r$ or
- $t \rhd u$ is a β-reduction step, so $t = @(\lambda x{:}\alpha.t_l, t_r) \rhd t_l[x/t_r] = u$.

All those properties but the last one are somehow standard. The last one demands reductions of application to operate argument-wise or be a β-reduction step. This property is specific for the $\rhd$ relation being union of HORPO and β-reduction relation as we will use it in Section 7.

Let us recall that we did not assume variables to be computable. Variables of a base type are computable due to the definition of computability and (P_4). Variables of a functional type are computable by property (C_3) presented in the following lemma which forbids us to prove the following computability properties (C_1), (C_2) and (C_3) separately.

Lemma 1. *For all terms $\Gamma \vdash t : \delta$, $\Delta \vdash u : \delta$ we prove that:*

(C_1) $t \in \mathbb{C}_\delta \implies t \in \mathcal{A}cc$
(C_2) $t \in \mathbb{C}_\delta \wedge t \rhd u \implies u \in \mathbb{C}_\delta$
(C_3) *if t-neutral then* $(\forall w : \delta \ . \ t \rhd w \implies w \in \mathbb{C}_\delta) \iff t \in \mathbb{C}_\delta$

Proof. Induction on type δ. Note that 'if' part of (C_3) is (C_2) so below we only prove the 'only if' part of this property.

- δ is a base type.
  (C_1) $t \in \mathbb{C}_\delta$ and δ is a base type so $t \in \mathcal{A}cc$ by the definition of computability.
  (C_2) $t \in \mathcal{A}cc$ by the same argument as in (C_1). $t \in \mathcal{A}cc$ and $t \rhd u$ hence $u \in \mathcal{A}cc$. By subject reduction for $\rhd$ (P_1), $u : \delta$, so $u \in \mathbb{C}_\delta$ by the definition of computability.
  (C_3) $t : \delta$ so to show $t \in \mathbb{C}_\delta$ we need to show $t \in \mathcal{A}cc$. But for every w such that $t \rhd w$ we have $w \in \mathbb{C}_\delta$ by assumption and hence $w \in \mathcal{A}cc$ by the definition of computability and thus $t \in \mathcal{A}cc$.
- $\delta = \alpha \to \beta$
  (C_1) Take variable $x : \alpha$ which is computable by induction hypothesis (C_3), as variables are not reducible (P_4). Now consider application $@(t, x)$ which is computable by the definition of computability. So $@(t, x) \in \mathcal{A}cc$ by induction hypothesis (C_1) and $t \in \mathcal{A}cc$ by monotonicity (P_7).
  (C_2) By the definition of computability $u \in \mathbb{C}_{\alpha \to \beta}$ if for every $s \in \mathbb{C}_\alpha$, $@(u, s) \in \mathbb{C}_\beta$. $@(t, s) \in \mathbb{C}_\beta$ by the definition of computability and $@(t, s) \rhd @(u, s)$ by monotonicity assumption (P_7). Finally we conclude $@(u, s) \in \mathbb{C}_\beta$ by induction hypothesis (C_2).
  (C_3) By the definition of computability $t \in \mathbb{C}_{\alpha \to \beta}$ if for every $s \in \mathbb{C}_\alpha$, $@(t, s) \in \mathbb{C}_\beta$. By induction hypothesis for (C_1), $s \in \mathcal{A}cc$ so we continue by well-founded inner induction on s with respect to $\rhd$.
  $@(t, s) : \beta$ is neutral so we can apply induction hypothesis for (C_3) and we are left to show that all reducts of $@(t, s)$ are computable. We do case analysis using (P_9). Since t is neutral and hence is not abstraction, we can exclude the β-reduction case and we are left with the following cases:

- $@(t, s) \rhd @(t', s)$ with $t \rhd t'$. Then t' is computable as so is every reduct of t and application of two computable terms is computable by the definition of computability.
- $@(t, s) \rhd @(t, s')$ with $s \rhd s'$. We observe that $s' \in \mathbb{C}$ by induction hypothesis for (C_2) and since $s \rhd s'$ we apply inner induction hypothesis to conclude $(t, s') \in \mathbb{C}_\beta$.
- $@(t, s) \rhd @(t', s')$ with $t \rhd t'$ and $s \rhd s'$. Every reduct of t is computable so $t' \in \mathbb{C}_{\alpha \to \beta}$. By induction hypothesis for (C_2) $s' \in \mathbb{C}_\alpha$. Again application of two computable terms is computable.

A simple consequence of (C_3) and (P_4) is $(\mathbf{C_4})$: all variables are computable. The last computability property involves abstractions.

Lemma 2 ($\mathbf{C_5}$). *Consider abstraction $(\lambda x : \alpha.t) : \alpha \to \beta$. If for every $u \in \mathbb{C}_\alpha$, $t[x/u] \in \mathbb{C}_\beta$ then $(\lambda x : \alpha.t) \in \mathbb{C}_{\alpha \to \beta}$.*

Proof. By the definition of computability $\lambda x : \alpha.t$ is computable if for every $s \in \mathbb{C}_\alpha$, $@(\lambda x : \alpha.t, s) \in \mathbb{C}_\beta$. Note that $t \in \mathbb{C}$ by assumption because $t = t[x/x]$ and variables are computable by (C_4). So by (C_1) both $t \in \mathcal{A}cc$ and $s \in \mathcal{A}cc$ and we proceed by induction on a pair of computable terms (t, s) with respect to the ordering $\gg = (\rhd, \rhd)_{lex}$. Now, since $@(\lambda x : \alpha.t, s)$ is neutral, by (C_3) we are left to show that all its reducts are computable. Let us continue by considering possible reducts of this application using (P_9). So we have $@(\lambda x : \alpha.t, s) \rhd u$ and the following cases to consider:

- $u = t[x/s]$. $u \in \mathbb{C}$ by the assumption.
- $u = @(\lambda x : \alpha.t, s')$ with $s \rhd s'$. $u \in \mathbb{C}$ by induction hypothesis for $(t, s') \ll (t, s)$.
- $u = @(w, s)$ with $\lambda x : \alpha.t \rhd w$. By (P_8) we know that the reduction is in the abstraction body of $\lambda x : \alpha.t$ so in fact $w = \lambda x : \alpha.t'$ with $t \rhd t'$. We conclude computability of u by induction hypothesis for $(t', s) \ll (t, s)$.
- $u = @(w, s')$ with $\lambda x : \alpha.t \rhd w$ and $s \rhd s'$. As in above case, by (P_8) we observe that $w = \lambda x : \alpha.t'$ with $t \rhd t'$ and we conclude computability of u by induction hypothesis for $(t', s') \ll (t, s)$.

6.3   Computability in Coq

Computability turned out to be by far the most difficult part of the development. In its first version ([10]) computability properties were stated as axioms. Making development axiom-free and proving all computability properties turned out to be a very laborious task after which the size of Coq scripts tripled.

Strictly speaking in terms of script size, the part of the formalization dealing with computability accounts for only slightly more than 5%. However, as those properties are at the heart of proofs concerning HORPO relation, providing proofs for them triggered many other developments.

This difficulty can be partially explained by the real complexity of the computability predicate proof method. Other factors that contributed to making this task difficult include:

– the fact that algebraic terms were encoded using pure $\lambda^\rightarrow$ terms,
– the necessity of defining computability modulo equivalence relation on terms.

For the clarity of presentation those issues are left implicit in the computability proofs presented in this section but in Coq proofs all of them had to be taken care of. Another difference is the use of de Bruijn indices [4] in the formalization to represent terms.

7   HORPO

In this section we present the core of this work: the results concerning the higher-order recursive path ordering (HORPO). We begin by presenting the definition of HORPO in 7.1, then some of its properties in 7.2 and its main property – well-foundedness – in 7.3. We conclude this section in 7.4 where we make some remarks about the formalization of HORPO.

7.1   Definition of HORPO

As indicated in the introduction the subject of our formalization is a slight variant of HORPO as presented in [8]. We begin by first presenting our version of the definition and then we discuss the differences comparing to the original definition by Jouannaud and Rubio.

Definition 2 (The higher-order recursive path ordering, $\succ$). *Assume a well-founded order $\triangleright$ on the set of function symbols, called a precedence. We define HORPO relation $\succ$ on terms and in this definition by $\succeq$ we denote reflexive closure of HORPO (that is $\succeq \equiv \succ \cup =$) and by $\succ_{mul}$ its multiset extension.*
$$\Gamma \vdash t : \delta \succ \Gamma \vdash u : \delta \text{ iff one of the following holds:}$$

$(\mathbf{H_1})$ $t = f(t_1, \ldots, t_n)$, $\exists i \in \{1, \ldots, n\}$. $t_i \succeq u$,
$(\mathbf{H_2})$ $t = f(t_1, \ldots, t_n)$, $u = g(u_1, \ldots, u_k)$, $f \triangleright g$, $t \succ\succ \{u_1, \ldots u_k\}$,
$(\mathbf{H_3})$ $t = f(t_1, \ldots, t_n)$, $u = f(u_1, \ldots, u_k)$, $\{\{t_1, \ldots t_n\}\} \succ_{mul} \{\{u_1, \ldots, u_k\}\}$,
$(\mathbf{H_4})$ $@(u_1, \ldots, u_k)$ is a partial-left flattening of u, $t \succ\succ \{u_1, \ldots u_k\}$,
$(\mathbf{H_5})$ $t = @(t_l, t_r)$, $u = @(u_l, u_r)$, $\{\{t_l, t_r\}\} \succ_{mul} \{\{u_l, u_r\}\}$,
$(\mathbf{H_6})$ $t = \lambda x{:}\alpha.t'$, $u = \lambda x{:}\alpha.u'$, $t' \succ u'$

where $\succ\succ$ is a relation between a term and a set of terms, defined as:
$t = f(t_1, \ldots, t_k) \succ\succ \{u_1, \ldots, u_n\}$ iff $\forall i \in \{1, \ldots, n\}$. $t \succ u_i \vee (\exists j . t_j \succeq u_i)$.

Note that, following Jouannaud and Rubio, we do not prove HORPO to be an ordering. In the following sections we will prove its well-foundedness and thus its transitive closure will be a well-founded ordering. There are three major differences between our definition and the definition from [8].

First let us note that in our variant only terms of equal types can be compared whereas in the original definition this restriction is weaker and it is possible to compare terms of equivalent types, where equivalence of types is a congruence generated by equating all sorts (in other words two types are equivalent if they have the same arrow structure). The reason for strengthening this assumption is that allowing to reduce between different sorts poses some technical difficulties. In [8] this problem was solved by extending the typing rules with the congruence rule which presence is basically equivalent to collapsing all sorts and which allows typing terms that normally would be ill-typed due to a sort clash. Our goal was to use $\lambda^{\rightarrow}$ in its purest form as a meta-language and hence we decided not to do that. Note however that this remark is relevant only for many-sorted signatures as for one sorted signatures type equality and type equivalence coincide.

The second difference is that the original definition of HORPO uses status and allows arguments of function symbols to be compared either lexicographically or as multisets, depending on the status, whereas we allow only for comparing arguments of functions as multisets. This choice was made simply to avoid dealing with status. Multiset comparison has been chosen as posing more difficulties. An extension with status and possibility of comparing arguments lexicographically should be relatively easy.

Finally we use different definition of multiset ordering. The property (M_1) will be crucial in lemmas preceding proof of well-foundedness of HORPO and for alternative definition of multiset extension only its weaker variant holds. This was the source of flaw in the approach of Jouannaud and Rubio we mentioned in Section 2. For more details see [11].

7.2   Properties of HORPO

In this section we will prove some properties of HORPO.

Lemma 3. *HORPO is stable under substitution, that is:* $t \succ u \implies t\gamma \succ u\gamma$

Proof. Induction on pair (t, u) ordered by $(\sqsubset, \sqsubset)_{lex}$ followed by case analysis on $t \succ u$.

(H_1) $t = f(t_1, \ldots, t_n)$ and $t_i \succeq u$ for some $i \in \{1, \ldots, n\}$. But then $t\gamma = f(t_1\gamma, \ldots, t_n\gamma) \succ u\gamma$ by (H_1) since $t_i\gamma \succeq u\gamma$ by the induction hypothesis.

(H_2) $t = f(t_1, \ldots, t_n)$, $u = g(u_1, \ldots, u_k)$, $f \rhd g$ and $t \succ\succ \{u_1, \ldots, u_k\}$. But then to get $t\gamma \succ u\gamma$ by (H_2) we only need to show $t\gamma \succ\succ \{u_1\gamma, \ldots, u_k\gamma\}$. For every $i \in \{1, \ldots, k\}$ we have $t \succ u_i \vee (\exists j . t_j \succeq u_i)$. In either case we have $t\gamma \succ u_i\gamma$ or $t_j\gamma \succeq u_i\gamma$ by the induction hypothesis.

(H_3) $t = f(t_1, \ldots, t_n)$, $u = f(u_1, \ldots, u_k)$ and $\{\{t_1, \ldots t_n\}\} \succ_{mul} \{\{u_1, \ldots, u_k\}\}$ but then $\{\{t_1\gamma, \ldots t_n\gamma\}\} \succ_{mul} \{\{u_1\gamma, \ldots, u_k\gamma\}\}$ since for all $i \in \{1, \ldots, n\}$, $j \in \{1, \ldots, k\}$, $t_i \succ u_j$ implies $t_i\gamma \succ u_j\gamma$ by the induction hypothesis. So we get $t\gamma \succ u\gamma$ by (H_3).

(H_4) $@(u_1, \ldots, u_k)$ is a partial flattening of u and $t \succ\succ \{u_1, \ldots u_k\}$. We use the same partial flattening for $u\gamma$ and get $t\gamma \succ\succ \{u_1\gamma, \ldots, u_k\gamma\}$ with the same argument as in case (H_2). We conclude $t\gamma \succ u\gamma$ by (H_4).

(H_5) $t = @(t_l, t_r)$, $u = @(u_l, u_r)$ and $\{\{t_l, t_r\}\} \succ_{mul} \{\{u_l, u_r\}\}$. Type considerations show that $t_l \succeq u_l$, $t_r \succeq u_r$ and $t_l \succ u_l \vee t_r \succ u_r$. By induction hypothesis on $(t_l\gamma, u_l\gamma)$ and $(t_r\gamma, u_r\gamma)$ we conclude $\{\{t_l\gamma, t_r\gamma\}\} \succ_{mul} \{\{u_l\gamma, u_r\gamma\}\}$ and hence $t\gamma \succ u\gamma$ by (H_5).

(H_6) $t = \lambda x\!:\!\alpha.t'$, $u = \lambda x\!:\!\alpha.u'$ and $t' \succ u'$. But then $t\gamma = \lambda x\!:\!\alpha.t'\gamma$, $u\gamma = \lambda x\!:\!\alpha.u'\gamma$ and $t'\gamma \succ u'\gamma$ by the induction hypothesis. So $t\gamma \succ u\gamma$ by (H_6).

Lemma 4. *HORPO is monotonous, that is:* $u \succ u' \implies t[u]_p \succ t[u']_p$.

Proof. The proof proceeds by induction on p and essentially uses the following observations:

- if $w_r \succ w'_r$ then $@(w_l, w_r) \succ @(w_l, w'_r)$ by (H_5).
- if $w_l \succ w'_l$ then $@(w_l, w_r) \succ @(w'_l, w_r)$ by (H_5).
- if $w \succ w'$ then $f(\ldots, w, \ldots) \succ f(\ldots, w', \ldots)$ by (H_3).
- if $w \succ w'$ then $\lambda x\!:\!\alpha.w \succ \lambda x\!:\!\alpha.w'$ by (H_6).

So we presented proofs for computability properties (P_6) and (P_7). Properties (P_1), (P_2), (P_4), (P_8) and (P_9) are direct from the definition of HORPO. (P_3) is easy using induction on pair of terms ordered by $(\sqsubset, \sqsubset)_{lex}$. For ($P_5$) more detailed presentation of equivalence relation $\sim$ is needed and we refer the interested reader to [11] for details.

We conclude this section with a result that is not present in [8], namely a proof of the fact that $\succ$ is decidable.

Theorem 1. *Given terms t and u the problem whether $t \succ u$ is decidable.*

Proof. Induction on the pair (t, u) ordered by $(\sqsubset, \sqsubset)_{lex}$ followed by a case analysis on t.

- $t = x$. Variables are in normal forms with respect to $\succ$ so we cannot have $x \succ u$.
- $t = @(t_l, t_r)$. Only (H_5) is applicable if $u = @(u_l, u_r)$ and for that, taking typing consideration into account, it is required that $t_l \succeq u_l$, $t_r \succeq u_r$ and $t_l \succ u_l \vee u_l \succ u_r$ all of which is decidable by induction hypothesis.
- $t = \lambda x\!:\!\alpha.t_b$. Only ($H_6$) is applicable for $u = \lambda x\!:\!\alpha.u_b$ and it is required that $t_b \succ u_b$ which we can decide by induction hypothesis.
- $t = f(t_1, \ldots, t_n)$. We have several cases to consider corresponding to application of different clauses of HORPO:
  - (H_1): for every $i \in \{1, \ldots, n\}$ we check whether $t_i \succeq u$ by application of induction hypothesis.
  - (H_2): u needs to be of the shape $u = g(u_1, \ldots, u_k)$ with $f \rhd g$ (we assume precedence to be decidable). We need to check whether $t \succ\!\!\succ \{u_1, \ldots, u_k\}$. So for every $i \in \{1, \ldots, k\}$ we check whether $t \succ u_i$ or $t_j \succ u_i$ for some $j \in \{1, \ldots, n\}$. Typing consideration are helpful in immediately discarding of many cases.
  - (H_3): comparison between all arguments of t and u is decidable by induction hypothesis so to conclude whether multisets of arguments can be compared we use (M_3).
  - (H_4): we consider all the possible partial flattenings $@(u_1, \ldots, u_k)$ of u (bounded by the size of u) and for each of them we check whether $t \succ\!\!\succ \{u_1, \ldots, u_k\}$ in the same way as in the (H_2) case.

7.3   Well-Foundedness of HORPO

In this section we present the proof of well-foundedness of $\succ \cup \to_\beta$. This relation will play important role in this section so let us abbreviate it by $\leadsto \equiv \succ \cup \to_\beta$. For the proof we will use the computability predicate proof method due to Tait and Girard (as in [8]) which was discussed in Section 6.

Note that we will use computability with respect to $\leadsto$ and for that we need to prove the properties (P_1)-(P_9) for $\leadsto$. We proved (P_6) and (P_7) for $\succ$ in Section 7.2. For the remaining properties of $\succ$, all properties for $\to_\beta$ (which are easy and standard) we refer to [11]. All those properties easily generalize to the union if they hold for the components.

The crucial lemma states that if function arguments are computable then so is the function application. We first need an auxiliary lemma for which the proof is easy and can be found in [11].

Lemma 5. *For any $t = f(t_1, \ldots, t_n)$ and $u = g(u_1, \ldots, u_k)$ if $t \gg \{u_1, \ldots u_k\}$ and $\forall i \in \{1, \ldots, n\} . t_i \in \mathbb{C}$ and $\forall j \in \{1, \ldots, k\} . t \leadsto u_j \implies u_j \in \mathbb{C}$ then $\forall j \in \{1, \ldots, k\} . u_j \in \mathbb{C}$.*

Lemma 6. *If $t_1, \ldots, t_n \in \mathbb{C}$ then $t = f(t_1, \ldots, t_n) \in \mathbb{C}$.*

Proof. The proof proceeds by well-founded induction on the pair of a function symbol and a multiset of computable terms, $(f, \{\{t_1, \ldots, t_n\}\})$, ordered lexicographically by $(\rhd, \leadsto_{mul})_{lex}$. Note that all terms in the multiset are computable and hence, by (C_1), strongly normalizable. So $(\rhd, \leadsto_{mul})_{lex}$ is well-founded by (M_2) and (L_1) which justifies the induction argument.

Since t is neutral we apply (C_3) and we are left to show that for arbitrary u, such that $t \leadsto u$, $u \in \mathbb{C}$. We will show that by inner induction on the structure of u. We continue by case analysis on $t \leadsto u$. The first case corresponds to a beta-reduction step and the following ones to applications of the clauses (H_1), (H_2), (H_3) and (H_4) of the HORPO definition. Note that the clauses (H_5) and (H_6) are not applicable.

(β) Let $t \to_\beta u$. The $\to_\beta$ step is in one of the arguments, so for some j we have $u = f(t_1, \ldots t'_j, \ldots t_n)$ with $t_j \to_\beta t'_j$. For every i, $t_i \in \mathbb{C}$ by assumption and $t'_j \in \mathbb{C}$ by (C_2) so we conclude $u \in \mathbb{C}$ by the outer induction hypothesis.

(H_1) $t_i \succeq u$ for some $i \in \{1, \ldots, n\}$. By assumption $t_i \in \mathbb{C}$ so $u \in \mathbb{C}$ by (C_2).

(H_2) $u = g(u_1, \ldots u_k)$ with $f \rhd g$. All $u_i \in \mathbb{C}$ for $1 \leq i \leq k$ by Lemma 5 and since $(f, \{\{t_1, \ldots t_n\}\})$ $(\rhd, \leadsto_{mul})_{lex}$ $(g, \{\{u_1, \ldots, u_k\}\})$ we conclude that $u \in \mathbb{C}$ by the outer induction hypothesis.

(H_3) $u = f(u_1, \ldots u_k)$ with $\{\{t_1, \ldots, t_n\}\} \succ_{mul} \{\{u_1, \ldots, u_k\}\}$. We can conclude $u \in \mathbb{C}$ by the outer induction hypothesis if we can prove that $u_i \in \mathbb{C}$ for $1 \leq i \leq k$. For arbitrary i, by (M_1) we get $t_j \succeq u_i$ for some j and since $t_j \in \mathbb{C}$ by assumption we conclude $u_i \in \mathbb{C}$ by (C_2).

(H_4) $@(u_1, \ldots, u_k)$ is some left-partial flattening of u and $t \gg \{u_1, \ldots, u_k\}$. By Lemma 5 we get $u_i \in \mathbb{C}$ for $1 \leq i \leq k$ and hence $u \in \mathbb{C}$.

The next step is to show that the application of a computable substitution gives computable term, where we define computable substitution as a substitution containing in its domain only computable terms.

Lemma 7. *We say that $\gamma = [x_1/u_1, \ldots, x_n/u_n]$ is a* computable substitution *if for every $i \in \{1, \ldots, n\}$, $u_i \in \mathbb{C}$. Let γ be computable substitution. Then for any term t, $t\gamma \in \mathbb{C}$.*

Proof. We proceed by induction on the structure of term t.

- $t = x$. If $x \in Dom(\gamma)$ then $\gamma = [\ldots, x/u, \ldots]$ and $t\gamma = u$ but $u \in \mathbb{C}$ since γ is computable. Otherwise $t\gamma = x \in \mathbb{C}$ as variables are computable (C_4).
- $t = f(t_1, \ldots, t_n)$ so $t\gamma = f(t_1\gamma, \ldots, t_n\gamma)$. We apply Lemma 6 and we are left to show that for $i \in \{1, \ldots, n\}$, $t_i\gamma \in \mathbb{C}$ which easily follows from the induction hypothesis.
- $t = @(t_l, t_r)$ and $t\gamma = @(t_l\gamma, t_r\gamma)$. Both $t_l\gamma$ and $t_r\gamma$ are computable by the induction hypothesis so $t\gamma \in \mathbb{C}$ by the definition of computability.
- $t = \lambda x{:}\alpha.t_b$ so $t\gamma = \lambda x{:}\alpha.t_b\gamma$. By application of ($C_5$) we are left to show that $t_b\gamma[x/u] \in \mathbb{C}$ for any $u \in \mathbb{C}_\alpha$. But $t_b\gamma[x/u] = t_b(\gamma \cup [x/u])$ since $x \notin Dom(\gamma)$. Since $\gamma \cup [x/u]$ is a computable substitution as so is γ and $u \in \mathbb{C}$, we can conclude $t\gamma \in \mathbb{C}$ by the induction hypothesis.

Now we are ready to present the main theorem stating that the union of HORPO and THE β-reduction relation of THE simply typed λ-calculus, is a well-founded relation on terms.

Theorem 2. *The relation $\rightsquigarrow$ is well-founded.*

Proof. We need to show that $t \in \mathcal{A}cc$ for arbitrary t. Consider an empty substitution ϵ, which is computable by definition. We also have $t = t\epsilon$ so we conclude $t \in \mathbb{C}$ by Lemma 7 and then $t \in \mathcal{A}cc$ by (C_1).

7.4   HORPO in Coq

The definition of HORPO and proof of its well-foundedness are main subjects of this work, however only less than 10% of Coqscripts is devoted to them. The development of this part was going rather smoothly. Mostly problematic were the computability properties but once they were there the proof of well-foundedness and other properties of HORPO could be accomplished with relative ease following closely the proofs from [8,9].

   The definition of HORPO is slightly complicated and that is because $\succ$, $\succ_{mul}$, $\succeq$ and $\succ\!\!\succ$ all need to be combined in one mutually inductive definition as all of them refer to each other.

8   Conclusions

We presented a formalization of the higher-order recursive path ordering in the theorem prover Coq. This development took two years and resulted in a rather big, complete, axiom-free Coq formalization that is a part of the CoLoR project.

The development contains three parts that are completely autonomous: simply typed lambda calculus, multisets and the multiset ordering. These parts will be submitted as Coq contributions so that hopefully they will be used in other developments. In fact the formalization of multisets and multiset ordering already have been used by Coupet-Grimal and Delobel in their recent formalization of RPO [5]. As for the main results concerning HORPO they are integral part of CoLoR library and hopefully will stimulate more developments in the area of higher-order rewriting.

Acknowledgements. The author would like to thank the anonymous reviewers for their useful comments.

References

1. H. P. Barendregt. Lambda calculi with types. *Handbook of logic in computer science (vol. II)*, pages 117–309, 1992.
2. U. Berger, S. Berghofer, P. Letouzey, and H. Schwichtenberg. Program extraction from normalization proofs. *Studia Logica*, 2005. Special issue, to appear.
3. S. Berghofer. A constructive proof of Higman's lemma in Isabelle. 3085:66–82, 2004.
4. N. G. de Bruijn. Lambda-calculus notation with nameless dummies: a tool for automatic formula manipulation with application to the Church-Rosser theorem. *Indag. Math.*, 34(5):381–392, 1972.
5. Solange Coupet-Grimal and William Delobel. A Constructive Axiomatization of the Recursive Path Ordering. Research report 28-2006, LIF, France, 2006.
6. N. Dershowitz. Orderings for term-rewriting systems. *Theor. Comput. Sci.*, 17:279–301, 1982.
7. J.-P. Jouannaud and M. Okada. Executable higher order algebraic specification languages. In *LICS '91*, 350–361, 1991.
8. J.-P. Jouannaud and A. Rubio. The higher-order recursive path ordering. In *LICS '99*, 402–411, Italy, 1999.
9. J.-P. Jouannaud and A. Rubio. Higher-order recursive path orderings 'à la carte'. 2001.
10. A. Koprowski. Well-foundedness of the higher-order recursive path ordering in Coq. TI-IR-004, Vrije Universiteit, The Netherlands, 2004. Master's Thesis.
11. A. Koprowski. Coq formalization of the higher-order recursive path ordering. Technical report in CS-Report series, Eindhoven Univ. of Tech.
12. F. Leclerc. Termination proof of term rewriting systems with the multiset path ordering: A complete development in the system Coq. In TLCA '95, 902:312–327, 1995.
13. C. Murthy. Extracting constructive content from classical proofs. 1990. PhD Thesis.
14. H. Persson. Type theory and the integrated logic of programs. 1999. PhD Thesis.
15. F. van Raamsdonk. *Term Rewriting Systems*, volume 55 of *Cambridge Tracts in TCS*, chapter 11, pages 588–668. Cambridge University Press, 2003.

Dealing with Non-orientable Equations in Rewriting Induction

Takahito Aoto

Research Institute of Electrical Communication, Tohoku University, Japan
aoto@nue.riec.tohoku.ac.jp

Abstract. Rewriting induction (Reddy, 1990) is an automated proof method for inductive theorems of term rewriting systems. Reasoning by the rewriting induction is based on the noetherian induction on some reduction order. Thus, when the given conjecture is not orientable by the reduction order in use, any proof attempts for that conjecture fails; also conjectures such as a commutativity equation are out of the scope of the rewriting induction because they can not be oriented by any reduction order. In this paper, we give an enhanced rewriting induction which can deal with non-orientable conjectures. We also present an extension which intends an incremental use of our enhanced rewriting induction.

1 Introduction

Properties of programs are often proved by induction on the data structures such as natural numbers or lists. Such properties are called inductive properties of programs. Inductive properties are indispensable in formal treatments of programs. Thus automated reasoning of inductive properties is appreciated in techniques such as the program verification and the program transformation.

Term rewriting systems (TRSs) is a computational model based on equational logic. Equational inductive properties of TRSs are called inductive theorems, and automated reasoning methods for inductive theorems have been investigated many years [3,4,7,8,9,10,11,12,15]. In this paper, we extend rewriting induction proposed by Reddy [12], which is one of such inductive theorem proving methods.

The rewriting induction falls in a category of implicit induction methods; in implicit induction, induction scheme is not specified explicitly—such methods are different from explicit induction methods that stem from [5]. Historically, the implicit induction method has been investigated mainly in the context of inductionless induction [7,8,9,11,15]. Usually inductionless induction methods require (kinds of) the Church-Rosser property; while in the rewriting induction, the termination property is needed instead—Koike and Toyama [10] revealed that the rewriting induction[1] and the inductionless induction have different underlying principles. In this context, the underlying principle of (the inductive theorem proving part of) the inductive theorem prover SPIKE [3,4] can be also

---

[1] Renaming the original "term rewriting induction" [6,12] to "rewriting induction" is proposed by them.

F. Pfenning (Ed.): RTA 2006, LNCS 4098, pp. 242–256, 2006.

classified as a rewriting induction method. The rewriting induction is also useful as a program synthesis [6,13].

Inductive proofs by the rewriting induction are based on the noetherian induction on some reduction order. Thus, when the given conjecture is not orientable by the reduction order in use, any proof attempt for that conjecture fails; also conjectures such as a commutativity equation are out of the scope of the rewriting induction because they can not be oriented by any reduction order.

To overcome this defect, several approaches have been proposed. One is to use rewriting modulo equations [12]. Another is to use ordered rewriting technique [2,6] which rewrites a term by possibly non-oriented equations when it simplifies (w.r.t. some ordering). The former appears only in a short remark in [12], and, as far as the author knows, the idea is not explored since then. The latter approach has been embodied in the inductive theorem prover SPIKE. In this paper, we present an enhanced rewriting induction designed following the first approach.

In our enhanced rewriting induction, a reduction order whose equational classes are "coarser" is more suitable to prove non-oriented conjectures. On the other hand, such a reduction order may fail to handle some equations orientable by other reduction orders. This observation leads us to introduce incremental rewriting induction in which already-proved lemmas can be applied more easily.

The rest of the paper is organized as follows. After fixing basic notations (Section 2), we review the principle and the procedure of the rewriting induction (Section 3). In Section 4, we give an enhanced rewriting induction that can deal with non-orientable conjectures and show its correctness. In Section 5, we introduce incremental rewriting induction which intends an incremental use of the enhanced rewriting induction. In Section 6, we conclude our result and compare our approach and the ordered rewriting approach.

2   Preliminaries

Let us fix some notations in *abstract reduction systems (ARSs)*. Let $\to$ be a binary relation on a set A. The reflexive transitive closure (transitive closure, symmetric closure, equivalence closure) of $\to$ is denoted by $\xrightarrow{*}$ ($\xrightarrow{+}$, $\leftrightarrow$, $\xleftrightarrow{*}$, respectively). The relation $\to$ is well-founded (denoted by $\mathrm{SN}(\to)$) when there exists no infinite chain $a_0 \to a_1 \to \cdots$. An element $a \in A$ is said to be *normal* when there is no $b \in A$ such that $a \to b$. The set of normal elements is denoted by $\mathrm{NF}(\to)$. The union $\to_i \cup \to_j$ of two binary relations $\to_i$ and $\to_j$ is abbreviated as $\to_{i \cup j}$. The composition is denoted by $\circ$. We denote by $\to_i / \to_j$ the relation defined by $\xleftrightarrow{*}_j \circ \to_i \circ \xleftrightarrow{*}_j$. The relation $\to_i / \to_j$ is abbreviated as $\to_{i/j}$. We assume / associates stronger than $\cup$; /, $\cup$ associate stronger than closure operations so that, for example, $\xleftrightarrow{*}_{1\cup2}$ stands for the equivalence closure of $\to_1 \cup \to_2$.

We next introduce notations on term rewriting used in this paper. (See [1,14] for details.) The sets of (arity-fixed) function symbols and variables are denoted by $\mathcal{F}$ and V, respectively. $\mathrm{T}(\mathcal{F}, V)$ is the set of terms over $\mathcal{F}, V$. We use $\equiv$ to denote the syntactical equality on terms. The set of variables contained in

t is denoted by $V(t)$. $\mathrm{root}(t)$ is the *root symbol* of a term t. The *domain* of a substitution σ is denoted by $\mathrm{dom}(\sigma)$. A term $\sigma(t)$ is called an *instance* of the term t; $\sigma(t)$ is also written as $t\sigma$. We denote by $\mathrm{mgu}(s,t)$ the *most general unifier* of terms s, t. A pair $l \to r$ of terms satisfying conditions (1) $\mathrm{root}(l) \in \mathcal{F}$; (2) $V(r) \subseteq V(l)$ is said to be a *rewrite rule*. A *term rewriting system (TRS)* is a set of rewrite rules. When the underlying set of function symbols is not clear, we refer to a pair $\langle \mathcal{F}, \mathcal{R} \rangle$ as a TRS—however, we assume that the set of function symbols are those appearing in rewrite rules in this paper. The *rewrite relation* of a TRS $\mathcal{R}$ is denoted by $s \to_{\mathcal{R}} t$. An *equation* $l \doteq r$ is just a pair $\langle l, r \rangle$ of terms in $\mathrm{T}(\mathcal{F}, V)$. When we write $l \doteq r$, we do not distinguish $\langle l, r \rangle$ and $\langle r, l \rangle$.

Function symbols that are roots of some lhs of rewrite rules are called *defined function symbols*; we write $\mathcal{D}_{\mathcal{R}}$ the set $\{\mathrm{root}(l) \mid l \to r \in \mathcal{R}\}$ of defined function symbols (of a TRS $\mathcal{R}$). When $\mathcal{R}$ is obvious from its context, we omit the subscript $\mathcal{R}$. The set of defined symbols appearing in a term t is denoted by $\mathcal{D}(t)$. The set $\mathcal{C} = \mathcal{F} \setminus \mathcal{D}$ of function symbols is the set of *constructor symbols*. Terms in $\mathrm{T}(\mathcal{C}, V)$ are said to be *constructor terms*; substitution σ such that $\sigma(x) \in \mathrm{T}(\mathcal{C}, V)$ for any $x \in \mathrm{dom}(\sigma)$ is called a *constructor substitution*. A term of the form $f(c_1, \ldots, c_n)$ for some $f \in \mathcal{D}$ and $c_1, \ldots, c_n \in \mathrm{T}(\mathcal{C}, V)$ is said to be *basic*. We write $u \trianglelefteq s$ to express that u is a subterm of s. The set $\{u \trianglelefteq s \mid \exists f \in \mathcal{D}. \, \exists c_1, \ldots, c_n \in \mathrm{T}(\mathcal{C}, V). \, u \equiv f(c_1, \ldots, c_n)\}$ of basic subterms of s is written as $\mathcal{B}(s)$.

A term t is said to be *ground* when $V(t) = \emptyset$. $\mathrm{T}(\mathcal{F})$ is the set of ground terms. When $t\sigma \in \mathrm{T}(\mathcal{F})$, $t\sigma$ is called a *ground instance* of t. Ground instances of rewrite rules, equations, etc. are defined similarly. A *ground substitution* is a substitution σ_g such that $\sigma_g(x) \in \mathrm{T}(\mathcal{F})$ for any $x \in \mathrm{dom}(\sigma_g)$. A TRS $\mathcal{R}$ is said to be *quasi-reducible* if no ground basic term is normal. In this paper, we assume w.l.o.g. that $t\sigma_g$ is ground (i.e. $V(t) \subseteq \mathrm{dom}(\sigma_g)$) when we speak of an instance $t\sigma_g$ of t by a ground substitution σ_g. An *inductive theorem* of a TRS $\mathcal{R}$ is an equation that is valid on $\mathrm{T}(\mathcal{F})$, that is, $s \doteq t$ is an inductive theorem when $s\sigma_g \overset{*}{\leftrightarrow}_{\mathcal{R}} t\sigma_g$ holds for any ground instance $s\sigma_g \doteq t\sigma_g$.

A relation R on $\mathrm{T}(\mathcal{F}, V)$ is said to be *closed under substitution* when $s\,R\,t \Rightarrow s\sigma\,R\,t\sigma$ for any substitution σ; *closed under context* when $s\,R\,t \Rightarrow C[s]\,R\,C[t]$ for any context C. A *reduction order* is a well-founded partial order that is closed under substitution and context. A quasi-order $\succsim$ is a *reduction quasi-order* when it is closed under substitution and context and its strict part $\succ = \succsim \setminus \precsim$ is a reduction order. We write the relation $\succsim \cap \precsim$ as $\approx$.

3 Rewriting Induction

Rewriting induction (RI, for short) is an automated inductive theorem proving method proposed by Reddy [12]. The inference system of rewriting induction deals with a set E of equations and a set H of rewrite rules. Intuitively, E is a set of equations to be proved and H is a set of induction hypotheses and theorems already proved. In Figure 1, we list the (downward) inference rules of the rewriting induction. Here $\uplus$ denotes the disjoint union. We note that the direction of each equation is not distinguished. $\mathcal{R}$ and $>$ are a TRS and a

Simplify

$$\frac{\langle E \uplus \{s \doteq t\}, \ H \rangle}{\langle E \cup \{s' \doteq t\}, \ H \rangle} \quad s \rightarrow_{\mathcal{R} \cup H} s'$$

Delete

$$\frac{\langle E \uplus \{s \doteq s\}, \ H \rangle}{\langle E, \ H \rangle}$$

Expand

$$\frac{\langle E \uplus \{s \doteq t\}, \ H \rangle}{\langle E \cup \mathrm{Expd}_u(s, t), \ H \cup \{s \rightarrow t\} \rangle} \quad u \in \mathcal{B}(s), s > t$$

Fig. 1. Inference rules of the rewriting induction

reduction order given as inputs. The set $\mathrm{Expd}_u(s, t)$ of equations is defined like this:

$$\mathrm{Expd}_u(s, t) = \{C[r]\sigma \doteq t\sigma \mid s \equiv C[u], \ \sigma = \mathrm{mgu}(u, l), \ l \rightarrow r \in \mathcal{R}, \ l\text{:basic}\}$$

The following property of Expd will be used later.

Lemma 1 (property of Expd). *Let $\mathcal{R}$ be a quasi-reducible TRS and $u \in \mathcal{B}(s)$. Then*

(1) $s\sigma_g \rightarrow_{\mathcal{R}} \circ \leftrightarrow_{\mathrm{Expd}_u(s,t)} t\sigma_g$ for any ground constructor substitution σ_g;

(2) $v \leftrightarrow_{\mathrm{Expd}_u(s,t)} w \Rightarrow v \overset{*}{\leftrightarrow}_{\mathcal{R} \cup \{s \doteq t\}} w$.

Proof. (1) Since u is basic and σ_g is a ground constructor substitution, $u\sigma_g$ is a basic ground term. Thus, by the quasi-reducibility of $\mathcal{R}$, there exists $l \rightarrow r \in \mathcal{R}$ such that $u\sigma_g$ is an instance of l. W.l.o.g. we may assume $V(l) \cap V(s) = \emptyset$ and thus by extending σ_g one can let $u\sigma_g \equiv l\sigma_g$. Then σ_g is a constructor unifier of u and l and thus we have $\sigma_g = \theta_g \circ \sigma$ for some constructor substitution θ_g, where $\sigma = \mathrm{mgu}(u, l)$. Then by letting $s \equiv C[u]$, we have $s\sigma_g \equiv C[u]\sigma_g \equiv C\sigma_g[u\sigma\theta_g] \equiv C\sigma_g[l\sigma\theta_g] \rightarrow_{\mathcal{R}} C\sigma_g[r\sigma\theta_g] \equiv C[r]\sigma\theta_g \leftrightarrow_{\mathrm{Expd}_u(s,t)} t\sigma\theta_g \equiv t\sigma_g$. (2) Let $v \leftrightarrow_{\mathrm{Expd}_u(s,t)} w$. Then $v \equiv \hat{C}[C[r]\sigma\hat{\sigma}]$, $w \equiv \hat{C}[t\sigma\hat{\sigma}]$ (or $w \equiv \hat{C}[C[r]\sigma\hat{\sigma}]$, $v \equiv \hat{C}[t\sigma\hat{\sigma}]$) for some context $\hat{C}$ and substitution $\hat{\sigma}$, where $\sigma = \mathrm{mgu}(u, l)$, $s \equiv C[u]$, $l \rightarrow r \in \mathcal{R}$. Then we have $v \equiv \hat{C}[C[r]\sigma\hat{\sigma}] \leftarrow_{\mathcal{R}} \hat{C}[C[l]\sigma\hat{\sigma}] \equiv \hat{C}[C\sigma[l\sigma]\hat{\sigma}] \equiv \hat{C}[C\sigma[u\sigma]\hat{\sigma}] \equiv \hat{C}[C[u]\sigma\hat{\sigma}] \equiv \hat{C}[s\sigma\hat{\sigma}] \leftrightarrow_{\{s \doteq t\}} \equiv \hat{C}[t\sigma\hat{\sigma}] \equiv w$. $\qquad\square$

Definition 1 (rewriting induction). *We write $\langle E, H \rangle \rightsquigarrow_{\mathrm{RI}} \langle E', H' \rangle$ when $\langle E', H' \rangle$ is obtained from $\langle E, H \rangle$ by applying one of the inference rules of Figure 1. The reflexive transitive closure of $\rightsquigarrow_{\mathrm{RI}}$ is denoted by $\overset{*}{\rightsquigarrow}_{\mathrm{RI}}$. We sometimes put superscripts ^{s,d,e} to indicate which inference rule is used.*

The rewriting induction procedure starts by putting conjectures into E and letting $H = \emptyset$. Then the procedure rewrites $\langle E, H \rangle$ by applying one of the inference rules. If it eventually becomes of the form $\langle \emptyset, H' \rangle$ then the procedure returns "success"—this means that the conjectures are inductive theorems of $\mathcal{R}$. On the other hand, when none of the rules are applicable, it reports "failure", or it also may run forever ("divergence"), which means the rewriting induction fails to prove that the conjectures are inductive theorems.

Koike and Toyama [10] revealed that the underlying principle of rewriting induction can be formulated in terms of ARSs as below. The proof is by the noetherian induction on $>$. Later, we will give a proof of a more general theorem.

Proposition 1 (principle of rewriting induction [10]). *Let* $\to_1$, $\to_2$ *be binary relations on a set* A. *Let* $>$ *be a well-founded partial order on* A. *Suppose*

$$(i) \quad \to_{1\cup2} \subseteq >$$
$$(ii) \quad \to_2 \subseteq \to_1 \circ \overset{*}{\to}_{1\cup2} \circ \overset{*}{\leftarrow}_{1\cup2}.$$

Then $\overset{*}{\leftrightarrow}_1 = \overset{*}{\leftrightarrow}_{1\cup2}$.

The following proposition states the correctness of the rewriting induction. The proof basically proceeds by applying Proposition 1 to binary relations $\to_{\mathcal{R}}$ and $\to_H$ on the set $\mathrm{T}(\mathcal{F})$ of ground terms. Later, we will give a proof of a more general theorem.

Proposition 2 (correctness of rewriting induction [12]). *Let* $\mathcal{R}$ *be a quasi-reducible TRS,* E *a set of equations,* $>$ *a reduction order satisfying* $\mathcal{R} \subseteq >$. *If there exists a set* H *such that* $\langle E, \emptyset \rangle \overset{*}{\leadsto}_{\mathrm{RI}} \langle \emptyset, H \rangle$ *then equations in* E *are inductive theorems of* $\mathcal{R}$.

Example 1 (rewriting induction). Let $\mathcal{R}$ and E be a TRS and a set of equations given as below.

$$\mathcal{R} = \left\{ \begin{array}{l} \mathsf{plus}(0, y) \quad \to y \\ \mathsf{plus}(\mathsf{s}(x), y) \to \mathsf{s}(\mathsf{plus}(x, y)) \end{array} \right\}$$

$$E = \{\, \mathsf{plus}(\mathsf{plus}(x, y), z) \doteq \mathsf{plus}(x, \mathsf{plus}(y, z)) \,\}$$

Let $>$ be a lexicographic path order [1] based on precedence $\mathsf{plus} > \mathsf{s} > 0$. Below we show how the rewriting induction for proving E proceeds based on the TRS $\mathcal{R}$ and the reduction order $>$.

$$\langle \{\, \mathsf{plus}(\mathsf{plus}(x, y), z) \doteq \mathsf{plus}(x, \mathsf{plus}(y, z)) \,\}, \{\} \rangle$$

$$\leadsto_{\mathrm{RI}}^{e} \left\langle \begin{array}{l} \left\{ \begin{array}{l} \mathsf{plus}(y_0, z) \doteq \mathsf{plus}(0, \mathsf{plus}(y_0, z)) \\ \mathsf{plus}(\mathsf{s}(\mathsf{plus}(x_1, y_1)), z) \doteq \mathsf{plus}(\mathsf{s}(x_1), \mathsf{plus}(y_1, z)) \end{array} \right\} \\ \{\, \mathsf{plus}(\mathsf{plus}(x, y), z) \to \mathsf{plus}(x, \mathsf{plus}(y, z)) \,\} \end{array} \right\rangle$$

$$\leadsto_{\mathrm{RI}}^{s} \leadsto_{\mathrm{RI}}^{s} \leadsto_{\mathrm{RI}}^{s} \left\langle \begin{array}{l} \left\{ \begin{array}{l} \mathsf{plus}(y_0, z) \doteq \mathsf{plus}(y_0, z) \\ \mathsf{s}(\mathsf{plus}(\mathsf{plus}(x_1, y_1), z)) \doteq \mathsf{s}(\mathsf{plus}(x_1, \mathsf{plus}(y_1, z))) \end{array} \right\} \\ \{\, \mathsf{plus}(\mathsf{plus}(x, y), z) \to \mathsf{plus}(x, \mathsf{plus}(y, z)) \,\} \end{array} \right\rangle$$

$$\leadsto_{\mathrm{RI}}^{s} \leadsto_{\mathrm{RI}}^{d} \leadsto_{\mathrm{RI}}^{d} \langle \{\}, \{\, \mathsf{plus}(\mathsf{plus}(x, y), z) \to \mathsf{plus}(x, \mathsf{plus}(y, z)) \,\} \rangle$$

The procedure ends in the form $\langle \emptyset, H \rangle$. Thus from Proposition 2 it follows that the equation in E is an inductive theorem of $\mathcal{R}$.

4   Proving Non-orientable Conjectures

A key of the rewriting induction is the *Expand* rule. But *Expand* rule is applicable only to the equation that can be oriented by the input reduction order. Thus, when the given conjecture is not orientable by the given reduction order, the proof of that conjecture always fails.

Example 2 (failure of rewriting induction). Let $\mathcal{R}$ be a TRS for the addition of natural numbers.

$$\mathcal{R} = \left\{ \begin{array}{l} \mathsf{plus}(0, y) \quad \rightarrow y \\ \mathsf{plus}(\mathsf{s}(x), y) \rightarrow \mathsf{s}(\mathsf{plus}(x, y)) \end{array} \right\}$$

The following equation e expresses the commutativity of addition.

$$e = \quad \mathsf{plus}(x, y) \doteq \mathsf{plus}(y, x)$$

The equation e is an inductive theorem of $\mathcal{R}$. However, because neither $\mathsf{plus}(x, y) > \mathsf{plus}(y, x)$ nor $\mathsf{plus}(y, x) > \mathsf{plus}(x, y)$ holds, the rewriting induction procedure starting with $\langle \{e\}, \emptyset \rangle$ stops immediately having no rules to apply.

To deal with non-orientable equations, Reddy proposed to use $\rightarrow_{\mathcal{R}}/\rightarrow_H$ instead of $\rightarrow_{\mathcal{R}} \cup \rightarrow_H$ (Remark 14 in [12]); however, he does not seem to elaborate on this. In fact, a naive extension seems to lead unsound reasoning—this is illustrated by the following proposition obtained by modifying Propsition 1 suitably for $\rightarrow_{\mathcal{R}}/\rightarrow_H$.

Conjecture 1 (incorrect conjecture). Let $\rightarrow_1, \rightarrow_2$ be binary relations on a set A. Let $\succsim$ be a well-founded quasi-order on A. Suppose

$$\begin{array}{ll} (i) & \rightarrow_1 \subseteq \succ \\ (ii) & \rightarrow_2 \subseteq \approx \\ (iii) & \rightarrow_2 \subseteq \rightarrow_1 \circ \xrightarrow{*}_{1/2} \circ \xleftarrow{*}_{1/2}. \end{array}$$

Then $\xleftrightarrow{*}_1 = \xleftrightarrow{*}_{1 \cup 2}$.

Example 3 (a counterexample to the Conjecture 1). Consider a set $A = \{a, b, c\}$ and relations $\rightarrow_1 = \{\langle a, b \rangle\}$ and $\rightarrow_2 = \{\langle a, c \rangle\}$ on A. Let $\succsim$ be a quasi-order such that $c \approx a \succ b$. Then conditions (i),(ii) clearly hold. Since $a \rightarrow_1 b \leftarrow_1 a \leftrightarrow_2 c$, condition (iii) holds also. But we have $c \xleftrightarrow{*}_{1 \cup 2} b$ and $c \xcancel{\leftrightarrow}^{*}_1 b$.

In Figure 2, we list the inference rules in which $\mathcal{R} \cup H$ is replaced by $\mathcal{R}/H$. This inference system is not sound as the following example shows.

Example 4 (incorrect inference). Let $\mathcal{R}$ be a TRS for the append of two lists:

$$\mathcal{R} = \left\{ \begin{array}{l} \mathsf{app}(\mathsf{nil}, ys) \quad\quad\quad \rightarrow ys \\ \mathsf{app}(\mathsf{cons}(x, xs), ys) \rightarrow \mathsf{cons}(x, \mathsf{app}(xs, ys)) \end{array} \right\}.$$

The append operation is not commutative, hence

$$\mathsf{app}(xs, ys) \doteq \mathsf{app}(ys, xs) \tag{1}$$

Simplify

$$\frac{\langle E \uplus \{s \doteq t\},\ H\rangle}{\langle E \cup \{s' \doteq t\},\ H\rangle} \quad s \to_{R/H} s'$$

Delete

$$\frac{\langle E \uplus \{s \doteq t\},\ H\rangle}{\langle E,\ H\rangle} \quad s \overset{*}{\leftrightarrow}_H t$$

Expand

$$\frac{\langle E \uplus \{s \doteq t\},\ H\rangle}{\langle E \cup \mathrm{Expd}_u(s,t),\ H \cup \{s \doteq t\}\rangle} \quad u \in \mathcal{B}(s), s \approx t$$

Fig. 2. Inference rules with rewriting modulo equations(not sound)

is not an inductive theorem of $\mathcal{R}$. However, by taking $\succsim$ as a recursive path order [1] based on the precedence $\mathsf{app} \succ \mathsf{cons} \succ \mathsf{nil}$, the inference of modified rewriting induction successfully proves the conjecture (1).

In *Expand* rule in Figure 2, for $v \doteq w$ in $\mathrm{Expd}_u(s,t)$, only v is "smaller" than $s\sigma$ while w is "just as big" as $s\sigma$. Hence, application of the inductive hypothesis to w is unsound. This observation suggests a new kind of *Expand* rule for non-orientable equations (*Expand2*, below), which expands both lhs and rhs of the equation.

Simplify

$$\frac{\langle E \uplus \{s \doteq t\},\ H,\ G\rangle}{\langle E \cup \{s' \doteq t\},\ H,\ G\rangle} \quad s \to_{(\mathcal{R}\cup H)/G} s'$$

Delete

$$\frac{\langle E \uplus \{s \doteq t\},\ H,\ G\rangle}{\langle E,\ H,\ G\rangle} \quad s \overset{*}{\leftrightarrow}_G t$$

Expand

$$\frac{\langle E \uplus \{s \doteq t\},\ H,\ G\rangle}{\langle E \cup \mathrm{Expd}_u(s,t),\ H \cup \{s \to t\},\ G\rangle} \quad u \in \mathcal{B}(s), s \succ t$$

Expand2

$$\frac{\langle E \uplus \{s \doteq t\}, H, G\rangle}{\langle E \cup \mathrm{Expd2}_{u,v}(s,t), H, G \cup \{s \doteq t\}\rangle} \quad u \in \mathcal{B}(s), v \in \mathcal{B}(t), s \approx t$$

Fig. 3. Inference rules of eRI

Here, $\mathrm{Expd2}_{u,v}(s,t)$ is defined like this:

$$\mathrm{Expd2}_{u,v}(s,t) = \bigcup \left\{ \mathrm{Expd}_{v\sigma}(t\sigma, s') \mid \langle s', t\sigma\rangle \in \mathrm{Expd}_u(s,t) \right\}$$

Definition 2 (enhanced rewriting induction). *We write* $\langle E, H, G\rangle \rightsquigarrow_{\mathrm{eRI}} \langle E', H', G'\rangle$ *when* $\langle E', H', G'\rangle$ *is obtained from* $\langle E, H, G\rangle$ *by applying one of the inference rules of Figure 3. The reflexive transitive closure of* $\rightsquigarrow_{\mathrm{eRI}}$ *is denoted by* $\overset{*}{\rightsquigarrow}_{\mathrm{eRI}}$. *We sometimes put superscripts* ^{s,d,e,e2} *to indicate which inference rule is used.*

Example 5 (application of Expand2 rule). Let

$$\mathcal{R} = \left\{ \begin{array}{l} \mathsf{plus}(0, y) \quad \to y \\ \mathsf{plus}(\mathsf{s}(x), y) \to \mathsf{s}(\mathsf{plus}(x, y)) \end{array} \right\},$$

$s \doteq t = \mathsf{plus}(x, y) \doteq \mathsf{plus}(y, x)$. Then we have

$$\mathrm{Expd2}_{s,t}(s, t) = \left\{ \begin{array}{l} 0 \doteq 0 \\ \mathsf{s}(x_1) \doteq \mathsf{s}(\mathsf{plus}(x_1, 0)) \\ \mathsf{s}(\mathsf{plus}(x_3, \mathsf{s}(y_3))) \doteq \mathsf{s}(\mathsf{plus}(y_3, \mathsf{s}(x_3))) \end{array} \right\}.$$

Note that an equation $\mathsf{s}(\mathsf{plus}(x_2, 0)) \doteq \mathsf{s}(x_2)$ which is also included in $\mathrm{Expd2}_{s,t}(s, t)$ is omitted, since this equation is same as the second one (as an equation).

Lemma 2 (property of Expd2). *Let $\mathcal{R}$ be a quasi-reducible TRS and let $u \in \mathcal{B}(s)$ and $v \in \mathcal{B}(t)$. Then*

(1) $s\sigma_g \to_{\mathcal{R}} \circ \leftrightarrow_{\mathrm{Expd2}_{u,v}(s,t)} \circ \leftarrow_{\mathcal{R}} t\sigma_g$ for any ground constructor substitution σ_g;

(2) $q \leftrightarrow_{\mathrm{Expd2}_{u,v}(s,t)} w \Rightarrow q \overset{*}{\leftrightarrow}_{\mathcal{R} \cup \{s \doteq t\}} w$.

Proof. (1) Let $l \to r \in \mathcal{R}$, l: basic, $\sigma = \mathrm{mgu}(u, l)$, $s \equiv C[u]$. Then, by Lemma 1 (1), $s\sigma_g \to_{\mathcal{R}} C[r]\sigma\theta_g \leftrightarrow_{\{C[r]\sigma \doteq t\sigma\}} t\sigma\theta_g \equiv t\sigma_g$ for some $\langle C[r]\sigma, t\sigma \rangle \in \mathrm{Expd}_u(s, t)$ and constructor substitution θ_g. Since σ is a constructor substitution and $v \in \mathcal{B}(t)$, we know $v\sigma \in \mathcal{B}(t\sigma)$. Thus applying Lemma 1 (1) once again, we have $t\sigma\theta_g \to_{\mathcal{R}} \circ \leftrightarrow_{\mathrm{Expd}_{v\sigma}(t\sigma, C[r]\sigma)} C[r]\sigma\theta_g$. Thus $s\sigma_g \to_{\mathcal{R}} C[r]\sigma\theta_g \leftrightarrow_{\mathrm{Expd2}_{u,v}(s,t)} \circ \leftarrow_{\mathcal{R}} t\sigma\theta_g \equiv t\sigma_g$. (2) Suppose $q \leftrightarrow_{\mathrm{Expd2}_{u,v}(s,t)} w$. Then, by definition, there exists $\langle s', t\sigma \rangle \in \mathrm{Expd}_u(s, t)$ such that $q \leftrightarrow_{\mathrm{Expd}_{v\sigma}(t\sigma, s')} w$, where $\sigma = \mathrm{mgu}(l, u)$, $l \to r \in \mathcal{R}$, l: basic. Then σ is a constructor substitution and thus $v\sigma \in \mathcal{B}(t\sigma)$. Then by Lemma 1 (2) $q \overset{*}{\leftrightarrow}_{\mathcal{R} \cup \{s' \doteq t\sigma\}} w$. Therefore we have $q \overset{*}{\leftrightarrow}_{\mathcal{R} \cup \mathrm{Expd}_u(s,t)} w$. By applying Lemma 1 (2) once again, we have $q \overset{*}{\leftrightarrow}_{\mathcal{R} \cup \{s \doteq t\}} w$. $\square$

The soundness of the enhanced rewriting induction is based on the following alternative principle.

Lemma 3 (principle of enhanced rewriting induction). *Let $\to_i$ ($1 \le i \le 3$) be binary relations on a set A, and $\succsim$ be a well-founded quasi-order on A. Suppose*

$$(i) \quad \to_{1 \cup 2} \subseteq \succ$$
$$(ii) \quad \to_3 \subseteq \approx$$
$$(iii) \quad \to_2 \subseteq \to_1 \circ \overset{*}{\to}_{(1 \cup 2)/3} \circ \overset{*}{\leftrightarrow}_3 \circ \overset{*}{\leftarrow}_{(1 \cup 2)/3}$$
$$(iv) \quad \to_3 \subseteq \to_1 \circ \overset{*}{\to}_{(1 \cup 2)/3} \circ \overset{*}{\leftrightarrow}_3 \circ \overset{*}{\leftarrow}_{(1 \cup 2)/3} \circ \leftarrow_1$$
$$(v) \quad \forall x, y \in \mathrm{NF}(\to_{(1 \cup 2)/3}). \ (x \overset{*}{\leftrightarrow}_3 y \Rightarrow x = y).$$

Then $\overset{*}{\leftrightarrow}_1 = \overset{*}{\leftrightarrow}_{1 \cup 2 \cup 3}$.

Proof. It suffices to show $\supseteq$. For this, we first show by noetherian induction on $\succ$ that

$$\text{for any } x \in A \ [\forall y \in A. \ (x \overset{*}{\to}_{(1 \cup 2)/3} y \Rightarrow x \overset{*}{\leftrightarrow}_1 y)] \tag{2}$$

holds.

(Base Step) Then $x = y$ and thus $x \overset{*}{\leftrightarrow}_1 y$ trivially holds.

(Induction Step) The case when $x \in \mathrm{NF}(\to_{(1\cup2)/3})$ follows immediately; so, suppose $x \overset{*}{\leftrightarrow}_3 u \to_{1\cup2} v \overset{*}{\leftrightarrow}_3 z \overset{*}{\to}_{(1\cup2)/3} y$. Since $x \succ z$, it follows

$$z \overset{*}{\leftrightarrow}_1 y \tag{3}$$

by induction hypothesis.

We now claim that

$$a \to_3 b \Rightarrow a \overset{*}{\leftrightarrow}_1 b \text{ for any } a, b \precsim x \tag{4}$$

We note that it immediately follows from this that

$$x \overset{*}{\leftrightarrow}_1 u \text{ and } v \overset{*}{\leftrightarrow}_1 z \tag{5}$$

Suppose $a \to_3 b$. Then by condition (iv) we have

$$a \to_1 c \overset{*}{\to}_{(1\cup2)/3} c' \overset{*}{\leftrightarrow}_3 d' \overset{*}{\leftarrow}_{(1\cup2)/3} d \leftarrow_1 b$$

for some c, c', d, d'. By induction hypothesis we have $c \overset{*}{\leftrightarrow}_1 c'$ and $d' \overset{*}{\leftrightarrow}_1 d$. If $c', d' \in \mathrm{NF}(\to_{(1\cup2)/3})$ then the claim follows from the condition (v). So, suppose $c' \notin \mathrm{NF}(\to_{(1\cup2)/3})$. Then by conditions (i),(ii), $c' \overset{+}{\to}_{(1\cup2)/3} n$ for some $n \in \mathrm{NF}(\to_{(1\cup2)/3})$. Then we have $d' \overset{*}{\leftrightarrow}_3 c' \overset{+}{\to}_{(1\cup2)/3} n$ and hence $d \overset{*}{\to}_{(1\cup2)/3} n$. Thus we have

$$a \to_1 c \overset{*}{\to}_{(1\cup2)/3} n \overset{*}{\leftarrow}_{(1\cup2)/3} d \leftarrow_1 b$$

Then by the induction hypothesis $a \to_1 c \overset{*}{\leftrightarrow}_1 n \overset{*}{\leftrightarrow}_1 d \leftarrow_1 b$ follows. Since the case $d' \notin \mathrm{NF}(\to_{(1\cup2)/3})$ is shown similarly, the claim (4) has been shown.

It remains to show

$$u \overset{*}{\leftrightarrow}_1 v \tag{6}$$

The case when $u \to_1 v$ is trivial. So, suppose $u \to_2 v$. By condition (iii), we have

$$u \to_1 w \overset{*}{\to}_{(1\cup2)/3} w' \overset{*}{\leftrightarrow}_3 v' \overset{*}{\leftarrow}_{(1\cup2)/3} v$$

Since $x \succ w, v$, we have $w \overset{*}{\leftrightarrow}_1 w'$ and $v' \overset{*}{\leftrightarrow}_1 v$ by induction hypothesis. Then, as above, one can suppose w.l.o.g. $w', v' \in \mathrm{NF}(\to_{(1\cup2)/3})$. Thus the claim follows from the condition (v). Thus by (3), (5), (6), the proof of the claim (2) has been completed.

Next we show $\to_3 \subseteq \overset{*}{\leftrightarrow}_1$. This can be proved same as the proof of the claim (4), except that this time we use already proved claim (2) instead of the induction hypothesis.

Finally, the statement of the lemma follows from the fact $\overset{*}{\to}_{1\cup2\cup3} \subseteq \overset{*}{\to}_{(1\cup2)/3} \cup \overset{*}{\leftrightarrow}_3$. $\qquad\square$

Below we prove the correctness of the enhanced rewriting induction. In remaining lemmata in this section, we assume that the TRS $\mathcal{R}$ is quasi-reducible and that $\precsim$ is a reduction quasi-order satisfying $\mathcal{R} \subseteq \succ$.

Lemma 4 (invariance). *Let* $\langle E_n, H_n, G_n \rangle \leadsto_{\text{eRI}} \langle E_{n+1}, H_{n+1}, G_{n+1} \rangle$. *Then* $\overset{*}{\leftrightarrow}_{\mathcal{R} \cup E_n \cup H_n \cup G_n} = \overset{*}{\leftrightarrow}_{\mathcal{R} \cup E_{n+1} \cup H_{n+1} \cup G_{n+1}}$ *on* $T(\mathcal{F})$.

Proof. Use Lemma 1 (2) and Lemma 2 (2). □

Lemma 5 (property of E_n). *Let* $\langle E_n, H_n, G_n \rangle \overset{*}{\leadsto}_{\text{eRI}} \langle \emptyset, H^\sharp, G^\sharp \rangle$. *Then* $\leftrightarrow_{E_n} \subseteq \overset{*}{\rightarrow}_{(\mathcal{R} \cup H^\sharp)/G^\sharp} \circ \overset{*}{\leftrightarrow}_{G^\sharp} \circ \overset{*}{\leftarrow}_{(\mathcal{R} \cup H^\sharp)/G^\sharp}$ *on* $T(\mathcal{F})$.

Proof. By induction on the length of $\langle E_n, H_n, G_n \rangle \overset{*}{\leadsto}_{\text{eRI}} \langle \emptyset, H^\sharp, G^\sharp \rangle$. □

Lemma 6 (property of $H^\sharp$). *Let* $\langle E_n, H_n, G_n \rangle \overset{*}{\leadsto}_{\text{eRI}} \langle \emptyset, H^\sharp, G^\sharp \rangle$. *Then* $\rightarrow_{H^\sharp} \subseteq \rightarrow_{\mathcal{R}} \circ \overset{*}{\rightarrow}_{(\mathcal{R} \cup H^\sharp)/G^\sharp} \circ \overset{*}{\leftrightarrow}_{G^\sharp} \circ \overset{*}{\leftarrow}_{(\mathcal{R} \cup H^\sharp)/G^\sharp}$ *on* $T(\mathcal{F})$.

Proof. Suppose $s \rightarrow t \in H^\sharp$, $s_g \rightarrow_{\{s \rightarrow t\}} t_g$, and let $s_g \equiv C_g[s\sigma_g]$, $t_g \equiv C_g[t\sigma_g]$.

It suffices to consider the case when $\sigma_g(x) \in \text{NF}(\rightarrow_{\mathcal{R}})$ for any $x \in V(s)$. For, by $\text{SN}(\rightarrow_{\mathcal{R}})$, there exists a substitution $\hat{\sigma}_g$ such that $\sigma_g(x) \overset{*}{\rightarrow}_{\mathcal{R}} \hat{\sigma}_g(x) \in \text{NF}(\rightarrow_{\mathcal{R}})$ for any $x \in V(s)$. Thus if once we have shown $C_g[s\hat{\sigma}_g] \rightarrow_{\mathcal{R}} \circ \overset{*}{\rightarrow}_{(\mathcal{R} \cup H^\sharp)/G^\sharp} \circ \overset{*}{\leftrightarrow}_{G^\sharp} \circ \overset{*}{\leftarrow}_{(\mathcal{R} \cup H^\sharp)/G^\sharp} C_g[t\hat{\sigma}_g]$, then by $C_g[s\sigma_g] \overset{*}{\rightarrow}_{\mathcal{R}} C_g[s\hat{\sigma}_g]$, $C_g[t\sigma_g] \overset{*}{\rightarrow}_{\mathcal{R}} C_g[t\hat{\sigma}_g]$, we would have $C_g[s\sigma_g] \rightarrow_{\mathcal{R}} \circ \overset{*}{\rightarrow}_{(\mathcal{R} \cup H^\sharp)/G^\sharp} \circ \overset{*}{\leftrightarrow}_{G^\sharp} \circ \overset{*}{\leftarrow}_{(\mathcal{R} \cup H^\sharp)/G^\sharp} C_g[t\sigma_g]$.

Thus let us suppose that $\sigma_g(x) \in \text{NF}(\rightarrow_{\mathcal{R}})$ for any $x \in V(s)$. Then by the quasi-reducibility of $\mathcal{R}$, σ_g is a constructor substitution. For some n, we have $\langle E_0, \emptyset, \emptyset \rangle \overset{*}{\leadsto}_{\text{eRI}} \langle E_n, H_n, G_n \rangle \leadsto_{\text{eRI}} \langle E_{n+1}, H_{n+1}, G_{n+1} \rangle \overset{*}{\leadsto}_{\text{eRI}} \langle \emptyset, H^\sharp, G^\sharp \rangle$, $H_{n+1} = H_n \cup \{s \rightarrow t\}$ where $E_n = E \uplus \{s \doteq t\}$, $E_{n+1} = E \cup \text{Expd}_u(s,t)$, $u \in \mathcal{B}(s)$, $s \succ t$. Then by Lemma 1 (1) we have $s\sigma_g \rightarrow_{\mathcal{R}} \circ \leftrightarrow_{\text{Expd}_u(s,t)} t\sigma_g$. Thus $s_g \equiv C_g[s\sigma_g] \rightarrow_{\mathcal{R}} \circ \leftrightarrow_{E_{n+1}} C_g[t\sigma_g] \equiv t_g$. Therefore, by Lemma 5, $s_g \rightarrow_{\mathcal{R}} \circ \overset{*}{\rightarrow}_{(\mathcal{R} \cup H^\sharp)/G^\sharp} \circ \overset{*}{\leftrightarrow}_{G^\sharp} \circ \overset{*}{\leftarrow}_{(\mathcal{R} \cup H^\sharp)/G^\sharp} t_g$. □

Lemma 7 (property of $G^\sharp$). *Let* $\langle E_n, H_n, G_n \rangle \overset{*}{\leadsto}_{\text{eRI}} \langle \emptyset, H^\sharp, G^\sharp \rangle$. *Then* $\leftrightarrow_{G^\sharp} \subseteq \rightarrow_{\mathcal{R}} \circ \overset{*}{\rightarrow}_{(\mathcal{R} \cup H^\sharp)/G^\sharp} \circ \overset{*}{\leftrightarrow}_{G^\sharp} \circ \overset{*}{\leftarrow}_{(\mathcal{R} \cup H^\sharp)/G^\sharp} \circ \leftarrow_{\mathcal{R}}$ *on* $T(\mathcal{F})$.

Proof. Similar to Lemma 6 using Lemma 2 (1) instead of Lemma 1 (1). □

Theorem 1 (correctness of enhanced rewriting induction). *Let* $\mathcal{R}$ *be a quasi-reducible TRS, E a set of equations, $\succsim$ a reduction quasi-order satisfying $\mathcal{R} \subseteq \succ$. If there exist sets H, G (of rewrite rules and of equations, respectively) such that $\langle E, \emptyset, \emptyset \rangle \overset{*}{\leadsto}_{\text{eRI}} \langle \emptyset, H, G \rangle$, then equations in E are inductive theorems of $\mathcal{R}$.*

Proof. By repeatedly applying Lemma 4, from $\langle E, \emptyset, \emptyset \rangle \overset{*}{\leadsto}_{\text{eRI}} \langle \emptyset, H, G \rangle$, it follows that $\overset{*}{\leftrightarrow}_{E \cup \mathcal{R}} = \overset{*}{\leftrightarrow}_{\mathcal{R} \cup H \cup G}$ on $T(\mathcal{F})$. Therefore it suffices to show $\overset{*}{\leftrightarrow}_{\mathcal{R} \cup H \cup G} = \overset{*}{\leftrightarrow}_{\mathcal{R}}$ holds on $T(\mathcal{F})$. We apply Lemma 3 for $A = T(\mathcal{F})$, $\rightarrow_1 = \rightarrow_{\mathcal{R}}$, $\rightarrow_2 = \rightarrow_H$, and $\rightarrow_3 = \rightarrow_G$.

By $\mathcal{R} \cup H \subseteq \succ$, $G \subseteq \approx$, the conditions (i),(ii) of Lemma 3 hold. The condition (iii) holds by Lemma 6 and (iv) by Lemma 7. Also by the side conditions of *Expand2* rule, we have $\mathcal{B}(s) \neq \emptyset$ and $\mathcal{B}(t) \neq \emptyset$ for any $s \doteq t \in G$. Thus by the quasi-reducibility of $\mathcal{R}$, $s_g \overset{+}{\leftrightarrow}_G t_g \Rightarrow s_g, t_g \notin \text{NF}(\rightarrow_{\mathcal{R}})$ for any s_g, t_g. Hence the condition (v) holds. Therefore, by Lemma 3, $\overset{*}{\leftrightarrow}_{\mathcal{R}} = \overset{*}{\leftrightarrow}_{\mathcal{R} \cup H \cup G}$. □

Example 6 (enhanced rewriting induction). Let $\mathcal{R}$ and E be as follows:

$$\mathcal{R} = \left\{ \begin{array}{l} \mathsf{plus}(0, y) \;\;\;\; \to y \\ \mathsf{plus}(\mathsf{s}(x), y) \to \mathsf{s}(\mathsf{plus}(x, y)) \end{array} \right\}$$

$$E = \left\{ \mathsf{plus}(x, y) \doteq \mathsf{plus}(y, x) \right\}$$

Let $\succsim$ be a recursive path order based on the precedence $\mathsf{plus} \succ \mathsf{s} \succ 0$. Then the eRI works as follows.

$$\langle \left\{ \mathsf{plus}(x, y) \doteq \mathsf{plus}(y, x) \right\}, \{\}, \{\} \rangle$$

$$\rightsquigarrow^{e2} \left\langle \begin{array}{l} \left\{ \begin{array}{l} 0 \doteq 0, \; \mathsf{s}(x_1) \doteq \mathsf{s}(\mathsf{plus}(x_1, 0)) \\ \mathsf{s}(\mathsf{plus}(x_2, \mathsf{s}(y_2))) \doteq \mathsf{s}(\mathsf{plus}(y_2, \mathsf{s}(x_2))) \end{array} \right\} \\ \{\}, \left\{ \mathsf{plus}(x, y) \doteq \mathsf{plus}(y, x) \right\} \end{array} \right\rangle$$

$$\rightsquigarrow^d \rightsquigarrow^s \rightsquigarrow^d \left\langle \begin{array}{l} \left\{ \mathsf{s}(\mathsf{plus}(x_2, \mathsf{s}(y_2))) \doteq \mathsf{s}(\mathsf{plus}(y_2, \mathsf{s}(x_2))) \right\} \\ \{\}, \left\{ \mathsf{plus}(x, y) \doteq \mathsf{plus}(y, x) \right\} \end{array} \right\rangle$$

$$\rightsquigarrow^s \rightsquigarrow^s \rightsquigarrow^d \langle \{\}, \{\}, \left\{ \mathsf{plus}(x, y) \doteq \mathsf{plus}(y, x) \right\} \rangle$$

By Theorem 1, the equation in E is an inductive theorem of $\mathcal{R}$.

5   Incremental Proofs by Rewriting Induction

In the enhanced rewriting induction, a reduction order whose equational classes are "coarser" is more suitable to prove non-oriented conjectures. On the other hand, such a reduction order may fail to handle some equations orientable by other reduction orders.

Example 7 (handling multiple conjectures). Let $\mathcal{R}$ and E be as follows:

$$\mathcal{R} = \left\{ \begin{array}{l} \mathsf{plus}(0, y) \;\;\;\; \to y \\ \mathsf{plus}(\mathsf{s}(x), y) \to \mathsf{s}(\mathsf{plus}(x, y)) \\ \mathsf{times}(0, y) \;\;\;\; \to 0 \\ \mathsf{times}(\mathsf{s}(x), y) \to \mathsf{plus}(\mathsf{times}(x, y), y) \end{array} \right\}$$

$$E = \left\{ \begin{array}{l} \mathsf{plus}(x, y) \;\;\;\;\;\;\;\;\;\;\; \doteq \mathsf{plus}(y, x) \\ \mathsf{plus}(x, \mathsf{plus}(y, z)) \doteq \mathsf{plus}(\mathsf{plus}(x, y), z) \\ \mathsf{times}(x, y) \;\;\;\;\;\;\;\;\;\; \doteq \mathsf{times}(y, x) \end{array} \right\}$$

The recursive path order can handle the commutativity equations but not the associativity equation. To the contrary, the lexicographic path order can handle the associativity equations but not the commutativity equations. As we will see, we need both commutativity and associativity of plus to prove the commutativity of times and thus eRI can not handle the commutativity of times.

This observation leads us to introduce *incremental rewriting induction* in which already-proved lemmas can be applied more easily. The incremental rewriting induction can employ different reduction orders in each phase so that it can be also benefited from variations of reduction orders.

We first formulate abstract principle of incremental rewriting induction. The proof is similar to that of Lemma 3.

Lemma 8 (principle of incremental rewriting induction). *Let $\to_i$ $(1 \le i \le 4)$ be a relation on a set A, and $\succsim$ be a well-founded quasi-order on A. Suppose*

(i) $\quad \to_{1 \cup 2} \subseteq \succ$

(ii) $\quad \to_3 \subseteq \approx$

(iii) $\quad \to_4 \subseteq \overset{*}{\leftrightarrow}_1 \cap \succsim$

(iv) $\quad \to_2 \subseteq \to_1 \circ \overset{*}{\to}_{((1 \cup 2)/3) \cup 4} \circ (\overset{*}{\leftrightarrow}_3 \cup \overset{*}{\leftrightarrow}_1) \circ \overset{*}{\leftarrow}_{((1 \cup 2)/3) \cup 4}$

(v) $\quad \to_3 \subseteq \to_1 \circ \overset{*}{\to}_{((1 \cup 2)/3) \cup 4} \circ (\overset{*}{\leftrightarrow}_3 \cup \overset{*}{\leftrightarrow}_1) \circ \overset{*}{\leftarrow}_{((1 \cup 2)/3) \cup 4} \circ \leftarrow_1$

(vi) $\quad \forall x, y \in \mathrm{NF}(\to_{(1 \cup 2)/3}).\ (x \overset{*}{\leftrightarrow}_3 y \Rightarrow x = y)$.

Then $\overset{*}{\leftrightarrow}_1 = \overset{*}{\leftrightarrow}_{1 \cup 2 \cup 3 \cup 4}$.

In Figure 4, we list inference rules designed based on this abstract principle.

Definition 3 (incremental rewriting induction). *We write $\langle E, H, G \rangle \rightsquigarrow_{\mathrm{iRI}}$ $\langle E', H', G' \rangle$ when $\langle E', H', G' \rangle$ is obtained from $\langle E, H, G \rangle$ by applying one of the inference rules of Figure 4. The reflexive transitive closure of $\rightsquigarrow_{\mathrm{iRI}}$ is denoted by $\overset{*}{\rightsquigarrow}_{\mathrm{iRI}}$. We put superscripts $s, s2, d, d2, e, e2$ to indicate which inference rule is used.*

The correctness of the incremental rewriting induction is proved similarly to the enhanced rewriting induction by putting $\to_1 = \to_\mathcal{R}$, $\to_2 = \to_H$, $\to_3 = \to_G$, and $\to_4 = \overset{*}{\leftrightarrow}_{\mathcal{R} \cup \mathcal{E}} \cap \succsim$.

Simplify
$$\frac{\langle E \uplus \{s \doteq t\},\ H,\ G \rangle}{\langle E \cup \{s' \doteq t\},\ H,\ G \rangle} \quad s \to_{(\mathcal{R} \cup H)/G} s'$$

Simplify2
$$\frac{\langle E \uplus \{s \doteq t\},\ H,\ G \rangle}{\langle E \cup \{s' \doteq t\},\ H,\ G \rangle} \quad s \overset{*}{\leftrightarrow}_{\mathcal{R} \cup \mathcal{E}} s',\ s \succsim s'$$

Delete
$$\frac{\langle E \uplus \{s \doteq t\},\ H,\ G \rangle}{\langle E,\ H,\ G \rangle} \quad s \overset{*}{\leftrightarrow}_G t$$

Delete2
$$\frac{\langle E \uplus \{s \doteq t\},\ H,\ G \rangle}{\langle E,\ H,\ G \rangle} \quad s \overset{*}{\leftrightarrow}_{\mathcal{R} \cup \mathcal{E}} t$$

Expand
$$\frac{\langle E \uplus \{s \doteq t\},\ H,\ G \rangle}{\langle E \cup \mathrm{Expd}_u(s, t),\ H \cup \{s \to t\},\ G \rangle} \quad u \in \mathcal{B}(s), s \succ t$$

Expand2
$$\frac{\langle E \uplus \{s \doteq t\}, H, G \rangle}{\langle E \cup \mathrm{Expd2}_{u,v}(s, t), H, G \cup \{s \doteq t\} \rangle} \quad u \in \mathcal{B}(s), v \in \mathcal{B}(t), s \approx t$$

Fig. 4. Inference rules of iRI

Theorem 2 (correctness of incremental rewriting induction). *Let $\mathcal{R}$ be a quasi-reducible TRS, $E, \mathcal{E}$ sets of equations, $\succsim$ a reduction quasi-order satisfying $\mathcal{R} \subseteq \succ$. Suppose equations in $\mathcal{E}$ are inductive theorems of $\mathcal{R}$. If there exist sets H, G (of rewrite rules and of equations, respectively) such that $\langle E, \emptyset, \emptyset \rangle \overset{*}{\leadsto}_{\mathrm{iRI}} \langle \emptyset, H, G \rangle$, then equations in E are inductive theorems of $\mathcal{R}$.*

Example 8 (incremental rewriting induction). Let $\mathcal{R}$, $\mathcal{E}$ and E be as below. We may suppose that equations in $\mathcal{E}$ has been already proved (Examples 1, 6).

$$\mathcal{R} = \begin{cases} \mathsf{plus}(0, y) & \to y \\ \mathsf{plus}(\mathsf{s}(x), y) & \to \mathsf{s}(\mathsf{plus}(x, y)) \\ \mathsf{times}(0, y) & \to 0 \\ \mathsf{times}(\mathsf{s}(x), y) & \to \mathsf{plus}(\mathsf{times}(x, y), y) \end{cases}$$

$$\mathcal{E} = \begin{cases} \mathsf{plus}(x, \mathsf{plus}(y, z)) \doteq \mathsf{plus}(\mathsf{plus}(x, y), z) \\ \mathsf{plus}(x, y) \qquad\qquad \doteq \mathsf{plus}(y, x) \end{cases}$$

$$E = \{\, \mathsf{times}(x, y) \doteq \mathsf{times}(y, x) \,\}$$

Then the incremental rewriting induction by the recursive path order based on precedence $\mathsf{times} \succ \mathsf{plus} \succ \mathsf{s} \succ 0$ proceeds as follows:

$$\langle \{\, \mathsf{times}(x, y) \doteq \mathsf{times}(y, x) \,\}, \{\}, \{\} \rangle$$

$\leadsto_{\mathrm{iRI}}^{e2}$

$$\left\langle \begin{cases} 0 \doteq 0 \\ \mathsf{plus}(\mathsf{times}(x_1, 0), 0) \doteq 0 \\ \mathsf{plus}(\mathsf{times}(x_1, \mathsf{s}(y_1)), \mathsf{s}(y_1)) \doteq \mathsf{plus}(\mathsf{times}(y_1, \mathsf{s}(x_1)), \mathsf{s}(x_1)) \end{cases} \right\rangle$$
$$\{\}, \{\, \mathsf{times}(x, y) \doteq \mathsf{times}(y, x) \,\}$$

$\leadsto_{\mathrm{iRI}}^{d} \leadsto_{\mathrm{iRI}}^{s} \leadsto_{\mathrm{iRI}}^{s} \leadsto_{\mathrm{iRI}}^{d}$

$$\left\langle \{\, \mathsf{plus}(\mathsf{times}(x_1, \mathsf{s}(y_1)), \mathsf{s}(y_1)) \doteq \mathsf{plus}(\mathsf{times}(y_1, \mathsf{s}(x_1)), \mathsf{s}(x_1)) \,\} \right\rangle$$
$$\{\}, \{\, \mathsf{times}(x, y) \doteq \mathsf{times}(y, x) \,\}$$

$\leadsto_{\mathrm{iRI}}^{s} \leadsto_{\mathrm{iRI}}^{s2} \leadsto_{\mathrm{iRI}}^{s} \leadsto_{\mathrm{iRI}}^{s} \leadsto_{\mathrm{iRI}}^{s2} \leadsto_{\mathrm{iRI}}^{s}$

$$\left\langle \{\, \mathsf{s}(\mathsf{plus}(y_1, \mathsf{plus}(\mathsf{times}(x_1, y_1), x_1))) \doteq \mathsf{s}(\mathsf{plus}(x_1, \mathsf{plus}(\mathsf{times}(x_1, y_1), y_1))) \,\} \right\rangle$$
$$\{\}, \{\, \mathsf{times}(x, y) \doteq \mathsf{times}(y, x) \,\}$$

$\leadsto_{\mathrm{iRI}}^{d2}$

$$\langle \{\}, \{\}, \{\, \mathsf{times}(x, y) \doteq \mathsf{times}(y, x) \,\} \rangle$$

Thus the commutativity of times has been proved.

6   Conclusion

We have presented an extension of the rewriting induction that can deal with conjectures not orientable by the given reduction order. We gave inference rules of the enhanced rewriting induction and proved its correctness. We have also present incremental rewriting induction in which already-proved lemmas can be applied more easily.

Our approach to deal with non-orientable equations is based on the rewriting modulo equations originally suggested in [12]. Another approach is to use ordered rewriting technique [2,6]. The latter approach is embodied in the inductive theorem prover SPIKE [3,4]. Below we list some results of inductive theorem proving of non-orientable conjectures (in purely equational setting) by SPIKE and our inference systems. It appears that results of these two approaches are quite different even in simple examples. In particular, our system can not directly deal with conjectures that are incomparable in the given reduction quasi-order, although SPIKE can deal with such conjectures directly. On the other hand, our system successfully handle commutativity equations that are hard for SPIKE.

(many-sorted) conjectures	SPIKE	Enhanced RI	Incremental RI
$\max(x, y) \doteq \max(y, x)$	√	√	√
$\mathsf{minus}(\mathsf{minus}(x, y), z) \doteq \mathsf{minus}(\mathsf{minus}(x, z), y)$	√	√	√
$\mathsf{len}(\mathsf{app}(xs, ys)) \doteq \mathsf{len}(\mathsf{app}(ys, xs))$	√	√	√
$\mathsf{len}(\mathsf{qrev}(xs, ys)) \doteq \mathsf{len}(\mathsf{qrev}(ys, xs))$	×	√	√
$\mathsf{plus}(x, \mathsf{plus}(y, z)) \doteq \mathsf{plus}(y, \mathsf{plus}(x, z))$	√	×	×
$\left\{ \begin{array}{l} \mathsf{plus}(x, y) \doteq \mathsf{plus}(y, x) \\ \mathsf{plus}(x, \mathsf{plus}(y, z)) \doteq \mathsf{plus}(\mathsf{plus}(x, y), z) \\ \mathsf{plus}(x, \mathsf{plus}(y, z)) \doteq \mathsf{plus}(y, \mathsf{plus}(x, z)) \end{array} \right\}$	√	×	√
$\left\{ \begin{array}{l} \mathsf{plus}(x, \mathsf{plus}(y, z)) \doteq \mathsf{plus}(\mathsf{plus}(x, y), z) \\ \mathsf{plus}(x, y) \doteq \mathsf{plus}(y, x) \\ \mathsf{times}(x, y) \doteq \mathsf{times}(y, x) \end{array} \right\}$	×	×	√
$\left\{ \begin{array}{l} \mathsf{plus}(x, \mathsf{plus}(y, z)) \doteq \mathsf{plus}(\mathsf{plus}(x, y), z) \\ \mathsf{plus}(x, y) \doteq \mathsf{plus}(y, x) \\ \mathsf{sum}(\mathsf{app}(xs, ys)) \doteq \mathsf{sum}(\mathsf{app}(ys, xs)) \end{array} \right\}$	×	×	√

To see the difference, we show how the process proving the commutativity of times proceeds in SPIKE. First by expansion rule, it produces

$$0 \doteq \mathsf{times}(0, 0) \tag{7}$$
$$\mathsf{plus}(\mathsf{times}(x_1, 0), 0) \doteq \mathsf{times}(0, \mathsf{s}(x_1)) \tag{8}$$
$$0 \doteq \mathsf{times}(\mathsf{s}(x_1), 0) \tag{9}$$
$$\mathsf{plus}(\mathsf{times}(x_2, \mathsf{s}(x_1)), \mathsf{s}(x_1)) \doteq \mathsf{times}(\mathsf{s}(x_1), \mathsf{s}(x_2)) \tag{10}$$

In the presence of commutativity and associativity equations for plus (as proved lemmas), the successive simplification procedure works as follows:

$(7) \Rightarrow 0 \doteq 0 \Rightarrow$ *deleted*
$(8) \Rightarrow \mathsf{plus}(\mathsf{times}(0, x_1), 0) \doteq \mathsf{times}(0, \mathsf{s}(x_1)) \Rightarrow 0 \doteq 0 \Rightarrow$ *deleted*

$(9) \Rightarrow 0 \doteq \mathsf{plus}(\mathsf{times}(x_1, 0), 0)$
$(10) \Rightarrow \mathsf{plus}(\mathsf{times}(\mathsf{s}(x_1), x_2), \mathsf{s}(x_1)) \doteq \mathsf{times}(\mathsf{s}(x_1), \mathsf{s}(x_1))$
$\quad\Rightarrow \mathsf{plus}(\mathsf{plus}(\mathsf{times}(x_1, x_2), x_2), \mathsf{s}(x_1)) \doteq \mathsf{plus}(\mathsf{times}(x_1, \mathsf{s}(x_2)), \mathsf{s}(x_2))$
$\quad\Rightarrow \mathsf{plus}(\mathsf{times}(x_1, x_2), \mathsf{plus}(x_2, \mathsf{s}(x_1))) \doteq \mathsf{plus}(\mathsf{times}(x_1, \mathsf{s}(x_2)), \mathsf{s}(x_2))$

Thus, it results in a two elements set of remaining conjectures. On the other hand, in our procedure, as shown in Example 8, one expansion and successive simplifications successfully eliminate all equations.

Acknowledgments

Thanks are due to anonymous referees for helpful comments and suggestions. This work was partially supported by a grant from Japan Society for the Promotion of Science, No. 17700002.

References

1. F. Baader and T. Nipkow. *Term Rewriting and All That*. Cambridge University Press, 1998.
2. L. Bachmair, N. Dershowitz, and D. A. Plaisted. Completion without failure. In *Resolution of Equations in Algebraic Structure*, volume 2, pages 1–30. Academic Press, 1989.
3. A. Bouhoula. Automated theorem proving by test set induction. *Journal of Symbolic Computation*, 23:47–77, 1997.
4. A. Bouhoula, E. Kounalis, and M. Rusinowitch. Automated mathematical induction. *Journal of Logic and Computation*, 5(5):631–668, 1995.
5. R. S. Boyer and J. S. Moore. *A Computational Logic*. Academic Press, 1979.
6. N. Dershowitz and U. S. Reddy. Deductive and inductive synthesis of equational programs. *Journal of Symbolic Computation*, 15:467–494, 1993.
7. G. Huet and J.-M. Hullot. Proof by induction in equational theories with constructors. *Journal of Computer and System Sciences*, 25(2):239–266, 1982.
8. J.-P. Jouannaud and E. Kounalis. Automatic proofs by induction in theories without constructors. *Information and Computation*, 82:1–33, 1989.
9. D. Kapur, P. Narendran, and H. Zhang. Automating inductionless induction using test sets. *Journal of Symbolic Computation*, 11(1–2):81–111, 1991.
10. H. Koike and Y. Toyama. Inductionless induction and rewriting induction. *Computer Software*, 17(6):1–12, 2000. In Japanese.
11. D. R. Musser. On proving inductive properties of abstract data types. In *Proc. of the 7th Annual ACM Symposium on Principles of Programming Languages*, pages 154–162. ACM Press, 1980.
12. U. S. Reddy. Term rewriting induction. In *Proc. of the 10th International Conference on Automated Deduction*, volume 449 of *LNAI*, pages 162–177. Springer-Verlag, 1990.
13. K. Sakamoto, T. Aoto, and Y. Toyama. Fusion transformation based on rewriting induction. In *Proc. of the JSSST 21th Annual Conference*, 2B-3, 2004. In Japanese.
14. Terese. *Term Rewriting Systems*. Cambridge University Press, 2003.
15. Y. Toyama. How to prove equivalence of term rewriting systems without induction. *Theoretical Computer Science*, 90(2):369–390, 1991.

TPA: Termination Proved Automatically

Adam Koprowski

Technical University of Eindhoven
P.O. Box 513, 5600 MB, Eindhoven, The Netherlands
A.Koprowski@tue.nl

Abstract. TPA is a tool for proving termination of term rewrite systems (TRSs) in a fully automated fashion. The distinctive feature of TPA is the support for relative termination and the use of the technique of semantic labelling with natural numbers. Thanks to the latter, TPA is capable of delivering automated termination proofs for some difficult TRSs for which all other tools fail.

1 Introduction

Termination is an important concept in term rewriting and, although in general undecidable, several techniques have been developed for proving termination. The present focus in this area is on automation of the proof searching process. A number of tools have been developed to facilitate this and they are capable of finding proofs that are often fairly complicated and unlikely to be found by hand. To stimulate those developments an annual competition is being organized in which all the participating tools compete on a set of termination problems [1].

The tool TPA developed by the author is such tool aiming at proving termination of TRSs in an automatic fashion. What makes it different from all the other tools is the support for relative termination and the use of semantic labelling with natural numbers.

TPA took part in the termination competition of 2005. Only after three months of author's work, participating among well-established tools like C*i*ME [5], TTT [11] or AProVE [8], each developed by a group of people, it got the 3rd place there. Moreover, it succeeded in providing termination proofs for some systems for which all other tools failed. Two of such systems we will see in Figure 2. In particular it can solve the well-known SUBST [10] example encoding process of substitution in combinatory categorical logic for which showing termination was considered to be difficult. For a more throughout benchmark of TPA performance compared to other tools we refer to results of [1].

The following techniques are used in TPA:

- polynomial interpretations [14],
- recursive path order [6] (also over infinite signatures [13]),
- semantic labelling [16] (also with natural numbers [13]),
- dependency pairs [3],
- dummy elimination [7] and
- reduction of right hand sides [18].

F. Pfenning (Ed.): RTA 2006, LNCS 4098, pp. 257–266, 2006.

TPA is freely available and can be downloaded from its web page:

http://www.win.tue.nl/tpa

It is written in Objective Caml (OCaml)[1] and is available in native code for Linux platforms and as a byte code which can be run on any platform for which OCaml interpreter is available.

2   Preliminaries

For a signature Σ and a set of variables $\mathcal{V}$ we denote the set of terms over Σ and $\mathcal{V}$ by $\mathcal{T}(\Sigma, \mathcal{V})$. We denote by $\mathsf{Var}(t)$ a set of variables occurring in term t. A *rewrite rule* is a pair (ℓ, r), written $\ell \to r$ with $\ell, r \in \mathcal{T}(\Sigma, \mathcal{V})$, $\ell \notin \mathcal{V}$, $\mathsf{Var}(r) \subseteq \mathsf{Var}(l)$. A *term rewriting system* (TRS) is a set of rewrite rules. The *rewrite relation* $\to_R$ for a TRS R is defined as $s \to_R t$ if there exists a rewrite rule $\ell \to r \in R$, a substitution σ and a context C such that $s = C[l\sigma]$ and $t = C[r\sigma]$. A TRS R is called *terminating* ($\mathsf{SN}(R)$) is there is no infinite reduction $t_1 \to_R t_2 \to_R \ldots$, that is when $\to_R$ is a well-founded relation. For two relations R, S we define $R/S \equiv S^* \cdot R \cdot S^*$ and we say that *relative termination* between R and S holds if $\mathsf{SN}(\to_R / \to_S)$. We will then say that R terminates relative to S. Note that this is equivalent to lack of infinite reduction in $R \cup S$ with infinitely many R-steps. For relative termination problems we will refer to the rules from R as *strict rules* and to the rules from S as *non-strict rules*.

3   Motivation

With that many available tools for proving termination automatically it seems natural to ask why creating yet another one? There are three main reasons why TPA has been developed and they are listed below in no particular order.

- **Semantic labelling with natural numbers.** Semantic labelling is a transformational technique for proving termination of TRSs [16]. Its variant with the model over two or three element sets is used in some tools. However the infinite model variants were considered not to be suitable for automation so far. Indeed they pose some difficulties as the labelled TRS and its signature are usually infinite.

  The author's hope was that automation of semantic labelling over infinite sets (natural numbers in particular) can be accomplished and that it will be a fruitful technique for proving termination of TRSs. In fact TPA started as a prototype to verify this conjecture. The experiment turned out to be rather successful [13] and the prototype grew into a tool on its own. We will treat semantic labelling with natural numbers in some more detail in Section 4.4.

---

[1] http://www.ocaml.org

- **Relative termination.** In our opinion the notion of relative termination is very natural. First of all termination is the special case of relative termination as we have $\mathsf{SN}(R) \equiv \mathsf{SN}(R/\emptyset)$. The concept of relative termination is also very closely related to the top termination and hence to the dependency pairs method [3].

  Giesl and Zantema proposed a method for verification of liveness properties using rewriting techniques [9]. In continuation of this line of research Koprowski and Zantema extended the framework to deal with fairness and there relative termination turned out to be a very natural and needed concept [12].

  Lucas and Meseguer conducted research on termination of concurrent systems under fairness assumptions [15]. In their setting they again use the notion of relative termination to establish the property of fair-termination. They used TPA for proving relative termination.

  Still up till now the support for relative termination was very limited. TORPA [17] incorporates this notion but is limited to string rewriting. TEPARLA is another tool capable of proving relative termination but it is not actively developed anymore. So being able to deal with relative termination problems for TRSs was one of the motivations for developing TPA.

- **CoLoR.** CoLoR[2] stands for *Coq library on rewriting and termination*. It is an initiative aiming at proving theoretical results from term rewriting in the theorem prover Coq. The ultimate goal is to (automatically) transform the termination proof candidates produced by termination tools into formal Coq proofs certifying termination.

  This will involve some cooperation between tool authors and CoLoR developers. Presently the output produced by termination tools is just an informal, textual description of the termination proof. Certification of such results will require more explicit output. The idea within the CoLoR project is to provide a description of the proof output format in XML. Every tool being able to generate output in this common proof format would potentially be able to employ CoLoR to certify its proofs.

  The author is involved in the CoLoR project and having his own tool is helpful in bridging the gap between those two communities and propagating the idea of formal verification of termination proofs. Hence it is the author's hope that TPA will be the first tool capable of producing formally certified termination proofs.

4   Overview

We begin with a few remarks about the implementation of TPA in Section 4.1. Then in Section 4.2 we give a description of our approach to proving termination. Finally in the following sections we briefly present the theory behind techniques used in TPA as well as some details of the way in which they are used.

---

[2] http://color.loria.fr

4.1   Implementation

TPA is implemented in Objective Caml (OCaml). It is written almost fully in a functional programming style with only few fragments using imperative features of OCaml. It does not use any third-party libraries and the source code of TPA consists of about 10,000 lines of OCaml code. It is equipped with command line interface. It is available in native code for Linux platform and as OCaml byte code that can be run on any platform supported by OCaml (that includes all the major platforms).

4.2   TPA Approach to Proving Termination

First we introduce the transformation used by TPA to eliminate all the function symbols of arity greater than 2.

Definition 1. *Given TRS R over Σ we define the transformed TRS $\overline{R}$ over signature*

$$\Sigma' = \{f_i \mid f \in \Sigma,\ 1 \leq i \leq \mathsf{arity}(f) - 1\} \cup \{f \mid f \in \Sigma, \mathsf{arity}(f) \leq 2\}$$

to consist of the rules $\overline{R} = \{\overline{\ell} \to \overline{r} \mid \ell \to r \in R\}$, where $\overline{\cdot}$ is defined as follows:

$$\overline{\overline{x} = x}$$

$$\frac{}{\overline{f(t_1,\ldots,t_n)} = f(\overline{t_1},\ldots,\overline{t_n})} \qquad\qquad \textit{if } \mathsf{arity}(f) \leq 2$$

$$\overline{f(t_1,\ldots,t_n)} = f_1(\overline{t_1}, f_2(\overline{t_2},\ldots, f_{n-1}(\overline{t_{n-1}}, \overline{t_n}))) \qquad \textit{if } \mathsf{arity}(f) > 2$$

The main result concerning this transformation states that $\mathsf{SN}(R) \iff \mathsf{SN}(\overline{R})$. The ($\Leftarrow$) is crucial in this application and is easy to show. The ($\Rightarrow$), implying completeness, of the transformation is more involved but can be shown using Aoto's theorem [2].

Note that in the transformed system only constants, unary and binary symbols may occur; a fact that is heavily used in TPA as we shall see. On the other hand the original structure of the TRS is somehow obscured by the transformation which may in some cases make termination arguments more difficult.

TPA tries to use a modular approach to proving termination. Given an input TRS it first applies the transformation from Definition 1 and then it tries to apply different tactics according to the *prover configuration* that we will describe later on. As soon as one of the tactics succeeds, resulting in simplification of the TRS, the whole procedure is repeated until finally termination is proved, all tactics fail or the maximum specified search time is used and TPA stops due to timeout. For this modular approach the following theorem from [17] is essential.

Theorem 1. *Let R, S, R' and S' be TRSs such that:*

- $R \cup S = R' \cup S'$ and $R \cap S = R' \cap S' = \emptyset$,
- $\mathsf{SN}(R'/S')$ and $\mathsf{SN}((R \cap S')/(S \cap S'))$.

Then $\mathsf{SN}(R/S)$.

The *prover configuration* is a description of the way TPA should employ different tactics in order to prove termination. In the present version of the tool the configuration is embedded in the source code but it is only a matter of providing the user interface to let the user specify custom configurations. Below is a snippet from the OCaml source code of TPA with a slightly simplified description of the prover configuration.

```
type proverStep = ElimDummy
                | ReduceRHS
                | PolyInt of PolyInterp.configuration * bool
                | SemLabBool of SemLabBool.semLabCfg
                | SemLabNat of PolyInterp.configuration * proverConfig
                | Rpo of Rpo.configuration
                | DP of proverConfig
                | Parallel of proverConfig list
and proverConfig = proverStep list
```

proverConfig corresponds to the prover configuration and consists of a list of steps that should be taken successively in an attempt to prove termination. The possible steps include:

- ElimDummy: dummy elimination transformation [7].
- ReduceRHS: reduction of right hand sides [18].
- PolyInt: polynomial interpretations [14] (see Section 4.3)
- SemLabBool: semantic labelling with boolean values [16] (see Section 4.4).
- SemLabNat: semantic labelling with natural numbers [13] (see Section 4.4).
- Rpo: recursive path order [6] (see Section 4.5).
- DP: a simple variant of dependency pairs method [3] (see Section 4.6).
- Parallel: this meta-tactic allows to try few different approaches in parallel using threads. As soon as one of them succeeds the other are abandoned and the termination procedure continues with the initial configuration.

Every transformational technique (like semantic labelling or dependency pair method) is parameterized by the configuration to be used after the transformation. Hence such technique is considered to be successful if the TRS can be simplified by applying transformation and all those techniques (for semantic labelling resulting system is subject to un-labelling at the end of this procedure).

Note that the proof search procedure in TPA is fully deterministic. The only non-deterministic behavior can occur due to the use of threads for parallel computations.

The present configuration of TPA is as follows:

- simple transformations: dummy elimination and reduction of right sides are tried,
- then a simple argument of counting different symbols in left and right hand sides of the rules is tried (which corresponds to the use of polynomial interpretations with identity and successor as the only interpretation functions).
- then an attempt is made at direct termination proof with RPO.

- If all of the above fail or cannot be applied anymore then the following tactics are all executed in parallel:
  - semantic labelling with natural numbers,
  - semantic labelling with booleans with two different sets of default interpretations and also with recursive labelling (see Section 4.4),
  - polynomial interpretations method with an extended set of interpretations and
  - the dependency pairs approach.

Now we will describe the main techniques used in TPA in some more detail.

4.3   Polynomial Interpretations

The idea of proving termination with polynomial interpretations goes back to [14], [4]. We briefly present the theory for this approach.

Let Σ be a signature and let A be a non-empty set. Now for every function symbol $f \in \Sigma$ of arity n fix an *interpretation function* $f_A : A^n \to A$. Let $\alpha : \mathcal{V} \to A$ be an assignment to variables. We define the term evaluation $[\cdot]^\alpha : \mathcal{T}(\Sigma, \mathcal{V}) \to A$:

$$[x]^\alpha = \alpha(x) \qquad [f(t_1, \ldots, t_n)]^\alpha = f_A([t_1]^\alpha, \ldots, [t_n]^\alpha)$$

Theorem 2. *Let A be a non-empty set, $>$ be a well-founded order on A. Let $f_A : A^n \to A$ be strictly monotone for every $f : \Sigma$. Let R, S be disjoint TRSs over Σ. If $\forall \alpha \; [\ell]^\alpha > [r]^\alpha$ for every $\ell \to r \in R$ and $\forall \alpha \; [\ell]^\alpha \geq [r]^\alpha$ for every $\ell \to r \in S$ then $\mathsf{SN}(R/S)$.*

This approach of monotone algebras goes back to Lankford. If for the set A we take $\mathbb{N} \setminus \{0, 1\}$, for $>$ usual comparison on $\mathbb{N}$ and all f_A are polynomials then this approach is called *polynomial interpretations*.

TPA uses a fixed (per arity) set of polynomial interpretations. To further limit the search space for every arity there is a default interpretation and in the search procedure only limited number (being a parameter of the search procedure) of deviations from those standard interpretations is allowed. The standard interpretations are 2 for constants, $\lambda x.x$ for unary symbols and $\lambda xy.x + y - 2$ for binary symbols (corresponding to zero, identity and summation in $\mathbb{N}$).

4.4   Semantic Labelling

The technique of semantic labelling goes back to Zantema [16]. We present its variant that is used in TPA.

Let A be a non-empty set. For every $f \in \Sigma$ fix an interpretation function $f_A : A^n \to A$, where n is the arity of f. We define the term evaluation function $[\cdot]$ as in Section 4.3. We define the extended signature $\Sigma_L = \{f_{l_1, \ldots, l_n} \mid f \in \Sigma; \; l_1, \ldots, l_n \in A\}$ and the labelling function on terms $lab : \mathcal{T}(\Sigma, \mathcal{V}) \times A^{\mathcal{V}} \to \mathcal{T}(\Sigma, \mathcal{V})$ as:

$$lab(x, \alpha) = x$$
$$lab(f(t_1, \ldots, t_n), \alpha) = f_{[t_1]^\alpha, \ldots, [t_n]^\alpha}(lab(t_1, \alpha), \ldots, lab(t_n, \alpha))$$

For TRS R we define its labelled version over Σ_L as $\mathsf{lab}(R) = \{\mathsf{lab}(\ell) \to \mathsf{lab}(r) \mid \ell \to r \in R, \text{for all } \alpha : \mathcal{V} \to A\}$. We define Decr to be the TRS consisting of the rules: $f_{s_1,\dots,s_i,\dots,s_n}(x_1,\dots,x_n) \to f_{s_1,\dots,s_i',\dots,s_n}(x_1,\dots,x_n)$ for all $1 \le i \le n; s_1,\dots,s_n \in A; s_i' \in A$ and $s_i > s_i'$.

Now the main theorem for semantic labelling reads:

Theorem 3. *Let R, S be two disjoint TRSs over Σ. Let $>$ be a well-founded order on a non-empty set A. Let $f_A : A^n \to A$ for every $f \in \Sigma$ be weakly monotonic in all arguments and such that:*

$$[\ell]^\alpha \ge [r]^\alpha \text{ for all } \ell \to r \in R \cup S \text{ and all } \alpha : \mathcal{V} \to A$$

Then $\mathsf{SN}(R/S)$ iff $\mathsf{SN}(\mathsf{lab}(R)/(\mathsf{lab}(S) \cup \mathsf{Decr}))$

In a sense semantic labelling is a central technique for TPA. Its two variants are used: one rather standard where A is the set $\{0,1\}$ and another, presently implemented in no other tool except for TPA, where $A = \mathbb{N}$. Note that it is possible to label already labelled system and TPA makes use of this fact doing recursive labelling with boolean models.

The difficulty posed by using semantic labelling with $\mathbb{N}$ is that the labelled system has infinite signature and infinitely many rules. Hence all the standard techniques may need to be somehow adopted in order to be applied to such systems. In the present version of TPA two techniques are used in combination with semantic labelling with natural numbers: RPO and polynomial interpretations. For details about using RPO on systems with infinite signature we refer the reader to [13].

Polynomial interpretations are incorporated into this setting in a straightforward way. Denote by Φ_0, Φ_1 and Φ_2 sets of polynomial interpretations for constants, unary and binary symbols respectively. Constants get no label but for unary and binary symbols after labelling the following interpretations are used:

$$[f_i(x)] = \phi(x) \text{ or } \phi(x) + i \text{ with } \phi \in \Phi_1$$
$$[f_{i,j}(x,y)] = \phi(x,y) \text{ or } \phi(x,y) + i \text{ or } \phi(x,y) + j \text{ or } \phi(x,y) + i + j \text{ with } \phi \in \Phi_2$$

We will illustrate this approach on a simple relative termination problem.

Example 1. $R = \{T(I(x),y) \to T(x,y)\}$, $S = \{T(x,y) \to T(x,I(y))\}$.

There are two rules in this system: the rule from R allows to remove one I symbol from the left argument of T and the rule from S allows introducing this symbol in the right argument of T but it can be applied only finitely many times and thus $\mathsf{SN}(\to_R / \to_S)$. Figure 1 shows the proof produced by TPA for this system. TPA prints R-rules as `1 -> r` and S-rules as `1 ->= r`.

In Figure 2 we present two TRSs that can be solved by TPA with the use of semantic labelling with natural numbers and presently no other tool can deal with them. The first one is the SUBST TRS encoding the process of substitution in combinatory categorical logic and the second one is the GCD TRS encoding the computation of the greatest common divisor. For termination proofs of those systems generated by TPA we refer the reader to its homepage.

```
TPA v.1.0
Result: TRS is terminating
[1] TRS loaded from input file:
      (1) T(I(x),y) -> T(x,y)
      (2) T(x,y) ->= T(x,I(y))
[2] Label this TRS using following interpretation over N\{0,1}:
      [T(x,y)] = 2
      [I(x)] = x + 1
    This interpretation is a model and yields the following TRS:
      (1) T{i + 1,j}(I{i}(x),y) -> T{i,j}(x,y)
      (2) T{i,j}(x,y) ->= T{i,j + 1}(x,I{j}(y))
[3] Use the following polynomial interpretation:
      [T_{i,j}(x,y)] = x + y - 2 + i
      [I_{i}(x)] = x
    Remove rules with left hand side strictly bigger than right hand
    side: (1)
[4] Since there are no remaining strict rules, relative termination is
    proven.
```

Fig. 1. Proof generated by TPA for the system from Example 1. For the purpose of the presentation the proof has been made slightly less verbose.

$$
\begin{aligned}
\min(x,0) &\rightarrow 0 \\
\min(0,y) &\rightarrow 0 \\
\min(\mathsf{s}(x),\mathsf{s}(y)) &\rightarrow \mathsf{s}(\min(x,y)) \\
\max(x,0) &\rightarrow x \\
\max(0,y) &\rightarrow y \\
\max(\mathsf{s}(x),\mathsf{s}(y)) &\rightarrow \mathsf{s}(\max(x,y)) \\
-(x,0) &\rightarrow x \\
-(\mathsf{s}(x),\mathsf{s}(y)) &\rightarrow -(x,y) \\
\gcd(\mathsf{s}(x),\mathsf{s}(y)) &\rightarrow \gcd(-(\mathsf{s}(\max(x,y)), \\
& \quad \mathsf{s}(\min(x,y))),\mathsf{s}(\min(x,y)))
\end{aligned}
$$

$$
\begin{aligned}
\lambda(x) \circ y &\rightarrow \lambda(x \circ (1 \cdot (y \circ \uparrow))) \\
(x \cdot y) \circ z &\rightarrow (x \circ z) \cdot (y \circ z) \\
(x \circ y) \circ z &\rightarrow x \circ (y \circ z) \\
\mathsf{id} \circ x &\rightarrow x \\
1 \circ \mathsf{id} &\rightarrow 1 \\
\uparrow \circ \mathsf{id} &\rightarrow 1 \\
1 \circ (x \cdot y) &\rightarrow x \\
\uparrow \circ (x \cdot y) &\rightarrow y
\end{aligned}
$$

Fig. 2. TRSs: GCD(left) and SUBST (right)

4.5 RPO

Recursive path order (RPO) is an ordering introduced by Dershowitz [6].

Definition 2 (RPO). *Given an order* $\rhd$ *on function symbols called* precedence *and a status function* τ *associating every function symbol with either lexicographic or multiset status, we define the RPO ordering* $\succ_{rpo}$ *as follows:*

$s = f(s_1, \ldots, s_n) \succ_{rpo} g(t_1, \ldots, t_m) = t$ *iff one of the following holds:*

- $s_i \succeq_{rpo} t$ *for some* $1 \le i \le n$.
- $f \rhd g$ *and* $s \succ_{rpo} t_i$ *for all* $1 \le i \le m$
- $f = g$ *and* $(s_1, \ldots, s_n) \succ_{rpo}^{\tau(f)} (t_1, \ldots, t_m)$

Theorem 4. *If* $\rhd$ *is well-founded and* $\ell \succ_{rpo} r$ *for all* $\ell \rightarrow r \in R$ *then* $\mathsf{SN}(R)$.

In TPA RPO is implemented in this form with both lexicographic (left-to-right and right-to-left) and multiset statuses. However due to the transformation from Definition 1 the power of multiset status is limited and in the most recent config-uration it is not even used. To avoid extensive branching in TPA it is also possible to use a slightly weaker variant where the first clause of the RPO definition is used only if t can be embedded in s.

RPO in its standard variant is not very suitable for proving relative termina-tion since the equality part of the preorder it generates is rather small. Some experiments have been made to make it more suitable for proving relative ter-mination but they were not very successful.

Note that due to the use of semantic labelling with natural numbers also a variant of RPO for infinite signatures is used in TPA. See [13] for more details.

4.6 Dependency Pairs

The technique of dependency pairs was introduced in [3] and since then is a central technique for many termination provers. Due to that fact we decided to concentrate on other techniques and in TPA dependency pairs play only a minor role.

Definition 3 (Dependency pairs). *Let R be TRS over Σ. We split Σ into defined symbols $\Sigma_D = \{root(\ell) \mid \ell \to r \in R\}$ and constructor symbols $\Sigma_C = \Sigma \setminus \Sigma_C$. We extend signature Σ with fresh symbols for every defined symbol: $\Sigma' = \Sigma \cup \{\overline{f} \mid f \in \Sigma_D\}$.*

We define a TRS $\mathsf{DP}(R)$ over Σ' to represent dependency pairs of R:

$$\mathsf{DP}(R) = \{\overline{f}(s_1, \ldots, s_n) \to \overline{g}(t_1, \ldots, t_n) \mid g \in \Sigma_D \wedge$$
$$f(s_1, \ldots, s_n) \to C[g(t_1, \ldots, t_n)] \in R\}$$

The main theorem, as it is used in TPA, relates termination of R with relative termination of $\mathsf{DP}(R)/R$.

Theorem 5. *Let R be TRS over Σ. Then $\mathsf{SN}(R)$ iff $\mathsf{SN}(\mathsf{DP}(R)/R)$.*

So using the dependency pair method TPA transforms a termination problem to a, hopefully easier, relative termination problem. Moreover it computes a dependency graph approximation and treats its strongly connected components separately but it does not use more involved refinements of dependency pairs method such as argument filtering or narrowing.

5   Conclusions and Further Research

TPA is a new termination tool that got the 3rd place in the international ter-mination competition in 2005. It also proved that the approach of semantic labelling with natural numbers can be successfully used for proving termination automatically; using this technique TPA could prove termination of systems for which all other tools failed.

Clearly there is plenty of room for improvement and further research. For instance the idea of semantic labelling with natural numbers can be investigated further and techniques other than only RPO and polynomial interpretations can be adopted and applied on labelled systems. We also want to improve the proving capabilities of TPA in the subject of relative termination, either by developing new techniques devoted particularly to relative termination or by refining existing termination techniques. Finally we want to propagate the idea of certified termination by continuing developments in the CoLoR project.

References

1. The termination competition.
   http://www.lri.fr/~marche/termination-competition.
2. T. Aoto and T.Yamada. Termination of Simply Typed Term Rewriting by Translation and Labelling. In Proceedings of *RTA*, LNCS 2706:380–394, 2003.
3. T. Arts and J. Giesl. Termination of term rewriting using dependency pairs. *Theor. Comput. Sci.*, 236(1-2):133–178, 2000.
4. A. B. Cherifa and P. Lescanne. Termination of rewriting systems by polynomial interpretations and its implementation. *Sci. Comput. Program.*, 9(2):137–159, 1987.
5. E. Contejean, C. Marché, B. Monate, and X. Urbain. Proving termination of rewriting with CiME. 71–73, 2003. http://cime.lri.fr.
6. N. Dershowitz. Orderings for term-rewriting systems. *Theor. Comput. Sci.*, 17:279–301, 1982.
7. M. C. F. Ferreira and H. Zantema. Dummy elimination: Making termination easier. In Proceedings of *FCT*, LNCS 965:243–252, 1995.
8. J. Giesl, R. Thiemann, P. Schneider-Kamp, and S. Falke. Automated termination proofs with AProVE. In Proceedings of *RTA*, LNCS 3091:210–220, 2004.
9. J. Giesl and H. Zantema. Liveness in rewriting. In Proceedings of *RTA*, LNCS 2706:321–336, 2003.
10. T. Hardin and A. Laville. Proof of termination of the rewriting system SUBST on CCL. *Theor. Comput. Sci.*, 46(2-3):305–312, 1986.
11. N. Hirokawa and A. Middeldorp. Tyrolean termination tool. In Proceedings of *RTA*, LNCS 3467:175–184, 2005.
12. A. Koprowski and H. Zantema. Proving liveness with fairness using rewriting. In Proceedings of *FroCoS*, LNCS 3717:232–247, 2005.
13. A. Koprowski and H. Zantema. Automation of recursive path ordering for infinite labelled rewrite systems. In J.Harrison, editor, *IJCAR '06*, LNCS, 2006.
14. D. S. Lankford. On proving term rewriting systems are noetherian. Tech. Rep. MTP-3, Louisiana Tech. Univ., Ruston, 1979.
15. S. Lucas and J. Meseguer. Termination of fair computations in term rewriting. In Proceedings of *LPAR*, LNCS 3835:184–198, 2005.
16. H. Zantema. Termination of term rewriting by semantic labelling. *Fundam. Inform.*, 24(1/2):89–105, 1995.
17. H. Zantema. Torpa: Termination of rewriting proved automatically. In Proceedings of *RTA*, LNCS 3091:95–104, 2004.
18. H. Zantema. Reducing right-hand sides for termination. In *Processes, Terms and Cycles*, LNCS 3838:173–197, 2005.

RAPT: A Program Transformation System Based on Term Rewriting

Yuki Chiba and Takahito Aoto

Research Institute of Electrical Communication, Tohoku University, Japan
{chiba, aoto}@nue.riec.tohoku.ac.jp

Abstract. Chiba et al. (2005) proposed a framework of program transformation by template based on term rewriting in which correctness of the transformation is verified automatically. This paper describes RAPT (Rewriting-based Automated Program Transformation system) which implements this framework.

1 Introduction

Chiba et al. [4] proposed a framework of program transformation by template based on term rewriting in which correctness of the transformation is verified automatically. In their framework, programs and program schemas are given by term rewriting systems (TRS, for short) and TRS patterns. A program transformation template consists of input and output TRS patterns and a hypothesis which is a set of equations the input TRS has to satisfy to guarantee the correctness of transformation.

This paper describes RAPT (Rewriting-based Automated Program Transformation system) which implements this framework. RAPT transforms a many-sorted TRS according to a specified program transformation template. Based on the rewriting induction proposed by Reddy [14], RAPT automatically verifies whether the input TRS satisfies the hypothesis of the transformation template. It also verifies conditions imposed to the input TRS and generated TRS by utilizing standard techniques in term rewriting. Thus, presupposing the program transformation template is *developed* [4], the correctness of the transformation is automatically verified so that the transformation keeps the relationship between initial ground terms and their normal forms.

2 Transformation by Templates

Let $\mathscr{P}$ be a set of (arity-fixed) pattern variables (disjoint from the set $\mathscr{F}$ of function symbols and the set $\mathscr{V}$ of variables). A *pattern* is a term with pattern variables. A *TRS pattern* $\mathcal{P}$ is a set of rewriting rules over patterns. A *hypothesis* $\mathcal{H}$ is a set of equations over patterns. A *transformation template* (or just *template*) is a triple $\langle \mathcal{P}, \mathcal{P}', \mathcal{H} \rangle$ of two TRS patterns $\mathcal{P}$, $\mathcal{P}'$ and a hypothesis $\mathcal{H}$.

F. Pfenning (Ed.): RTA 2006, LNCS 4098, pp. 267–276, 2006.

The following template $\langle \mathcal{P}, \mathcal{P}', \mathcal{H} \rangle$ describes a well-known transformation from the recursive form to the iterative (tail-recursive) form:

$$
\mathcal{P} \begin{cases}
f(a) & \to b \\
f(c(u,v)) & \to g(e(u), f(v)) \\
g(b,u) & \to u \\
g(d(u,v),w) & \to d(u, g(v,w))
\end{cases}
$$

$$
\mathcal{P}' \begin{cases}
f(u) & \to f_1(u, b) \\
f_1(a, u) & \to u \\
f_1(c(u,v), w) & \to f_1(v, g(w, e(u))) \\
g(b, u) & \to u \\
g(d(u,v), w) & \to d(u, g(v,w))
\end{cases}
$$

$$
\mathcal{H} \begin{cases}
g(b, u) & \approx g(u, b) \\
g(g(u,v), w) & \approx g(u, g(v,w))
\end{cases}
$$

Here, the symbols $f, a, b, g, \ldots$ are pattern variables.

To achieve the program transformation by templates, we need a mechanism to specify how a template is applied to a concrete TRS. For this we use a notion of *term homomorphism* [4]. If we match the TRS pattern $\mathcal{P}$ to a concrete TRS $\mathcal{R}$ with a term homomorphism φ, we obtain a generated TRS $\mathcal{R}'$ by applying φ to the TRS pattern $\mathcal{P}'$ (Figure 1). A matching algorithm to find all (most general) term homomorphisms φ satisfying $\mathcal{R} = \varphi(\mathcal{P})$ from a given TRS $\mathcal{R}$ and a TRS pattern $\mathcal{P}$ is presented in [4].

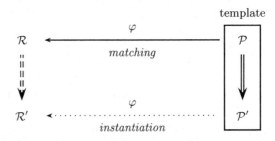

Fig. 1. TRS transformation

Definition 1 ([4]). *Let $\langle \mathcal{P}, \mathcal{P}', \mathcal{H} \rangle$ be a template. A TRS $\mathcal{R}$ is transformed into $\mathcal{R}'$ by $\langle \mathcal{P}, \mathcal{P}', \mathcal{H} \rangle$ if there exists a term homomorphism φ such that $\mathcal{R} = \varphi(\mathcal{P})$ and $\mathcal{R}' = \varphi(\mathcal{P}')$.*

The following TRS $\mathcal{R}_{sum}$ computes the summation of a list using a recursive call.

$$
\mathcal{R}_{sum} \begin{cases}
\mathsf{sum}([\,]) & \to 0 \\
\mathsf{sum}(x : y) & \to +(x, \mathsf{sum}(y)) \\
+(0, x) & \to x \\
+(\mathsf{s}(x), y) & \to \mathsf{s}(+(x, y))
\end{cases}
$$

The following term homomorphism φ is used to transform the TRS $\mathcal{R}_{sum}$.

$$\varphi = \begin{cases} \mathsf{f} \mapsto \mathsf{sum}(\square_1), & \mathsf{b} \mapsto 0, \\ \mathsf{g} \mapsto +(\square_1, \square_2), & \mathsf{c} \mapsto \square_1{:}\square_2, \\ \mathsf{f}_1 \mapsto \mathsf{sum1}(\square_1, \square_2), & \mathsf{d} \mapsto \mathsf{s}(\square_2), \\ \mathsf{a} \mapsto [\,], & \mathsf{e} \mapsto \square_1 \end{cases}$$

Applying φ to $\mathcal{P}'$, we get the following output TRS $\mathcal{R}'_{sum}$.

$$\mathcal{R}'_{sum} \begin{cases} \mathsf{sum}(x) & \to \mathsf{sum1}(x, 0) \\ \mathsf{sum1}([\,], x) & \to x \\ \mathsf{sum1}(x{:}y, z) & \to \mathsf{sum1}(y, +(z, x)) \\ +(0, x) & \to x \\ +(\mathsf{s}(x), y) & \to \mathsf{s}(+(x, y)) \end{cases}$$

$\mathcal{R}'_{sum}$ computes the summation of a list more efficiently without the recursion.

3 Design of **RAPT**

We assume that the set $\mathscr{F}$ of function symbols is divided into disjoint two sets: the set $\mathscr{F}_\mathrm{d}$ of *defined function symbols* and the set $\mathscr{F}_\mathrm{c}$ of *constructor symbols*. The following is a sufficient condition to guarantee the correctness of the transformation from a TRS $\mathcal{R}$ on $\mathscr{G}$ to a TRS $\mathcal{R}'$ on $\mathscr{G}'$ by a template $\langle \mathcal{P}, \mathcal{P}', \mathcal{H} \rangle$ through a term homomorphism φ (Theorem 2 of [4]):

- $\mathcal{R}$ is a left-linear confluent constructor system,
- $\langle \mathcal{P}, \mathcal{P}', \mathcal{H} \rangle$ is a developed template,
- φ is a CS-homomorphism,
- equations in $\varphi(\mathcal{H})$ are inductive consequences of $\mathcal{R}$ for $\mathscr{G}$,
- $\mathcal{R}$ is sufficiently complete for $\mathscr{G}$, and
- $\mathcal{R}'$ is sufficiently complete for $\mathscr{G}'$,

where $\mathscr{F}_\mathrm{c} \subseteq \mathscr{G}, \mathscr{G}' \subseteq \mathscr{F}$.

A key property of our framework is sufficient completeness, which has to be satisfied by input and output TRSs [4]. Sufficient completeness is checked in RAPT by the decidable necessary and sufficient condition for terminating TRSs [9,11], and thus currently the target of program transformation by RAPT is limited to terminating TRSs. A simple procedure to check confluence is also available for terminating TRSs [1].

RAPT uses *rewriting induction* [14], in which termination plays an essential role, to verify that the instantiated hypotheses of transformation template are inductive consequences of the input TRS. Since RAPT handles only terminating TRSs, rewriting induction is integrated keeping the whole system simple. Other inductive proving methods [2,5] also can be possibly incorporated.

For the termination checking, RAPT detects a possible compatible precedence for the lexicographic path ordering (LPO) [1]. The obtained reduction ordering is used as a basis of rewriting induction. Other methods to verify termination of TRSs [1] may well be incorporated.

```
FUNCTIONS                       INPUT
  sum: List -> Nat;             ?f(?a()) -> ?b();
  cons: Nat * List -> List;     ?f(?c(u,v)) -> ?g(?e(u),?f(v));
  nil: List;                    ?g(?b(),u) -> u;
  +: Nat * Nat -> Nat;          ?g(?d(u,v),w) -> ?d(u,?g(v,w))
  s: Nat -> Nat;
  0: Nat                        OUTPUT
                                ?f(u) -> ?f1(u,?b());
                                ?f1(?a(),u) -> u;
RULES                           ?f1(?c(u,v),w) -> ?f1(v,?g(w,?e(u)));
  sum(nil()) -> 0();            ?g(?b(),u) -> u;
  sum(cons(x,ys)) -> +(x,sum(ys));  ?g(?d(u,v),w) -> ?d(u,?g(v,w))
  +(0(), x) -> x;
  +(s(x),y) -> s(+(x,y))        HYPOTHESIS
                                ?g(?b(),u) = ?g(u,?b());
                                ?g(?g(u,v),w) = ?g(u,?g(v,w))
```

Fig. 2. Specification of input TRS and transformation template

4   Implementation

4.1   Specification of Input TRS and Transformation Template

Inputs of RAPT are a many-sorted TRS and a transformation template. The
input TRS is specified by the following sections.

1. FUNCTIONS: function symbols with sort declaration.
2. RULES: rewrite rules over many-sorted terms.

The transformation template $\langle \mathcal{P}, \mathcal{P}', \mathcal{H} \rangle$ is specified by the following sections.

1. INPUT: rewrite rules of $\mathcal{P}$ over patterns,
2. OUTPUT: rewrite rules of $\mathcal{P}'$ over patterns,
3. HYPOTHESIS: equations of $\mathcal{H}$ over patterns.

Figure 2 shows the many-sorted TRS $\mathcal{R}_{sum}$ and the template $\langle \mathcal{P}, \mathcal{P}', \mathcal{H} \rangle$ which
appear in Section 2 prepared as an input to RAPT: rules, equations and sort
declarations are separated by ";"; pattern variables are preceded by "?"; and to
distinguish variables from constants, the latter are followed by "()" .

4.2   Implementation Details

RAPT is implemented using SML/NJ. The source code of RAPT consists of
about 5,000 lines.

The TRS transformation and the verification of its correctness are conducted
in RAPT in 6 phases. In Figure 3, we describe these phases and dependencies
among each phases. Solid arrows represent data flow and dotted arrows explain
how information obtained in each phase is used.

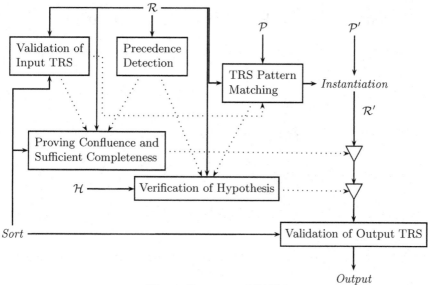

Fig. 3. Overview of RAPT

If these 6 phases are successfully passed then RAPT produces output TRSs. The correctness of the transformation is guaranteed, provided the transformation template is developed. RAPT can also report summaries of program transformation in a readable format (Figure 4).

We now explain operations of each phases briefly.

1. Validation of input TRS. In this phase, RAPT checks whether the input TRS is left-linear and well-typed, and from rewrite rules divides function symbols into defined function symbols and constructor symbols and checks whether the input TRS is a constructor system. The information of function symbols will be used in Phases 3 and 4.

2. Precedence detection. In this phase, RAPT checks the input TRS is terminating by LPO and (if it is the case) detects a precedence. The suitable precedence (if there exists one) for LPO is computed based on the LPO constraint solving algorithm described in [7].

3. Proving confluence and sufficient completeness. In this phase, RAPT proves whether the input TRS is confluent and sufficiently complete. This makes use of the information of constructor symbols detected at Phase 1 and the fact that the input TRS is left-linear and terminating verified at Phases 1 and 2, respectively. For confluence, it is checked whether all critical pairs are joinable. For sufficient completeness, quasi-reducibility of the TRS is checked; this part is based on the (many-sorted extension of) complement algorithm introduced in [10] that computes the complement of a substitution.

Summary of Program Transformation

reported by RAPT

February 21, 2006

Transformation Template:

$$\mathcal{P} \begin{cases} f(a) & \to b \\ f(c(u,v)) \to g(e(u,v), f(v)) \end{cases}$$

$$\mathcal{P}' \begin{cases} f(u) & \to f1(u,b) \\ f1(a,u) & \to u \\ f1(c(u,v),w) \to f1(v, g(w, e(u,v))) \end{cases}$$

$$\mathcal{H} \begin{cases} g(b,u) & \approx u \\ g(u,b) & \approx u \\ g(g(u,v),w) \approx g(u, g(v,w)) \end{cases}$$

Input TRS:

$$\mathcal{R} \begin{cases} \mathsf{rev}(\mathsf{nil}) & \to \mathsf{nil} \\ \mathsf{rev}(\mathsf{cons}(x,ys)) & \to \mathsf{app}(\mathsf{rev}(ys), \mathsf{cons}(x, \mathsf{nil})) \\ \mathsf{app}(\mathsf{nil}, x) & \to x \\ \mathsf{app}(\mathsf{cons}(x,y), z) \to \mathsf{cons}(x, \mathsf{app}(y,z)) \end{cases}$$

Termination of $\mathcal{R}$ is checked by LPO with the precedence $\{\mathsf{rev} > \mathsf{app}, \mathsf{rev} > \mathsf{nil}, \mathsf{rev} > \mathsf{cons}, \mathsf{app} > \mathsf{cons}\}$. The set of critical pairs of $\mathcal{R}$ is $\{\}$.

A solution of matching (CS-homomorphisms):

$$\varphi = \begin{cases} \mathsf{b} \mapsto \mathsf{nil} \\ \mathsf{a} \mapsto \mathsf{nil} \\ \mathsf{e} \mapsto \mathsf{cons}(\square_1, \mathsf{nil}) \\ \mathsf{g} \mapsto \mathsf{app}(\square_2, \square_1) \\ \mathsf{c} \mapsto \mathsf{cons}(\square_1, \square_2) \\ \mathsf{f} \mapsto \mathsf{rev}(\square_1) \end{cases}$$

The instantiation of hypothesis:

$$\varphi(\mathcal{H}) \begin{cases} \mathsf{app}(u, \mathsf{nil}) & \approx u \\ \mathsf{app}(\mathsf{nil}, u) & \approx u \\ \mathsf{app}(w, \mathsf{app}(v,u)) \approx \mathsf{app}(\mathsf{app}(w,v), u) \end{cases}$$

Output TRS:

$$\mathcal{R}' \begin{cases} \mathsf{rev}(u) & \to \mathsf{f1}(u, \mathsf{nil}) \\ \mathsf{f1}(\mathsf{nil}, u) & \to u \\ \mathsf{f1}(\mathsf{cons}(u,v), w) & \to \mathsf{f1}(v, \mathsf{cons}(u,w)) \\ \mathsf{app}(\mathsf{nil}, x) & \to x \\ \mathsf{app}(\mathsf{cons}(x,y), z) \to \mathsf{cons}(x, \mathsf{app}(y,z)) \end{cases}$$

Fig. 4. Example of a program transformation report

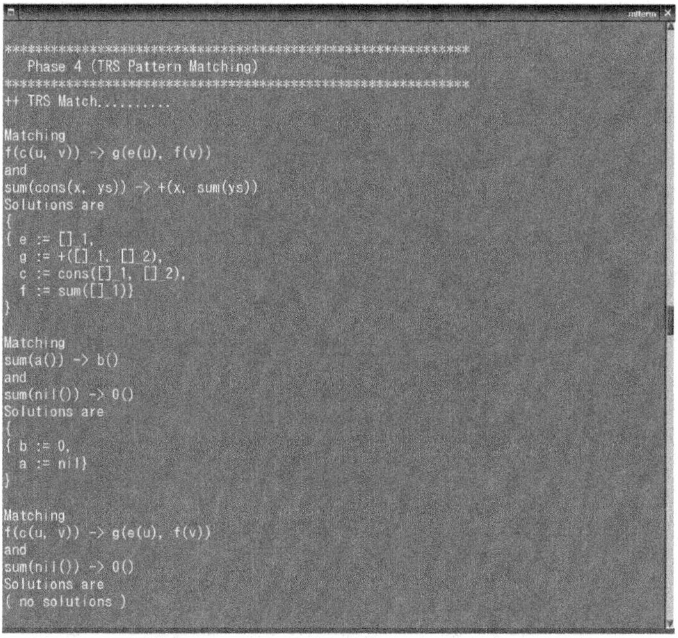

Fig. 5. Snapshot of TRS pattern matching

4. TRS pattern matching. In this phase, RAPT finds a combination of rewrite rules to apply the transformation and the term homomorphism which instantiates the input pattern TRS to these rewrite rules; the matching algorithm in [4] is used in this part. Using information of function symbols detected in Phase 1, it is also checked whether this term homomorphism is a CS-homomorphism. Pattern matching of rewrite rules are carried out in order, and use the information of matching solutions to limit next rewrite rules to perform the pattern match. Since solving the patten matching of main function usually gives information which subfunctions are used in sequel, this heuristics performs the TRS matching relatively well. Visually, consider the case when $\mathcal{P} = \{p_i(x) \to p_{i-1}(x) \mid 1 \leq i \leq 9\} \cup \{p_0(x) \to a\}$ and $\mathcal{R} = \{f_i(x) \to f_{i-1}(x) \mid 1 \leq i \leq 9\} \cup \{f_0(x) \to 0\}$ where the number of all possible combinations of rewrite rules becomes $10! = 3,628,800$ while the number of matching performed becomes $\sum_{i=0}^{10} i = 55$.

5. Verification of hypothesis. In this phase, RAPT checks whether the input TRS satisfies the hypothesis part of the template. This is done by (1) instantiating the hypotheses through the term homomorphism found at Phase 4 and (2) proving they are inductive consequences of the input TRS, using rewriting induction. The latter uses LPO with the precedence detected at Phase 2.

6. Validation of output TRS. In this phase, RAPT checks whether the output TRS is (1) terminating, (2) left-linear, (3) type consistent, and (4) sufficiently complete. In (3), because the pattern TRS $\mathcal{P}'$ for the output may

Table 1. Experimental result

Template I	TRSs	Template II	TRSs
$\left.\begin{array}{ll} f(a) & \rightarrow b \\ f(c(u,v)) & \rightarrow g(e(u),f(v)) \\ g(b,u) & \rightarrow u \\ g(d(u,v),w) & \rightarrow d(u,g(v,w)) \end{array}\right\}$, $\left.\begin{array}{ll} f(u) & \rightarrow f_1(u,b) \\ f_1(a,u) & \rightarrow u \\ f_1(c(u,v),w) \rightarrow f_1(v,g(w,e(u))) \\ g(b,u) & \rightarrow u \\ g(d(u,v),w) \rightarrow d(u,g(v,w)) \end{array}\right\}$, $\left.\begin{array}{l} g(b,u) \approx g(u,b) \\ g(g(u,v),w) \approx g(u,g(v,w)) \end{array}\right\}$	3	$\left.\begin{array}{ll} f(a) & \rightarrow b \\ f(c(u,v)) & \rightarrow g(f(v),e(u)) \\ g(b,u) & \rightarrow u \\ g(d(u,v),w) & \rightarrow d(u,g(v,w)) \end{array}\right\}$, $\left.\begin{array}{ll} f(u) & \rightarrow f_1(u,b) \\ f_1(a,u) & \rightarrow u \\ f_1(c(u,v),w) \rightarrow f_1(v,g(e(u),w)) \\ g(b,u) & \rightarrow u \\ g(d(u,v),w) \rightarrow d(u,g(v,w)) \end{array}\right\}$, $\left.\begin{array}{l} g(b,u) \approx g(u,b) \\ g(g(u,v),w) \approx g(u,g(v,w)) \end{array}\right\}$	3

Template III	TRSs	Template IV	TRSs
$\left.\begin{array}{ll} f(a) & \rightarrow b \\ f(c(u,v)) \rightarrow g(e(u,v),f(v)) \end{array}\right\}$, $\left.\begin{array}{ll} f(u) & \rightarrow f_1(u,b) \\ f_1(a,u) & \rightarrow u \\ f_1(c(u,v),w) \rightarrow f_1(v,g(w,e(u,v))) \end{array}\right\}$, $\left.\begin{array}{l} g(b,u) \approx u \\ g(u,b) \approx u \\ g(g(u,v),w) \approx g(u,g(v,w)) \end{array}\right\}$	11	$\left.\begin{array}{ll} f(x,y,z) & \rightarrow g(h(x,y),z) \\ g(a,y) & \rightarrow b(u) \\ g(c(x,y),z) & \rightarrow e(x,g(y,z)) \\ h(a,y) & \rightarrow r(y) \\ h(c(x,y),z) & \rightarrow c(d(x),h(y,z)) \end{array}\right\}$, $\left.\begin{array}{ll} f(a,y,z) & \rightarrow g(r(y),z) \\ f(c(x,y),z,w) \rightarrow e(d(x),f(y,z,w)) \\ g(a,y) & \rightarrow b(u) \\ g(c(x,y),z) & \rightarrow e(x,g(y,z)) \\ h(a,y) & \rightarrow r(y) \\ h(c(x,y),z) & \rightarrow c(d(x),h(y,z)) \end{array}\right\}$, $\{\}$	8

contain a pattern variable not occurring in the pattern TRS $\mathcal{P}$ for the input, types may be unknown for some of function symbols in $\mathcal{R}'$. Therefore, we need to infer the type information together with the type consistency check. (4) is proved based on the fact the output TRS is terminating which is verified at (1) using LPO.

5   Experiments

We have checked operations of RAPT using several templates. Table 1 describes some of transformation templates and numbers of TRSs succeeded in transformation by each template. Template I is the one which appears in Section 2. This template represents a well-known transformation from recursive programs to iterative programs. A same kind of transformation is also described by Template II. The main difference between Template I and II is the right-hand side of second rule of input parts. In our experiments, there exist TRSs which cannot be transformed by one of these templates but can be done by the other. Template III is the one which overcomes this difference; unchanged rewrite rules of input and output TRS patterns are removed and rewrite rules which are necessary to develop the template are pushed into the hypothesis. Template IV represents another transformation known as fusion or deforestation [16]. RAPT performs transformations of these examples in less than 100 msec.

6   Concluding Remarks

Program transformation techniques have been widely investigated in various fields [3,12,13,16]. This paper describes the system RAPT, which implements the program transformation based on term rewriting introduced in [4]. RAPT transforms a term rewriting system according to a specified program transformation template and automatically verifies correctness of the transformation. We have described the design and implementation of RAPT. An experimental result for several templates has been shown.

Another framework of program transformation by templates is the one based on lambda calculus [6,8,15]. MAG system [6,15] is a program transformation system based on this framework. MAG supports transformations which include modification of expressions, matching with a help of hypothesis; its target also includes higher-order programs. RAPT does not handle such refinements, and cannot deal with most of transformations appearing in [15]. The advantage of RAPT against MAG lies on the approach to the verification of hypothesis. Since the correctness of transformation by MAG system is based on Huet and Lang's original framework [8], users are usually need to verify the hypothesis. In contrast, RAPT proves the hypothesis automatically without help of users.

Besides the limitation of the theoretical framework, several limitations are imposed in the current implementation of RAPT:

- RAPT handles only terminating TRSs. In fact, termination of input and output TRSs are not required in the theoretical framework on which RAPT is based. The main reason to limit its target to terminating TRSs is to reduce checking of sufficient completeness to that of quasi-reducibility, which can be easily verified.
- RAPT allows only confluent TRSs for input. Theoretically, not confluence but ground confluence is sufficient. Replacing confluence checking by ground confluence checking might enlarge the scope of input programs.
- RAPT implements only a naive rewriting induction. Thus, incorporating lemma discovery mechanism and other inductive theorem proving methods may largely enhance the power of inductive theorem proving. Since verification of the hypothesis of template is an important part of the correctness verification, enhancing this part will increase the flexibility of the program transformation.

Extending RAPT to make more flexible transformation possible remains as a future work.

Acknowledgments

Thanks are due to anonymous referees for useful comments and advices. The authors also thank Yoshihito Toyama for valuable comments and discussions. This work was partially supported by a grant from Japan Society for the Promotion of Science, No. 17700002.

References

1. F. Baader and T. Nipkow. *Term Rewriting and All That*. Cambridge University Press, 1998.
2. A. Bundy. The automation of proof by mathematical induction. In *Handbook of Automated Reasoning*, chapter 13, pages 845–911. Elsevier and MIT Press, 2001.
3. R.M. Burstall and J. Darlington. A transformation system for developing recursive programs. *Journal of the ACM*, 24(1):44–67, 1977.
4. Y. Chiba, T. Aoto, and Y. Toyama. Program transformation by templates based on term rewriting. In *Proceedings of the 7th ACM-SIGPLAN International Conference on Principles and Practice of Declarative Programming (PPDP 2005)*, pages 59–69. ACM Press, 2005.
5. H. Comon. Inductionless induction. In *Handbook of Automated Reasoning*, chapter 14, pages 913–962. Elsevier and MIT Press, 2001.
6. O. de Moor and G. Sittampalam. Higher-order matching for program transformation. *Theoretical Computer Science*, 269:135–162, 2001.
7. N. Hirokawa and A. Middeldorp. Tsukuba termination tool. In *Proceedings of the 14th International Conference on Rewriting Techniques and Applications*, volume 2706 of *LNCS*, pages 311–320. Springer-Verlag, 2003.
8. G. Huet and B. Lang. Proving and applying program transformations expressed with second order patterns. *Acta Informatica*, 11:31–55, 1978.
9. D. Kapur, P. Narendran, and H. Zhang. On sufficient-completeness and related properties of term rewriting systems. *Acta Informatica*, 24(4):395–415, 1987.
10. A. Lazrek, P. Lescanne, and J. J. Thiel. Tools for proving inductive equalities, relative completeness, and ω-completeness. *Information and Computation*, 84:47–70, 1990.
11. T. Nipkow and G. Weikum. A decidability result about sufficient-completeness of axiomatically specified abstract data types. In *Proceedings of the 6th GI-Conference on Theoretical Computer Science*, volume 145 of *LNCS*, pages 257–268. Springer-Verlag, 1983.
12. R. Paige. Future directions in program transformations. *ACM Computing Surveys*, 28(4es):170, 1996.
13. H. Partsch and R. Steinbrüggen. Program transformation systems. *ACM Computing Surveys*, 15(3):199–236, 1983.
14. U. S. Reddy. Term rewriting induction. In *Proceedings of the 10th International Conference on Automated Deduction*, volume 449 of *LNAI*, pages 162–177, 1990.
15. G. Sittampalam. *Higher-Order Matching for Program Transformation*. PhD thesis, Magdalen College, 2001.
16. P. Wadler. Deforestation: transforming programs to eliminate trees. *Theoretical Computer Science*, 73:231–248, 1990.

The CL-Atse Protocol Analyser

Mathieu Turuani

Loria-INRIA, Vandoeuvre-lès-Nancy, France
turuani@loria.fr

Abstract. This paper presents an overview of the CL-Atse tool, an efficient and versatile automatic analyser for the security of cryptographic protocols. CL-Atse takes as input a protocol specified as a set of rewriting rules (IF format, produced by the AVISPA compiler), and uses rewriting and constraint solving techniques to model all reachable states of the participants and decide if an attack exists w.r.t. the Dolev-Yao intruder. Any state-based security property can be modelled (like secrecy, authentication, fairness, etc...), and the algebraic properties of operators like xor or exponentiation are taken into account with much less limitations than other tools, thanks to a complete modular unification algorithm. Also, useful constraints like typing, inequalities, or shared sets of knowledge (with set operations like removes, negative tests, etc...) can also be analysed.

1 Introduction

Designing secure communication systems in open environments such as the Internet is a challenging task, which heavily relies on cryptographic protocols. However, severe attacks have been discovered on protocols even assuming perfect cryptographic primitives. Also, a complete manual analysis of a security protocol is usually a very difficult work. Therefore, many decision procedures have been proposed to decide security properties of protocols w.r.t. a bounded number of sessions [1,7,16,15] in the so called Dolev-Yao model of intruder [13], the dominating formal security model in this line of research (see [14] for an overview of the early history of protocol analysis). In particular, among the different approaches the symbolic ones [15,10,12] have proved to be very effective on standard benchmarks [11] and discovered new flaws on several protocols.

The main design goals of CL-Atse[1] are modularity and performance. These two features proved crucial for i) easily extending the class of protocols that can be analysed (modularity) and ii) obtaining results for a large number of protocol sessions (performance). This appeared to be very useful for analysing protocols from the AVISPA [2] project in which CL-Atse is involved since a few years (with OFMC [5], SATMC [3] and TA4SP [6]), as well as for the RNTL Prouvé project that CL-Atse joined recently. The CL-Atse tool can be freely used, either by binary download on the CL-Atse web page[2], or through on-line execution on the AVISPA web page[3].

The protocol analysis methods of CL-Atse have their roots in the generic knowledge deduction rules from casrul [10] and AVISPA. However, a lot of optimisations and

---

[1] CL-Atse stands for Constraint-Logic-based ATtack SEarcher.
[2] http://www.loria.fr/equipes/cassis/softwares/AtSe/
[3] http://www.avispa-project.org/web-interface/

F. Pfenning (Ed.): RTA 2006, LNCS 4098, pp. 277–286, 2006.

major extensions have been integrated in the tool, like preprocessing of the protocol specifications of extensions to manage the algebraic properties of operators like xor[4] or exponentiation. In practice, the main characteristics of CL-Atse are:

- A general protocol language: CL-Atse can analyse any protocol specified as a set of IF rewriting rules (no restriction). The following figure shows the standard process of protocol analysis using the AVISPA tools, from a specification in HLPSL (role-based, same idea as strands) to any of the four tools available at the moment.

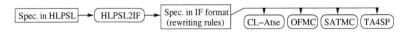

- Flexibility and modularity: CL-Atse structure allows easy integration of new deduction rules and operator properties. In particular, CL-Atse integrates an optimised version of the well-known Baader & Schulz unification algorithm [4], with modules for xor, exponentiation, and associative pairing. To our knowledge, CL-Atse is the only protocol analysis tool that includes complete unification algorithms for xor and exponentiation, with no limitation on terms or intruder operations.
- Efficiency: CL-Atse takes advantage of many optimisations, like simplification and re-writing of the input specification, or optimisations of the analysis method.
- Expressive language for security goals: CL-Atse can analyse any user-defined state-based property specified in AVISPA IF format.

Since protocol security is undecidable for unbounded number of sessions, the analysis is restricted to a fixed but arbitrary large number of sessions (or loops, specified by the user). Other tools provide different features. The closest to CL-Atse are:

The OFMC tool [5], also part of AVISPA, solves the same problem as CL-Atse except that loops and sessions are iterated indefinitely. However, OFMC proposes a different method to manage algebraic properties of operators: instead of hard-coding these properties in the tool, a language of operator properties is provided to the user. Equality modulo theories is solved through modular rewriting instead of direct unification with state-of-the-art algorithms for CL-Atse. However, since this language covers all theories, termination is only obtained by specifying bounds on message depths and number of intruder operations used to create new terms. Hence, completeness cannot be ensured. CL-Atse does not provide such flexibility on properties, but it also does not have any limitation for the theories it can handle (xor, exponentiation, etc...). Moreover, thanks to modularity in the unification algorithm and in knowledge deduction rules, it is quite easy to include new algebraic (or cryptographic) properties directly in the tool. Also, CL-Atse seems to be much faster than OFMC (see Section 3.3).

The Corin-Etalle [12] constraint-based system, which improves upon one developed by Millen & Schmatikov, relies on an expressive syntax based on strands and some efficient semantics to analyse and validate security protocols. Here, strands are extended to allow any agent to perform explicit checks (i.e. equality test over terms). This makes a quite expressive syntax for modelling protocols, that is however subsumed by IF rules. Moreover, to our knowledge no implementation for xor and exponential is provided.

---

[4] We specially thank Max Tuengal who largely contributed to the integration of xor in CL-Atse.

2   The Internal of CL-Atse

We now describe how CL-Atse models protocols and states, and how these objects are used in analysis methods. We start with term signature in CL-Atse used to model messages sent by parties (honest or malicious):

$$Term = \mathcal{A}tom \mid \mathcal{V}ar \mid \mathcal{T}erm.\mathcal{T}erm \mid \{\mathcal{T}erm\}^s_{\mathcal{T}erm} \mid \{\mathcal{T}erm\}^a_{\mathcal{T}erm}$$
$$\mid inv(\mathcal{T}erm) \mid \mathcal{T}erm \oplus \mathcal{T}erm \mid Exp(\mathcal{T}erm, \mathcal{P}roduct)$$
$$\mathcal{P}roduct = (\mathcal{T}erm)^{\pm 1} \mid (\mathcal{T}erm)^{\pm 1} \times \mathcal{P}roduct$$

Terms can be atoms, variables, concatenations (or pairing), and symmetric or asymmetric encryption (marked by s or a). Also, $inv(k)$ is the inverse of k for asymmetric encryption[5]. The $\oplus$ and $Exp(..)$ operators are presented in Section 3.1, and model the xor and exponentiation operators.

The intruder capabilities in CL-Atse match the Dolev-Yao model [13], extended for xor and exponentiation as in [8,9]. Following the formalism of [16,8,9], we write $Forge(E)$ for the infinite set of messages that the intruder can generate from a set of ground terms E. In particular, the intruder can compose pairs, encryption, xor and exponentiation terms, and decompose pairs, encryption (if possible), etc...

As usual, (ground) substitutions are (ground) term assignments to variables. See [9] for a discussion about how to rewrite a protocol specification to avoid products as variable values. Moreover, allowing agents to make tests of quadratic residuosity for the exponential is an easy extension of CL-Atse planned for near future.

2.1   Protocol and System State in CL-Atse

For performance issues, various algorithms are implemented in CL-Atse to simplify and optimise the input protocol specification, and also to guide the protocol analysis. However, these methods require working on a protocol specification with some special features. Listing these would be quite technical, but the most important ones are the fact that all protocol steps and roles must be local to only one participant, and that CL-Atse must eliminate all honest agents' knowledge by converting them into a small set of equality and inequality constraints over terms with global variables. This allows CL-Atse to compute closures of the participant's or intruder knowledge, unforgeable terms, sets or facts, and to optimise each role instance accordingly (preprocessing). The way CL-Atse converts an IF file is out of the scope of this paper.

An execution trace in CL-Atse is built over (protocol) steps and states, and represents the list of state changes when running a list of steps, starting from the initial state. The basic objects used by CL-Atse are defined as follows.

A system state in CL-Atse is a symbolic representation of an infinite number of "real" (i.e. ground) states. Since honest agent's states have been converted into constraints, only the intruder state is relevant in the definition of states, here. Formally:

$$state = Subst, Sets, ToDec, Known$$
$$ToDec = (\mathcal{T}erm, \mathcal{T}erm)^*$$
$$Known = H(\mathcal{V}ar) \triangleright Known \mid D(\mathcal{T}erm) \triangleright Known \mid \epsilon$$

---

[5] If k is a (random) term, then $inv(k)$ exists but is unknown to every agent.

with $Subst$ a (partial) substitution, $Sets$ a list of facts $t \in set$, with $\{t, set\} \subset \mathcal{T}erm$, saying that in this state the element t is present in the set $named\ set$, and $ToDec$ a list of opportunities of knowledge deductions: if $(m, k) \in ToDec$, then the intruder will get m as soon as he will be able to forge k. Typically, a knowledge $\{m\}_k^s$ creates an entry in $ToDec$ if k in not known at the point $\{m\}_k^s$ is obtained. Finally, $Known$ is a list of elementary 'D'ecomposed knowledge $D(t)$, and 'H'ypothesis $H(v)$ (i.e. variable constraints), ordered by creation time in the execution trace. For example:

$$Known = H(x) \triangleright H(y) \triangleright D(\{z\}_k^s) \triangleright H(z) \triangleright D(a) \triangleright D(b) \triangleright \epsilon$$

means that the intruder knows $\{z\}_k^s, a, b$ but must forge the value of z from $\{a, b\}$, and the values of x and y from $\{\{z\}_k^s, a, b\}$. We denote $E_{|D}$ the set of terms t such that $D(t) \in E$. Naturally, a symbolic state as above models the infinite set of ground states $\sigma(Known_{|D}, Sets)$ such that σ is a ground instance of $Subst$ and $\sigma(v) \in Forge(\sigma(F_{|D}))$, with $Known = E \triangleright H(v) \triangleright F$. The analysis methods of CL-Atse use rewriting of symbolic states in order to filter or update the set of ground states that it represents.

A protocol step in CL-Atse represents an elementary reaction of an agent: when receiving the message rcv, and provided that a list $CtrList$ of constraints $u = v$ or $u \neq v$ over terms and a list $SetTests$ of constraints $t \in set$ or $t \notin set$ over sets are satisfied, the agent sends a message snd as a response and executes a list $SetOperations$ of add or remove operations over sets and set elements. That is:

$$step = iknows(rcv) \ \& \ CtrList \ \& \ SetTests$$
$$\Rightarrow iknows(snd) \ \& \ SetOperations$$

Note that IF facts are converted into constraints over sets. The semantic of ground step execution is defined as usual: given some intruder knowledge E, a populated list of named sets, and a ground substitution σ, if $\sigma(rcv) \in Forge(E)$, if $\sigma(u) = \sigma(v)$ or $\sigma(s) \neq \sigma(t)$ for any constraint $u = v$ or $s \neq t$ in $CtrList$, and if $\sigma(t) \in \sigma(set)$ (resp. $\sigma(t) \notin \sigma(set)$) for any test $t \in set$ (resp. $t \notin set$) in $SetTests$, then $\sigma(snd)$ is added to E and all add or remove operations in $SetOperations$ are performed modulo σ.

A role in CL-Atse is a tree-structured set of roles that captures the non-determinism of the execution of IF rules. Formally:

$$role = Step(step, role) \mid Choice(role\ list) \mid EndRole$$

where $Choice$ describes an agent's choice point, i.e. from that point only one role in $role\ list$ may be run, like in $A|B$. Thanks to equality and inequality constraints, this may model pattern matching. Moreover, thread creation is supported through tokens. For example, $A.(B\|C)$ is modeled by 3 roles A, B, C where A send tok_1 at its last step and B and C wait for tok_1 in their first steps. Same for confluence $((B\|C).D)$, with a pool of tokens.

A security property in CL-Atse is modeled as the negation of a list of attack states, defined as follows:

$$attack\_states = (iknows(rcv) \ \& \ CtrList \ \& \ SetTests)^*$$

with the same definitions as for *step*. An attack is found when at least one ground state of a symbolic system state satisfies the constraints of one of the attack states. This definition of security failure is quite versatile since it allows the user to use any IF facts (self-made or not) to define any property based on states adapted to his protocol. Standard properties like secrecy or authentication are naturally supported and an implementation of temporal security properties is planned for very near future. For example, fairness[6] in a two-party contract-signing protocol may be coded by:

$$fairness\_atk = (i, Alice, \text{text}) \in play\_together \& iknows(ctr(\text{text}))$$
$$\& ctr(\text{text}) \notin ctr\_list(Alice) \& Alice \in finished$$

with text a contract text, $ctr(\text{text})$ a term representing the valid signed contract, and $play\_together$, $ctr\_list(Alice)$ and $finished$ user-defined facts representing the lists of initiated sessions, contracts of *Alice*, and terminated agent's roles. Similar for *Alice* playing with an honest agent.

A protocol specification in CL-Atse is simply a set of instantiated roles (one for each participant) plus an initial state and a set of attack states:

$$protocol = RoleList, InitState, attack\_state$$

2.2 Protocol Simplifications and Optimisations

During the AVISPA project, it became increasingly clearer that two important ingredients that may contribute to the efficiency of the CL-Atse tool would be protocol simplification strategies and optimisation operations on the protocol specification. Therefore, without neglecting the importance of efficiency for the analysis algorithm, some important efforts were devoted to the two axes of protocol simplification and optimisation.

Protocol simplifications reduce the overall size of the protocol, and specifically the number of steps, by merging as many steps together as possible, or at least marking them to be executed as soon, or as late, as possible. A step marked to be run as soon as possible will be run in any trace immediately after its parent step. Since these marks or merges put very restrictive constraints on the step interleaving, they greatly reduce CL-Atse computing time (the analysis is necessarily exponential in the number of unmarked steps). However, CL-Atse can only take such decisions when it can automatically build a proof that it would not void the insecurity of the protocol, i.e., that if the protocol was flawed then necessarily at least one attack remains. To do so, CL-Atse builds various protocol-dependant objects like a set of unforgeable terms (atoms, keys, etc.. that the intruder cannot create in any execution). Then, given a protocol step, CL-Atse tests its elements for possibilities of merging (or marking).

To do so, set tests and operations are checked for possibilities of being executed as soon, or as late, as possible. For example, if *set* is a set name unforgeable by the intruder, then an operation that removes the term t from *set* can be performed as soon as possible when either there exists no set operation that add t' to set' or tests $t' \in set'$ in any step of other roles that may be run before this operation; or the operation is useless

---

[6] It intuitively requires that whenever a participant obtains a valid contract, there is a way for it's partner to also obtain one.

or impossible (similar tests). Similar criteria are evaluated for each set operation or test, and for both as soon, or as late, as possible executions.

Also, all other step elements are tested in similar ways, as well as attack states: for example, marking a step to be run as soon as possible requires that if any attack state is validated, then it is also validated either before the previous step, or after the current step if no constraints could prevent running this step. When all tests are successful, the step is marked and another one is analysed.

Optimisations: Protocol optimisations aim at rewriting some parts of the protocol in order to accelerate the search for attacks. The acceleration can be significant, but the protocol structure can be changed. The idea is to track all possible origins of ciphertexts that the intruder must send but cannot create himself (i.e. necessarily obtained from an agent). By building an exhaustive list of origins for such terms, CL-Atse can reduce the future work of the analysis algorithm by unifying these terms with each of their possible origins and generate choice points accordingly. Analysis acceleration comes from a reduction of possible redundancy in step execution. Moreover, this strategy also fixes the moment when steps holding such cipher terms must be run in the analysis. The same must also be done on the awaited sets that the intruder cannot create himself (same idea). For example, if we have some protocol steps

$$step_1 = iknows(\{m\}_k) \Rightarrow ....$$
$$step_2 = ... \Rightarrow iknows(\{m'\}_{k'}) ...$$
$$step_3 = ... \Rightarrow iknows(\{m''\}_{k''}) ...$$

where CL-Atse computes that k is unforgeable by the intruder, and that $step_2$ and $step_3$ are the only origins of $\{m\}_k$, then these steps may be replaced by:

$$step_4 = Choice(step_5, step_6)$$
$$step_5 = iknows(token_1) \& equal(\{m\}_k, \{m'\}_{k'}) \Rightarrow ....$$
$$step_6 = iknows(token_2) \& equal(\{m\}_k, \{m''\}_{k''}) \Rightarrow ....$$
$$step_7 = ... \Rightarrow iknows(\{m'\}_{k'}, token_1) ...$$
$$step_8 = ... \Rightarrow iknows(\{m''\}_{k''}, token_2) ...$$

The big difference is that only atoms are now awaited in $step_5$ and $step_6$. This gives us the chance to optimise their execution (when possible) by running these steps immediately as soon as $token_1$ or $token_2$ is added to the intruder knowledge. This strategy allows CL-Atse to analyse rapidly some protocols that it could not analyse otherwise.

3   The Analysis Method

As said before, the analysis algorithm implemented in CL-Atse follows the general ideas developed in the AVISPA Project, that is, to symbolically execute the protocol in any possible step ordering. We saw in the previous section some of the important optimisations of CL-Atse for step interleaving above this generic method. Moreover, in order to perform this exploration of all possible execution traces, the analysis algorithm relies on two major components: a (generic) unification algorithm modulo the properties of the operators, like xor or exponentiation, that provide all term-specific computations;

and the management of states and constraints when running a protocol step. This modular structure allowed us to code the tool extensions required by the AVISPA project (like sets, properties, typing, etc..) in a direct and natural way. We now present the two major components of CL-Atse and the analysis method.

3.1 Modular Unification (with Xor and Exponentiation)

The unification module provides a (generic) complete unification algorithm modulo the algebraic or cryptographic properties of the CL-Atse operators (encryption, xor, exp, pair, etc..), as well as related algorithms like term purification or normalisation. From a general point of view, the problem that must be decided here is: given a (partial) substitution σ and two terms u and v, generate a list of most general unifiers $Mgu_{u,v}^{\sigma} = \{\sigma_1', .., \sigma_p'\}$ of u and v that validate σ, i.e.:

$$\forall \sigma' \in Mgu_{u,v}^{\sigma}, \quad \sigma'(u) = \sigma'(v) \text{ and } \sigma'(Var) = \sigma'(\sigma(Var))$$

Since mgu(s) are used to generate new system states, a great care must be taken to generate a list of mgu as small as possible. The latest implementation of CL-Atse manages the properties of the xor operator, the exponentiation, and the associative concatenation. To manage these properties, the tool unifies terms thanks to an implementation of an optimised version of the well-known Baader & Schulz unification algorithm [4], which splits the unification problem into smaller unification problems, one for each theory. Therefore, the unification algorithm is very modular, and we consider that it would be reasonably difficult to add new operator properties to the previous ones. Currently, we have:

The xor operator: Denoted $\oplus$, this is an associative $(a \oplus (b \oplus c) = (a \oplus b) \oplus c)$ and commutative $(a \oplus b = b \oplus a)$ operator equipped with a unit element $(a \oplus 0 = a)$ and nilpotent $(a \oplus a = 0)$;
The exponentiation: Denoted $Exp(g, a)$, it represents g^a in some fixed group of prime order. Also, the product $\times$ on exponents models the multiplication in the corresponding (abelian) multiplicative group. Properties include inverse $(a \times a^{-1} = 1)$, commutativity $(a \times b = b \times a)$, normalisation $(Exp(Exp(g, M), N) = Exp(g, M \times N))$, ...
The associative concatenation: it represents the basic bit string concatenation, without any header giving the splitting position: in this case, associativity models the chance (or a risk) that an agent will not cut the concatenation correctly when parsing it. Naturally, a non-associative pairing operator is also provided.

3.2 The Kernel: Running a Protocol Step

The second foundational element of the protocol analysis is the kernel module, which aims at running a protocol step on a symbolic system state by adding new constraints, reducing them to elementary constraints, testing their validity, etc... All these operations are described as rewriting rules and follow carefully the IF semantics. For performance issues all these rules are directly implemented in the tool as operations on constraints. Therefore, adding new intruder deduction rules requires to implement them in the tool. However, the recent extensions to algebraic properties proved that the tool is sufficiently modular to make such integration quite easy. In particular, the rewriting rules described below correspond to matching in the tool very precisely.

Hypothesis reductions: We call non-reduced a hypothesis $H(t)$ where t is not a variable. This is the received message of a protocol step. Assume that $s = (E \triangleright H(t) \triangleright F \triangleright \epsilon, td, set, \sigma)$ is a system state where $t \notin Var$ and F reduced already. Then, we reduce $H(t)$ depending on t with rewriting rules on $E \triangleright H(t) \triangleright F \triangleright \epsilon$ (and σ). For example:

- $E \triangleright H(u, v) \triangleright F \longrightarrow E \triangleright H(u) \triangleright H(v) \triangleright F$;
- $E \triangleright H(t) \triangleright F \triangleright D(t') \triangleright G \longrightarrow \sigma'(E \triangleright F \triangleright D(t') \triangleright G)$ with $\sigma' \in Mgu^\sigma_{t,t'}$;
- $E \triangleright H(\{t\}_k^{s \ or \ a}) \triangleright F \longrightarrow E \triangleright H(t) \triangleright H(k) \triangleright F$ if k in not unforgeable;

These rules model respectively the creation of a pair, the redirection of a known message, and the creation of a cipher. Also, similar but more complex rules allow us to construct xor or exponentiation terms, by enumerating the possibilities available to an adversary for constructing such terms, like building an xor by combining xor and non-xor terms. If defined, σ' is the new state's substitution. These rules are naturally non-deterministic (create a set of states), and are iterated until all variables are reduced.

Knowledge deductions: We increase $Known$ with $K(t)$, $t \in Term$, for new non-decomposed 'K'nowledge (sent message in a protocol step), and $T(t)$ for a knowledge being processed (i.e. 'T'emporary). Reducing an $Known$ containing some $K(t)$ is done in two steps, similar to those for hypothesis. That is, the processing of a new knowledge follows this scheme:

$$\ldots \triangleright K(t) \triangleright \ldots \xrightarrow{\text{decompose } K(t)} \ldots \triangleright T(t) \triangleright \ldots \xrightarrow{\text{analyse } ToDec \text{ with } t} \ldots \triangleright D(t) \triangleright \ldots$$

The first set of rules decompose any $K(t)$. For example:

- $E \triangleright K(u, v) \triangleright F \longrightarrow E \triangleright K(u) \triangleright K(v) \triangleright F$; $\quad E \triangleright K(t) \triangleright F \longrightarrow E \triangleright T(t) \triangleright F$;
- $E \triangleright K(\{m\}_k^{s \ or \ a}) \triangleright F \longrightarrow E \triangleright K(m) \triangleright T(\{m\}_k^{s \ or \ a}) \triangleright H(k') \triangleright F$ with $k' = inv(k)$ (for asymmetric encryption) or $k' = k$ (for symmetric encryption);

These rules model respectively the decomposition of a pair, the fact that a term may not be decomposable, and the decryption of a cipher. Rules in CL-Atse include various optimisations and variations of the techniques described above (like state filtering depending on key availability, ...). Moreover, rules for $\oplus$ or Exp are also included (to get a from $a \oplus t$, or g from $Exp(g, M)$). The second set of rules can analyse $ToDec$ to add or remove deduction opportunities depending on $T(..)$. That is, assuming that $Known = E \triangleright T(t) \triangleright F$, we:

- Add (m, k) to $ToDec$ when $t = \{m\}_k$;
- Remove $\{(m'_i, k'_i)\}_{i \in 1..n}$ from $ToDec$ when we can reduce the hypothesis $H(k'_1) \triangleright .. \triangleright H(k'_n) \triangleright E \triangleright D(t) \triangleright F$ to some G such that $D(t)$ is used at least once for each k'_i, and create a new state with $K(m') \triangleright G$. This is again non-deterministic. Also, create a new state with $E \triangleright D(t) \triangleright F$, in case no k'_i may be computed.

These rules, too, are significantly optimised in CL-Atse. Moreover, the last rule guarantee that we won't ever build k' is a way that has already been tried before, which is critical for tool performance.

Other operations: To run a protocol step, we need to perform other operations on states than the two above, like adding (and validating) new equality or inequality constraints, managing sets, etc.. Since they are quite straightforward and coded in a similar way as the two above, they are not detailed here.

3.3   Search for Attacks

Using the previously described kernel module, we are now able to run a protocol step on a system state and get the resulting set of new states. Therefore, we can easily explore all possible runs of a protocol by iteratively running steps in any possible ordering, starting from the initial state. Moreover, we reduce step interleaving by using the step marking described in the simplification and optimisation Section 2.2. Finally, each time a protocol step is run, we test the non-satisfiability of each attack state.

Performances: The analysis algorithm of CL-Atse gives very good performances in practice, as shown in the small benchmark table that follows. Times are computing times of the latest versions (feb. 20, 2006) of OFMC and CL-Atse, and protocol specifications are taken from AVISPA. Note also that (2) is CL-Atse without some optimisations. The "Timeout" for QoS in that case is due to an explosion of the number of states. Both binaries and on-line tool execution are available (see introduction for URLs).

Protocol Name Alg. theory	Result	OFMC	CL-Atse	CL-Atse[2]
ASW - Abort part	Secrecy failure	3.94s	**0.03s**	0.16s
EAP with Archie method	Safe	0.70s	**0.07s**	5.94s
EAP TTLS with CHAP	Safe	1.27s	**0.18s**	0.19s
Fair Zhou-Gollmann	Auth. failure	Timeout	**0.13s**	0.13s
Fair Zhou-Gollmann (fixed)	Safe	7.65s	**4.57s**	5.34
IKEv2 with MAC auth. - **Exp.**	Safe	20.29s	**7.62s**	7.62s
Kerberos, cross-realm ver. - **Exp.**	Safe	5.83s	**0.42s**	0.42s
Kerberos, forwardable tickets - **Exp.**	Safe	15.40s	**0.14s**	0.15s
Purpose Built Keys protocol	Auth. failure	0.35s	**0.00s**	0.00s
PEAP with MS-CHAP auth.	Safe	14.25s	**0.18s**	0.18s
Next Steps In Signaling, QoS	Safe	15.53s	**0.86s**	Timeout
SET - Purchase Request	Secrecy failure	1.17s	**0.14s**	0.15s
Diameter Session Init. Prot.	Safe	1.80s	**0.01s**	0.02s
SPEKE, with strong pwd. - **Exp.**	Safe	2.75s	**0.04s**	0.04s
SSH Transport Layer Prot. - **Exp.**	Safe	33.96s	**2.12s**	2.16s

4   Conclusion

As mentioned before, the analysis algorithm implemented in CL-Atse proposes a solution to the NP-Complete protocol insecurity problem w.r.t. a bounded number of sessions, and with (or without) the algebraic or cryptographic properties of operators, like xor, exponentiation, or associative pairing. The methods of CL-Atse include many important optimisations for step interleaving, either by preprocessing or by optimised data structures and deduction rules. This allows CL-Atse to reduce redundancies and limit the overall number of elementary actions needed at each step (performance). Moreover, the tool proved to be sufficiently flexible to support major improvements and extensions of the past few years (modularity). For example, extensions to inequalities, set operations, state-based properties, or typing required only little recoding of previous works. Also, while the recent implementation of the Baader & Schulz unification required a significant amount of work, the extension of CL-Atse with new operator properties, like Cipher block chaining, is now largely facilitated, as well as planned extensions to temporal security properties of heuristics for unbounded analysis.

References

1. R. Amadio, D. Lugiez, and V. Vanackère. On the symbolic reduction of processes with cryptographic functions. *Theor. Comput. Sci.*, 290(1):695–740, 2003.
2. The AVISPA Team. The Avispa Tool for the automated validation of internet security protocols and applications. In *Proceedings of CAV 2005, Computer Aided Verification*, LNCS 3576, Springer Verlag.
3. A. Armando, L. Compagna. An Optimized Intruder Model for SAT-based Model-Checking of Security Protocols. In *Proceedings of the Workshop on Automated Reasoning for Security Protocol Analysis (ARSPA 2004)*, ENTCS 125(1):91-108, 2005.
4. F. Baader and K.U. Schulz. Unification in the Union of Disjoint Equational Theories: Combining Decision Procedures. In *Journal of Symbolic Computing*. 21(2): 211-243 (1996).
5. D. Basin, S. Mödersheim, L. Viganò. OFMC: A symbolic model checker for security protocols. In *International Journal of Information Security* 4(3):181–208, 2005.
6. Y. Boichut, P.-C. Héam, O. Kouchnarenko. Automatic Verification of Security Protocols Using Approximations. *INRIA Research Report*, October 2005.
7. M. Boreale. Symbolic trace analysis of cryptographic protocols. In *Proceedings of the 28th ICALP'01*, LNCS 2076, pages 667–681. Springer-Verlag, Berlin, 2001.
8. Y. Chevalier, R. Küsters, M. Rusinowitch, and M. Turuani. An NP decision procedure for protocol insecurity with xor. In *Proceedings of LICS 2003*, 2003.
9. Y. Chevalier, R. Küsters, M. Rusinowitch, and M. Turuani. Deciding the Security of Protocols with Diffie-Hellman Exponentiation and Products in Exponents. In *Proceedings of the Foundations of Software Technology and Theoretical Computer Science (FSTTCS'03)*, LNCS 2914, Springer-Verlag, December 2003.
10. Y. Chevalier and L. Vigneron. A Tool for Lazy Verification of Security Protocols. In *Proceedings of the Automated Software Engineering Conference (ASE'01)*. IEEE CSP, 2001.
11. J. Clark and J. Jacob. A Survey of Authentication Protocol Literature: Version 1.0, 17. Nov. 1997. URL: www.cs.york.ac.uk/~jac/papers/drareview.ps.gz.
12. R. Corin and S. Etalle. An improved constraint-based system for the verification of security protocols. In *SAS*, LNCS 2477:326–341, Springer-Verlag, 2002.
13. D. Dolev and A.C. Yao. On the Security of Public-Key Protocols. *IEEE Transactions on Information Theory*, 29(2):198–208, 1983.
14. C. Meadows. Open issues in formal methods for cryptographic protocol analysis. In *Proceedings of DISCEX 2000*, pages 237–250. IEEE Computer Society Press, 2000.
15. J. Millen and V. Shmatikov. Symbolic protocol analysis with products and Diffie-Hellman exponentiation. In *Proceedings of the 16th IEEE Computer Security Foundations Workshop (CSFW'03)*, pages 47–61, 2003.
16. M. Rusinowitch and M. Turuani. Protocol Insecurity with Finite Number of Sessions is NP-complete. In *14th IEEE Computer Security Foundations Workshop (CSFW-14)*, pages 174–190, 2001.

SLOTHROP: Knuth-Bendix Completion with a Modern Termination Checker

Ian Wehrman, Aaron Stump, and Edwin Westbrook[*]

Dept. of Computer Science and Engineering
Washington University in St. Louis
St. Louis, MO USA

Abstract. A Knuth-Bendix completion procedure is parametrized by a reduction ordering used to ensure termination of intermediate and resulting rewriting systems. While in principle any reduction ordering can be used, modern completion tools typically implement only Knuth-Bendix and path orderings. Consequently, the theories for which completion can possibly yield a decision procedure are limited to those that can be oriented with a single path order.

In this paper, we present a variant on the Knuth-Bendix completion procedure in which no ordering is assumed. Instead we rely on a modern termination checker to verify termination of rewriting systems. The new method is correct if it terminates; the resulting rewrite system is convergent and equivalent to the input theory. Completions are also not just ground-convergent, but fully convergent. We present an implementation of the new procedure, SLOTHROP, which automatically obtains such completions for theories that do not admit path orderings.

1 Introduction

A Knuth-Bendix completion procedure is a technique for solving the word problem for a finite set of identities. In this procedure, the user provides the set of identities as well as a reduction order on terms.

Using *unfailing completion* [3], ground-convergent completions can be discovered even when no compatible reduction order exists. However, successful completions may contain unoriented equations along with oriented rewrite rules. The resulting system is convergent only for ground terms and often contains more rules than a fully convergent completion.

Nonetheless, many theories are easily oriented by a few useful classes of reduction orderings, such as Knuth-Bendix and recursive path orderings (KBO and RPO, respectively). The wide applicability of KBO and RPO has led to the success of completion procedures. While in principle any reduction ordering can be used, modern completion tools like WALDMEISTER [9] typically implement only these two classes of orderings. (However, the tool CiME also implements polynomial orderings [4].) Consequently, the theories for which completion can

---

[*] This work was partially supported by NSF grant CCF-0448275.

F. Pfenning (Ed.): RTA 2006, LNCS 4098, pp. 287–296, 2006.

possibly yield a decision procedure in the form of a convergent rewrite system are limited to those with completions that admit such a path order.

In this paper, we present a new variant on the standard Knuth-Bendix completion procedure in which no ordering is explicitly provided. Instead a constraint rewriting system is constructed during execution, and its reduction relation is used for an ordering. Termination of the constraint system then implies termination of the intermediate rewrite systems. In the implementation, we rely on modern termination-checking methods to verify the termination of the constraint systems.

The new method is correct if it terminates; the resulting rewrite system is convergent and equivalent to the input theory. In addition, the completions are not just ground-convergent, but fully convergent. The method is also complete for finite executions in that if there exists a successful, finite execution of a standard Knuth-Bendix completion procedure with some reduction ordering, then an equivalent execution in the modified system exists.

We begin in Sect. 2 with a presentation of standard Knuth-Bendix completion and statements of correctness. In Sect. 3 we introduce our new completion variant in which the reduction order is left implicit and prove its correctness. In Sect. 4, we present an implementation of the new procedure called SLOTHROP, and in Sect. 5 we discuss its performance and results, including convergent completions automatically obtained for the first time. Finally in Sect. 6, we discuss areas of future interest with respect to the new technique.

2   Knuth-Bendix Completion

In this section, we review the basic definition and properties of completion procedures. We use standard notation for terms and term rewriting systems, as presented in [1].

A *Knuth-Bendix completion procedure* [8] is an algorithm that takes as input a reduction ordering $>$ and a finite set of equations E and attempts to produce a decision procedure for the word problem for E in the form of a rewriting system. Completion algorithms attempt to construct a convergent rewriting system R that is equivalent to E (i.e., with the same equational theory, $\overset{*}{\leftrightarrow}_E = \overset{*}{\leftrightarrow}_R$) by generating a possibly infinite sequence of intermediate rewriting systems which yield approximations of the equational theory of E.

Bachmair formulated Knuth-Bendix completion as an equational inference system [2]. We refer to this standard system as $\mathcal{C}$ because it serves as the basis of a correctness condition for our refinement of the procedure. The rules of the inference system $\mathcal{C}$ are shown in Fig. 1. A *deduction* of $\mathcal{C}$, written $(E, R) \vdash_{\mathcal{C}} (E', R')$, consists of finite sets of identities E, E' and rewrite systems R, R'. A *finite execution* γ of the system $\mathcal{C}$ is the pair $(E_0, \emptyset)$ followed by a finite sequence of deductions

$$(E_0, \emptyset) \vdash_{\mathcal{C}} (E_1, R_1) \cdots \vdash_{\mathcal{C}} (E_n, R_n),$$

where E_0 is the input theory provided by the user, and each deduction results from an application of one of the inference rules of $\mathcal{C}$. (We consider only finite

$$\begin{array}{lll} \text{ORIENT:} & \dfrac{(E \cup \{s \doteq t\}, R)}{(E, R \cup \{s \to t\})} & \text{if } s > t \\[2ex] \text{DEDUCE:} & \dfrac{(E, R)}{(E \cup \{s = t\}, R)} & \text{if } s \leftarrow_R u \to_R t \\[2ex] \text{DELETE:} & \dfrac{(E \cup \{s = s\}, R)}{(E, R)} & \\[2ex] \text{SIMPLIFY:} & \dfrac{(E \cup \{s \doteq t\}, R)}{(E \cup \{u \doteq t\}, R)} & \text{if } s \to_R u \\[2ex] \text{COMPOSE:} & \dfrac{(E, R \cup \{s \to t\})}{(E, R \cup \{s \to u\})} & \text{if } t \to_R u \\[2ex] \text{COLLAPSE:} & \dfrac{(E, R \cup \{s \to t\})}{(E \cup \{v = t\}, R)} & \text{if } s \xrightarrow{\exists}_R v \end{array}$$

Fig. 1. Standard Knuth-Bendix Completion ($\mathcal{C}$)

executions of $\mathcal{C}$ in this paper; the infinite case is discussed briefly in Sect. 6.) The *length* of γ, written $|\gamma|$, is the number of deductions in γ. A finite execution γ of $\mathcal{C}$ *succeeds* if $E_{|\gamma|} = \emptyset$ and $R_{|\gamma|}$ is a convergent rewrite system equivalent to E as described above; otherwise it *fails*. Elsewhere ([2], [1]), $\mathcal{C}$ is proved correct in that any successful, finite execution γ results in a convergent rewrite system $R_{|\gamma|}$ equivalent to the input identities E.

The main difficulty with the standard completion procedure is in finding an appropriate reduction order. Choosing a suitable RPO, KBO or polynomial interpretation (the only options available in known tools) is difficult even for experienced users, and for many theories no such path ordering exists. In the next section, we solve this problem with a variant on the standard completion procedure which discovers a suitable reduction ordering without input from the user.

3   Completion with Termination Checking

We now present a modification of the standard Knuth-Bendix completion procedure. The primary difference is that no reduction order is explicitly provided as input, only a finite set of identities. Lacking any specific reduction order to guide the search, we preserve termination of each intermediate rewrite system R_i by ensuring that some reduction order $\succ_i$ compatible with R_i exists. The orders $\succ_i$ are constructed using terminating rewrite systems C_i, specifically as the transitive closure of the reduction relation on C_i, written $\xrightarrow{+}_{C_i}$. This relation is a well-founded order exactly when the system C_i is terminating. While in the standard system $\mathcal{C}$ a rule $s \to t$ is added by ORIENT to R_i only if $s > t$ with the user-specified reduction order, in the modified system the rule is added only if the addition of $s \to t$ to C_i preserves termination. Of course, deciding termination is not possible in general. In Sect. 4, we discuss how this test is accomplished in practice.

Figure 2 provides the inference rules for a modification of the standard completion procedure, which we refer to as system $\mathcal{A}$. A deduction of $\mathcal{A}$, written $(E, R, C) \vdash_{\mathcal{A}} (E', R', C')$, consists of identities E, E' and rewrite systems R, R' as in standard completion, and finite *constraint* rewrite systems C, C' new to $\mathcal{A}$. A finite execution α of the system $\mathcal{A}$ is the triple $(E_0, \emptyset, \emptyset)$ followed by a finite sequence of deductions

$$(E_0, \emptyset, \emptyset) \vdash_{\mathcal{A}} (E_1, R_1, C_1) \cdots \vdash_{\mathcal{A}} (E_n, R_n, C_n),$$

with E_0 the set of input identities and where each deduction results from an application of one inference rule from $\mathcal{A}$. We consider only finite executions of $\mathcal{A}$; infinite executions are discussed in Sect. 6. We write $|\alpha|$ to denote the length of the sequence. An execution α of $\mathcal{A}$ is *equivalent* to an execution γ of $\mathcal{C}$ when the intermediate equations and rewrite systems are the same at each step. A finite execution α of system $\mathcal{A}$ succeeds when $E_{|\alpha|} = \emptyset$ and $R_{|\alpha|}$ is a convergent rewrite system equivalent to E. Because we only consider finite executions, every execution that does not succeed fails.

ORIENT:	$\dfrac{(E \cup \{s \doteq t\}, R, C)}{(E, R \cup \{s \to t\}, C \cup \{s \to t\})}$	if $C \cup \{s \to t\}$ terminates
DEDUCE:	$\dfrac{(E, R, C)}{(E \cup \{s = t\}, R, C)}$	if $s \leftarrow_R u \to_R t$
DELETE:	$\dfrac{(E \cup \{s = s\}, R, C)}{(E, R, C)}$	
SIMPLIFY:	$\dfrac{(E \cup \{s \doteq t\}, R, C)}{(E \cup \{u \doteq t\}, R, C)}$	if $s \to_R u$
COMPOSE:	$\dfrac{(E, R \cup \{s \to t\}, C)}{(E, R \cup \{s \to u\}, C)}$	if $t \to_R u$
COLLAPSE:	$\dfrac{(E, R \cup \{s \to t\}, C)}{(E \cup \{v = t\}, R, C)}$	if $s \overset{\sqsupset}{\to}_R v$

Fig. 2. Modified Knuth-Bendix Completion ($\mathcal{A}$)

The rules DEDUCE, DELETE, SIMPLIFY, COMPOSE and COLLAPSE of $\mathcal{A}$ are identical to those of $\mathcal{C}$, except for the presence of the constraint system C which is carried unmodified from antecedent to consequent. The critical difference between $\mathcal{A}$ and $\mathcal{C}$ is in the definition of the ORIENT rule. In the standard system $\mathcal{C}$, an identity $s \doteq t$ of E is added to R as rule $s \to t$ only when $s > t$ for the given reduction order. In the modified system $\mathcal{A}$, we add the rule $s \to t$ to R only when the augmented constraint system $C \cup \{s \to t\}$ is terminating. The system $\mathcal{A}$ accepts as input only the finite set of identities E; no reduction order is explicitly provided.

We now state correctness of $\mathcal{A}$ for finite executions (partial correctness). The proof proceeds by showing that $\mathcal{A}$ simulates a standard Knuth-Bendix completion procedure $\mathcal{C}$. For each finite execution α of $\mathcal{A}$, we construct an execution

γ of C with an equivalent sequence of deductions as α. The constraint systems are used to show that any finite execution is equivalent to one which uses the single order induced by the final constraint system. This is important because completion is not generally correct when reduction orders are changed during execution, even if each is compatible with the immediate intermediate rewrite system [10]. The induced order is $\xrightarrow{+}_{C_{|\alpha|}}$, the transitive closure of the reduction relation of the final constraint system $C_{|\alpha|}$.

Theorem 1 (Partial Correctness of $\mathcal{A}$). *Let α be a finite execution of the system $\mathcal{A}$. Then there exists an equivalent execution γ of C using reduction order $\xrightarrow{+}_{C_{|\alpha|}}$.*

3.1 Partial Completeness

We now show a limited form of completeness for our procedure with respect to standard Knuth-Bendix completion. Namely, for any successful execution of the standard completion procedure C there exists a corresponding execution of the modified procedure $\mathcal{A}$ with the same deductions. This shows that our method can at least construct decision procedures for those theories that are decidable by the standard method. In Sect. 5, we give an example of a theory for which our method constructs a convergent completion that, to the authors' knowledge, cannot be automatically constructed by any tool that implements C due to inability to specify an appropriate reduction order.

Theorem 2 (Partial Completeness of $\mathcal{A}$). *For any finite execution γ of C with reduction order $>$, there exists an equivalent execution α of $\mathcal{A}$. Furthermore, $\xrightarrow{+}_{C_{|\gamma|}} \subseteq >$.*

Proof. By induction on γ. The beginning execution is $\gamma = (E_0, \emptyset)$, which translates to $\alpha = (E_0, \emptyset, \emptyset)$. Otherwise, $\gamma = \gamma' \vdash_C (E_k, R_k)$ and by IH there exists α' that satisfies the claim for γ', and also $\to_{C_{k-1}} \subseteq >$. Let $C_k = C_{k-1}$ if the final deduction is the result of any rule except ORIENT, and $C_k = C_{k-1} \cup \{s \to t\}$ otherwise, with $\{s \to t\} = R_k - R_{k-1}$. We claim $\alpha = \alpha' \vdash_{\mathcal{A}} (E_k, R_k, C_k)$ and show this is a correct execution of $\mathcal{A}$. This is trivial for rules other than ORIENT, since their side conditions do not mention the constraint systems. Otherwise, $s > t$ and $\xrightarrow{+}_{C_{k-1}} \subseteq >$ which implies $\xrightarrow{+}_{C_k} \subseteq >$. This in turn implies that C_k is terminating because its rules are compatible with the reduction order $>$.

This theorem demonstrates the existence of a successful execution of $\mathcal{A}$ for every successful finite execution of C. But note that that the rule ORIENT in $\mathcal{A}$ can orient an equation $s = t$ in either direction when both $C \cup \{s \to t\}$ and $C \cup \{t \to s\}$ are terminating systems. Consequently, an execution of $\mathcal{A}$ as defined above will fail if a poor decision is made during orientation. The ability to construct a successful execution relies on a non-deterministic orientation choice. Deterministically, an execution of $\mathcal{A}$ becomes a binary tree in which each node is an instance of the rule ORIENT. In practice, we must *search* for a successful execution. We ensure discovery of such an execution (corresponding to a path from the root $(E_0, \emptyset)$) by fairly advancing each of the individual executions.

4   Implementation

We have implemented our modified Knuth-Bendix completion procedure in a 7000-line Ocaml program called SLOTHROP.[1] The implementation is based on a particular completion strategy developed and proved correct by Huet [7], and later by Bachmair using the inference system $\mathcal{C}$ [2]. The implementation itself is originally based on an ML implementation of Huet's algorithm by Baader and Nipkow [1] and makes use of data structures programmed by Filliâtre [5].

The main technical challenge in the implementation is with the ORIENT rule. As is well known, determining whether or not a term rewriting system terminates is undecidable in general. However, modern termination-checking tools, such as APROVE [6], succeed in proving many systems terminating or nonterminating with almost alarming success. In our implementation, we take advantage of this success and use APROVE as an oracle to answer queries about the termination of constraint rewriting systems in each orientation step. If APROVE fails to prove a system terminating or nonterminating, we treat it as a nonterminating system and delay its treatment. However, the array of techniques used by APROVE to show termination includes recursive path orders among many others, so there is little difficulty recognizing the termination of systems compatible with such an order. Furthermore, since APROVE is able to prove termination of systems that are not compatible with a path order, SLOTHROP can find convergent completions of theories other modern completion tools (e.g., WALDMEISTER) cannot. One example of such a theory is given in Sect. 5.

Integrating a separate termination checker also provides separation-of-concerns benefits for theorem proving. As the power and speed of the APROVE tool, so does SLOTHROP. This also provides the opportunity to leverage other termination checkers with different properties (e.g., one which is faster but less powerful might be useful for simple theories).

Another important aspect of our implementation is the manner in which different branches of executions are explored. When APROVE determines that some equation $s = t$ can be oriented either way, both branches are explored. Implementation of this exploration is critical to performance. The binary tree of executions is potentially infinite, and branches whenever orderings exist that are compatible with both orientations of an equation.

A breadth-first search of the branches is sufficient for partial completeness; if there is some successful finite execution corresponding to a branch on the tree, it will eventually be expanded. In practice, however, this strategy spends too much time in uninteresting areas of the search space, and prevents SLOTHROP from finding completions for any but the most modest theories in a reasonable amount of time. A more effective strategy is a best-first search in which the next execution to advance is chosen based on a cost function defined by

$$\mathrm{cost}(E, R, C) = \mathrm{size}(C) + \mathrm{size}(E) + \mathrm{size}(\Gamma(R)),$$

---

[1] SLOTHROP is available online at `http://cl.cse.wustl.edu/` on the software page.

where $\Gamma(R)$ denotes the set of all nontrivial critical pairs of R. With this strategy, size(C) can be thought of as the cost to reach the current intermediate step in the execution and size(E) + size($\Gamma(R)$) as a heuristic estimate for the cost to find a convergent completion.

In some cases, APROVE is unable to prove termination of a system with either orientation of a particular equation. Here, we do not discard the system entirely, but attempt to orient other equations in hope that the previously unorientable equation will simplify into an orientable (or trivial) one. In the current implementation of SLOTHROP, treatment of such systems is delayed until others are explored which can be proved terminating or nonterminating. This heuristic is suitable for simple systems, but better ones are needed for more difficult theories.

5   Performance and New Results

SLOTHROP is capable of completing a variety of theories fully automatically in a modest amount of time. For example, the standard 10-rule completion of the group axioms is discovered in under 3 seconds on a modern desktop PC. On the way to this completion, it encounters 27 orientations, roughly half of which are not trivially nonterminating and must be verified with APROVE. On the execution branch that leads to a completion, however, only two orientation steps are required. SLOTHROP automatically completes the theory of groups plus a single endomorphism (GE_1) in under 10 seconds, requiring about 100 calls to APROVE. A large theory with 21 equations corresponding to propositional proof simplification rules [12] is considerably more difficult to complete because of the number of orientations. Nonetheless, SLOTHROP does find a completion without user intervention after about 7 hours and 3000 calls to APROVE.

The majority of SLOTHROP's running time is spent waiting for calls to APROVE. Although we have encountered many examples of rewriting systems which APROVE can show terminating after a prohibitively long amount of time, in practice we have found that it is uncommon for such difficult systems to appear on the branch of a successful execution. Most calls to APROVE that occur on successful branches return in under 2 seconds. Completeness of SLOTHROP can be exchanged for performance enhancements by calling APROVE with a short timeout. The above completions were obtained with a 5-second timeout.

Since SLOTHROP is not restricted to a given reduction ordering, it can also search for multiple completions of a given theory. For example, it finds two

$1 * x = x$	$x^{-1} * x = 1$	$(x * y) * z = x * (y * z)$
$f(x * y) = f(x) * f(y)$	$g(x * y) = g(x) * g(y)$	$f(x) * g(y) = g(y) * f(x)$

Fig. 3. The Theory of Two Commuting Group Endomorphisms (CGE_2)

completions of the basic group axioms corresponding to both orientations of the associativity rule. It also finds four completions of GE_1 corresponding to the orientations of the associativity endomorphism rules. It also discovers a number of other larger completions of the same theory in which endomorphism are oriented differently depending on the context.

Additionally, a convergent completion can be obtained by SLOTHROP for the theory of two commuting group endomorphisms (CGE_2), shown in Fig. 5. The reader may verify that no RPO or KBO is compatible with the theory (in particular, the final commutativity rule). A completion was recently obtained for the first time by hand [11] — rules derived from critical pairs were manually oriented, local confluence checked, and termination of the resulting system verified by APROVE.

Using unfailing completion [3], WALDMEISTER is able to complete CGE_2 as well, but constructs a larger system which is ground-confluent only — i.e, it contains equations as well as rewrite rules. This system is often less helpful than a small convergent completion, for example, in characterizing the normal forms of the system for algebraic proof mining [12]. Furthermore, WALDMEISTER does not appear to be able to find this ground-convergent completion fully automatically; a carefully selected Knuth-Bendix ordering (given in [11]) must be provided. SLOTHROP is able to find the convergent completion with no input from the user other than the theory itself. (This still takes more than an hour, however, even using the heuristic described in Sect. 4.)

6   Conclusion and Future Work

We have presented a new variant on Knuth-Bendix completion which does not require the user to provide a reduction ordering to orient identities. The procedure is correct and complete, but only for finite executions. An implementation of the procedure, called SLOTHROP, can find convergent completions for a number of interesting theories without any input from the user, including one (CGE_2) which cannot be obtained by any existing tool.

A primary goal of future work is to increase the efficiency of SLOTHROP. Basic heuristic search techniques have made the algorithm feasible for many theories, but it is still prohibitively slow for large theories — completion of the CGE_3 has not yet been achieved. The performance of SLOTHROP also does not approach that of well-tuned equational theorem provers such as WALDMEISTER for most tasks. Modern search and learning techniques, e.g. as developed for SAT, may be applicable to the search for a convergent completion. Finally, we would like to explore extensions to termination checking techniques to allow proofs to be constructed incrementally. This may significantly decrease the amortized time to prove a series of term rewriting systems terminating, since SLOTHROP tends to make a number of successive calls on rewrite systems whose rules form increasing chains.

Infinite Executions. While the provided argument for completeness carries to infinite executions essentially unmodified, it is not the case that all non-failing runs are successful. In particular, termination of the infinite union of the intermediate constraint systems does not follow from termination of the individual systems; this is because in general the union of an infinite number of finite, terminating rewrite systems is not itself terminating. For example, consider the family of (string) rewriting systems $R_j = \cup_{0 \leq i \leq j} \{fg^i f \to fg^{i+1}\}$. For any $k \in \mathbb{N}$ it is easy to see that $\cup_{0 \leq j \leq k} R_j$ is terminating. But it is not the case that $\cup_{j \in \mathbb{N}} R_j$ is terminating, for it contains the infinite derivation $ff \to fgf \to fggf \to \cdots$.

Instead, it must be shown in a proof of correctness for the infinite case that *some* successful branch of execution always exists, and that it will always be found in the search for a completion. The authors believe the modified procedure to be correct in the infinite case, making it usable as a semidecision procedure for theories. We have a proof sketch of correctness for the infinite case of the system $\mathcal{A}$, and a complete proof is in progress.

Acknowledgements. The authors are especially appreciative of Peter Schneider-Kamp and the APROVE team for their willingness to make changes to their system necessary for integration with SLOTHROP. We thank Stephan Falke for testing SLOTHROP and providing a theory that requires delayed treatment of unorientable equations. We also thank Li-Yang Tan and the referees for helpful comments on drafts.

References

1. F. Baader and T. Nipkow. *Term Rewriting and All That.* Cambridge University Press, 1998.
2. L. Bachmair. *Canonical Equational Proofs.* Progress in Theoretical Computer Science. Birkhäuser, 1991.
3. L. Bachmair, N. Dershowitz, and D.A. Plaisted. Completion Without Failure. In *Resolution of Equations in Algebraic Structures*, volume 2, Rewriting Techniques, pages 1–30. Academic Press, 1989.
4. Evelyne Contejean, Claude Marché, and Xavier Urbain. CiME3, 2004. Available at http://cime.lri.fr/.
5. Jean-Christophe Filliâtre. Ocaml data structures. Available at http://www.lri.fr/~filliatr/software.en.html.
6. J. Giesl, R. Thiemann, P. Schneider-Kamp, and S. Falke. Automated Termination Proofs with AProVE. In V. van Oostrom, editor, *the 15th International Conference on Rewriting Techniques and Applications*, pages 210–220. Springer, 2004.
7. G. Huet. A Complete Proof of Correctness of the Knuth-Bendix Completion Algorithm. *Journal of Computer and System Science*, 23(1):11–21, 1981.
8. D. Knuth and P. Bendix. Simple Word Problems in Universal Algebras. In J. Leech, editor, *Computational Problems in Abstract Algebra*, pages 263–297. Pergamon Press, 1970.
9. B. Löchner and T. Hillenbrand. The Next Waldmeister Loop. In A. Voronkov, editor, *18th International Conference on Automated Deduction*, pages 486–500, 2002.

10. A. Sattler-Klein. About Changing the Ordering During Knuth-Bendix Completion. In P. Enjalbert E. W. Mayr and K. W. Wagner, editors, *Symposium on Theoretical Aspects of Computer Science*, volume 775 of *LNCS*, pages 176–186. Springer, 1994.

11. A. Stump and B. Löchner. Knuth-Bendix Completion of Theories of Commuting Group Endomorphisms. *Information Processing Letters*, 2006. To appear.

12. I. Wehrman and A. Stump. Mining Propositional Simplification Proofs for Small Validating Clauses. In A. Armando and A. Cimatti, editors, *3rd International Workshop on Pragmatics of Decision Procedures in Automated Reasoning*, 2005.

Automated Termination Analysis for Haskell: From Term Rewriting to Programming Languages⋆

Jürgen Giesl, Stephan Swiderski, Peter Schneider-Kamp, and René Thiemann

LuFG Informatik 2, RWTH Aachen, Ahornstr. 55, 52074 Aachen, Germany
{giesl, swiderski, psk, thiemann}@informatik.rwth-aachen.de

Abstract. There are many powerful techniques for automated termination analysis of term rewriting. However, up to now they have hardly been used for real programming languages. We present a new approach which permits the application of existing techniques from term rewriting in order to prove termination of programs in the functional language Haskell. In particular, we show how termination techniques for ordinary rewriting can be used to handle those features of Haskell which are missing in term rewriting (e.g., lazy evaluation, polymorphic types, and higher-order functions). We implemented our results in the termination prover AProVE and successfully evaluated them on existing Haskell-libraries.

1 Introduction

We show that termination techniques for term rewrite systems (TRSs) are also useful for termination analysis of programming languages like Haskell. Of course, any program can be translated into a TRS, but in general, it is not obvious how to obtain TRSs *suitable for existing automated termination techniques*. Adapting TRS-techniques for termination of Haskell is challenging for the following reasons:

- Haskell has a *lazy evaluation* strategy. However, most TRS-techniques ignore such evaluation strategies and try to prove that *all* reductions terminate.
- Defining equations in Haskell are handled from top to bottom. In contrast for TRSs, *any* rule may be used for rewriting.
- Haskell has polymorphic types, whereas TRSs are untyped.
- In Haskell-programs with infinite data objects, only certain functions are terminating. But most TRS-methods try to prove termination of *all* terms.
- Haskell is a *higher-order* language, whereas most automatic termination techniques for TRSs only handle first-order rewriting.

There are only few techniques for automated termination analysis of functional programs. Methods for first-order languages with strict evaluation strategy were developed in [5,10,16]. For higher-order languages, [1,3,17] study how to ensure termination by typing and [15] defines a restricted language where all evaluations terminate. A successful approach for automated termination proofs for a small Haskell-like language was developed in [11]. (A related technique is [4], which handles outermost evaluation of untyped first-order rewriting.) How-

---

⋆ Supported by the Deutsche Forschungsgemeinschaft DFG under grant GI 274/5-1.

F. Pfenning (Ed.): RTA 2006, LNCS 4098, pp. 297–312, 2006.

ever, these are all "stand-alone" methods which do not allow the use of modern termination techniques from term rewriting. In our approach we build upon the method of [11], but we adapt it in order to make TRS-techniques applicable.[1]

We recapitulate Haskell in Sect. 2 and introduce our notion of "termination". To analyze termination, our method first generates a corresponding *termination graph* (similar to the "termination tableaux" in [11]), cf. Sect. 3. But in contrast to [11], then our method transforms the termination graph into *dependency pair problems* which can be handled by existing techniques from term rewriting (Sect. 4). Our approach in Sect. 4 can deal with any termination graph, whereas [11] can only handle termination graphs of a special form ("without crossings"). We implemented our technique in the termination prover AProVE [9], cf. Sect. 5.

2   Haskell

We now give the syntax and semantics for a subset of Haskell which only uses certain easy patterns and terms (without "λ"), and function definitions without conditions. Any Haskell-program (without type classes and built-in data structures)[2] can automatically be transformed into a program from this subset [14].[3] For example, in our implementation lambda abstractions are removed by replacing every Haskell-term "$\setminus t_1...t_n \rightarrow t$" with the free variables $x_1, \ldots x_m$ by "$f\,x_1 \ldots x_m$". Here, f is a new function symbol with the defining equation $f\,x_1...x_m\,t_1...t_n = t$.

2.1   Syntax of Haskell

In our subset of Haskell, we only permit user-defined data structures such as

data Nats = Z | S Nats   data List a = Nil | Cons a (List a)

These data-declarations introduce two *type constructors* Nats and List of arity 0 and 1, respectively. So Nats is a type and for every type τ, "List τ" is also a type representing lists with elements of type τ. Moreover, there is a pre-defined binary type constructor $\rightarrow$ for function types. Since Haskell's type system is polymorphic, it also has *type variables* like a which stand for any type.

For each type constructor like Nats, a data-declaration also introduces its *data constructors* (e.g., Z and S) and the types of their arguments. Thus, Z has arity 0 and is of type Nats and S has arity 1 and is of type Nats $\rightarrow$ Nats.

Apart from data-declarations, a program has function declarations. Here, "from x" generates the infinite list of numbers starting with x and "take $n\,xs$" returns the first n elements of xs. The type of from is "List Nats" and take has type "Nats $\rightarrow$ (List a) $\rightarrow$ (List a)" where $\tau_1 \rightarrow \tau_2 \rightarrow \tau_3$ stands for $\tau_1 \rightarrow (\tau_2 \rightarrow \tau_3)$.

---

[1] Alternatively, one could simulate Haskell's evaluation strategy by *context-sensitive rewriting* (CSR), cf. [6]. But termination of CSR is hard to analyze automatically.

[2] See Sect. 5 for an extension to type classes and pre-defined data structures.

[3] Of course, it would be possible to restrict ourselves to programs from an even smaller "core"-Haskell subset. However, this would not simplify the subsequent termination analysis any further. In contrast, the resulting programs would usually be less readable, which would make interactive termination proofs harder.

$$\text{from } x = \text{Cons } x \, (\text{from } (\text{S } x))$$

$$\begin{aligned}
\text{take } Z \; xs &= \text{Nil} \\
\text{take } n \; \text{Nil} &= \text{Nil} \\
\text{take } (\text{S } n) \, (\text{Cons } x \; xs) &= \text{Cons } x \, (\text{take } n \; xs)
\end{aligned}$$

In general, function declarations have the form "$f \, \ell_1 \ldots \ell_n = r$". The function symbols f on the "outermost" position of left-hand sides are called *defined*. So the set of function symbols is the disjoint union of the (data) constructors and the defined function symbols. All defining equations for f must have the same number of arguments n (called f's *arity*). The right-hand side r is an arbitrary *term*, whereas $\ell_1, \ldots, \ell_n$ are special terms, so-called *patterns*. Moreover, the left-hand side must be *linear*, i.e., no variable may occur more than once in "$f \, \ell_1 \ldots \ell_n$".

The set of *terms* is the smallest set containing all variables, function symbols, and *well-typed* applications $(t_1 \, t_2)$ for terms t_1 and t_2. As usual, "$t_1 \, t_2 \, t_3$" stands for "$((t_1 \, t_2) \, t_3)$". The set of *patterns* is the smallest set with all variables and terms "$c \, t_1 \ldots t_n$" where c is a constructor of arity n and $t_1, \ldots, t_n$ are patterns.

The positions of t are $Pos(t) = \{\varepsilon\}$ if t is a variable or function symbol. Otherwise, $Pos(t_1 \, t_2) = \{\varepsilon\} \cup \{1\, \pi \mid \pi \in Pos(t_1)\} \cup \{2\, \pi \mid \pi \in Pos(t_2)\}$. As usual, we define $t|_\varepsilon = t$ and $(t_1 \, t_2)|_{i\, \pi} = t_i|_\pi$. The *head* of t is $t|_{1^n}$ where n is the maximal number with $1^n \in Pos(t)$. So the head of $t = \text{take } n \; xs$ (i.e., "$(\text{take } n) \; xs$") is $t|_{11} = \text{take}$.

2.2 Operational Semantics of Haskell

Given an underlying program, for any term t we define the position $\mathbf{e}(t)$ where the next evaluation step has to take place due to Haskell's outermost strategy. So in most cases, $\mathbf{e}(t)$ is the top position ε. An exception are terms "$f \, t_1 \ldots t_n \, t_{n+1} \ldots t_m$" where $\text{arity}(f) = n$ and $m > n$. Here, f is applied to too many arguments. Thus, one considers the subterm "$f \, t_1 \ldots t_n$" at position 1^{m-n} to find the evaluation position. The other exception is when one has to evaluate a subterm of $f \, t_1 \ldots t_n$ in order to check whether a defining f-equation $\ell = r$ will then become applicable on top position. We say that an equation $\ell = r$ from the program is *feasible* for a term t and define the corresponding *evaluation position* $\mathbf{e}_\ell(t)$ w.r.t. ℓ if either

(a) ℓ matches t (then we define $\mathbf{e}_\ell(t) = \varepsilon$),   or
(b) for the leftmost outermost position π where $\text{head}(\ell|_\pi)$ is a constructor and where $\text{head}(\ell|_\pi) \neq \text{head}(t|_\pi)$, the symbol $\text{head}(t|_\pi)$ is defined or a variable. Then $\mathbf{e}_\ell(t) = \pi$.

Since Haskell considers the order of the program's equations, t is evaluated below the top (on position $\mathbf{e}_\ell(t)$) whenever (b) holds for the *first* feasible equation $\ell = r$ (even if an evaluation with a *subsequent* defining equation would be possible at top position). Thus, this is no ordinary leftmost outermost evaluation strategy.

Definition 1 (Evaluation Position $\mathbf{e}(t)$). *For any term t, we define*

$$\mathbf{e}(t) = \begin{cases} 1^{m-n}\, \pi, & \textit{if } t = f \, t_1 \ldots t_n \, t_{n+1} \ldots t_m, \; f \textit{ is defined, } m > n = \text{arity}(f), \\ & \textit{and } \pi = \mathbf{e}(\, f \, t_1 \ldots t_n \,) \\ \mathbf{e}_\ell(t)\, \pi, & \textit{if } t = f \, t_1 \ldots t_n, \, f \textit{ is defined, } n = \text{arity}(f), \textit{ there are feasible} \\ & \textit{equations for } t \textit{ (the first is ``}\ell = r\textit{''), } \mathbf{e}_\ell(t) \neq \varepsilon, \textit{ and } \pi = \mathbf{e}(t|_{\mathbf{e}_\ell(t)}) \\ \varepsilon, & \textit{otherwise} \end{cases}$$

If $t = \text{take } u \, (\text{from } m)$ and $s = \text{take } (\text{S } n) \, (\text{from } m)$, then $t|_{\mathbf{e}(t)} = u$ and $s|_{\mathbf{e}(s)} = \text{from } m$.

We now present Haskell's operational semantics by defining the *evaluation relation* $\to_H$. For any term t, it performs a rewrite step on position $\mathbf{e}(t)$ using the *first* applicable defining equation of the program. So terms like "$x\,Z$" or "take Z" are normal forms: If the head of t is a variable or if a symbol is applied to too few arguments, then $\mathbf{e}(t) = \varepsilon$ and no rule rewrites t at top position. Moreover, a term $s = f\,s_1\ldots s_m$ with a defined symbol f and $m \geq \text{arity}(f)$ is a normal form if no equation in the program is feasible for s. If $\text{head}(s|_{\mathbf{e}(s)})$ is a defined symbol g, then we call s an *error term* (i.e., then g is not "completely" defined).

For terms $t = c\,t_1 \ldots t_n$ with a constructor c of arity n, we also have $\mathbf{e}(t) = \varepsilon$ and no rule rewrites t at top position. However, here we permit rewrite steps below the top, i.e., $t_1, \ldots, t_n$ may be evaluated with $\to_H$. This corresponds to the behavior of Haskell-interpreters like Hugs which evaluate terms until they can be displayed as a string. To transform data objects into strings, Hugs uses a function "show". This function can be generated automatically for user-defined types by adding "deriving Show" behind the data-declarations. This show-function would transform every data object "$c\,t_1 \ldots t_n$" into the string consisting of "c" and of show t_1, ..., show t_n. Thus, show would require that all arguments of a term with a constructor head have to be evaluated.

Definition 2 (Evaluation Relation $\to_H$). *We have $t \to_H s$ iff either*

(1) t rewrites to s on the position $\mathbf{e}(t)$ using the first equation of the program whose left-hand side matches $t|_{\mathbf{e}(t)}$,   or
(2) $t = c\,t_1 \ldots t_n$ for a constructor c of arity n, $t_i \to_H s_i$ for some $1 \leq i \leq n$, and $s = c\,t_1 \ldots t_{i-1}\,s_i\,t_{i+1} \ldots t_n$

For example, we have the infinite evaluation from $m \to_H$ Cons m (from (S m)) $\to_H$ Cons m (Cons (S m) (from (S m))) $\to_H \ldots$ On the other hand, the following evaluation is finite: take (S Z) (from m) $\to_H$ take (S Z) (Cons m (from (S m))) $\to_H$ Cons m(take Z (from (S m))) $\to_H$ Cons m Nil.

The reason for permitting non-ground terms in Def. 1 and 2 is that our termination method in Sect. 3 evaluates Haskell *symbolically*. Here, variables stand for arbitrary *terminating* terms. Def. 3 introduces our notion of termination.

Definition 3 (H-Termination). *A ground term t is H-terminating iff*

(a) t does not start an infinite evaluation $t \to_H \ldots$,
(b) if $t \to_H^* (f\,t_1 \ldots t_n)$ for a defined function symbol f, $n < \text{arity}(f)$, and the term t' is H-terminating, then $(f\,t_1 \ldots t_n\,t')$ is also H-terminating,   and
(c) if $t \to_H^* (c\,t_1 \ldots t_n)$ for a constructor c, then $t_1, \ldots, t_n$ are also H-terminating.

A term t is H-terminating iff $t\sigma$ is H-terminating for all substitutions σ with H-terminating ground terms (of the correct types). These substitutions σ may also introduce new defined function symbols with arbitrary defining equations.

So a term is only H-*terminating* if all its applications to H-terminating terms H-terminate, too. Thus, "from" is not H-terminating, as "from Z" has an infinite evaluation. But "take u (from m)" is H-terminating: when instantiating u and m by H-terminating ground terms, the resulting term has no infinite evaluation.

To illustrate that one may have to add defining equations to examine H-termination, consider the function nonterm of type Bool $\to$ (Bool $\to$ Bool) $\to$ Bool:

$$\text{nonterm True } x = \text{True} \qquad \text{nonterm False } x = \text{nonterm } (x \text{ True}) x \qquad (1)$$

The term "nonterm False x" is not H-terminating: one obtains an infinite evaluation if one instantiates x by the function mapping all arguments to False. In full Haskell, such functions can of course be represented by lambda terms and indeed, "nonterm False $(\backslash y \to \text{False})$" starts an infinite evaluation.

3   From Haskell to Termination Graphs

Our goal is to prove H-termination of a *start term* t. By Def. 3, H-*termination* of t implies that $t\sigma$ is H-terminating for all substitutions σ with H-terminating ground terms. Thus, t represents a (usually infinite) set of terms and we want to prove that they are all H-terminating. Without loss of generality, we can restrict ourselves to normal ground substitutions σ, i.e., substitutions where $\sigma(x)$ is a ground term in normal form w.r.t. $\to_H$ for all variables x in t.

Regard the start term $t = \text{take } u \, (\text{from } m)$. A naive approach would be to consider the defining equations of all needed functions (i.e., take and from) as rewrite rules. However, this disregards Haskell's lazy evaluation strategy. So due to the non-terminating rule for "from", we would fail to prove H-termination of t.

Therefore, our approach starts evaluating the start term a few steps. This gives rise to a so-called *termination graph*. Instead of transforming defining Haskell-equations into rewrite rules, we then transform the termination graph into rewrite rules. The advantage is that the initial evaluation steps in this graph take the evaluation strategy and the types of Haskell into account and therefore, this is also reflected in the resulting rewrite rules.

To construct a termination graph for the start term t, we begin with the graph containing only one single node, marked with t. Similar to [11], we then apply *expansion rules* repeatedly to the leaves of the graph in order to extend it by new nodes and edges. As usual, a *leaf* is a node with no outgoing edges. We have obtained a *termination graph* for t if no expansion rules is applicable to its leaves anymore. Afterwards, we try to prove H-termination of all terms occurring in the termination graph, cf. Sect. 4. We now describe our five expansion rules intuitively using Fig. 1. Their formal definition is given in Def. 4.

When constructing termination graphs, the goal is to *evaluate* terms. However, $t = \text{take } u \, (\text{from } m)$ cannot be evaluated with $\to_H$, since it has a variable u on its evaluation position $\mathbf{e}(t)$. The evaluation can only continue if we know how u is going to be instantiated. Therefore, the first expansion

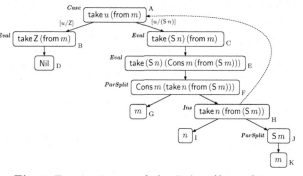

Fig. 1. Termination graph for "take u (from m)"

rule is called **Case Analysis** (or "**Case**", for short). It adds new child nodes where u is replaced by all terms of the form $(c\,x_1 \ldots x_n)$. Here, c is a constructor of the appropriate type and $x_1, \ldots, x_n$ are fresh variables. The edges to these children are labelled with the respective substitutions $[u/(c\,x_1 \ldots x_n)]$. In our example, u is a variable of type Nats. Therefore, the **Case**-rule adds two child nodes B and C to our initial node A, where u is instantiated by Z and by (S n), respectively. Since the children of A were generated by the **Case**-rule, we call A a "**Case**-node". Every node in the graph has the following property: If all its children are marked with H-terminating terms, then the node itself is also marked by a H-terminating term. Indeed, if the terms in nodes B and C are H-terminating, then the term in node A is H-terminating as well.

Now the terms in nodes B and C can indeed be evaluated. Therefore, the **Evaluation**-rule ("**Eval**") adds the nodes D and E resulting from one evaluation step with $\rightarrow_H$. Moreover, E is also an **Eval**-node, since its term can be evaluated further to the term in node F. So the **Case**- and **Eval**-rule perform a form of *narrowing* that respects the evaluation strategy and the types of Haskell.

The term Nil in node D cannot be evaluated and therefore, D is a leaf of the termination graph. But the term "Cons m (take n (from (S m)))" in node F may be evaluated further. Whenever the head of a term is a constructor like Cons or a variable,[4] then evaluations can only take place on its arguments. We use a **Parameter Split**-rule ("**ParSplit**") which adds new child nodes with the arguments of such terms. Thus, we obtain the nodes G and H. Again, H-termination of the terms in G and H obviously implies H-termination of the term in node F.

The node G remains a leaf since its term m cannot be evaluated further for any normal ground instantiation. For node H, we could continue by applying the rules **Case**, **Eval**, and **ParSplit** as before. However, in order to obtain finite graphs (instead of infinite trees), we also have an **Instantiation**-rule ("**Ins**"). Since the term in node H is an *instance* of the term in node A, one can draw an *instantiation edge* from the instantiated term to the more general term (i.e., from H to A). We depict instantiation edges by dashed lines. These are the only edges which may point to already existing nodes (i.e., one obtains a tree if one removes the instantiation edges from a termination graph).

To guarantee that the term in node H is H-terminating whenever the terms in its child nodes are H-terminating, the **Ins**-rule has to ensure that one only uses instantiations with H-terminating terms. In our example, the variables u and m of node A are instantiated with the terms n and (S m), respectively. Therefore, in addition to the child A, the node H gets two more children I and J marked with n and (S m). Finally, the **ParSplit**-rule adds J's child K, marked with m.

Now we consider a different start term, viz. "take". If a defined function has "too few" arguments, then by Def. 3 we have to apply it to additional H-terminating arguments in order to examine H-termination. Therefore, we have a **Variable Expansion**-rule ("**VarExp**") which would add a child marked with "take x" for a fresh variable x. Another application of **VarExp** gives "take $x\,xs$". The remaining termination graph is constructed by the rules discussed before.

---

[4] The reason is that "$x\,t_1 \ldots t_n$" H-terminates iff the terms $t_1, \ldots, t_n$ H-terminate.

Definition 4 (Termination Graph). *Let G be a graph with a leaf marked with the term t. We say that G can be* expanded *to G' (denoted "$G \Rightarrow G'$") if G' results from G by adding new **child** nodes marked with the elements of $\mathbf{ch}(t)$ and by adding edges from t to each element of $\mathbf{ch}(t)$. Only in the **Ins**-rule, we also permit to add an edge to an already existing node, which may then lead to cycles. All edges are marked by the identity substitution unless stated otherwise.*

___**Eval:**___ $\mathbf{ch}(t) = \{\tilde{t}\}$, *if $t = (f\, t_1 \ldots t_n)$, f is a defined symbol, $n \geq \text{arity}(f)$, $t \rightarrow_H \tilde{t}$*

___**Case:**___ $\mathbf{ch}(t) = \{t\sigma_1, \ldots, t\sigma_k\}$, *if $t = (f\, t_1 \ldots t_n)$, f is a defined function symbol, $n \geq \text{arity}(f)$, $t|_{\mathbf{e}(t)}$ is a variable x of type "$d\,\tau_1 \ldots \tau_m$" for a type constructor d, the type constructor d has the data constructors c_i of arity n_i (where $1 \leq i \leq k$), and $\sigma_i = [x/(c_i\, x_1 \ldots x_{n_i})]$ for fresh pairwise different variables $x_1, \ldots, x_{n_i}$. The edge from t to $t\sigma_i$ is marked with the substitution σ_i.*

___**VarExp:**___ $\mathbf{ch}(t) = \{t\,x\}$, *if $t = (f\, t_1 \ldots t_n)$, f is a defined function symbol, $n < \text{arity}(f)$, x is a fresh variable*

___**ParSplit:**___ $\mathbf{ch}(t) = \{t_1, \ldots, t_n\}$ *if $t = (c\, t_1 \ldots t_n)$, c is a constructor or variable, $n > 0$*

___**Ins:**___ $\mathbf{ch}(t) = \{s_1, \ldots, s_m, \tilde{t}\}$, *if $t = (f\, t_1 \ldots t_n)$, t is not an error term, f is a defined symbol, $n \geq \text{arity}(f)$, $t = \tilde{t}\sigma$ for some term $\tilde{t}$, $\sigma = [x_1/s_1, \ldots, x_m/s_m]$. Moreover, either $\tilde{t} = (x\,y)$ for fresh variables x and y, or $\tilde{t}$ is an **Eval**-node, or $\tilde{t}$ is a **Case**-node and all paths starting in $\tilde{t}$ reach an **Eval**-node or a leaf with an error term after traversing only **Case**-nodes.[5] The edge from t to $\tilde{t}$ is called an* instantiation edge.

*If the graph already contained a node marked with $\tilde{t}$, then we permit to reuse this node in the **Ins**-rule. So in this case, instead of adding a new child marked with $\tilde{t}$, one may add an edge from t to the already existing node $\tilde{t}$.*

Let G_t be the graph with a single node marked with t and no edges. G is a termination graph *for t iff $G_t \Rightarrow^* G$ and G is in normal form w.r.t. $\Rightarrow$.*

If one disregards **Ins**, then for each leaf there is at most one rule applicable.[6] However, the **Ins**-rule introduces indeterminism. Instead of applying the **Case**-rule on node A in Fig. 1, we could also apply **Ins** and generate an instantiation edge to a new node with $\tilde{t} = (\text{take}\, u\, ys)$. Since the instantiation is $[ys/(\text{from}\, m)]$, node A would get an additional child node marked with the non-H-terminating term $(\text{from}\, m)$. Then our approach in Sect. 4 which tries to prove H-termination of *all* terms in the termination graph would fail, whereas it succeeds for the graph in Fig. 1. Therefore, in our implementation we developed a heuristic for constructing termination graphs which tries to avoid unnecessary applications of **Ins** (since applying **Ins** means that one has to prove H-termination of more terms).

An instantiation edge to $\tilde{t} = (x\,y)$ is needed to get termination graphs for functions like tma which are applied to "too many" arguments in recursive calls.

$$\text{tma}\,(S\,m) = \text{tma}\,m\,m \tag{2}$$

---

[5] This ensures that every cycle of the graph contains at least one **Eval**-node.

[6] No rule is applicable to leaves with variables, constructors of arity 0, or error terms.

Here, tma has the type Nats → a. We obtain the termination graph in Fig. 2. After applying **Case** and **Eval**, we result in "tma $m\,m$" in node D which is not an instance of the start term "tma n" in node A. Of course, we could continue with **Case** and **Eval** infinitely often, but to obtain a termination graph, at some point we need to apply the **Ins**-rule. Here, the only possibility is to regard $t = (\text{tma}\,m\,m)$ as an instance of the term $\tilde{t} = (x\,y)$. Thus, we obtain an instantiation edge to the new node E. As the instantiation is $[x/(\text{tma}\,m), y/m]$, we get additional child nodes F and G marked with "tma m" and m, respectively. Now we can "close" the graph, since "tma m" is an instance of the start term "tma n" in node A. So the

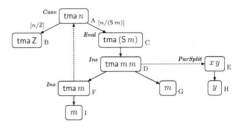

Fig. 2. Termination graph for "tma n"

instantiation edge to the special term $(x\,y)$ is used to remove "superfluous" arguments (i.e., it permits to go from "tma $m\,m$" in node D to "tma m" in node F). Thm. 5 shows that by the expansion rules of Def. 4 one can always obtain normal forms.[7]

Theorem 5 (Existence of Termination Graphs). *The relation* ⇒ *is normalizing, i.e., for any term t there exists a termination graph.*

4   From Termination Graphs to DP Problems

Now we present a method to prove H-termination of all terms in a termination graph. To this end, we want to use existing techniques for termination analysis of term rewriting. One of the most popular techniques for TRSs is the *dependency pair* (DP) method [2]. In particular, the DP method can be formulated as a general framework which permits the integration and combination of *any* termination technique for TRSs [7]. This *DP framework* operates on so-called *DP problems* $(\mathcal{P}, \mathcal{R})$. Here, $\mathcal{P}$ and $\mathcal{R}$ are TRSs that may also have rules $\ell \to r$ where r contains extra variables not occurring in ℓ. $\mathcal{P}$'s rules are called *dependency pairs*. The goal of the DP framework is to show that there is no infinite *chain*, i.e., no infinite reduction $s_1\sigma_1 \to_{\mathcal{P}} t_1\sigma_1 \to_{\mathcal{R}}^* s_2\sigma_2 \to_{\mathcal{P}} t_2\sigma_2 \to_{\mathcal{R}}^* \dots$ where $s_i \to t_i \in \mathcal{P}$ and σ_i are substitutions. In this case, the DP problem $(\mathcal{P}, \mathcal{R})$ is called *finite*. See [7] for an overview on techniques to prove finiteness of DP problems.[8]

Instead of transforming termination graphs into TRSs, the information available in the termination graph can be better exploited if one transforms these

---

[7] All proofs can be found at http://aprove.informatik.rwth-aachen.de/eval/Haskell/.

[8] In the DP literature, one usually does not regard rules with extra variables on right-hand sides, but almost all existing termination techniques for DPs can also be used for such rules. (For example, finiteness of such DP problems can often be proved by eliminating the extra variables by suitable *argument filterings* [2].)

graphs into DP problems, cf. the end of this section. In this way, we also do not have to impose any restrictions on the form of the termination graph (as in [11] where one can only analyze certain start terms which lead to termination graphs "without crossings"). Then finiteness of the resulting DP problems implies H-termination of all terms in the termination graph.

Note that termination graphs still contain higher-order terms (e.g., applications of variables to other terms like "$x\,y$" and partial applications like "take u"). Since most methods and tools for automated termination analysis only operate on first-order TRSs, we translate higher-order terms into *applicative* first-order terms containing just variables, constants, and a binary symbol ap for function application. So terms like "$x\,y$", "take u", and "take $u\,xs$" are transformed into the first-order terms $\mathsf{ap}(x, y)$, $\mathsf{ap}(\mathsf{take}, u)$, and $\mathsf{ap}(\mathsf{ap}(\mathsf{take}, u), xs)$, respectively. As shown in [8], the DP framework is well suited to prove termination of applicative TRSs automatically. To ease readability, in the remainder we will not distinguish anymore between higher-order and corresponding applicative first-order terms, since the conversion between these two representations is obvious.

Recall that if a node in the termination graph is marked with a non-H-terminating term, then one of its children is also marked with a non-H-terminating term. Hence, every non-H-terminating term corresponds to an infinite path in the termination graph. Since a termination graph only has finitely many nodes, infinite paths have to end in a cycle. Thus, it suffices to prove H-termination for all terms occurring in cycles resp. in *strongly connected components (SCCs)* of the termination graph. Moreover, one can analyze H-termination separately for each SCC. Here, an SCC is a maximal subgraph G' of the termination graph such that for all nodes n_1 and n_2 in G' there is a non-empty path from n_1 to n_2 traversing only nodes of G'. (In particular, there must also be a non-empty path from every node to itself in G'.) The termination graph for "take u (from m)" in Fig. 1 has just one SCC with the nodes A, C, E, F, H. The following definition is needed to extract dependency pairs from SCCs of the termination graph.

Definition 6 (DP Path). *Let G' be an SCC of a termination graph containing a path from a node marked with s to a node marked with t. We say that this path is a DP path if it does not traverse instantiation edges, if s has an incoming instantiation edge in G', and if t has an outgoing instantiation edge in G'.*

So in Fig. 1, the only DP path is A, C, E, F, H. Since every infinite path has to traverse instantiation edges infinitely often, it also has to traverse DP paths infinitely often. Therefore, we generate a dependency pair for each DP path. If there is no infinite chain with these dependency pairs, then no term corresponds to an infinite path, i.e., then all terms in the graph are H-terminating.

More precisely, whenever there is a DP path from a node marked with s to a node marked with t and the edges of the path are marked with $\sigma_1, \ldots, \sigma_m$, then we generate the dependency pair $s\sigma_1 \ldots \sigma_m \to t$. In Fig. 1, the first edge of the DP path is labelled with the substitution $[u/(\mathsf{S}\,n)]$ and all remaining edges are labelled with the identity. Thus, we generate the dependency pair

$$\mathsf{take}\,(\mathsf{S}\,n)\,(\mathsf{from}\,m) \to \mathsf{take}\,n\,(\mathsf{from}\,(\mathsf{S}\,m)). \qquad (3)$$

The resulting DP problem is $(\mathcal{P}, \mathcal{R})$ where $\mathcal{P} = \{(3)\}$ and $\mathcal{R} = \varnothing$.[9] Automated termination tools can easily show that this DP problem is finite. Hence, the start term "take u (from m)" is H-terminating in the original Haskell-program.

Similarly, finiteness of the DP problem $(\{\text{tma}\,(\text{S}\,m) \rightarrow \text{tma}\,m\}, \varnothing)$ for the start term "tma n" from Fig. 2 is also easy to prove automatically.

A slightly more challenging example is obtained by replacing the last take-rule by the following two rules, where p computes the predecesor function.

$$\text{take}\,(\text{S}\,n)\,(\text{Cons}\,x\,xs) = \text{Cons}\,x\,(\text{take}\,(\text{p}\,(\text{S}\,n))\,xs) \qquad \text{p}\,(\text{S}\,x) = x \qquad (4)$$

Now the resulting termination graph can be obtained from the graph in Fig. 1 by replacing the subgraph starting with node F by the subgraph in Fig. 3.

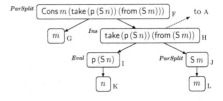

We want to construct an infinite chain whenever the termination graph contains a non-H-terminating term. In this case, there also exists a DP path with first node s such that s is not H-terminating.

Fig. 3. Subtree at node F of Fig. 1

So there is a normal ground substitution σ where $s\sigma$ is not H-terminating either. There must be a DP path from s to a term t labelled with the substitutions $\sigma_1, \ldots, \sigma_m$ such that σ is an instance of $\sigma_1 \ldots \sigma_m$ and such that $t\sigma$ is also not H-terminating.[10] So the first step of the desired corresponding infinite chain is $s\sigma \rightarrow_{\mathcal{P}} t\sigma$. The node t has an outgoing instantiation edge to a node $\tilde{t}$ which starts another DP path. So to continue the construction of the infinite chain in the same way, we now need a non-H-terminating instantiation of $\tilde{t}$ with a normal ground substitution. Obviously, $\tilde{t}$ matches t by some matcher τ. But while $\tilde{t}\tau\sigma$ is not H-terminating, the substitution $\tau\sigma$ is not necessarily a normal ground substitution. The reason is that t and hence τ may contain defined symbols.

This is also the case in our example. The only DP path is A, C, E, F, H which would result in the dependency pair $\text{take}\,(\text{S}\,n)\,(\text{from}\,m) \rightarrow t$ with $t = \text{take}\,(\text{p}\,(\text{S}\,n))\,(\text{from}\,(\text{S}\,m))$. Now t has an instantiation edge to node A with $\tilde{t} = \text{take}\,u\,(\text{from}\,m)$. The matcher is $\tau = [u/(\text{p}\,(\text{S}\,n)), m/(\text{S}\,m)]$. So $\tau(u)$ is not normal.

In this example, the problem can be avoided by already evaluating the right-hand sides of dependency pairs as much as possible. So instead of a dependency pair $s\sigma_1 \ldots \sigma_m \rightarrow t$ we now generate the dependency pair $s\sigma_1 \ldots \sigma_m \rightarrow \mathbf{ev}(t)$. For a node marked with t, $\mathbf{ev}(t)$ is the term reachable from t by traversing only *Eval*-nodes. So in our example $\mathbf{ev}(\text{p}\,(\text{S}\,n)) = n$, since node I is an *Eval*-node with an

---

[9] Def. 11 will explain how to generate $\mathcal{R}$ in general.

[10] To ease the presentation, we require that user-defined data structures (base types) may not be "empty". (But our approach can easily be extended to "empty" structures as well.) Then we may restrict ourselves to substitutions σ where all subterms of $\sigma(x)$ with base type have a constructor as head, for all variables x in s. This ensures that for every *Case*-node in the DP path, one child corresponds to the instantiation σ. To obtain a *ground* term $t\sigma$, we extend the substitution σ appropriately to the variables in t that do not occur in s. These variables were introduced by *VarExp*.

edge to node K. Moreover, $\mathbf{ev}(t)$ can also evaluate subterms of t if t is an **Ins**-node or a **ParSplit**-node with a constructor as head. We obtain $\mathbf{ev}(\mathsf{S}\,m) = \mathsf{S}\,m$ for node J and $\mathbf{ev}(\mathsf{take}\,(\mathsf{p}\,(\mathsf{S}\,n))\,(\mathsf{from}\,(\mathsf{S}\,m))) = \mathsf{take}\,n\,(\mathsf{from}\,(\mathsf{S}\,m))$ for node H. Thus, the resulting DP problem is again $(\mathcal{P}, \mathcal{R})$ with $\mathcal{P} = \{(3)\}$ and $\mathcal{R} = \varnothing$.

To see how $\mathbf{ev}(t)$ must be defined for **ParSplit**-nodes where $\mathrm{head}(t)$ is a variable, we regard the function nonterm again, cf. (1). In the termination graph for the start term "nonterm $b\,x$", we obtain a DP path from the node with the start term to a node with "nonterm $(x\,\mathsf{True})\,x$" labelled with the substitution $[b/\mathsf{False}]$. So the resulting DP problem only contains the dependency pair "nonterm $\mathsf{False}\,x \;\rightarrow\; \mathbf{ev}(\mathsf{nonterm}\,(x\,\mathsf{True})\,x)$". If we would define $\mathbf{ev}(x\,\mathsf{True}) = x\,\mathsf{True}$, then $\mathbf{ev}$ would not modify the term "nonterm $(x\,\mathsf{True})\,x$". But then the resulting DP problem would be finite and one could falsely prove H-termination. (The reason is that the DP problem contains no rule to transform any instance of "$x\,\mathsf{True}$" to False.) But as discussed in Sect. 3, x can be instantiated by arbitrary H-terminating functions and then, "$x\,\mathsf{True}$" can evaluate to any term. Therefore, $\mathbf{ev}$ must replace terms like "$x\,\mathsf{True}$" by fresh variables.

Definition 7 (ev). *Let G be a termination graph with a node t.[11] Then*

$$\mathbf{ev}(t) = \begin{cases} t, & \textit{if } t \textit{ is a leaf, a } \textbf{Case}\textit{-node, or a } \textbf{VarExp}\textit{-node} \\ x, & \textit{if } t \textit{ is } \textbf{ParSplit}\textit{-node, } \mathrm{head}(t) \textit{ is a variable, and } x \textit{ is a fresh variable} \\ \mathbf{ev}(\tilde{t}), & \textit{if } t \textit{ is an } \textbf{Eval}\textit{-node with child } \tilde{t} \\ \tilde{t}[x_1/\mathbf{ev}(t_1), \ldots, x_n/\mathbf{ev}(t_n)], & \textit{if } t = \tilde{t}[x_1/t_1, \ldots, x_n/t_n] \textit{ and either} \\ & \quad t \textit{ is an } \textbf{Ins}\textit{-node with the children } t_1, \ldots, t_n, \tilde{t} \textit{ or} \\ & \quad t \textit{ is a } \textbf{ParSplit}\textit{-node, and } \tilde{t} = (c\,x_1 \ldots x_n) \textit{ for a constructor } c \end{cases}$$

Our goal was to construct an infinite chain whenever s is the first node in a DP path and $s\sigma$ is not H-terminating for a normal ground substitution σ. As discussed before, there is a DP path from s to t such that the chain starts with $s\sigma \rightarrow_{\mathcal{P}} \mathbf{ev}(t)\sigma$ and such that $t\sigma$ and hence $\mathbf{ev}(t)\sigma$ is also not H-terminating. The node t has an instantiation edge to some node $\tilde{t}$. Thus $t = \tilde{t}[x_1/t_1, \ldots, x_n/t_n]$ and $\mathbf{ev}(t) = \tilde{t}[x_1/\mathbf{ev}(t_1), \ldots, x_n/\mathbf{ev}(t_n)]$. In order to continue the construction of the infinite chain, we need a non-H-terminating instantiation of $\tilde{t}$ with a normal ground substitution. Clearly, if $\tilde{t}$ is instantiated by the substitution $[x_1/\mathbf{ev}(t_1)\sigma, \ldots, x_n/\mathbf{ev}(t_n)\sigma]$, then it is again not H-terminating. However, the substitution $[x_1/\mathbf{ev}(t_1)\sigma, \ldots, x_n/\mathbf{ev}(t_n)\sigma]$ is not necessarily normal. The problem is that $\mathbf{ev}$ does not perform those evaluations that correspond to instantiation edges and to edges from **Case**-nodes. Therefore, we now generate DP problems which do not just contain dependency pairs $\mathcal{P}$, but they also contain all rules $\mathcal{R}$ which might be needed to evaluate $\mathbf{ev}(t_i)\sigma$ further. Then we obtain $s\sigma \rightarrow_{\mathcal{P}} \mathbf{ev}(t)\sigma \rightarrow_{\mathcal{R}}^* \tilde{t}\sigma'$ for a normal ground substitution σ'. Since $\tilde{t}$ is again the first node in a DP path, now this construction of the chain can be continued in the same way infinitely many times. Hence, we obtain an infinite chain.

As an example, we replace the equation for p in (4) by the following two defining equations:

$$\mathsf{p}\,(\mathsf{S}\,\mathsf{Z}) = \mathsf{Z} \qquad\qquad \mathsf{p}\,(\mathsf{S}\,x) = \mathsf{S}\,(\mathsf{p}\,x) \tag{5}$$

---

[11] To simplify the presentation, we identify nodes with the terms they are labelled with.

In the termination graph for "take u (from m)" from Fig. 1 and 3, the node I would now be replaced by the subtree in Fig. 4. So I is now a ***Case***-node. Thus, instead of (3) we obtain the dependency pair

$$\text{take} (S\, n)\, (\text{from}\, m) \rightarrow \text{take} (p\, (S\, n))\, (\text{from}\, (S\, m)), \tag{6}$$

since now **ev** does not modify its right-hand side anymore (i.e., $\mathbf{ev}(p\, (S\, n)) = p\, (S\, n)$). Hence, now the resulting DP problem must contain all rules $\mathcal{R}$ that might be used to evaluate $p\, (S\, n)$ when instantiated by σ.

So for any term t, we want to detect rules that might be needed to evaluate $\mathbf{ev}(t)\sigma$ further for normal ground substitutions σ. To this end, we first compute the set $\mathbf{con}(t)$ of those terms that are reachable from t, but where the computation of **ev** stopped. So $\mathbf{con}(t)$ contains all terms which might give rise to further **con**tinuing evaluations that are not captured by **ev**. To compute $\mathbf{con}(t)$, we traverse all paths starting in t. If we reach a ***Case***-node s, we stop traversing

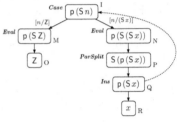

Fig. 4. Subtree at node I of Fig. 3

this path and insert s into $\mathbf{con}(t)$. Moreover, if we traverse an instantiation edge to some node $\tilde{t}$, we also stop and insert $\tilde{t}$ into $\mathbf{con}(t)$. So in the example of Fig. 4, we obtain $\mathbf{con}(p\, (S\, n)) = \{p\, (S\, n)\}$, since I is now a ***Case***-node. If we started with the term $t = \text{take} (S\, n)\, (\text{from}\, m)$ in node C, then we would reach the ***Case***-node I and the node A which is reachable via an instantiation edge. So $\mathbf{con}(t) = \{p\, (S\, n), \text{take}\, u\, (\text{from}\, m)\}$. Finally, **con** also stops at ***VarExp***-nodes (they are in normal form w.r.t. $\rightarrow_H$) and at ***ParSplit***-nodes whose head is a variable (since **ev** already "approximates" their result by fresh variables).

Definition 8 (con). *Let G be a termination graph with a node t. Then*

$$\mathbf{con}(t) = \begin{cases} \varnothing, & \text{if } t \text{ is a leaf, a } \textbf{VarExp-}, \text{ or a } \textbf{ParSplit}\text{-node with variable head} \\ \{t\}, & \text{if } t \text{ is a } \textbf{Case}\text{-node} \\ \{\tilde{t}\} \cup \mathbf{con}(t_1) \cup \ldots \cup \mathbf{con}(t_n), & \text{if } t \text{ is an } \textbf{Ins}\text{-node with the} \\ \quad \text{children } t_1, \ldots, t_n, \tilde{t} \text{ and an instantiation edge from } t \text{ to } \tilde{t} \\ \bigcup_{t' \text{ child of } t} \mathbf{con}(t'), & \text{otherwise} \end{cases}$$

Now we can define how to extract a DP problem $\mathbf{dp}_{G'}$ from every SCC G' of the termination graph. As mentioned, we generate a dependency pair $s\sigma_1 \ldots \sigma_m \rightarrow \mathbf{ev}(t)$ for every DP path from s to t labelled with $\sigma_1, \ldots, \sigma_m$ in G'. If $t = \tilde{t}[x_1/t_1, \ldots, x_n/t_n]$ has an instantiation edge to $\tilde{t}$, then the resulting DP problem must contain all rules that can be used to reduce the terms in $\mathbf{con}(t_1) \cup \ldots \cup \mathbf{con}(t_n)$. For any term s, let $\mathbf{rl}(s)$ be the rules that can be used to reduce $s\sigma$ for normal ground substitutions σ. We will give the definition of $\mathbf{rl}$ afterwards.

Definition 9 (dp). *For a termination graph containing an SCC G', we define* $\mathbf{dp}_{G'} = (\mathcal{P}, \mathcal{R})$*. Here, $\mathcal{P}$ and $\mathcal{R}$ are the smallest sets such that*

- "$s\sigma_1 \ldots \sigma_m \to \mathbf{ev}(t)$" $\in \mathcal{P}$ and
- $\mathbf{rl}(q) \subseteq \mathcal{R}$,

whenever G' contains a DP path from s to t labelled with $\sigma_1, \ldots, \sigma_m$, $t = \tilde{t}[x_1/t_1, \ldots, x_n/t_n]$ has an instantiation edge to $\tilde{t}$, and $q \in \mathbf{con}(t_1) \cup \ldots \cup \mathbf{con}(t_n)$.

In our example with the start term "$\mathsf{take}\, u\, (\mathsf{from}\, m)$" and the p-equations from (5), the termination graph in Fig. 1, 3, and 4 has two SCCs G_1 (consisting of the nodes A, C, E, F, H) and G_2 (consisting of I, N, P, Q). Finiteness of the two DP problems $\mathbf{dp}_{G_1}$ and $\mathbf{dp}_{G_2}$ can be proved independently. The SCC G_1 only has the DP path from A to H leading to the dependency pair (6). So we obtain $\mathbf{dp}_{G_1} = (\{(6)\}, \mathcal{R}_1)$ where $\mathcal{R}_1$ contains $\mathbf{rl}(q)$ for all $q \in \mathbf{con}(\mathsf{p}\,(\mathsf{S}\,n)) = \{\mathsf{p}\,(\mathsf{S}\,n)\}$. Thus, $\mathcal{R}_1 = \mathbf{rl}(\mathsf{p}\,(\mathsf{S}\,n))$. The SCC G_2 only has the DP path from I to Q. Hence, $\mathbf{dp}_{G_2} = (\mathcal{P}_2, \mathcal{R}_2)$ where $\mathcal{P}_2$ consists of the dependency pair "$\mathsf{p}\,(\mathsf{S}\,(\mathsf{S}\,x)) \to \mathsf{p}\,(\mathsf{S}\,x)$" (since $\mathbf{ev}(\mathsf{p}\,(\mathsf{S}\,x)) = \mathsf{p}\,(\mathsf{S}\,x)$) and $\mathcal{R}_2$ contains $\mathbf{rl}(q)$ for all $q \in \mathbf{con}(x) = \varnothing$, i.e., $\mathcal{R}_2 = \varnothing$. Thus, finiteness of $\mathbf{dp}_{G_2}$ can easily be proved automatically.

For every term s, we now show how to extract a set of rules $\mathbf{rl}(s)$ such that every evaluation of $s\sigma$ for a normal ground substitution σ corresponds to a reduction with $\mathbf{rl}(s)$.[12] The only expansion rules which transform terms into "equal" ones are **Eval** and **Case**. This leads to the following definition.

Definition 10 (Rule Path). *A path from a node marked with s to a node marked with t is a rule path if s and all other nodes on the path except t are **Eval**- or **Case**-nodes and t is no **Eval**- or **Case**-node. So t may also be a leaf.*

In Fig. 4, there are two rule paths starting in node I. The first one is I, M, O (since O is a leaf) and the other is I, N, P (since P is a **ParSplit**-node).

While DP paths give rise to dependency pairs, rule paths give rise to rules. Therefore, if there is a rule path from s to t labelled with $\sigma_1, \ldots, \sigma_m$, then $\mathbf{rl}(s)$ contains the rule $s\sigma_1 \ldots \sigma_m \to \mathbf{ev}(t)$. In addition, $\mathbf{rl}(s)$ must also contain all rules required to evaluate $\mathbf{ev}(t)$ further, i.e., all rules in $\mathbf{rl}(q)$ for $q \in \mathbf{con}(t)$.[13]

Definition 11 (rl). *For a node labelled with s, $\mathbf{rl}(s)$ is the smallest set with*

- "$s\sigma_1 \ldots \sigma_m \to \mathbf{ev}(t)$" $\in \mathbf{rl}(s)$ and
- $\mathbf{rl}(q) \subseteq \mathbf{rl}(s)$,

whenever there is rule path from s to t labelled with $\sigma_1, \ldots, \sigma_m$, and $q \in \mathbf{con}(t)$.

For the start term "$\mathsf{take}\, u\, (\mathsf{from}\, m)$" and the p-equations from (5), we obtained the DP problem $\mathbf{dp}_{G_1} = (\{6\}, \mathbf{rl}(\mathsf{p}\,(\mathsf{S}\,n)))$. Here, $\mathbf{rl}(\mathsf{p}\,(\mathsf{S}\,n))$ consists of

$$\mathsf{p}\,(\mathsf{S}\,\mathsf{Z}) \to \mathsf{Z} \qquad \text{(due to the rule path from I to O)} \qquad (7)$$

$$\mathsf{p}\,(\mathsf{S}\,(\mathsf{S}\,x)) \to \mathsf{S}\,(\mathsf{p}\,(\mathsf{S}\,x)) \qquad \text{(due to the rule path from I to P),} \qquad (8)$$

---

[12] More precisely, $s\sigma \to_{\mathsf{H}}^* q$ implies $s\sigma \to_{\mathbf{rl}(s)}^* q'$ for a term q' which is "at least as evaluated" as q (i.e., one can evaluate q further to q' if one also permits evaluation steps below or beside the evaluation position).

[13] So if $t = \tilde{t}[x_1/t_1, \ldots, x_n/t_n]$ has an instantiation edge to $\tilde{t}$, then here we also include all rules of $\mathbf{rl}(\tilde{t})$, since $\mathbf{con}(t) = \{\tilde{t}\} \cup \mathbf{con}(t_1) \cup \ldots \cup \mathbf{con}(t_n)$. In contrast, for the definition of $\mathbf{dp}$ in Def. 9 we only regard the rules $\mathbf{rl}(q)$ for $q \in \mathbf{con}(t_1) \cup \ldots \cup \mathbf{con}(t_n)$, whereas the evaluations of $\tilde{t}$ are captured by the dependency pairs.

as **ev** does not modify the right-hand sides of (7) and (8). Moreover, the require-
ment "$\mathbf{rl}(q) \subseteq \mathbf{rl}(\mathsf{p}\,(\mathsf{S}\,n))$ for all $q \in \mathbf{con}(\mathsf{Z})$ and all $q \in \mathbf{con}(\mathsf{S}\,(\mathsf{p}\,(\mathsf{S}\,x)))$" does not
add further rules. The reason is that $\mathbf{con}(\mathsf{Z}) = \varnothing$ and $\mathbf{con}(\mathsf{S}\,(\mathsf{p}\,(\mathsf{S}\,x))) = \{\mathsf{p}\,(\mathsf{S}\,n)\}$.
Now finiteness of $\mathbf{dp}_{G_1} = (\{6\}, \{(7), (8)\})$ is also easy to show automatically.

Finally, consider the following program which leads to the graph in Fig. 5.

$$\mathsf{f}\,x = \mathsf{applyToZero}\,\mathsf{f} \qquad\qquad \mathsf{applyToZero}\,x = x\,\mathsf{Z}$$

This example shows that one also has to traverse edges resulting from **VarExp**
when constructing dependency pairs. Otherwise one would falsely prove H-termination. Since the only DP path goes from node A to F, we obtain the DP problem
$(\{\mathsf{f}\,x \to \mathsf{f}\,y\}, \mathcal{R})$ with $\mathcal{R} = \mathbf{rl}(y) = \varnothing$. This problem is not finite (and indeed, "$\mathsf{f}\,x$"
is not H-terminating). In contrast, the definition of **rl** stops at **VarExp**-nodes.

The example also illustrates that **rl** and **dp**
handle instantiation edges differently, cf. Foot-
note 13. Since there is a rule path from A to B, we
would obtain $\mathbf{rl}(\mathsf{f}\,x) = \{\mathsf{f}\,x \to \mathsf{applyToZero}\,\mathsf{f}\} \cup$
$\mathbf{rl}(\mathsf{applyToZero}\,x)$, since $\mathbf{con}(\mathsf{applyToZero}\,\mathsf{f}) =$
$\mathsf{applyToZero}\,x$. So for the construction of **rl** we
also have to include the rules resulting from
nodes like C which are only reachable by instan-
tiation edges.[14] We obtain $\mathbf{rl}(\mathsf{applyToZero}\,x) =$
$\{\mathsf{applyToZero}\,x \to z\}$, since $\mathbf{ev}(x\,\mathsf{Z}) = z$ for a
fresh variable z. The following theorem states
the soundness of our approach.

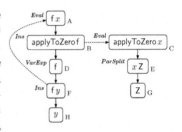

Fig. 5. Termination graph for "$\mathsf{f}\,x$"

Theorem 12 (Soundness). *Let G be termination graph. If the DP problem*
$\mathbf{dp}_{G'}$ *is finite for all SCCs G' of G, then all nodes t in G are H-terminating.*[15]

While we transform termination graphs into DP problems, it would also be possible to transform termination graphs into TRSs instead and then prove termination of the resulting TRSs. However, this approach has several disadvantages.
For example, if the termination graph contains a **VarExp**-node or a **ParSplit**-
node with a variable as head, then we would result in rules with extra variables
on right-hand sides and thus, the resulting TRSs would never be terminating.
In contrast, a DP problem $(\mathcal{P}, \mathcal{R})$ with extra variables in $\mathcal{P}$ and $\mathcal{R}$ can still be
finite, since dependency pairs from $\mathcal{P}$ are only be applied on top positions in
chains and since $\mathcal{R}$ need not be terminating for finite DP problems $(\mathcal{P}, \mathcal{R})$.

5   Extensions, Implementation, and Experiments

We presented a technique for automated termination analysis of Haskell which
works in three steps: First, it generates a termination graph for the given start

---

[14] This is different in the definition of **dp**. Otherwise, we would have $\mathcal{R} = \mathbf{rl}(y) \cup \mathbf{rl}(\mathsf{f}\,x)$.
[15] Instead of $\mathbf{dp}_{G'} = (\mathcal{P}, \mathcal{R})$, for H-termination it suffices to prove finiteness of $(\mathcal{P}^\sharp, \mathcal{R})$.
Here, $\mathcal{P}^\sharp$ results from $\mathcal{P}$ by replacing each rule $f(t_1, ..., t_n) \to g(s_1, ..., s_m)$ in $\mathcal{P}$ by
$f^\sharp(t_1, ..., t_n) \to g^\sharp(s_1, ..., s_m)$, where $f^\sharp$ and $g^\sharp$ are fresh "*tuple*" function symbols [2].

term. Then it extracts DP problems from the termination graph. Finally, one uses existing methods from term rewriting to prove finiteness of these DP problems.

To ease readability, we did not regard Haskell's *type classes* and built-in data structures in the preceding sections. However, our approach easily extends to these concepts [14]. To deal with type classes, we use an additional *Case*-rule in the construction of termination graphs, which instantiates type variables by all instances of the corresponding type class. Built-in data structures like Haskell's lists and tuples simply correspond to user-defined types with a different syntax. To deal with integers, we transform them into a notation with the constructors Pos and Neg (which take arguments of type Nats) and provide pre-defined rewrite rules for integer operations like addition, subtraction, etc. Floating-point numbers can be handled in a similar way (e.g., by representing them as fractions).

We implemented our approach in the termination prover APrOVE [9]. It accepts the full Haskell 98 language defined in [12] and we successfully evaluated our implementation with standard Haskell-libraries from the Hugs-distribution such as Prelude, Monad, List, FiniteMap, etc. To access the implementation via a web interface, for details on our experiments, and for the proofs of all theorems, see http://aprove.informatik.rwth-aachen.de/eval/Haskell/.

We conjecture that term rewriting techniques are also suitable for termination analysis of other kinds of programming languages. In [13], we recently adapted the dependency pair method in order to prove termination of *logic* programming languages like Prolog. In future work, we intend to examine the use of TRS-techniques for *imperative* programming languages as well.

References

1. A. Abel. Termination checking with types. *RAIRO - Theoretical Informatics and Applications*, 38(4):277–319, 2004.
2. T. Arts and J. Giesl. Termination of term rewriting using dependency pairs. *Theoretical Computer Science*, 236:133–178, 2000.
3. G. Barthe, M. J. Frade, E. Giménez, L. Pinto, and T. Uustalu. Type-based termination of recursive definitions. *Math. Structures in Comp. Sc.*, 14(1):1–45, 2004.
4. O. Fissore, I. Gnaedig, and H. Kirchner. Outermost ground termination. In *Proc. WRLA '02*, ENTCS 71, 2002.
5. J. Giesl. Termination analysis for functional programs using term orderings. In *Proc. SAS' 95*, LNCS 983, pages 154–171, 1995.
6. J. Giesl and A. Middeldorp. Transformation techniques for context-sensitive rewrite systems. *Journal of Functional Programming*, 14(4):379–427, 2004.
7. J. Giesl, R. Thiemann, and P. Schneider-Kamp. The dependency pair framework: Combining techniques for automated termination proofs. In *Proc. LPAR '04*, LNAI 3452, pages 301–331, 2005.
8. J. Giesl, R. Thiemann, and P. Schneider-Kamp. Proving and disproving termination of higher-order functions. In *Proc. FroCoS '05*, LNAI 3717, pp. 216-231, 2005.
9. J. Giesl, P. Schneider-Kamp, and R. Thiemann. APrOVE 1.2: Automatic termination proofs in the DP framework. In *Proc. IJCAR '06*, LNAI, 2006. To appear.
10. C. S. Lee, N. D. Jones, and A. M. Ben-Amram. The size-change principle for program termination. In *Proc. POPL '01*, pages 81–92. ACM Press, 2001.

11. S. E. Panitz and M. Schmidt-Schauss. TEA: Automatically proving termination of programs in a non-strict higher-order functional language. In *Proc. SAS '97*, LNCS 1302, pages 345–360, 1997.
12. S. Peyton Jones (ed.). Haskell 98 *Languages and Libraries: The revised report.* Cambridge University Press, 2003.
13. P. Schneider-Kamp, J. Giesl, A. Serebrenik, and R. Thiemann. Automated termination analysis for logic programs by term rewriting. In *Proc. LOPSTR '06*, LNCS, 2006. To appear.
14. S. Swiderski. Terminierungsanalyse von Haskellprogrammen. Diploma Thesis, RWTH Aachen, 2005. See http://aprove.informatik.rwth-aachen.de/eval/Haskell/.
15. A. Telford and D. Turner. Ensuring termination in ESFP. *Journal of Universal Computer Science*, 6(4):474–488, 2000.
16. C. Walther. On proving the termination of algorithms by machine. *Artificial Intelligence*, 71(1):101–157, 1994.
17. H. Xi. Dependent types for program termination verification. *Higher-Order and Symbolic Computation*, 15(1):91–131, 2002.

Predictive Labeling

Nao Hirokawa and Aart Middeldorp

Institute of Computer Science
University of Innsbruck
6020 Innsbruck, Austria
{nao.hirokawa, aart.middeldorp}@uibk.ac.at

Abstract. Semantic labeling is a transformation technique for proving the termination of rewrite systems. The semantic part is given by a quasi-model of the rewrite rules. In this paper we present a variant of semantic labeling in which the quasi-model condition is only demanded for the usable rules induced by the labeling. Our variant is less powerful in theory but maybe more useful in practice.

1 Introduction

Numerous methods are available for proving the termination of term rewrite systems, ranging from simplification orders like the Knuth-Bendix order [10], polynomial interpretations [12,3], and path orders [4,9], via transformation methods like semantic labeling [18] and the dependency pair method [1], to recent methods based on results from automata theory [5,6].

In this paper we revisit the semantic labeling method of Zantema [18]. Invented back in 1995, only recently the method has become available in tools that aim to prove termination automatically. Zantema implemented a version with a binary (quasi-)model in his termination prover TORPA [19] for string rewrite systems. The termination prover TPA [11] developed by Koprowski for term rewrite systems, additionally employs natural numbers as semantics and labels. As shown by the performance of TPA in the TRS category of the 2005 termination competition,[1] this is surprisingly powerful.

We present a variant of semantic labeling which comes with less constraints on the part of the semantics. More precisely, our variant does not require that all rewrite rules of the rewrite system that we want to prove terminating need to be considered when checking the quasi-model condition. To make the discussion more concrete, let us consider the following example.

Example 1. Consider the TRS $\mathcal{R}$ consisting of the following rewrite rules:

$$\mathsf{fact}(0) \to \mathsf{s}(0) \qquad\qquad 0 + y \to y$$
$$\mathsf{fact}(\mathsf{s}(x)) \to \mathsf{fact}(\mathsf{p}(\mathsf{s}(x))) \times \mathsf{s}(x) \qquad\qquad \mathsf{s}(x) + y \to \mathsf{s}(x + y)$$
$$\mathsf{p}(\mathsf{s}(0)) \to 0 \qquad\qquad 0 \times y \to 0$$
$$\mathsf{p}(\mathsf{s}(\mathsf{s}(x))) \to \mathsf{s}(\mathsf{p}(\mathsf{s}(x))) \qquad\qquad \mathsf{s}(x) \times y \to (x \times y) + y$$

---

[1] http://www.lri.fr/~marche/termination-competition/2005

F. Pfenning (Ed.): RTA 2006, LNCS 4098, pp. 313–327, 2006.

This is the leading example from [18] extended with the rule $\mathsf{fact}(0) \rightarrow \mathsf{s}(0)$ and recursive rules for addition and multiplication. These additional rules cause no problems for the "standard" semantic labeling proof, which employs natural numbers as semantics and as labels for the function symbol fact, using the natural interpretations $0_\mathbb{N} = 0$, $\mathsf{s}_\mathbb{N}(x) = x + 1$, $\mathsf{p}_\mathbb{N}(x) = \max\{x - 1, 0\}$, $x +_\mathbb{N} y = x + y$, $x \times_\mathbb{N} y = x \times y$, $\mathsf{fact}_\mathbb{N}(x) = x!$ and the labeling function $\mathsf{fact}_\ell(x) = x$. Note that the resulting algebra is a model of the rewrite rules of $\mathcal{R}$. By replacing the two rules

$$\mathsf{fact}(0) \rightarrow \mathsf{s}(0) \qquad\qquad \mathsf{fact}(\mathsf{s}(x)) \rightarrow \mathsf{fact}(\mathsf{p}(\mathsf{s}(x))) \times \mathsf{s}(x)$$

with the infinitely many rules

$$\mathsf{fact}_0(0) \rightarrow \mathsf{s}(0) \qquad \mathsf{fact}_{i+1}(\mathsf{s}(x)) \rightarrow \mathsf{fact}_i(\mathsf{p}(\mathsf{s}(x))) \times \mathsf{s}(x) \quad (\forall\, i \geqslant 0)$$

the labeled TRS $\mathcal{R}_{\mathsf{lab}}$ is obtained. The rules of this TRS are oriented from left to right by the lexicographic path order induced by the well-founded precedence

$$\mathsf{fact}_{i+1} > \mathsf{fact}_i > \cdots > \mathsf{fact}_0 > \times > + > \mathsf{p} > \mathsf{s}$$

and hence $\mathcal{R}_{\mathsf{lab}}$ is terminating. The soundness of semantic labeling guarantees that $\mathcal{R}$ is terminating, too.

Semantic labeling requires that the algebra defining the semantics is a (quasi-) model of all rewrite rules of the TRS that we want to prove terminating. This entails that we need to define semantics for all function symbols occurring in the TRS. In the variant we present in this paper, we need to define the semantics of the function symbols that appear below a function symbol that we want to label as well as the function symbols that depend on them, and the (quasi-)model condition is required only for the rules that define these function symbols. In our example, the interpretations of the function symbols $+$, $\times$, and fact may be ignored. Furthermore, the (quasi-)model condition needs to be checked for the two rules

$$\mathsf{p}(\mathsf{s}(0)) \rightarrow 0 \qquad\qquad \mathsf{p}(\mathsf{s}(\mathsf{s}(x))) \rightarrow \mathsf{s}(\mathsf{p}(\mathsf{s}(x)))$$

only. We prove that this is sound provided an additional condition on the algebras that may be used in connection with our variant of semantic labeling is imposed. This condition makes our variant less powerful in theory but maybe more useful in practice. Our variant is certainly more difficult to prove correct since the standard proof of transforming a presupposed infinite rewrite sequence into an infinite labeled rewrite sequence will not work without further ado due to a lack of semantic information. In the correctness proof we predict this missing information, which is why we call our variant *predictive* labeling.

The remainder of the paper is organized as follows. In the next section we recapitulate the formal definition of semantic labeling. In Section 3 we present our main result. Some more examples are presented in Section 4 and we conclude with mentioning some open issues in Section 5.

2   Preliminaries

We assume that the reader is familiar with term rewriting [2,14]. Let $\mathcal{R}$ be a TRS over a signature $\mathcal{F}$ and let $\mathcal{A} = (A, \{f_\mathcal{A}\}_{f\in\mathcal{F}})$ be an $\mathcal{F}$-algebra. A *labeling* ℓ for $\mathcal{A}$ consists of sets of labels $L_f \subseteq A$ for every $f \in \mathcal{F}$ together with mappings $\ell_f \colon A^n \to L_f$ for every n-ary function symbol $f \in \mathcal{F}$ with $L_f \neq \varnothing$. In examples we have $L_f = A$ whenever $L_f \neq \varnothing$. The labeled signature $\mathcal{F}_{\mathsf{lab}}$ consists of n-ary function symbols f_a for every n-ary function symbol $f \in \mathcal{F}$ and label $a \in L_f$ together with all function symbols $f \in \mathcal{F}$ such that $L_f = \varnothing$. The mapping ℓ_f determines the label of the root symbol f of a term $f(t_1, \ldots, t_n)$ based on the values of the arguments $t_1, \ldots, t_n$. Let $\mathcal{V}$ be the set of variables. For every assignment $\alpha\colon \mathcal{V} \to A$ the mapping $\mathsf{lab}_\alpha \colon \mathcal{T}(\mathcal{F}, \mathcal{V}) \to \mathcal{T}(\mathcal{F}_{\mathsf{lab}}, \mathcal{V})$ is inductively defined as follows:

$$\mathsf{lab}_\alpha(t) = \begin{cases} t & \text{if } t \text{ is a variable,} \\ f(\mathsf{lab}_\alpha(t_1), \ldots, \mathsf{lab}_\alpha(t_n)) & \text{if } t = f(t_1, \ldots, t_n) \text{ and } L_f = \varnothing, \\ f_a(\mathsf{lab}_\alpha(t_1), \ldots, \mathsf{lab}_\alpha(t_n)) & \text{if } t = f(t_1, \ldots, t_n) \text{ and } L_f \neq \varnothing \end{cases}$$

where a denotes the label $\ell_f([\alpha]_\mathcal{A}(t_1), \ldots, [\alpha]_\mathcal{A}(t_n))$. The *labeled* TRS $\mathcal{R}_{\mathsf{lab}}$ over the signature $\mathcal{F}_{\mathsf{lab}}$ consists of the rewrite rules

$$\mathsf{lab}_\alpha(l) \to \mathsf{lab}_\alpha(r)$$

for all rules $l \to r \in \mathcal{R}$ and assignments $\alpha\colon \mathcal{V} \to A$.

Theorem 2 (Zantema [18]). *Let $\mathcal{R}$ be a TRS. Let the algebra $\mathcal{A}$ be a non-empty model of $\mathcal{R}$ and let ℓ be a labeling for $\mathcal{A}$. The TRS $\mathcal{R}$ is terminating if and only if the TRS $\mathcal{R}_{\mathsf{lab}}$ is terminating.*   □

The condition that $\mathcal{A}$ is a model is somewhat restrictive. A stronger (in the sense that more terminating TRSs can be transformed into TRSs that can be proved terminating by simple methods) result is obtained by equipping $\mathcal{A}$ with a well-founded order such that all algebra operations and all labeling functions are weakly monotone in all coordinates.

A well-founded weakly monotone $\mathcal{F}$-algebra $(\mathcal{A}, >)$ consists of a non-empty $\mathcal{F}$-algebra $\mathcal{A} = (A, \{f_\mathcal{A}\}_{f\in\mathcal{F}})$ and a well-founded order $>$ on the carrier A of $\mathcal{A}$ such that every algebra operation is weakly monotone in all coordinates, i.e., if $f \in \mathcal{F}$ has arity $n \geqslant 1$ then

$$f_\mathcal{A}(a_1, \ldots, a_i, \ldots, a_n) \geqslant f_\mathcal{A}(a_1, \ldots, b, \ldots, a_n)$$

for all $a_1, \ldots, a_n, b \in A$ and $i \in \{1, \ldots, n\}$ with $a_i > b$. The relation $\geqslant_\mathcal{A}$ on $\mathcal{T}(\mathcal{F}, \mathcal{V})$ is defined as follows: $s \geqslant_\mathcal{A} t$ if $[\alpha]_\mathcal{A}(s) \geqslant [\alpha]_\mathcal{A}(t)$ for all assignments α. We say that $(\mathcal{A}, >)$ is a quasi-model of a TRS $\mathcal{R}$ if $\mathcal{R} \subseteq \geqslant_\mathcal{A}$.

A labeling ℓ for $\mathcal{A}$ is called weakly monotone if all labeling functions ℓ_f are weakly monotone in all coordinates. The TRS $\mathcal{D}ec$ consists of all rewrite rules

$$f_a(x_1, \ldots, x_n) \to f_b(x_1, \ldots, x_n)$$

with f an n-ary function symbol, $a, b \in L_f$ such that $a > b$, and $x_1, \ldots, x_n$ pairwise different variables.

Theorem 3 (Zantema [18]). *Let $\mathcal{R}$ be a TRS, $(\mathcal{A}, >)$ a well-founded weakly monotone quasi-model for $\mathcal{R}$, and ℓ a weakly monotone labeling for $(\mathcal{A}, >)$. The TRS $\mathcal{R}$ is terminating if and only if the TRS $\mathcal{R}_{\mathsf{lab}} \cup \mathsf{Dec}$ is terminating.* □

In [13] it is shown how Theorem 3 can be used to transform any terminating TRS into a so-called *precedence terminating* TRS, which are defined as having the property that there exists a well-founded precedence $\sqsupset$ such that $\mathsf{root}(l) \sqsupset f$ for every rewrite rule $l \rightarrow r$ and every function symbol $f \in \mathcal{F}\mathsf{un}(r)$. This condition ensures that the rewrite rules can be oriented from left to right by the lexicographic path order induced by the precedence. Needless to say, this particular transformation is not effective.

We conclude this preliminary section with a simple but useful fact that underlies the dependency pair method [1]. This fact is used to obtain the main result presented in the next section. The easy proof can be found in [8]. Here $\mathcal{T}_\infty$ denotes the set of minimal non-terminating terms in $\mathcal{T}(\mathcal{F}, \mathcal{V})$, minimal in the sense that all arguments are terminating.

Lemma 4. *For every term $t \in \mathcal{T}_\infty$ there exists a rewrite rule $l \rightarrow r$, a substitution σ, and a non-variable subterm u of r such that $t \xrightarrow{>\epsilon}{}^* l\sigma \xrightarrow{\epsilon} r\sigma \trianglerighteq u\sigma$ and $l\sigma, u\sigma \in \mathcal{T}_\infty$.* □

In the following we do not use the fact that all steps in the rewrite sequence from t to $l\sigma$ take place below the root.

3   Predictive Labeling

Our aim is to weaken the quasi-model condition $\mathcal{R} \subseteq \geqslant_{\mathcal{A}}$ in Theorem 3 by replacing $\mathcal{R}$ with the *usable rules* of the labeling ℓ. The concept of usable rules originates from [1]. We extend the definition to labelings.

Definition 5. *For function symbols f and g we write $f \rhd_{\mathsf{d}} g$ if there exists a rewrite rule $l \rightarrow r \in \mathcal{R}$ such that $f = \mathsf{root}(l)$ and g is a defined function symbol in $\mathcal{F}\mathsf{un}(r)$. Let ℓ be a labeling and t a term. We define*

$$\mathcal{G}_\ell(t) = \begin{cases} \varnothing & \text{if } t \text{ is a variable,} \\ \mathcal{F}\mathsf{un}(t_1)^* \cup \cdots \cup \mathcal{F}\mathsf{un}(t_n)^* & \text{if } t = f(t_1, \ldots, t_n) \text{ and } L_f \neq \varnothing, \\ \mathcal{G}_\ell(t_1) \cup \cdots \cup \mathcal{G}_\ell(t_n) & \text{if } t = f(t_1, \ldots, t_n) \text{ and } L_f = \varnothing \end{cases}$$

where F^* denotes the set $\{g \mid f \rhd_{\mathsf{d}}^* g \text{ for some } f \in F\}$. Furthermore we define

$$\mathcal{G}_\ell(\mathcal{R}) = \bigcup_{l \rightarrow r \in \mathcal{R}} \mathcal{G}_\ell(l) \cup \mathcal{G}_\ell(r).$$

The set of usable rules *for ℓ is defined as $\mathcal{U}(\ell) = \{l \rightarrow r \in \mathcal{R} \mid \mathsf{root}(l) \in \mathcal{G}_\ell(\mathcal{R})\}$.*

In the following we simply write $\mathcal{G}_\ell$ for $\mathcal{G}_\ell(\mathcal{R})$.

Example 6. With respect to the TRS $\mathcal{R}$ and the labeling ℓ restricted to fact in Example 1 we have $\mathcal{G}_\ell = \{0, \mathsf{p}, \mathsf{s}\}$. Since 0 and s are constructors, $\mathcal{U}(\ell)$ consists of the two rules $\mathsf{p}(\mathsf{s}(0)) \to 0$ and $\mathsf{p}(\mathsf{s}(\mathsf{s}(x))) \to \mathsf{s}(\mathsf{p}(\mathsf{s}(x)))$ that define the function symbol p.

In our version of semantic labeling we require $\mathcal{U}(\ell) \subseteq \geqslant_\mathcal{A}$ instead of $\mathcal{R} \subseteq \geqslant_\mathcal{A}$. Moreover, we only need to define semantics for the function symbols in $\mathcal{G}_\ell$. Without further ado, this would be unsound, as can be seen from the following example.

Example 7. Consider the non-terminating TRS $\mathcal{R}$ (from [16])

$$\mathsf{f}(\mathsf{a}, \mathsf{b}, x) \to \mathsf{f}(x, x, x) \qquad \mathsf{g}(x, y) \to x \qquad \mathsf{g}(x, y) \to y$$

We want to distinguish the two occurrences of the function symbol f. This can be achieved by an algebra $\mathcal{A}$ consisting of the carrier $\{0, 1, a, b\}$ equipped with the well-founded order $> = \{(1, 0)\}$ and the interpretations $\mathsf{a}_\mathcal{A} = a$ and $\mathsf{b}_\mathcal{A} = b$, together with the weakly monotone labeling function

$$\ell_\mathsf{f}(x, y, z) = \begin{cases} 1 & \text{if } x = a \text{ and } y = b \\ 0 & \text{otherwise} \end{cases}$$

We have $\mathcal{G}_\ell = \{\mathsf{a}, \mathsf{b}\}$ and $\mathcal{U}(\ell) = \varnothing$. Obviously $\mathcal{U}(\ell) \subseteq \geqslant_\mathcal{A}$. The transformed TRS $\mathcal{R}_\mathsf{lab} \cup \mathcal{D}ec$

$$\mathsf{f}_1(\mathsf{a}, \mathsf{b}, x) \to \mathsf{f}_0(x, x, x) \qquad \mathsf{g}(x, y) \to x \qquad \mathsf{g}(x, y) \to y$$
$$\mathsf{f}_1(x, y, z) \to \mathsf{f}_0(x, y, z)$$

is terminating.

Definition 8. *Let $\mathcal{A} = (A, \{f_\mathcal{A}\}_{f \in \mathcal{F}})$ be an algebra equipped with a proper order $>$ on its carrier A. We say that $(\mathcal{A}, >)$ is a $\sqcup$-algebra if for all finite subsets $X \subseteq A$ there exists a least upper bound $\bigsqcup X$ of X in A. We denote $\bigsqcup \varnothing$ by $\bot$.*

Since every element of A is an upper bound of $\varnothing$, it follows that $\bot$ is the minimum element of A. This is used in the proof of Lemma 13 below. Note that the algebra in Example 7 is not a $\sqcup$-algebra as the set $\{a, b\}$ has no upper bound.

In the remainder of this section we assume that $\mathcal{R}$ is a *finitely branching* TRS over a signature $\mathcal{F}$, $(\mathcal{A}, >)$ with $\mathcal{A} = (A, \{f_\mathcal{A}\}_{f \in \mathcal{F}})$ a well-founded weakly monotone $\sqcup$-algebra, and ℓ a weakly monotone labeling for $(\mathcal{A}, >)$ such that $\mathcal{U}(\ell) \subseteq \geqslant_\mathcal{A}$ and $f_\mathcal{A}(a_1, \ldots, a_n) = \bot$ for all $f \notin \mathcal{G}_\ell$ and $a_1, \ldots, a_n \in A$. The latter assumption is harmless because function symbols in $\mathcal{F} \setminus \mathcal{G}_\ell$ are not involved when computing $\mathcal{R}_\mathsf{lab}$ or verifying $\mathcal{U}(\ell) \subseteq \geqslant_\mathcal{A}$.

The if-direction of Theorem 3 is proved in [18] by transforming a presupposed infinite rewrite sequence in $\mathcal{R}$ into an infinite rewrite sequence in $\mathcal{R}_\mathsf{lab} \cup \mathcal{D}ec$. This transformation is achieved by applying the labeling function $\mathsf{lab}_\alpha(\cdot)$ (for an arbitrary assignment α) to all terms in the infinite rewrite sequence of $\mathcal{R}$. The key property is that

$$\mathsf{lab}_\alpha(s) \to^+_{\mathcal{R}_\mathsf{lab} \cup \mathcal{D}ec} \mathsf{lab}_\alpha(t)$$

whenever $s \to_{\mathcal{R}} t$. In our setting this approach does not work since we lack sufficient semantic information to label arbitrary terms.

In the following definition an interpretation function $[\alpha]_{\mathcal{A}}^*(\cdot)$ is given for all terminating terms in $\mathcal{T}(\mathcal{F}, \mathcal{V})$ which provides more information than the standard interpretation function $[\alpha]_{\mathcal{A}}(\cdot)$. We write $\mathcal{SN}$ for the subset of $\mathcal{T}(\mathcal{F}, \mathcal{V})$ consisting of all terminating terms.

Definition 9. *Let $t \in \mathcal{SN}$ and α an assignment. We define the interpretation $[\alpha]_{\mathcal{A}}^*(t)$ inductively as follows:*

$$[\alpha]_{\mathcal{A}}^*(t) = \begin{cases} \alpha(x) & \text{if } t \text{ is a variable,} \\ f_{\mathcal{A}}([\alpha]_{\mathcal{A}}^*(t_1), \ldots, [\alpha]_{\mathcal{A}}^*(t_n)) & \text{if } t = f(t_1, \ldots, t_n) \text{ and } f \in \mathcal{G}_\ell, \\ \bigsqcup \{[\alpha]_{\mathcal{A}}^*(u) \mid t \to_{\mathcal{R}}^+ u\} & \text{if } t = f(t_1, \ldots, t_n) \text{ and } f \notin \mathcal{G}_\ell. \end{cases}$$

Note that the recursion in the definition of $[\alpha]_{\mathcal{A}}^*(\cdot)$ terminates because the union of $\to_{\mathcal{R}}^+$ and the proper superterm relation $\rhd$ is a well-founded relation on $\mathcal{SN}$. Further note that the operation $\bigsqcup$ is applied only to finite sets as $\mathcal{R}$ is assumed to be finitely branching. The definition of $[\alpha]_{\mathcal{A}}^*(t)$ can be viewed as a semantic version of a transformation of Gramlich [7, Definition 3], which is used for proving the modularity of collapsing extending $(\mathcal{C}_{\mathcal{E}})$ termination of finite branching TRSs. Here $\mathcal{R}$ is $\mathcal{C}_{\mathcal{E}}$-*terminating* if $\mathcal{R} \cup \{g(x, y) \to x, g(x, y) \to y\}$ with g a fresh function symbol is terminating. We remark that every $\bigsqcup$-algebra $(\mathcal{A}, >)$ satisfies $g(x, y) \geqslant_{\mathcal{A}} x$ and $g(x, y) \geqslant_{\mathcal{A}} y$ by taking the interpretation $g_{\mathcal{A}}(x, y) = \bigsqcup \{x, y\}$. Variations of Gramlich's definition have been more recently used in [17,8,15] to reduce the constraints originating from the dependency pair method.

The induced labeling function can be defined for terminating and for minimal non-terminating terms but not for arbitrary terms in $\mathcal{T}(\mathcal{F}, \mathcal{V})$.

Definition 10. *Let $t \in \mathcal{SN} \cup \mathcal{T}_\infty$ and α an assignment. We define the labeled term $\text{lab}_\alpha^*(t)$ inductively as follows:*

$$\text{lab}_\alpha^*(t) = \begin{cases} t & \text{if } t \text{ is a variable,} \\ f(\text{lab}_\alpha^*(t_1), \ldots, \text{lab}_\alpha^*(t_n)) & \text{if } L_f = \varnothing, \\ f_a(\text{lab}_\alpha^*(t_1), \ldots, \text{lab}_\alpha^*(t_n)) & \text{if } L_f \neq \varnothing \end{cases}$$

where $a = \ell_f([\alpha]_{\mathcal{A}}^*(t_1), \ldots, [\alpha]_{\mathcal{A}}^*(t_n))$.

We illustrate the above definitions on a concrete rewrite sequence with respect to the factorial example of the introduction.

Example 11. Consider the TRS $\mathcal{R}$ and the labeling ℓ restricted to fact of Example 1. We assume that $f_{\mathbb{N}}(x_1, \ldots, x_n) = 0$ for all function symbols $f \in \{\text{fact}, +, \times\}$ and all $x_1, \ldots, x_n \in \mathbb{N}$. Consider the rewrite sequence

$$\text{fact}(s(0) + \text{fact}(0)) \to \text{fact}(s(0 + \text{fact}(0))) \to \text{fact}(s(\text{fact}(0))) \to \text{fact}(s(s(0)))$$

and let α be an arbitrary assignment. (Since we deal with ground terms, the assignment does not matter.) We have

$$[\alpha]_{\mathbb{N}}^{*}(\mathsf{s}(0)) = 1$$

$$[\alpha]_{\mathbb{N}}^{*}(\mathsf{s}(\mathsf{s}(0))) = 2$$

$$[\alpha]_{\mathbb{N}}^{*}(\mathsf{fact}(0)) = [\alpha]_{\mathbb{N}}^{*}(0 + \mathsf{s}(0)) = \bigsqcup\{[\alpha]_{\mathbb{N}}^{*}(\mathsf{s}(0))\} = \bigsqcup\{1\} = 1$$

$$[\alpha]_{\mathbb{N}}^{*}(\mathsf{s}(\mathsf{fact}(0))) = \mathsf{s}_{\mathbb{N}}([\alpha]_{\mathbb{N}}^{*}(\mathsf{fact}(0))) = 1 + 1 = 2$$

$$[\alpha]_{\mathbb{N}}^{*}(0 + \mathsf{fact}(0)) = \bigsqcup\{[\alpha]_{\mathbb{N}}^{*}(\mathsf{fact}(0)), [\alpha]_{\mathbb{N}}^{*}(0 + \mathsf{s}(0)), [\alpha]_{\mathbb{N}}^{*}(\mathsf{s}(0))\} = \bigsqcup\{1\} = 1$$

$$[\alpha]_{\mathbb{N}}^{*}(\mathsf{s}(0 + \mathsf{fact}(0))) = \mathsf{s}_{\mathbb{N}}([\alpha]_{\mathbb{N}}^{*}(0 + \mathsf{fact}(0))) = 1 + 1 = 2$$

$$[\alpha]_{\mathbb{N}}^{*}(\mathsf{s}(0) + \mathsf{fact}(0)) = \bigsqcup\{\cdots\} = 2$$

and hence by applying $\mathsf{lab}_{\alpha}^{*}(\cdot)$ to all terms in the above rewrite sequence we obtain the sequence

$$\mathsf{fact}_2(\mathsf{s}(0) + \mathsf{fact}_0(0)) \to \mathsf{fact}_2(\mathsf{s}(0 + \mathsf{fact}_0(0))) \to \mathsf{fact}_2(\mathsf{s}(\mathsf{fact}_0(0)))$$
$$\to \mathsf{fact}_2(\mathsf{s}(\mathsf{s}(0)))$$

in $\mathcal{R}_{\mathsf{lab}}$.

The following lemma compares the predicted semantics of an instantiated terminating term to the original semantics of the uninstantiated term, in which the substitution becomes part of the assignment.

Definition 12. *Given an assignment α and a substitution σ such that $\sigma(x) \in \mathcal{SN}$ for all variables x, the assignment α_{σ}^{*} is defined as $[\alpha]_{\mathcal{A}}^{*} \circ \sigma$ and the substitution $\sigma_{\mathsf{lab}_{\alpha}^{*}}$ as $\mathsf{lab}_{\alpha}^{*} \circ \sigma$.*

Lemma 13. *If $t\sigma \in \mathcal{SN}$ then $[\alpha]_{\mathcal{A}}^{*}(t\sigma) \geqslant [\alpha_{\sigma}^{*}]_{\mathcal{A}}(t)$. If in addition $\mathcal{F}\mathsf{un}(t) \subseteq \mathcal{G}_{\ell}$ then $[\alpha]_{\mathcal{A}}^{*}(t\sigma) = [\alpha_{\sigma}^{*}]_{\mathcal{A}}(t)$.*

Proof. We use structural induction on t. If t is a variable then

$$[\alpha]_{\mathcal{A}}^{*}(t\sigma) = ([\alpha]_{\mathcal{A}}^{*} \circ \sigma)(t) = [\alpha_{\sigma}^{*}]_{\mathcal{A}}(t).$$

Suppose $t = f(t_1, \ldots, t_n)$. We distinguish two cases.

1. If $f \in \mathcal{G}_{\ell}$ then

$$\begin{aligned} [\alpha]_{\mathcal{A}}^{*}(t\sigma) &= f_{\mathcal{A}}([\alpha]_{\mathcal{A}}^{*}(t_1\sigma), \ldots, [\alpha]_{\mathcal{A}}^{*}(t_n\sigma)) \\ &\geqslant f_{\mathcal{A}}([\alpha_{\sigma}^{*}]_{\mathcal{A}}(t_1), \ldots, [\alpha_{\sigma}^{*}]_{\mathcal{A}}(t_n)) \\ &= [\alpha_{\sigma}^{*}]_{\mathcal{A}}(t) \end{aligned}$$

   where the inequality follows from the induction hypothesis (note that $t_i\sigma \in \mathcal{SN}$ for all $i = 1, \ldots, n$) and the weak monotonicity of $f_{\mathcal{A}}$. If $\mathcal{F}\mathsf{un}(t) \subseteq \mathcal{G}_{\ell}$ then $\mathcal{F}\mathsf{un}(t_i) \subseteq \mathcal{G}_{\ell}$ and the inequality is turned into an equality.
2. If $f \notin \mathcal{G}_{\ell}$ then $f_{\mathcal{A}}(a_1, \ldots, a_n) = \bot$ for all $a_1, \ldots, a_n \in A$ and thus

$$[\alpha]_{\mathcal{A}}^{*}(t\sigma) \geqslant \bot = [\alpha_{\sigma}^{*}]_{\mathcal{A}}(t)$$

   In this case $\mathcal{F}\mathsf{un}(t) \subseteq \mathcal{G}_{\ell}$ does not hold, so the second part of the lemma holds vacuously.   □

The next two lemmata do the same for labeled terms. Since the label of a function symbol depends on the semantics of its arguments, we can deal with minimal non-terminating terms.

Lemma 14. *Let $t\sigma \in \mathcal{SN} \cup \mathcal{T}_\infty$. If $\mathcal{F}un(t_1) \cup \cdots \cup \mathcal{F}un(t_n) \subseteq \mathcal{G}_\ell$ when $t = f(t_1, \ldots, t_n)$ then $\mathsf{lab}^*_\alpha(t\sigma) = \mathsf{lab}_{\alpha^*_\sigma}(t)\sigma_{\mathsf{lab}^*_\alpha}$.*

Proof. We use structural induction on t. If t is a variable then

$$\mathsf{lab}^*_\alpha(t\sigma) = t\sigma_{\mathsf{lab}^*_\alpha} = \mathsf{lab}_{\alpha^*_\sigma}(t)\sigma_{\mathsf{lab}^*_\alpha}.$$

Suppose $t = f(t_1, \ldots, t_n)$. The induction hypothesis yields

$$\mathsf{lab}^*_\alpha(t_i\sigma) = \mathsf{lab}_{\alpha^*_\sigma}(t_i)\sigma_{\mathsf{lab}^*_\alpha}$$

for $i = 1, \ldots, n$. We distinguish two cases.

1. If $L_f = \varnothing$ then

$$
\begin{aligned}
\mathsf{lab}^*_\alpha(t\sigma) &= f(\mathsf{lab}^*_\alpha(t_1\sigma), \ldots, \mathsf{lab}^*_\alpha(t_n\sigma)) \\
&= f(\mathsf{lab}_{\alpha^*_\sigma}(t_1)\sigma_{\mathsf{lab}^*_\alpha}, \ldots, \mathsf{lab}_{\alpha^*_\sigma}(t_n)\sigma_{\mathsf{lab}^*_\alpha}) \\
&= f(\mathsf{lab}_{\alpha^*_\sigma}(t_1), \ldots, \mathsf{lab}_{\alpha^*_\sigma}(t_n))\sigma_{\mathsf{lab}^*_\alpha} \\
&= \mathsf{lab}_{\alpha^*_\sigma}(f(t_1, \ldots, t_n))\sigma_{\mathsf{lab}^*_\alpha}.
\end{aligned}
$$

2. If $L_f \neq \varnothing$ then

$$
\begin{aligned}
\mathsf{lab}^*_\alpha(t\sigma) &= f_a(\mathsf{lab}^*_\alpha(t_1\sigma), \ldots, \mathsf{lab}^*_\alpha(t_n\sigma)) \\
&= f_a(\mathsf{lab}_{\alpha^*_\sigma}(t_1)\sigma_{\mathsf{lab}^*_\alpha}, \ldots, \mathsf{lab}_{\alpha^*_\sigma}(t_n)\sigma_{\mathsf{lab}^*_\alpha}) \\
&= f_a(\mathsf{lab}_{\alpha^*_\sigma}(t_1), \ldots, \mathsf{lab}_{\alpha^*_\sigma}(t_n))\sigma_{\mathsf{lab}^*_\alpha}
\end{aligned}
$$

and

$$\mathsf{lab}_{\alpha^*_\sigma}(t)\sigma_{\mathsf{lab}^*_\alpha} = f_b(\mathsf{lab}_{\alpha^*_\sigma}(t_1), \ldots, \mathsf{lab}_{\alpha^*_\sigma}(t_n))\sigma_{\mathsf{lab}^*_\alpha}$$

with $a = \ell_f([\alpha]^*_\mathcal{A}(t_1\sigma), \ldots, [\alpha]^*_\mathcal{A}(t_n\sigma))$ and $b = \ell_f([\alpha^*_\sigma]_\mathcal{A}(t_1), \ldots, [\alpha^*_\sigma]_\mathcal{A}(t_n))$. Because $\mathcal{F}un(t_i) \subseteq \mathcal{G}_\ell$, Lemma 13 yields $[\alpha]^*_\mathcal{A}(t_i\sigma) = [\alpha^*_\sigma]_\mathcal{A}(t_i)$, for all $i = 1, \ldots, n$. Hence $a = b$ and therefore $\mathsf{lab}^*_\alpha(t\sigma) = \mathsf{lab}_{\alpha^*_\sigma}(t)\sigma_{\mathsf{lab}^*_\alpha}$ as desired.   □

Lemma 15. *If $t\sigma \in \mathcal{SN} \cup \mathcal{T}_\infty$ then $\mathsf{lab}^*_\alpha(t\sigma) \to^*_{\mathcal{D}ec} \mathsf{lab}_{\alpha^*_\sigma}(t)\sigma_{\mathsf{lab}^*_\alpha}$.*

Proof. We use structural induction on t. If t is a variable then we obtain $\mathsf{lab}^*_\alpha(t\sigma) = \mathsf{lab}_{\alpha^*_\sigma}(t)\sigma_{\mathsf{lab}^*_\alpha}$ from Lemma 14. Suppose $t = f(t_1, \ldots, t_n)$. Note that $t_1, \ldots, t_n \in \mathcal{SN}$. The induction hypothesis yields $\mathsf{lab}^*_\alpha(t_i\sigma) \to^*_{\mathcal{D}ec} \mathsf{lab}_{\alpha^*_\sigma}(t_i)\sigma_{\mathsf{lab}^*_\alpha}$ for all $i = 1, \ldots, n$. We distinguish two cases.

1. If $L_f = \varnothing$ then

$$
\begin{aligned}
\mathsf{lab}^*_\alpha(t\sigma) = \ \ & f(\mathsf{lab}^*_\alpha(t_1\sigma), \ldots, \mathsf{lab}^*_\alpha(t_n\sigma)) \\
\to^*_{\mathcal{D}ec} \ & f(\mathsf{lab}_{\alpha^*_\sigma}(t_1)\sigma_{\mathsf{lab}^*_\alpha}, \ldots, \mathsf{lab}_{\alpha^*_\sigma}(t_n)\sigma_{\mathsf{lab}^*_\alpha}) \\
= \ \ & f(\mathsf{lab}_{\alpha^*_\sigma}(t_1), \ldots, \mathsf{lab}_{\alpha^*_\sigma}(t_n))\sigma_{\mathsf{lab}^*_\alpha} \\
= \ \ & \mathsf{lab}_{\alpha^*_\sigma}(f(t_1, \ldots, t_n))\sigma_{\mathsf{lab}^*_\alpha}.
\end{aligned}
$$

2. If $L_f \neq \varnothing$ then

$$\begin{aligned}
\mathsf{lab}_\alpha^*(t\sigma) &= f_a(\mathsf{lab}_\alpha^*(t_1\sigma), \ldots, \mathsf{lab}_\alpha^*(t_n\sigma)) \\
&\to_{\mathcal{D}ec}^* f_a(\mathsf{lab}_{\alpha_\sigma^*}(t_1)\sigma_{\mathsf{lab}_\alpha^*}, \ldots, \mathsf{lab}_{\alpha_\sigma^*}(t_n)\sigma_{\mathsf{lab}_\alpha^*})
\end{aligned}$$

and

$$\begin{aligned}
\mathsf{lab}_{\alpha_\sigma^*}(t)\sigma_{\mathsf{lab}_\alpha^*} &= f_b(\mathsf{lab}_{\alpha_\sigma^*}(t_1), \ldots, \mathsf{lab}_{\alpha_\sigma^*}(t_n))\sigma_{\mathsf{lab}_\alpha^*} \\
&= f_b(\mathsf{lab}_{\alpha_\sigma^*}(t_1)\sigma_{\mathsf{lab}_\alpha^*}, \ldots, \mathsf{lab}_{\alpha_\sigma^*}(t_n)\sigma_{\mathsf{lab}_\alpha^*})
\end{aligned}$$

with $a = \ell_f([\alpha]_\mathcal{A}^*(t_1\sigma), \ldots, [\alpha]_\mathcal{A}^*(t_n\sigma))$ and $b = \ell_f([\alpha_\sigma^*]_\mathcal{A}(t_1), \ldots, [\alpha_\sigma^*]_\mathcal{A}(t_n))$. Lemma 13 yields $[\alpha]_\mathcal{A}^*(t_i\sigma) \geqslant [\alpha_\sigma^*]_\mathcal{A}(t_i)$ for all $i = 1, \ldots, n$. Because the labeling function ℓ_f is weakly monotone in all its coordinates, $a \geqslant b$. If $a > b$ then $\mathcal{D}ec$ contains the rewrite rule $f_a(x_1, \ldots, x_n) \to f_b(x_1, \ldots, x_n)$ and thus (also if $a = b$)

$$f_a(\mathsf{lab}_{\alpha_\sigma^*}(t_1)\sigma_{\mathsf{lab}_\alpha^*}, \ldots, \mathsf{lab}_{\alpha_\sigma^*}(t_n)\sigma_{\mathsf{lab}_\alpha^*}) \to_{\mathcal{D}ec}^= \mathsf{lab}_{\alpha_\sigma^*}(t)\sigma_{\mathsf{lab}_\alpha^*}.$$

We conclude that $\mathsf{lab}_\alpha^*(t\sigma) \to_{\mathcal{D}ec}^* \mathsf{lab}_{\alpha_\sigma^*}(t)\sigma_{\mathsf{lab}_\alpha^*}$.   □

The next lemma states that the rewrite sequence in Lemma 15 is empty when t is a subterm of the right-hand side of a rule.

Lemma 16. *If $l \to r \in \mathcal{R}$ and $t \trianglelefteq r$ such that $t\sigma \in \mathcal{SN} \cup \mathcal{T}_\infty$ then $\mathsf{lab}_\alpha^*(t\sigma) = \mathsf{lab}_{\alpha_\sigma^*}(t)\sigma_{\mathsf{lab}_\alpha^*}$.*

Proof. We use structural induction on t. If t is a variable then we obtain

$$\mathsf{lab}_\alpha^*(t\sigma) = \mathsf{lab}_{\alpha_\sigma^*}(t)\sigma_{\mathsf{lab}_\alpha^*}$$

from Lemma 14. Suppose $t = f(t_1, \ldots, t_n)$. The induction hypothesis yields $\mathsf{lab}_\alpha^*(t_i\sigma) = \mathsf{lab}_{\alpha_\sigma^*}(t_i)\sigma_{\mathsf{lab}_\alpha^*}$ for all $i = 1, \ldots, n$. If $L_f = \varnothing$ then we obtain $\mathsf{lab}_\alpha^*(t\sigma) = \mathsf{lab}_{\alpha_\sigma^*}(t)\sigma_{\mathsf{lab}_\alpha^*}$ as in the proof of Lemma 14. If $L_f \neq \varnothing$ then $\mathcal{F}un(t_1) \cup \cdots \cup \mathcal{F}un(t_n) \subseteq \mathcal{G}_\ell$ by the definition of $\mathcal{G}_\ell$. Since $t_i\sigma \in \mathcal{SN}$, we have $t\sigma \in \mathcal{T}_\infty$ and therefore Lemma 14 yields $\mathsf{lab}_\alpha^*(t\sigma) = \mathsf{lab}_{\alpha_\sigma^*}(t)\sigma_{\mathsf{lab}_\alpha^*}$.   □

We are now ready for the key lemma, which states that rewrite steps between terminating and minimal non-terminating terms can be labeled.

Lemma 17. *Let $s, t \in \mathcal{SN} \cup \mathcal{T}_\infty$. If $s \to_\mathcal{R} t$ then $\mathsf{lab}_\alpha^*(s) \to_{\mathcal{R}_{\mathsf{lab}} \cup \mathcal{D}ec}^+ \mathsf{lab}_\alpha^*(t)$.*

Proof. Write $s = C[l\sigma]$ and $t = C[r\sigma]$. We use structural induction on the context C. If $C = \square$ then

$$\begin{aligned}
\mathsf{lab}_\alpha^*(s) = \mathsf{lab}_\alpha^*(l\sigma) &\to_{\mathcal{D}ec}^* \mathsf{lab}_{\alpha_\sigma^*}(l)\sigma_{\mathsf{lab}_\alpha^*} \\
&\to_{\mathcal{R}_{\mathsf{lab}}} \mathsf{lab}_{\alpha_\sigma^*}(r)\sigma_{\mathsf{lab}_\alpha^*} = \mathsf{lab}_\alpha^*(r\sigma)
\end{aligned}$$

using Lemmata 15 and 16. Let $C = f(s_1, \ldots, C', \ldots, s_n)$. The induction hypothesis yields $\mathsf{lab}_\alpha^*(C'[l\sigma]) \to_{\mathcal{R}_{\mathsf{lab}} \cup \mathcal{D}ec}^+ \mathsf{lab}_\alpha^*(C'[r\sigma])$. We distinguish two cases.

1. If $L_f = \varnothing$ then

$$
\begin{aligned}
\mathsf{lab}^*_\alpha(s) = & \quad f(\mathsf{lab}^*_\alpha(s_1), \ldots, \mathsf{lab}^*_\alpha(C'[l\sigma]), \ldots, \mathsf{lab}^*_\alpha(s_n)) \\
\to^+_{\mathcal{R}_{\mathsf{lab}} \cup \mathcal{D}ec} & \quad f(\mathsf{lab}^*_\alpha(s_1), \ldots, \mathsf{lab}^*_\alpha(C'[r\sigma]), \ldots, \mathsf{lab}^*_\alpha(s_n)) \\
= & \quad \mathsf{lab}^*_\alpha(t).
\end{aligned}
$$

2. If $L_f \neq \varnothing$ then

$$
\begin{aligned}
\mathsf{lab}^*_\alpha(s) = & \quad f_a(\mathsf{lab}^*_\alpha(s_1), \ldots, \mathsf{lab}^*_\alpha(C'[l\sigma]), \ldots, \mathsf{lab}^*_\alpha(s_n)) \\
\to^+_{\mathcal{R}_{\mathsf{lab}} \cup \mathcal{D}ec} & \quad f_a(\mathsf{lab}^*_\alpha(s_1), \ldots, \mathsf{lab}^*_\alpha(C'[r\sigma]), \ldots, \mathsf{lab}^*_\alpha(s_n))
\end{aligned}
$$

with

$$
a = \ell_f([\alpha]^*_{\mathcal{A}}(s_1), \ldots, [\alpha]^*_{\mathcal{A}}(C'[l\sigma]), \ldots, [\alpha]^*_{\mathcal{A}}(s_n))
$$

and

$$
\mathsf{lab}^*_\alpha(t) = f_b(\mathsf{lab}^*_\alpha(s_1), \ldots, \mathsf{lab}^*_\alpha(C'[r\sigma]), \ldots, \mathsf{lab}^*_\alpha(s_n))
$$

with

$$
b = \ell_f([\alpha]^*_{\mathcal{A}}(s_1), \ldots, [\alpha]^*_{\mathcal{A}}(C'[r\sigma]), \ldots, [\alpha]^*_{\mathcal{A}}(s_n))
$$

If we can show that

$$
[\alpha]^*_{\mathcal{A}}(C'[l\sigma]) \geqslant [\alpha]^*_{\mathcal{A}}(C'[r\sigma]) \tag{1}
$$

then $a \geqslant b$ by weak monotonicity of ℓ_f and thus

$$
f_a(\mathsf{lab}^*_\alpha(s_1), \ldots, \mathsf{lab}^*_\alpha(C'[r\sigma]), \ldots, \mathsf{lab}^*_\alpha(s_n)) \to^=_{\mathcal{D}ec} \mathsf{lab}^*_\alpha(t).
$$

We prove (1) by structural induction on C'.

(a) First assume that $C' = \square$. We distinguish two cases. If $\mathsf{root}(l\sigma) = \mathsf{root}(l) \in \mathcal{G}_\ell$ then $l \to r \in \mathcal{U}(\ell)$ and $\mathcal{F}\mathsf{un}(r) \subseteq \mathcal{G}_\ell$ according to the definition of $\mathcal{G}_\ell$. Hence

$$
[\alpha]^*_{\mathcal{A}}(l\sigma) \geqslant [\alpha^*_\sigma]_{\mathcal{A}}(l)
$$

by Lemma 13,

$$
[\alpha^*_\sigma]_{\mathcal{A}}(l) \geqslant [\alpha^*_\sigma]_{\mathcal{A}}(r)
$$

since $l \geqslant_{\mathcal{A}} r$ due to the assumption $\mathcal{U}(\ell) \subseteq \geqslant_{\mathcal{A}}$, and

$$
[\alpha^*_\sigma]_{\mathcal{A}}(r) = [\alpha]^*_{\mathcal{A}}(r\sigma)
$$

by another application of Lemma 13. The combination yields the desired $[\alpha]^*_{\mathcal{A}}(l\sigma) \geqslant [\alpha]^*_{\mathcal{A}}(r\sigma)$. If $\mathsf{root}(l\sigma) = \mathsf{root}(l) \notin \mathcal{G}_\ell$ then

$$
[\alpha]^*_{\mathcal{A}}(l\sigma) = \bigsqcup \{[\alpha]^*_{\mathcal{A}}(u) \mid l\sigma \to^+_{\mathcal{R}} u\}
$$

Because $l\sigma \to_{\mathcal{R}} r\sigma$, $[\alpha]^*_{\mathcal{A}}(r\sigma) \in \{[\alpha]^*_{\mathcal{A}}(u) \mid l\sigma \to^+_{\mathcal{R}} u\}$ and thus also in this case $[\alpha]^*_{\mathcal{A}}(l\sigma) \geqslant [\alpha]^*_{\mathcal{A}}(r\sigma)$.

(b) Next assume that $C' = g(u_1, \ldots, C'', \ldots, u_m)$. The induction hypothesis yields $[\alpha]_{\mathcal{A}}^*(C''[l\sigma]) \geqslant [\alpha]_{\mathcal{A}}^*(C''[r\sigma])$. If $g \in \mathcal{G}_\ell$ then

$$[\alpha]_{\mathcal{A}}^*(C'[l\sigma]) = g_{\mathcal{A}}([\alpha]_{\mathcal{A}}^*(u_1), \ldots, [\alpha]_{\mathcal{A}}^*(C''[l\sigma]), \ldots, [\alpha]_{\mathcal{A}}^*(u_m))$$

and

$$[\alpha]_{\mathcal{A}}^*(C'[r\sigma]) = g_{\mathcal{A}}([\alpha]_{\mathcal{A}}^*(u_1), \ldots, [\alpha]_{\mathcal{A}}^*(C''[r\sigma]), \ldots, [\alpha]_{\mathcal{A}}^*(u_m))$$

and thus $[\alpha]_{\mathcal{A}}^*(C'[l\sigma]) \geqslant [\alpha]_{\mathcal{A}}^*(C'[r\sigma])$ by the weak monotonicity of $g_{\mathcal{A}}$. If $g \notin \mathcal{G}_\ell$ then

$$[\alpha]_{\mathcal{A}}^*(C'[l\sigma]) = \bigsqcup \{[\alpha]_{\mathcal{A}}^*(u) \mid C'[l\sigma] \to_{\mathcal{R}}^+ u\}$$

Because $C'[l\sigma] \to_{\mathcal{R}} C'[r\sigma]$, $[\alpha]_{\mathcal{A}}^*(C'[r\sigma]) \in \{[\alpha]_{\mathcal{A}}^*(u) \mid C'[l\sigma] \to_{\mathcal{R}}^+ u\}$ and thus $[\alpha]_{\mathcal{A}}^*(C'[l\sigma]) \geqslant [\alpha]_{\mathcal{A}}^*(C'[r\sigma])$. □

We now have all the ingredients to prove the soundness of predictive labeling.

Theorem 18. *Let $\mathcal{R}$ be a TRS, $(\mathcal{A}, >)$ a well-founded weakly monotone $\sqcup$-algebra, and ℓ a weakly monotone labeling for $(\mathcal{A}, >)$ such that $\mathcal{U}(\ell) \subseteq \geqslant_{\mathcal{A}}$. If $\mathcal{R}_{\mathsf{lab}} \cup \mathcal{D}ec$ is terminating then $\mathcal{R}$ is terminating.*

Proof. According to Lemma 4 for every term $t \in \mathcal{T}_\infty$ there exist a rewrite rule $l \to r \in \mathcal{R}$, a substitution σ, and a subterm u of r such that

$$t \xrightarrow{>\epsilon}{}^* l\sigma \xrightarrow{\epsilon} r\sigma \trianglerighteq u\sigma$$

and $l\sigma, u\sigma \in \mathcal{T}_\infty$. Let α be an arbitrary assignment. We will apply lab_α^* to the terms in the above sequence. From Lemma 17 we obtain

$$\mathsf{lab}_\alpha^*(t) \to_{\mathcal{R}_{\mathsf{lab}} \cup \mathcal{D}ec}^* \mathsf{lab}_\alpha^*(l\sigma).$$

Since $r\sigma$ need not be an element of $\mathcal{T}_\infty$, we cannot apply Lemma 17 to the step $l\sigma \xrightarrow{\epsilon} r\sigma$. Instead we use Lemma 15 to obtain

$$\mathsf{lab}_\alpha^*(l\sigma) \to_{\mathcal{D}ec}^* \mathsf{lab}_{\alpha_\sigma^*}(l)\sigma_{\mathsf{lab}_\alpha^*}.$$

Since $\mathsf{lab}_{\alpha_\sigma^*}(l) \to \mathsf{lab}_{\alpha_\sigma^*}(r) \in \mathcal{R}_{\mathsf{lab}}$,

$$\mathsf{lab}_{\alpha_\sigma^*}(l)\sigma_{\mathsf{lab}_\alpha^*} \to_{\mathcal{R}_{\mathsf{lab}}} \mathsf{lab}_{\alpha_\sigma^*}(r)\sigma_{\mathsf{lab}_\alpha^*}.$$

Because u is a subterm of r, $\mathsf{lab}_{\alpha_\sigma^*}(r)\sigma_{\mathsf{lab}_\alpha^*} \trianglerighteq \mathsf{lab}_{\alpha_\sigma^*}(u)\sigma_{\mathsf{lab}_\alpha^*}$. Lemma 16 yields $\mathsf{lab}_{\alpha_\sigma^*}(u)\sigma_{\mathsf{lab}_\alpha^*} = \mathsf{lab}_\alpha^*(u\sigma)$. Putting everything together, we obtain

$$\mathsf{lab}_\alpha^*(t) \to_{\mathcal{R}_{\mathsf{lab}} \cup \mathcal{D}ec}^+ \cdot \trianglerighteq \mathsf{lab}_\alpha^*(u\sigma).$$

Now suppose that $\mathcal{R}$ is non-terminating. Then $\mathcal{T}_\infty$ is non-empty and thus there is an infinite sequence

$$t_1 \xrightarrow{>\epsilon}{}^* \cdot \xrightarrow{\epsilon} \cdot \trianglerighteq t_2 \xrightarrow{>\epsilon}{}^* \cdot \xrightarrow{\epsilon} \cdot \trianglerighteq t_3 \xrightarrow{>\epsilon}{}^* \cdot \xrightarrow{\epsilon} \cdot \trianglerighteq \cdots$$

By the above argument, this sequence is transformed into

$$\mathsf{lab}_\alpha^*(t_1) \to_{\mathcal{R}_{\mathsf{lab}} \cup \mathcal{D}\mathrm{ec}}^+ \cdot \unrhd \mathsf{lab}_\alpha^*(t_2) \to_{\mathcal{R}_{\mathsf{lab}} \cup \mathcal{D}\mathrm{ec}}^+ \cdot \unrhd \mathsf{lab}_\alpha^*(t_3) \to_{\mathcal{R}_{\mathsf{lab}} \cup \mathcal{D}\mathrm{ec}}^+ \cdot \unrhd \cdots$$

By introducing appropriate contexts, the latter sequence gives rise to an infinite rewrite sequence in $\mathcal{R}_{\mathsf{lab}} \cup \mathcal{D}\mathrm{ec}$, contradicting the assumption that the latter system is terminating. □

We conclude this section by showing that, due to the least upper bound condition on the algebras that may be used in connection with Theorem 18, predictive labeling does not succeed in transforming every terminating TRS into a precedence terminating TRS.

Example 19. The one-rule TRS $\mathcal{R} = \{\mathsf{f}(\mathsf{a}, \mathsf{b}, x) \to \mathsf{f}(x, x, x)\}$ is terminating but not precedence terminating. Suppose $\mathcal{R}$ can be transformed by predictive labeling into a precedence terminating TRS. This is only possible if the two occurrences of f get a different label. Let $\mathcal{A} = (A, \{\mathsf{f}_A, \mathsf{a}_A, \mathsf{b}_A\})$ be a well-founded weakly monotone $\sqcup$-algebra and ℓ a weakly monotone labeling such that $\mathcal{U}(\ell) \subseteq \geqslant_A$ and $\mathcal{R}_{\mathsf{lab}} \cup \mathcal{D}\mathrm{ec}$ is precedence terminating. Since $L_f \neq \varnothing$, the labeling function ℓ_f exists and we must have $\ell_f(\mathsf{a}_A, \mathsf{b}_A, x) \neq \ell_f(x, x, x)$ for all $x \in A$. Take $x = \bigsqcup\{\mathsf{a}_A, \mathsf{b}_A\}$. Since $x \geqslant \mathsf{a}_A$ and $x \geqslant \mathsf{b}_A$, we obtain

$$a = \ell_f(x, x, x) \geqslant \ell_f(\mathsf{a}_A, \mathsf{b}_A, x) = b$$

from the weak monotonicity of ℓ_f. Since we cannot have $a = b$, $a > b$ must hold. Hence $\mathcal{D}\mathrm{ec}$ contains the rule $\mathsf{f}_a(x, y, z) \to \mathsf{f}_b(x, y, z)$ whereas $\mathcal{R}_{\mathsf{lab}}$ contains the rule $\mathsf{f}_b(\mathsf{a}, \mathsf{b}, x) \to \mathsf{f}_a(x, x, x)$ It follows that $\mathcal{R}_{\mathsf{lab}} \cup \mathcal{D}\mathrm{ec}$ cannot be precedence terminating, contradicting our assumption.

4   Examples

In this section we present two more examples.

Example 20. Consider the TRS $\mathcal{R}$ consisting of the following rewrite rules:

1:	$x - 0$	$\to$	x	6:	$\mathsf{gcd}(0, y)$	$\to$	y
2:	$\mathsf{s}(x) - \mathsf{s}(y)$	$\to$	$x - y$	7:	$\mathsf{gcd}(\mathsf{s}(x), 0)$	$\to$	$\mathsf{s}(x)$
3:	$0 \leq y$	$\to$	true	8:	$\mathsf{gcd}(\mathsf{s}(x), \mathsf{s}(y))$	$\to$	$\mathsf{ifgcd}(y \leq x, \mathsf{s}(x), \mathsf{s}(y))$
4:	$\mathsf{s}(x) \leq 0$	$\to$	false	9:	$\mathsf{ifgcd}(\mathsf{true}, \mathsf{s}(x), \mathsf{s}(y))$	$\to$	$\mathsf{gcd}(x - y, \mathsf{s}(y))$
5:	$\mathsf{s}(x) \leq \mathsf{s}(y)$	$\to$	$x \leq y$	10:	$\mathsf{ifgcd}(\mathsf{false}, \mathsf{s}(x), \mathsf{s}(y))$	$\to$	$\mathsf{gcd}(y - x, \mathsf{s}(x))$

We use the interpretations

$$0_\mathbb{N} = \mathsf{true}_\mathbb{N} = \mathsf{false}_\mathbb{N} = \leq_\mathbb{N}(x, y) = 0 \qquad \mathsf{s}_\mathbb{N}(x) = x + 1 \qquad -_\mathbb{N}(x, y) = x$$

and the labeling

$$\ell_{\mathsf{gcd}}(x, y) = x + y \qquad \ell_{\mathsf{ifgcd}}(x, y, z) = y + z$$

We have $\mathcal{G}_\ell = \{0, \mathsf{s}, \mathsf{true}, \mathsf{false}, \leq, -\}$ and thus $\mathcal{U}(\ell) = \{1, 2, \ldots, 5\}$. One easily checks that $\mathcal{U}(\ell) \subseteq \geqslant_\mathbb{N}$. The TRS $\mathcal{R}_{\mathsf{lab}}$ consists of the rewrite rules

$$
\begin{aligned}
1: & & x - 0 &\rightarrow x \\
2: & & \mathsf{s}(x) - \mathsf{s}(y) &\rightarrow x - y \\
3: & & 0 \leq y &\rightarrow \mathsf{true} \\
4: & & \mathsf{s}(x) \leq 0 &\rightarrow \mathsf{false} \\
5: & & \mathsf{s}(x) \leq \mathsf{s}(y) &\rightarrow x \leq y \\
6': & & \mathsf{gcd}_j(0, y) &\rightarrow y \\
7': & & \mathsf{gcd}_{i+1}(\mathsf{s}(x), 0) &\rightarrow \mathsf{s}(x) \\
8': & & \mathsf{gcd}_{i+j+2}(\mathsf{s}(x), \mathsf{s}(y)) &\rightarrow \mathsf{ifgcd}_{i+j+2}(y \leq x, \mathsf{s}(x), \mathsf{s}(y)) \\
9': & \mathsf{ifgcd}_{i+j+2}(\mathsf{true}, \mathsf{s}(x), \mathsf{s}(y)) &\rightarrow \mathsf{gcd}_{i+j+1}(x - y, \mathsf{s}(y)) \\
10': & \mathsf{ifgcd}_{i+j+2}(\mathsf{false}, \mathsf{s}(x), \mathsf{s}(y)) &\rightarrow \mathsf{gcd}_{i+j+1}(y - x, \mathsf{s}(x))
\end{aligned}
$$

for all $i, j \geqslant 0$ and the TRS $\mathcal{D}ec$ consists of the rules $\mathsf{gcd}_i(x, y) \rightarrow \mathsf{gcd}_j(x, y)$ and $\mathsf{ifgcd}_i(x, y, z) \rightarrow \mathsf{ifgcd}_j(x, y, z)$ for all $i > j \geqslant 0$. Their union is oriented from left to right by the lexicographic path order induced by the well-founded precedence

$$\mathsf{ifgcd}_{i+1} > \mathsf{gcd}_{i+1} > \mathsf{ifgcd}_i > \mathsf{gcd}_i > \cdots > \mathsf{gcd}_0 > - > \leq > \mathsf{true} > \mathsf{false}$$

and hence terminating. Theorem 18 yields the termination of $\mathcal{R}$.

Example 21. Consider the TRS $\mathcal{R}$ consisting of the following rewrite rules:

$$
\begin{aligned}
1: && \mathsf{half}(0) &\rightarrow 0 & 4: && \mathsf{bits}(0) &\rightarrow 0 \\
2: && \mathsf{half}(\mathsf{s}(0)) &\rightarrow 0 & 5: && \mathsf{bits}(\mathsf{s}(x)) &\rightarrow \mathsf{s}(\mathsf{bits}(\mathsf{half}(\mathsf{s}(x)))) \\
3: & \mathsf{half}(\mathsf{s}(\mathsf{s}(x))) &\rightarrow \mathsf{s}(\mathsf{half}(x))
\end{aligned}
$$

We use the interpretations

$$0_\mathbb{N} = 0 \qquad \mathsf{s}_\mathbb{N}(x) = x + 1 \qquad \mathsf{half}_\mathbb{N}(x) = \max\{x - 1, 0\}$$

and the labeling $\ell_{\mathsf{bit}}(x) = x$. We have $\mathcal{G}_\ell = \{0, \mathsf{s}, \mathsf{half}\}$ and $\mathcal{U}(\ell) = \{1, 2, 3\}$. Clearly $\mathcal{U}(\ell) \subseteq \geqslant_\mathbb{N}$. The TRS $\mathcal{R}_{\mathsf{lab}}$ consists of the rewrite rules

$$
\begin{aligned}
1: && \mathsf{half}(0) &\rightarrow 0 & 4': && \mathsf{bits}_0(0) &\rightarrow 0 \\
2: && \mathsf{half}(\mathsf{s}(0)) &\rightarrow 0 & 5': && \mathsf{bits}_{i+1}(\mathsf{s}(x)) &\rightarrow \mathsf{s}(\mathsf{bits}_i(\mathsf{half}(\mathsf{s}(x)))) \\
3: & \mathsf{half}(\mathsf{s}(\mathsf{s}(x))) &\rightarrow \mathsf{s}(\mathsf{half}(x))
\end{aligned}
$$

for all $i \geqslant 0$ and the TRS $\mathcal{D}ec$ consists of the rules $\mathsf{bits}_i(x) \rightarrow \mathsf{bits}_j(x)$ for all $i > j \geqslant 0$. Their union is oriented from left to right by the lexicographic path order induced by the well-founded precedence

$$\mathsf{bits}_{i+1} > \mathsf{bits}_i > \cdots > \mathsf{bits}_0 > \mathsf{half} > \mathsf{s} > 0$$

and thus we conclude that $\mathcal{R}$ is terminating.

5   Conclusion

Predictive labeling (Theorem 18) is a variant of the quasi-model version of semantic labeling (Theorem 3). A natural question is whether the usable rules refinement is also applicable to the model version of semantic labeling (Theorem 2). This would be interesting not so much because we would get rid of the rewrite rules in $\mathcal{D}ec$ but especially because without the weak monotonicity condition more labeling functions are possible. The termination prover TORPA mentioned in the introduction implements both versions of semantic labeling for that reason. Unfortunately, it is not immediately clear how the definitions and proofs in Section 3 have to be modified in order to obtain a model version of predictive labeling.

Although the example at the end of Section 3 shows that predictive labeling is less powerful than semantic labeling, we believe that predictive labeling may be more useful when it comes to automation. First note that the algebras used in the implementations of semantic labeling mentioned in the introduction, $\{0, 1\}$ and $\mathbb{N}$ equipped with the standard order, are $\sqcup$-algebras. Second, the usable rules induced by the labeling are often a small subset of the set of all rewrite rules. Hence the possibility of finding a suitable interpretation increases while at the same time the search space decreases. The overhead of computing $\mathcal{U}(\ell)$ is negligible. Therefore we believe that termination provers that incorporate semantic labeling may benefit from our result. In the version of semantic labeling implemented in TPA [11] all function symbols are labeled. This entails that most rewrite rules are usable. As a consequence the termination proving power of TPA is only modestly increased by predictive labeling (Adam Koprowski, personal communication).

Acknowledgments

We thank Adam Koprowski, Georg Moser, and the anonymous referees for several suggestions to improve the presentation.

References

1. T. Arts and J. Giesl. Termination of term rewriting using dependency pairs. *Theoretical Computer Science*, 236:133–178, 2000.
2. F. Baader and T. Nipkow. *Term Rewriting and All That*. Cambridge University Press, 1998.
3. A. Ben Cherifa and P. Lescanne. Termination of rewriting systems by polynomial interpretations and its implementation. *Science of Computer Programming*, 9:137–159, 1987.
4. N. Dershowitz. Orderings for term-rewriting systems. *Theoretical Computer Science*, 17:279–301, 1982.
5. A. Geser, D. Hofbauer, and J. Waldmann. Match-bounded string rewriting. *Applicable Algebra in Engineering, Communication and Computing*, 15:149–171, 2004.

6. A. Geser, D. Hofbauer, J. Waldmann, and H. Zantema. On tree automata that certify termination of left-linear term rewriting systems. In *Proceedings of the 16th International Conference on Rewriting Techniques and Applications*, volume 3467 of *Lecture Notes Computer Science*, pages 353–367, 2005.

7. B. Gramlich. Generalized sufficient conditions for modular termination of rewriting. *Applicable Algebra in Engineering, Communication and Computing*, 5:131–158, 1994.

8. N. Hirokawa and A. Middeldorp. Dependency pairs revisited. In *Proceedings of the 16th International Conference on Rewriting Techniques and Applications*, volume 3091 of *Lecture Notes Computer Science*, pages 249–268, 2004.

9. S. Kamin and J.J. Lévy. Two generalizations of the recursive path ordering. Unpublished manuscript, University of Illinois, 1980.

10. D.E. Knuth and P. Bendix. Simple word problems in universal algebras. In J. Leech, editor, *Computational Problems in Abstract Algebra*, pages 263–297. Pergamon Press, 1970.

11. A. Koprowski. TPA: Termination proved automatically. In *Proceedings of the 17th International Conference on Rewriting Techniques and Applications*, Lecture Notes Computer Science, 2006. This volume.

12. D. Lankford. On proving term rewriting systems are Noetherian. Technical Report MTP-3, Louisiana Technical University, 1979.

13. A. Middeldorp, H. Ohsaki, and H. Zantema. Transforming termination by self-labelling. In *Proceedings of the 13th International Conference on Automated Deduction*, volume 1104 of *Lecture Notes in Artificial Intelligence*, pages 373–386, 1996.

14. Terese. *Term Rewriting Systems*, volume 55 of *Cambridge Tracts in Theoretical Computer Science*. Cambridge University Press, 2003.

15. R. Thiemann, J. Giesl, and P. Schneider-Kamp. Improved modular termination proofs using dependency pairs. In *Proceedings of the 2nd International Joint Conference on Automated Reasoning*, volume 3097 of *Lecture Notes in Artificial Intelligence*, pages 75–90, 2004.

16. Y. Toyama. Counterexamples to the termination for the direct sum of term rewriting systems. *Information Processing Letters*, 25:141–143, 1987.

17. X. Urbain. Modular & incremental automated termination proofs. *Journal of Automated Reasoning*, 32:315–355, 2004.

18. H. Zantema. Termination of term rewriting by semantic labelling. *Fundamenta Informaticae*, 24:89–105, 1995.

19. H. Zantema. TORPA: Termination of rewriting proved automatically. In *Proceedings of the 15th International Conference on Rewriting Techniques and Applications*, Lecture Notes Computer Science, pages 95–104, 2004.

Termination of String Rewriting
with Matrix Interpretations

Dieter Hofbauer[1] and Johannes Waldmann[2]

[1] Mühlengasse 16, D-34125 Kassel, Germany
dieter@theory.informatik.uni-kassel.de
[2] Hochschule für Technik, Wirtschaft und Kultur (FH) Leipzig
Fb IMN, PF 30 11 66, D-04251 Leipzig, Germany
waldmann@imn.htwk-leipzig.de

Abstract. A rewriting system can be shown terminating by an order-preserving mapping into a well-founded domain. We present an instance of this scheme for string rewriting where the domain is a set of square matrices of natural numbers, equipped with a well-founded ordering that is not total. The coefficients of the matrices can be found via a transformation to a boolean satisfiability problem. The matrix method also supports relative termination, thus it fits with the dependency pair method as well. Our implementation is able to automatically solve hard termination problems.

1   Introduction

To solve a problem in rewriting, one can try to translate it into a different domain that still allows to represent important aspects of rewriting, but at the same time provides new techniques. We consider string rewriting, so we need the concept of strings (concatenation of letters) and of replacement (rule application in context). This suggests a translation to *rings*. A ring has multiplicative structure, representing concatenation, and additive resp. subtractive structure, representing rule application, as the translation i maps a rewrite rule $\ell \to r$ to the ring element $i(\ell) - i(r)$.

We are especially interested in termination, so the ring has to be equipped with a suitable well-founded ordering such that for each rewrite step $u \to v$, we have $i(u) > i(v)$ for the respective images. This implies that the ring must be infinite, so the obvious example is the ordered ring of integers whose positive cone is well-founded. Indeed it is useful in termination proofs because it gives us proofs by weights (written multiplicatively). On the other hand, its usefulness is limited because multiplication is commutative: no integer weight function can show termination of e. g. $\{ab \to ba\}$. So we are looking for non-commutative rings, and the natural examples are rings of square integer matrices and this is indeed the main topic of the paper.

As an introductory example of our method, we show that the string rewriting system $\{aa \to aba\}$ is terminating, even relative to the system $\{b \to bb\}$. We interpret letters by square matrices

F. Pfenning (Ed.): RTA 2006, LNCS 4098, pp. 328–342, 2006.

$$i(a) = \begin{pmatrix} 1 & 1 \\ 1 & 0 \end{pmatrix}, \qquad i(b) = \begin{pmatrix} 1 & 0 \\ 0 & 0 \end{pmatrix}$$

and get $i(aa) - i(aba) = \begin{pmatrix} 2 & 1 \\ 1 & 1 \end{pmatrix} - \begin{pmatrix} 1 & 1 \\ 1 & 1 \end{pmatrix} = \begin{pmatrix} 1 & 0 \\ 0 & 0 \end{pmatrix}$ and $i(b) - i(bb) = \begin{pmatrix} 1 & 0 \\ 0 & 0 \end{pmatrix} - \begin{pmatrix} 1 & 0 \\ 0 & 0 \end{pmatrix} = \begin{pmatrix} 0 & 0 \\ 0 & 0 \end{pmatrix}$. It is easily verified that $i(xaay) - i(xabay) = i(x)(i(aa) - i(aba))i(y) \geq \begin{pmatrix} 1 & 0 \\ 0 & 0 \end{pmatrix}$ and $i(xby) - i(xbby) = i(x)(i(b) - i(bb))i(y) \geq \begin{pmatrix} 0 & 0 \\ 0 & 0 \end{pmatrix}$ for all context strings x and y. This implies relative termination, as will be justified and discussed in the remainder of the paper.

We describe interpretations into well-founded rings in Section 3 and specialize to the ring of matrices in Section 4. Examples are given in Section 5, among them two hard termination problems. Limitations of matrix interpretations are discussed in Section 6. We then describe how our method can be implemented and how it performs on the Termination Problem Data Base (TPDB) [12], a collection of termination problems that is used in the annual termination competitions, see http://www.lri.fr/~marche/termination-competition/. Finally, we relate our method to the monotone algebra point of view.

2   Notations and Preliminaries

A *string rewriting system* over an alphabet Σ is a relation $R \subseteq \Sigma^* \times \Sigma^*$, where a pair (ℓ, r) from R is usually referred to as a *rule* $\ell \to r$. The system R induces the *(one-step) rewrite relation* $\to_R = \{(x\ell y, xry) \mid x, y \in \Sigma^*, (\ell \to r) \in R\}$ on Σ^*. In this paper, all rewriting systems are finite. For more on strings and string rewriting see [2], for instance.

The transitive closure of a relation $\rho \subseteq A \times A$ is ρ^+, and ρ^* denotes its reflexive and transitive closure. The composition of two relations $\rho \subseteq A \times B$ and $\sigma \subseteq B \times C$ is $\rho \circ \sigma = \{(a, c) \mid \exists b \in B : (a, b) \in \rho, (b, c) \in \sigma\}$. For $\rho \subseteq A \times A$ and $\sigma \subseteq B \times B$, a mapping $f : A \to B$ is said to be *monotone* (or *order preserving*) if $(a, a') \in \rho$ implies $(f(a), f(a')) \in \sigma$.

A relation $\rho \subseteq A \times A$ is *terminating* (or *strongly normalizing*, or *well-founded*), written $\mathrm{SN}(\rho)$, if no infinite chain modulo ρ exists. For example, the usual order $>$ on $\mathbb{N}$ is terminating. For relations $\rho, \sigma \subseteq A \times A$ define ρ/σ by $\sigma^* \circ \rho \circ \sigma^*$. We say that ρ is *terminating relative to* σ if $\mathrm{SN}(\rho/\sigma)$. For example, $\mathrm{SN}(\rho/\sigma)$ for $\rho = \{(2n, n) \mid n > 0\}$ (division by 2) and $\sigma = \{(n, 3n) \mid n > 0\}$ (multiplication by 3) on $\mathbb{N}$. For string rewriting systems R and S we abbreviate $\mathrm{SN}(\to_R)$ and $\mathrm{SN}(\to_R/\to_S)$ by $\mathrm{SN}(R)$ and $\mathrm{SN}(R/S)$ respectively.

Relative termination is useful in step-wise proofs of termination: $\mathrm{SN}(R)$ can be inferred if there is some subset $R' \subseteq R$ with $\mathrm{SN}(R'/R)$ and $\mathrm{SN}(R \setminus R')$. The same idea of removal of rules works in proofs of relative termination: $\mathrm{SN}(R/S)$ holds if there are subsets $R' \subseteq R$ and $S' \subseteq S$ such that $\mathrm{SN}((R' \cup S')/(R \cup S))$ and $\mathrm{SN}((R \setminus R')/(S \setminus S'))$. For these and related concepts see [8,15], for instance.

A partial order $\geq$ is a reflexive, antisymmetric, and transitive relation. We write $>$ for $\geq \setminus =$, and the partial order is called well-founded if $>$ is well-founded. Throughout the paper, mappings $f : A_1 \times \cdots \times A_n \to A$ are extended to sets by $f(B_1, \ldots, B_n) = \{f(b_1, \ldots, b_n) \mid b_i \in B_i\}$ for $B_i \subseteq A_i$.

3   Interpretations into Well-Founded Rings

Interpretations into suitable well-founded partially ordered rings can be used for termination proofs. Before we consider matrix rings in Section 4, a slightly more general treatment of the underlying principle is given in this section.

Basically, interpretations are order preserving mappings into well-founded domains. The following lemma gives the simple general scheme for relative termination proofs that we are going to instantiate.

Lemma 1. *Let ρ and σ be relations on Σ^*. Let $(N, \geq)$ be a well-founded partial order, let $i : \Sigma^* \to N$ be a mapping. If i is order preserving both from (Σ^*, ρ^+) to $(N, >)$ and from (Σ^*, σ^*) to $(N, \geq)$, then ρ is terminating relative to σ.*

Proof. As we have $\geq \circ > \circ \geq \subseteq >$, the mapping i is also order preserving from $(\Sigma^*, \sigma^* \circ \rho \circ \sigma^*)$ to $(N, >)$. Therefore, an infinite chain modulo $\sigma^* \circ \rho \circ \sigma^*$ on Σ^* would induce an infinite chain modulo $>$ on N.   □

A *ring* $(D, 0, 1, +, \cdot)$ with domain D has two constants and two binary operations such that $(D, 0, +)$ is an Abelian group (implying the existence of an additive inverse $-d$ for each $d \in D$) and such that $(D, 1, \cdot)$ is a monoid; additionally, multiplication distributes over addition from both sides. Multiplication is neither required to be commutative nor invertible. As usual, subtraction $a - b$ stands for $a + (-b)$. The ring is *partially ordered by* $\geq$ if $(D, \geq)$ is a partial order which is compatible with the ring operations: for $a, b, c \in D$, $a \geq b \Rightarrow a + c \geq b + c$ and $a \geq b \wedge c \geq 0 \Rightarrow ac \geq bc \wedge ca \geq cb$.

The set $N = \{d \in D \mid d \geq 0\}$ of *positive* ring elements is said to be its *positive cone*, and $P = \{d \in D \mid d > 0\} = N \setminus \{0\}$ is its *strictly positive cone*, see [5]. Note that the ordering $\geq$ is already uniquely determined by the positive cone since $a \geq b$ is equivalent to $a - b \in N$. Analogously, $a > b$ is equivalent to $a - b \in P$. The ring is called *well-founded* if $>$ is well-founded on N. For $i \in \mathbb{N}$ and $A \subseteq D$ define A^i by $A^0 = \{1\}$ and $A^{i+1} = A^i \cdot A$, and let $A^* = \bigcup_{i \in \mathbb{N}} A^i$. Note that by definition we have $N + P = P + N = P$ and $N + N = N$. Further, $N \cdot P = P \cdot N = N = N \cdot N$, thus $N^* = N$, but in general we do not have $P \cdot P \subseteq P$ due to the existence of zero divisors.

A *(homomorphic ring) interpretation* of an alphabet Σ is a mapping $i : \Sigma \to D$ that is extended to a homomorphism from $(\Sigma^*, \epsilon, \cdot)$ to $(D, 1, \cdot)$ by $i(s_1 \cdot \ldots \cdot s_n) = i(s_1) \cdot \ldots \cdot i(s_n)$. We further extend i to a mapping from rules (i.e., pairs of strings) to D by $i(\ell \to r) = i(\ell) - i(r)$.

In order to apply ring interpretations for proving termination of rewriting, we want to ensure $i(x\ell y) > i(xry)$ for each rewrite step $x\ell y \to_R xry$, that is,

$$i(x\ell y) - i(xry) = i(x)i(\ell)i(y) - i(x)i(r)i(y) = i(x)\big(i(\ell) - i(r)\big)i(y) \in P. \quad (*)$$

Given the set of interpretations of letters $i(\Sigma) = A$, what is the set of admissible interpretations of rules $i(R) = B$? Inspecting condition $(*)$, it is obvious that $A^* B A^* \subseteq P$ is necessary. The largest such set B will be called the *core* for A.

Definition 1. *For $A \subseteq D$ define* $\text{core}(A) = \{d \in D \mid A^* d A^* \subseteq P\}$.

Example 1. For $A = \{(\begin{smallmatrix} 1 & 0 \\ 0 & 0 \end{smallmatrix})\}$ we have $\mathrm{core}(A) = \{d \mid d \geq (\begin{smallmatrix} 1 & 0 \\ 0 & 0 \end{smallmatrix})\}$, see Lemma 4.

Remark 1. For $A, B \subseteq D$ define left and right quotients by $A \backslash B = \{d \in D \mid Ad \subseteq B\}$ and $B/A = \{d \in D \mid dA \subseteq B\}$ respectively. Then we can write $\mathrm{core}(A) = A^* \backslash P / A^*$; note that $(A^* \backslash P)/A^* = A^* \backslash (P/A^*)$.

Remark 2. By definition, $\mathrm{core}(A)$ is the largest set B with $A^* B A^* \subseteq P$. On the other hand, for a given set B there is not necessarily a unique largest set A with $A^* B A^* \subseteq P$. To wit, for $B = \{(\begin{smallmatrix} 1 & 0 \\ 0 & 1 \end{smallmatrix})\}$ the sets $A = \{d \in N \mid d \geq (\begin{smallmatrix} 1 & 0 \\ 0 & 0 \end{smallmatrix})\}$ and $A' = \{d \in N \mid d \geq (\begin{smallmatrix} 0 & 0 \\ 0 & 1 \end{smallmatrix})\}$ are different maximal sets fulfilling that property.

In the following lemma, we collect a few basic properties for later reference.

Lemma 2. *(1)* $\mathrm{core}(A) \subseteq P$.
(2) $A^* \mathrm{core}(A) A^* = \mathrm{core}(A)$.
(3) For $A \subseteq N$, $\mathrm{core}(A) + N = \mathrm{core}(A)$.
(4) If $A_1 \subseteq A_2$, then $\mathrm{core}(A_2) \subseteq \mathrm{core}(A_1)$.
(5) For $A \subseteq N$, $\mathrm{core}(A + N) = \mathrm{core}(A)$.
(6) For $A \subseteq N$, $A^* \subseteq P$ if and only if $\mathrm{core}(A) \neq \emptyset$.

Proof. (1) By $\mathrm{core}(A) \subseteq A^* \mathrm{core}(A) A^* \subseteq P$. (2) This follows from $A^* A^* = A^*$. (3) For $c \in \mathrm{core}(A)$ and $n \in N$ we have $A^*(c + n)A^* \subseteq A^* c A^* + A^* n A^* \subseteq P + N = P$. (4) If $A_1 \subseteq A_2$, then for any $d \in D$ we have $A_1^* d A_1^* \subseteq A_2^* d A_2^*$, so $A_2^* d A_2^* \subseteq P$ implies $A_1^* d A_1^* \subseteq P$. (5) We have "$\subseteq$" by the previous item; the other inclusion holds since in a partially ordered ring, $a_i' \geq a_i$ implies $a_1' d a_2' \geq a_1 d a_2$. (6) If $A^* \subseteq P$, then $A^* \subseteq \mathrm{core}(A)$ by $A^* A^* A^* = A^* \subseteq P$, thus $\mathrm{core}(A) \neq \emptyset$. Conversely, $A^* \subseteq N$ and $A^* \nsubseteq P$ implies $0 \in A^*$. For $\mathrm{core}(A) \neq \emptyset$ we get the contradiction $\{0\} = 0 \mathrm{core}(A) 0 \subseteq A^* \mathrm{core}(A) A^* \subseteq P$. □

We always have $\mathrm{core}(A) \subseteq P$, but in general we do not have $\mathrm{core}(A) \subseteq A$, even for $A \subseteq P$ (see Lemma 4 in Section 4 for an example).

Remark 3. It is important to observe that adding elements to the potential range of interpretations of letters typically reduces the set of elements that can safely be chosen as interpretations of rules; this is what Lemma 2(4) says. We can, on the other hand, assume that the range of all interpretations is upward closed (i.e., if it contains a then it also contains every $a' \geq a$). This can be assumed without loss of generality for the interpretation of letters by Lemma 2(5), and it always holds for the interpretation of rules by Lemma 2(3).

Lemma 3. *Let R be a string rewriting system over Σ, and let $i : \Sigma \to D$ be an interpretation into a well-founded ring such that $i(\Sigma) \subseteq N$. Then i is*
(1) order preserving from $(\Sigma^*, \to_R^+)$ to $(D, >)$ if and only if $i(R) \subseteq \mathrm{core}(i(\Sigma))$,
(2) order preserving from $(\Sigma^*, \to_R^*)$ to $(D, \geq)$ if and only if $i(R) \subseteq N$.

Proof. Set $A = i(\Sigma)$. (1) Equivalence of $\forall x, y \in \Sigma^* \forall (\ell, r) \in R : i(x \ell y) > i(x r y)$ and $A^* i(R) A^* \subseteq P$ holds by $(*)$ above. (2) is proven analogously; note that $A^* i(R) A^* \subseteq N$ if and only if $i(R) \subseteq N$. □

Definition 2. *Let N be the positive cone of a well-founded ring. For $A \subseteq N$, an A-interpretation for a string rewriting system R over Σ is an interpretation $i : \Sigma \to N$ with $i(\Sigma) \subseteq A$ and $i(R) \subseteq \mathrm{core}(A)$.*

Lemma 3 together with Lemma 1 yields

Theorem 1. *Let N be the positive cone of a well-founded ring, let $A \subseteq N$, and let R and S be string rewriting systems over Σ. If there is an A-interpretation i for R with $i(S) \subseteq N$, then R is terminating relative to S. In particular, if there is an A-interpretation for R, then R is terminating.*

We remark that this termination proof method is complete in the sense that for all string rewriting systems R and S over Σ with $\mathrm{SN}(R/S)$, there is some well-founded ring and some subset A of its positive cone N such that an A-interpretation i for R with $i(S) \subseteq N$ exists. This ring can be chosen as the free semi-group ring over Σ (consisting of mappings $\Sigma^* \to \mathbb{Z}$), where the order $\geq$ is induced by the order $(\to_R/\to_S)^*$ on strings.

4   Matrix Interpretations

We now switch to the partially ordered ring of square matrices of a fixed dimension n over $\mathbb{Z}$, which is $D = \mathbb{Z}^{n \times n}$. Addition and multiplication are the usual matrix operations, and 0 and 1 are the zero and the identity matrix respectively. The order is defined component-wise: $d \geq d'$ if $\forall i, j : d_{i,j} \geq d'_{i,j}$. The positive ring elements are just the matrices over $\mathbb{N}$, thus the positive cone is $N = \mathbb{N}^{n \times n}$. The given partial order is indeed well-founded on the positive cone. For dimensions $n > 1$, this order is not total.

In order to apply Theorem 1 we need a set of matrices $A \subseteq N$ and an interpretation $i : \Sigma \to N$ such that $i(\Sigma) \subseteq A$, $i(R) \subseteq \mathrm{core}(A)$, and $i(S) \subseteq N$. In the sequel, we present two particular instances of this method, both parameterized by a set of matrix indices I. The one alternative is to choose $A = E_I$ with $\mathrm{core}(A) = P_I$, the other choice is $A = M_I$ with $\mathrm{core}(A) = M_I$, where E_I, P_I and M_I are simple "syntactically" defined subsets of N. Of course, further instances of the general scheme are conceivable.

Definition 3. *For non-empty $I \subseteq \{1, \ldots, n\}$ define subsets of $N = \mathbb{N}^{n \times n}$ by*

$$E_I = \{d \in N \mid \forall i \in I : d_{i,i} > 0\},$$
$$P_I = \{d \in N \mid \exists i \in I \exists j \in I : d_{i,j} > 0\},$$
$$M_I = \{d \in N \mid \forall i \in I \exists j \in I : d_{i,j} > 0\}.$$

For $1 \leq i \leq n$ let $P_i = P_{\{1,\ldots,i\}}$, and $P = P_n = N \setminus \{0\}$. Sets E_i, M_i, E, and M are defined analogously.

The following properties are all easily verified. By definition we have $1 \in E_I \subseteq M_I \subseteq P_I \subseteq P$. From $E_I^2 \subseteq E_I$ and $M_I^2 \subseteq M_I$ we obtain $E_I^* = E_I$ and $M_I^* = M_I$, but $P_I^2 \subseteq P_I$ does not hold since the matrix ring has zero divisors. Further, $E_I P_I = P_I E_I = P_I$, thus $E_I^* P_I E_I^* = P_I$. Note that $E_1 = P_1 = M_1$.

Lemma 4. *For non-empty $I \subseteq \{1, \ldots, n\}$,*

$$\mathrm{core}(E_I) = P_I, \tag{1}$$
$$\mathrm{core}(M_I) = M_I. \tag{2}$$

Proof. (1) Using quotients as defined in Remark 1, we have $P_I = P_I/E_I \subseteq P/E_I$ by $1 \in E_I$ and $P_I E_I = P_I \subseteq P$. For showing the inverse inclusion $P/E_I \subseteq P_I$, assume the existence of a matrix d with $dE_I \subseteq P$ and $d \notin P_I$, so $d_{i,j} = 0$ for $i, j \in I$. Define $e \in E_I$ by $e_{i,j} = 1$ for $i = j \in I$ and $e_{i,j} = 0$ otherwise. Then $de = 0 \notin P$, a contradiction. Symmetrically, we get $E_I \backslash P_I = P_I$. Thus by Remark 1, $\mathrm{core}(E_I) = E_I^* \backslash P/E_I^* = E_I \backslash P/E_I = E_I \backslash P_I = P_I$.

(2) Here we get $M_I = M_I/M_I \subseteq P/M_I$ by $1 \in M_I$ and $M_I^2 = M_I \subseteq P$, and symmetrically, $M_I = M_I \backslash M_I \subseteq M_I \backslash P$. For $M_I \backslash P \subseteq M_I$, again by way of contradiction, assume the existence of a matrix d with $M_I d \subseteq P$ and $d \notin M_I$. Then, for some $k \in I$, we have $\forall j \in I : d_{k,j} = 0$. Define $m \in M_I$ by $m_{i,j} = 1$ for $i \in I$, $j = k$, and $m_{i,j} = 0$ otherwise. Then $md = 0 \notin P$, contradiction. Finally, by Remark 1, $\mathrm{core}(M_I) = M_I^* \backslash P/M_I^* = M_I \backslash P/M_I = M_I/M_I = M_I$. □

Example 2. We claimed in the introduction that termination of the string rewriting system $\{aa \rightarrow aba\}$ relative to the system $\{b \rightarrow bb\}$ is shown by the interpretation

$$i(a) = \begin{pmatrix} 1 & 1 \\ 1 & 0 \end{pmatrix}, \qquad i(b) = \begin{pmatrix} 1 & 0 \\ 0 & 0 \end{pmatrix}.$$

This is an E_1-interpretation with $i(aa \rightarrow aba) = i(aa) - i(aba) = \begin{pmatrix} 2 & 1 \\ 1 & 1 \end{pmatrix} - \begin{pmatrix} 1 & 1 \\ 1 & 1 \end{pmatrix} = \begin{pmatrix} 1 & 0 \\ 0 & 0 \end{pmatrix} \in P_1$ and $i(b \rightarrow bb) = i(b) - i(bb) = 0 \in N$. Alternatively, the M_2-interpretation

$$i(a) = \begin{pmatrix} 1 & 1 \\ 1 & 0 \end{pmatrix}, \qquad i(b) = \begin{pmatrix} 0 & 1 \\ 0 & 1 \end{pmatrix}$$

with $i(aa \rightarrow aba) = i(aa) - i(aba) = \begin{pmatrix} 2 & 1 \\ 1 & 1 \end{pmatrix} - \begin{pmatrix} 2 & 0 \\ 1 & 0 \end{pmatrix} = \begin{pmatrix} 0 & 1 \\ 0 & 1 \end{pmatrix} \in M_2$, and where again $i(b \rightarrow bb) = 0$, serves the same purpose. Note that the latter interpretation is not E_I for any I.

Remark 4. Matrix interpretations have implicitly occurred in the literature before. For instance in [15, Section 6.2.5], termination of $R = \{aa \rightarrow aba\}$ is shown by an interpretation into a monotone algebra $\mathcal{A}$ with domain $\mathbb{N} \times \{0, 1\}$, where (after an argument swap)

$$a_{\mathcal{A}}(n, m) = (n + m, 1), \qquad b_{\mathcal{A}}(n, m) = (n, 0).$$

Equivalently, we can choose linear functions

$$a_{\mathcal{B}}(n, m, e) = (n + m, e, e), \qquad b_{\mathcal{B}}(n, m, e) = (n, 0, e)$$

in an algebra $\mathcal{B}$ with domain $\mathbb{N} \times \{0, 1\} \times \{1\} \subseteq \mathbb{N}^3$, corresponding to matrices

$$i(a) = \begin{pmatrix} 1 & 1 & 0 \\ 0 & 0 & 1 \\ 0 & 0 & 1 \end{pmatrix}, \qquad i(b) = \begin{pmatrix} 1 & 0 & 0 \\ 0 & 0 & 0 \\ 0 & 0 & 1 \end{pmatrix}.$$

This mapping i is in fact an $E_{\{1,3\}}$-interpretation for R, as

$$i(aa \to aba) = i(aa) - i(aba) = \begin{pmatrix} 1 & 1 & 1 \\ 0 & 0 & 1 \\ 0 & 0 & 1 \end{pmatrix} - \begin{pmatrix} 1 & 1 & 0 \\ 0 & 0 & 1 \\ 0 & 0 & 1 \end{pmatrix} = \begin{pmatrix} 0 & 0 & 1 \\ 0 & 0 & 0 \\ 0 & 0 & 0 \end{pmatrix} \in P_{\{1,3\}}.$$

Remark 5. Matrix interpretations are invariant under permutations, i.e., if i is an E_I-interpretation of dimension n for a system R, and if π is a permutation on $\{1, \ldots, n\}$, then there is also an $E_{\pi(I)}$-interpretation for R. First observe that for the corresponding permutation matrix p we have $E_{\pi(I)} = p^{-1} \cdot E_I \cdot p$ and $P_{\pi(I)} = p^{-1} \cdot P_I \cdot p$. Defining the $E_{\pi(I)}$-interpretation i_π by $i_\pi(s) = p^{-1} \cdot i(s) \cdot p$ for $s \in \Sigma$ we get $i_\pi(x) = p^{-1} \cdot i(x) \cdot p$ for $x \in \Sigma^*$. Thus, $i(R) \subseteq P_I$ implies $i_\pi(R) \subseteq P_{\pi(I)} = \mathrm{core}(E_{\pi(I)})$, so i_π is an $E_{\pi(I)}$-interpretation for R. The same considerations also apply to M_I. As a consequence we can replace an arbitrary index set I by the particular index set $\{1, \ldots, |I|\}$ without loss of generality.

Remark 6. In our implementation we use E_J-interpretations of dimension n for the particular two element index set $J = \{1, n\}$, see Section 7. In view of Remark 5 it is not difficult to see that a proof of $\mathrm{SN}(R/S)$ via some E_I-interpretation can be replaced by a sequence of such E_J-interpretations which successively remove the same rules.

Remark 7. It is decidable whether an arbitrary matrix interpretation $i : \Sigma \to N$ (i.e., not necessarily of type E_I or M_I) satisfies $i(R) \subseteq \mathrm{core}(i(\Sigma))$. In particular we can effectively determine a finite set $C \subseteq P$ such that $\mathrm{core}(i(\Sigma)) = \{d \geq c \mid c \in C\}$; note that a finite such set always exists by Lemma 2(3) and since the standard matrix ordering is a well-quasi-order.

5   Examples

Example 3. Zantema's System $\{a^2b^2 \to b^3a^3\}$ is a classical test case for (automated) termination methods (and therefore is problem z001 in TPDB). The matrix interpretation

$$i(a) = \begin{pmatrix} 1 & 0 & 0 & 1 & 0 \\ 0 & 0 & 0 & 2 & 0 \\ 0 & 0 & 0 & 1 & 0 \\ 0 & 0 & 1 & 0 & 0 \\ 0 & 0 & 0 & 0 & 1 \end{pmatrix}, \qquad i(b) = \begin{pmatrix} 1 & 0 & 0 & 0 & 0 \\ 0 & 0 & 1 & 2 & 1 \\ 0 & 1 & 0 & 0 & 0 \\ 0 & 0 & 0 & 0 & 0 \\ 0 & 0 & 0 & 0 & 1 \end{pmatrix}$$

of type $E_{\{1,5\}}$ shows termination, as

$$i(a^2b^2 \to b^3a^3) = \begin{pmatrix} 1 & 0 & 1 & 2 & 1 \\ 0 & 0 & 2 & 4 & 2 \\ 0 & 0 & 1 & 2 & 1 \\ 0 & 0 & 0 & 0 & 0 \\ 0 & 0 & 0 & 0 & 1 \end{pmatrix} - \begin{pmatrix} 1 & 0 & 1 & 2 & 0 \\ 0 & 0 & 2 & 1 & 2 \\ 0 & 0 & 0 & 2 & 1 \\ 0 & 0 & 0 & 0 & 0 \\ 0 & 0 & 0 & 0 & 1 \end{pmatrix} = \begin{pmatrix} 0 & 0 & 0 & 0 & 1 \\ 0 & 0 & 0 & 3 & 0 \\ 0 & 0 & 1 & 0 & 0 \\ 0 & 0 & 0 & 0 & 0 \\ 0 & 0 & 0 & 0 & 0 \end{pmatrix} \in P_{\{1,5\}}.$$

Note that exchanging a and b, and then transposing the matrices along the anti-diagonal is the identical mapping. This corresponds to taking the reversal of the system after exchanging letters, which indeed gives the original system.

Example 4. Zantema's "other" System $\{a^2 \to bc, b^2 \to ac, c^2 \to ab\}$ is TPDB problem z086 which became famous because no tool (and no tool author) could solve it in the previous termination competitions. It also appears as Problem #104 [17] in the RTA list of open problems. The following $E_{\{1,5\}}$-interpretation proves termination:

$$a = \begin{pmatrix} 1\,0\,0\,3\,1 \\ 0\,0\,1\,1\,1 \\ 0\,2\,0\,1\,0 \\ 0\,0\,0\,0\,0 \\ 0\,0\,0\,0\,1 \end{pmatrix}, \quad b = \begin{pmatrix} 1\,0\,2\,0\,0 \\ 0\,0\,1\,0\,0 \\ 0\,0\,2\,1\,2 \\ 0\,0\,0\,0\,0 \\ 0\,0\,0\,0\,1 \end{pmatrix}, \quad c = \begin{pmatrix} 1\,0\,0\,1\,1 \\ 0\,0\,1\,1\,3 \\ 0\,0\,0\,1\,0 \\ 0\,0\,2\,0\,0 \\ 0\,0\,0\,0\,1 \end{pmatrix}.$$

This interpretation has been published (in disguise) in [10], cf the remark at the end of Section 8. Again, note the symmetry by exchanging a and c and applying anti-transposal of the matrices resp. reversal of the system.

Remark 8. A termination criterion is not necessarily invariant under reversal, that is, the method may fail to show termination of a given system R, but may apply to its reversal $\mathrm{rev}(R) = \{\mathrm{rev}(\ell) \to \mathrm{rev}(r) \mid (\ell \to r) \in R\}$, where $\mathrm{rev}(a_1, \ldots, a_n) = a_n, \ldots, a_1$ for $a_i \in \Sigma$. For matrix interpretations, invariance under reversal is guaranteed if the class of matrices used is closed under transposal. This is due to the fact that $a \cdot b = (b^{\mathrm{T}} \cdot a^{\mathrm{T}})^{\mathrm{T}}$ for square matrices a and b, where a^{T} denotes the transpose of a. For a given interpretation $i : \Sigma \to A$ define $i^{\mathrm{T}} : \Sigma \to A^{\mathrm{T}}$ by $i^{\mathrm{T}}(c) = i(c)^{\mathrm{T}}$. Then $i(\mathrm{rev}(x)) = i^{\mathrm{T}}(x)^{\mathrm{T}}$ for $x \in \Sigma^*$, thus $i(\mathrm{rev}(\ell)) - i(\mathrm{rev}(r)) = i^{\mathrm{T}}(\ell)^{\mathrm{T}} - i^{\mathrm{T}}(r)^{\mathrm{T}} = (i^{\mathrm{T}}(\ell) - i^{\mathrm{T}}(r))^{\mathrm{T}}$, so $i(\mathrm{rev}(R)) \subseteq \mathrm{core}(A)$ implies $i^{\mathrm{T}}(R) \subseteq \mathrm{core}(A)^{\mathrm{T}}$. Note that $\mathrm{core}(A)^{\mathrm{T}} = \mathrm{core}(A^{\mathrm{T}})$, as $(A^{\mathrm{T}})^* = (A^*)^{\mathrm{T}}$ and $P^{\mathrm{T}} = P$. Therefore, if i is an A-interpretation for $\mathrm{rev}(R)$, then i^{T} is an A^{T}-interpretation for R. This shows that if A is closed under transposal, then there is an A-interpretation for R in case an A-interpretation for $\mathrm{rev}(R)$ exists. Finally note that all classes E_I and P_I are indeed closed under transposal, but that this does not hold for M_I.

Example 5. For $k \geq 0$, termination of the system $\{ab \to ba^k\}$ is proven by the E_2-interpretation $i(a) = \left(\begin{smallmatrix} 1 & 1 \\ 0 & 1 \end{smallmatrix}\right)$, $i(b) = \left(\begin{smallmatrix} 1 & 0 \\ 0 & k+1 \end{smallmatrix}\right)$, as $i(ab) - i(ba^k) = \left(\begin{smallmatrix} 1 & k+1 \\ 0 & k+1 \end{smallmatrix}\right) - \left(\begin{smallmatrix} 1 & k \\ 0 & k+1 \end{smallmatrix}\right) = \left(\begin{smallmatrix} 0 & 1 \\ 0 & 0 \end{smallmatrix}\right) \in P_2$.

Remark 9. All termination proofs by additive natural weights can be expressed as matrix interpretations with 2×2 matrices of type E_2. This is due to the fact that $(\mathbb{N}, +)$ is isomorphic to $(\{\left(\begin{smallmatrix} 1 & n \\ 0 & 1 \end{smallmatrix}\right) \mid n \in \mathbb{N}\}, \cdot)$. More generally, consider termination proofs by *linear interpretations*: Letters are interpreted by mappings $\lambda n.an + b$ on $\mathbb{N}$ with $a, b \in \mathbb{N}$ and $a \geq 1$, concatenation is interpreted by composition, and the proof obligation is $\forall n \in \mathbb{N} : i(\ell)(n) > i(r)(n)$ for $\ell \to r$ in R. This corresponds to matrix interpretations with matrices of the form $\left(\begin{smallmatrix} a & b \\ 0 & 1 \end{smallmatrix}\right)$.

Example 6. The system $R = \{aa \to bc, bb \to cd, b \to a, cc \to df, dd \to fff, d \to b, ff \to ga, gg \to a\}$, is TPDB problem z112. In the 2005 termination competition, it has been solved by Jambox [3] (being RFC-match-bounded by 4, witnessed by an automaton with 127 states) and Torpa [16] using additive weights $a \mapsto 16$, $b \mapsto 17$, $c \mapsto 15$, $d \mapsto 18$, $f \mapsto 12$, $g \mapsto 8$ to remove rules $\{bb \to cd, b \to a, d \to b\}$, and then another weighting to remove the remaining rules. It can be verified that integer weights ≥ 18 are necessary to remove any of the rules from R.

Additive weights correspond to matrices $\left(\begin{smallmatrix} 1 & * \\ 0 & 1 \end{smallmatrix}\right)$, see Remark 9, and we will now show that for a moderate increase in dimension one can sometimes get a drastic decrease in coefficients: The following is an $M_{\{1,4\}}$-interpretation that allows to remove rule $gg \to a$:

$$i(a) = \begin{pmatrix} 1 & 0 & 0 & 0 \\ 0 & 0 & 0 & 0 \\ 1 & 0 & 0 & 0 \\ 1 & 0 & 0 & 0 \end{pmatrix}, \quad i(b) = \begin{pmatrix} 1 & 0 & 0 & 0 \\ 0 & 0 & 0 & 0 \\ 1 & 0 & 0 & 0 \\ 1 & 0 & 0 & 0 \end{pmatrix}, \quad i(c) = \begin{pmatrix} 1 & 0 & 0 & 0 \\ 0 & 0 & 0 & 0 \\ 1 & 0 & 0 & 0 \\ 1 & 0 & 0 & 0 \end{pmatrix},$$

$$i(d) = \begin{pmatrix} 1 & 0 & 0 & 0 \\ 0 & 0 & 1 & 0 \\ 1 & 0 & 0 & 0 \\ 1 & 0 & 1 & 0 \end{pmatrix}, \quad i(f) = \begin{pmatrix} 1 & 0 & 0 & 0 \\ 1 & 0 & 0 & 0 \\ 0 & 0 & 0 & 0 \\ 1 & 1 & 0 & 0 \end{pmatrix}, \quad i(g) = \begin{pmatrix} 1 & 1 & 0 & 0 \\ 1 & 0 & 0 & 0 \\ 0 & 1 & 0 & 0 \\ 1 & 1 & 0 & 0 \end{pmatrix}.$$

The rest of the rules can be removed successively with additive weights ≤ 2.

6   Limitations

We will now discuss inherent limitations of the matrix method. We present different reasons why a terminating rewriting system does not have a termination proof via A-interpretations. These are: a particular shape of A, the dimension of A, and the plain fact that A consists of matrices.

Simple Termination. Simple termination of R over Σ is equivalent to $\mathrm{SN}(R \cup \Sigma \times \{\epsilon\})$. We claim that if there is an E-interpretation for a rewriting system R, then R is simply terminating. Recall that $E = \{d \mid d \geq 1\}$. Any E-interpretation i fulfills $i(\Sigma \times \{\epsilon\}) \subseteq N$ since $i(\epsilon) = 1$, so interpretation i shows $\mathrm{SN}(R/\Sigma \times \{\epsilon\})$, and $\mathrm{SN}(\Sigma \times \{\epsilon\})$ is obvious. This implies that for a non-simply terminating rewriting system as $\{aa \to aba\}$ we cannot have an E-interpretation. There are matrix interpretations for that system, see Example 2 and Remark 4, but they use E_1, M_2 or $E_{\{1,3\}}$ and not E.

Dimension restrictions. For dimension one, matrices are in fact scalars, and scalar multiplication is commutative. Therefore there can be no one-dimensional termination proof for $\{ab \to ba\}$, since any interpretation $i : \{a, b\} \to \mathbb{N}^{1 \times 1}$ verifies $i(ab) = i(ba)$. (Note that a "two-dimensional" proof is contained in Example 5.)

This is a general phenomenon: a matrix ring is not *free*. Depending on its dimension, certain polynomial identities hold. For 2×2-matrices A and B, we have that $[A, B]^2$ is a scalar multiple of the identity (where $[A, B]$ denotes the *commutator* $AB - BA$), therefore $[A, [B, C]^2] = 0$. In other words, $ABCBC + ACBCB + BCCBA + CBBCA = CBCBA + BCBCA + ABCCB + ACBBC$. This implies that there is no interpretation $i : \{a, b, c\} \to \mathbb{N}^{2 \times 2}$ that removes (by relative termination) a rule from $\{abcbc \to cbcba, acbcb \to bcbca, bccba \to abccb, cbbca \to acbbc\}$. Still this system is easily seen terminating because it is RFC-matchbounded. (We currently do not know of a matrix proof.) Similar polynomial identities are known [11] for matrix rings of any dimension, leading to similar counterexamples.

Exponentional derivational complexity. In a product of k matrices taken from a finite set A of matrices, the coefficients are bounded by an exponential function in k. This also bounds the derivational complexity of string rewriting systems for which we can obtain a termination proof via matrix interpretations.

For example, $\{ab \to baa, cb \to bbc\}$ has derivations of doubly exponential length, since $ab^k \to^* b^k a^{2^k}$, $c^k b \to^* b^{2^k} c^k$ and $ac^k b \to^* ab^{2^k} c^k \to^* b^{2^k} a^{2^{2^k}} c^k$, and by the fact that increasing the length of a string by n can only be the result of at least n rewrite steps. So this rewrite system cannot be proved terminating by a matrix interpretation. In this case, there still is a termination proof via matrix interpretations due to relative termination: we can first remove the rule $cb \to bbc$, and then the other one.

In contrast, consider the system $\{ab \to bca, cb \to bbc\}$, which is up to renaming the heart of TPDB problems z018 and z020. It has long derivations: We have $(cb)^k \to^* b^{2^{k+1}-2} c^k$, thus $ab^k \to^* (bc)^k a = b(cb)^{k-1} ca \to^* b^{2^k-1} c^k a$. This can be iterated, so from $a^k b$ we can start a derivation whose length is a tower of k exponentials and in which both rules are applied equally often. Therefore, none of the rules can be removed by a matrix interpretation. Note that this system is compatible with a recursive path order. It also has a termination proof via the dependency pair method combined with matrix interpretations, as discussed in Section 9.

7 Implementation

For automatically proving termination by step-wise removal of rules, the basic algorithm takes as input a description of a matrix set A and a nonempty rewriting system R, and it outputs a nonempty subset $R' \subseteq R$ and an A-interpretation i that proves $\mathrm{SN}(R'/R)$, or it fails.

No matter how the interpretation i will be found, its validity can be verified easily. Short of an exact solution, the obvious route to finding interpretations is to enumerate all, or randomly guess some, and then just check them. This is indeed feasible for small dimensions and small coefficients. For larger parameters, we tried genetic algorithms: candidate solutions are improved by evaluating, mutating and combining them. The difficult part is to find a good evaluation function, and we really do have none. So we need something more elaborate. In the following we describe the implementation in the current version of the automated termination prover Matchbox [13].

Restricting the matrix shape. The matrix method works with A-interpretations where $\mathrm{core}(A) \neq \emptyset$. We have given E_I- and M_I-interpretations as examples. We found experimental evidence (but no proof) that if there is an M_I-interpretation i that removes some rules, then there also is an E_J-interpretation j that removes some of these rules. Sometimes the dimension of j is larger than that of i. We are willing to pay this price because by Remark 6, then there is also an $E_{\{1,n\}}$-interpretation, so it is enough to search for matrices of that shape. We restrict this matrix class even more, as we found that in a lot of cases there is an interpretation into

$$T = \{d \mid d \in E_{\{1,n\}}, \text{ first column is } (1,0,\ldots,0)^{\mathrm{T}}, \text{ last row is } (0,\ldots,0,1)\}.$$

This shape is motivated by the transformation that introduced the third parameter in Remark 4, and it also occurs in Examples 3, 4 and Remark 9. Note that $\mathrm{core}(T) = P_{\{1,n\}}$, thus $\mathrm{core}(T) \cap (T - T) = \{d \mid d_{1,n} > 0\}$.

Formulating an integer constraint system. The conditions of Theorem 1 constitute a system of inequalities with the coefficients in the $n \times n$-matrices $i(\Sigma)$ as unknowns. The shape of T determines the first column and the last row, so we effectively have reduced the dimension by one, and there are only $(n-1)^2 \cdot |\Sigma|$ unknowns. Since the shape is respected by multiplication ($T^* \subseteq T$), in each product we only need to compute the upper right $(n-1) \times (n-1)$-submatrix.

The inequalities relate polynomials in the unknowns. For dimension $n = 2$, the polynomials are in fact linear functions, cf. Remark 9. For dimensions $n > 2$, the maximal degree of these polynomials is the maximal length of a side of a rewrite rule.

For each rule of the input system, we have $(n-1)^2$ inequalities. Each one relates polynomials with n^{l-1} monomials, where l is the length of the corresponding side of the corresponding rewrite rule. We can reduce the degree and size of these polynomials at the cost of introducing additional variables. In fact, we represent each product of two matrices by another matrix of shape T, introducing $(n-1)^2$ new variables. These new variables are constrained by equality relations. The resulting equalities and inequalities are quadratic.

Since matrix multiplication is associative, we can choose the grouping of the sub-products so as to minimize the number of additional variables. This is done by looking for factors that occur repeatedly in the sides of the rewrite rules. For $\{a^2b^2 \to b^3a^3\}$, e.g., we set $c = a^2$, $d = b^2$ and consider $\{cd \to (bd)(ca)\}$. This is most likely to be helpful for termination problems over small alphabets (but for a fixed problem size, problems over larger alphabets tend to be easier anyway).

Solving the constraint system. Since the constraints are nonlinear (for $n > 2$), we cannot hope for an algebraic algorithm that solves them exactly and efficiently. By putting a bound on the coefficients, we get a finite domain problem that can be solved by combinatorial methods.

The finite domain constraint system is translated into a formula in propositional logic. Each integer unknown is represented in unary or binary notation by a sequence of propositional variables. Relations between the variables are expressed by formulas. Then a state-of-the-art SAT solver (we use SatELiteGTI [7]) finds a satisfying assignment, from which we reconstruct the solution of the original system.

Flow control in the small. This is how we use the constraint solver to find the subset R' of rules to be removed from R: it solves the constraint system

Method/Prover	Number of proofs	Average time
matrix	106	18 s
matrix (no time limit)	124	—
matrix + RFC	138	6 s
Torpa-1.4 (2005)	138	0.04 s
Torpa-1.5 (2006)	144	0.1 s

Fig. 1. Performance on TPDB [12] (1 min default time limit)

corresponding to

$$i(\Sigma) \subseteq T \wedge i(R) \subseteq N \wedge \bigvee_{(\ell \to r) \in R} i(\ell \to r) \in \text{core}(T),$$

which implies that at least one rule can be removed. This also gives a nice safety measure: from the constraint solver we just get an assignment of the variables, and we re-compute all constraints to see which parts of the disjunction were true.

Flow control in the large. While this gives the general idea, quite some effort has to be invested to organize the repeated attempts in such a manner that all potentially successful parameter combinations are actually tried within the given time bound, where we have to consider also the time spent for failing attempts due to unsuccessful parameter settings. A good balance seems to be that we first take 4 seconds time to look for interpretations of dimension 2 with 6 bits for coefficients, then increase the dimension and the runtime while decreasing the bit width, arriving at dimension 5 with 3 bits and 25 seconds. This allows to prove both of Zantema's systems (Examples 3 and 4) within one minute on a standard personal computer.

Performance. The Termination Problem Data Base [12] (Version 2.0) contains 166 termination problems for string rewriting. Among these, 13 are relative termination problems, and 11 problems are non-terminating (on purpose).For 152 of these problems, at least one automated tool could prove termination in the 2005 competition. We run our prover Matchbox on this problem set. It finds 106 termination proofs via matrix interpretations (including 12 for relative termination) within the time limit of one minute per problem. This includes z001 and z086, see Examples 3 and 4. Using more time and more refined search methods (that are not completely automated) we found that a total of 124 problems can be solved by matrix interpretations. These numbers indicate that matrix interpretations are a very powerful stand-alone method for termination proofs, although our current implementation is slow compared to other provers.

The matrix method nicely combines with the RFC match-bound method [9], in the sense that it is often the case that either the whole system is RFC-match-bounded, or some rules can be removed by a matrix interpretation.By repeated application of these two approaches, Matchbox finds 138 termination proofs within the one minute time limit for each. We remark that this score just coincides with last year's winning score of Torpa [16].

8   Well-Founded Rings as Monotone Algebras

One particular instance of the matrix method (T-interpretations as defined in Section 7) has been extended to term rewriting [4]. Since the algebra of terms does not have a ring structure, ordered rings are replaced by "extended monotone algebras". In this section we show how these concepts are related for string rewriting.

In a Σ-algebra $\mathcal{A}$ with domain D, for every symbol $s \in \Sigma$ there is a mapping $s_\mathcal{A} : D \to D$, its *interpretation*. This induces an interpretation of strings by $\epsilon_\mathcal{A}(x) = x$ and, for $s \in \Sigma$ and $z \in \Sigma^*$, $(sz)_\mathcal{A}(x) = s_\mathcal{A}(z_\mathcal{A}(x))$. Assume that relations ρ and σ on D are given such that ρ is well-founded, $\rho \circ \sigma \subseteq \rho$, and for each $s \in \Sigma$, $s_\mathcal{A}$ is monotone modulo ρ and monotone modulo σ. Then $\mathcal{A}$ is called an *extended monotone algebra* in [4]. It is *strictly compatible* with a string rewriting system R over Σ if $\forall (\ell \to r) \in R \; \forall x \in D : \ell_\mathcal{A}(x) \, \rho \, r_\mathcal{A}(x)$, and it is *compatible* with R if $\forall (\ell \to r) \in R \; \forall x \in D : \ell_\mathcal{A}(x) \, \sigma \, r_\mathcal{A}(x)$.

Theorem 2 ([4]). *For string rewriting systems R and S the following properties are equivalent: (1) There is an extended monotone algebra that is strictly compatible with R and compatible with S. (2) R is terminating relative to S.*

Now let N be the positive cone of a well-founded ring, let $A \subseteq N$, and let again R and S be string rewriting systems over Σ. If there is an A-interpretation i for R with $i(S) \subseteq N$, Theorem 1 yields a proof of $SN(R/S)$ which can as well be formulated in the above cited monotone algebra framework:

Given a mapping $i : \Sigma \to A$, we build a Σ-algebra $\mathcal{A}$ with domain A^* as follows. The interpretation $s_\mathcal{A} : A^* \to A^*$ of $s \in \Sigma$ is defined by $s_\mathcal{A}(x) = i(s) \cdot x$; thus $z_\mathcal{A}(x) = i(z) \cdot x$ for $z \in \Sigma^*$ and $x \in A^*$. Apart from the usual strict matrix ordering $>$ and its reflexive closure $\geq$, we shall need the ordering $>_A$ on A^* which is defined by

$$x >_A y \quad \text{if} \quad x - y \in \text{core}(A).$$

Excluding the trivial cases $\Sigma = \emptyset$ or $R = \emptyset$, we can assume $A \neq \emptyset$ and $\text{core}(A) \neq \emptyset$, so by Lemma 2(6) we have $A \subseteq P$. Further, without loss of generality we have $A + N \subseteq A$ by Lemma 2(5), cf. Remark 3. Then choosing $\rho = >_A$ and $\sigma = \geq$, it is easy to establish

Proposition 1. *$\mathcal{A}$ is an extended monotone algebra, $\mathcal{A}$ is strictly compatible with R if and only if $i(R) \subseteq \text{core}(A)$, and $\mathcal{A}$ is compatible with S if and only if $i(S) \subseteq N$.*

From the above algebra of matrices $\mathbb{Z}^{d \times d}$ there is a homomorphism h into the algebra of vectors $\mathbb{Z}^d$ given by $h(m) = m \cdot (1, \ldots, 1)^T$, corresponding to constructing the column vector of row sums. Then an A-interpretation i corresponds to a Σ-algebra $\mathcal{C}$ with domain $\mathbb{Z}^d$ by $s_\mathcal{C}(x) = i(s) \cdot x$ (as above, but now x is a vector). For certain choices of A it is possible to define an ordering $>_A$, this time on vectors, such that $\mathcal{C}$ becomes an extended monotone algebra as above. E.g., define

$x >_{E_I} y$ by $x \geq y \wedge \exists i \in I : x_i > y_i$ and define $x >_{M_I} y$ by $x \geq y \wedge \forall i \in I : x_i > y_i$. For the set T from Section 7, take $x >_T y$ by $x \geq y \wedge x_1 > y_1$; this is the ordering we used for our termination proof of Example 4 in [10]. For $A = M_I^T$ there is no such ordering.

9 Discussion

Matrix interpretations and dependency pairs. The dependency pair method [1] infers termination of a rewriting system R over Σ from relative termination $SN(DP(R)/R)$, see [16]. For a string rewriting system R, the *marked* system is $DP(R) = \{x'u \rightarrow y'w \mid (xu, vyw) \in R, \ x, y \in \Sigma\}$ over an alphabet $\Sigma \cup \Sigma'$ where $\Sigma' = \{x' \mid x \in \Sigma\}$. Since the matrix method gives proofs for relative termination, it directly supports this basic version of the dependency pairs method. The marker symbols do in fact encode the idea that $DP(R)$ steps only happen at the left end (for terms: top position). As shown in [4], the matrix method can be adapted to relative top-termination, and it can be combined with refinements of the dependency pair approach.

Open problems and future work. One immediate problem with current implementations of the matrix method is that they are slow. Perhaps this could be improved by preprocessing the constraints. Performance is not a problem of the method itself—we have the strong feeling that more problems from the Termination Problem Database can be solved by matrix interpretations, and we already can construct some of them on paper—the matrices are large but sparse.

We want to investigate more closely the power of the matrix method. What other sets of matrices A with non-empty core(A) could be used in termination proofs? Explain the relationship between proofs via the sets E_I and M_I, and between M_I and $M_{I'}$ for $I \neq I'$.

By fixing matrix dimensions, we get a termination hierarchy where level n contains those systems that admit a matrix proof of dimension n. Is every level inhabited? Is it decidable whether a given system belongs to a given level? Is it for some fixed level? (It is for level 1, because (after taking logarithms) an interpretation can be found by linear programming. How about level 2?)

Another idea is to relate matrix interpretations to formal language theory. Matrix interpretations are in fact weighted finite automata [14]. The method of (RFC) match-bounds [9] also builds on weighted (annotated) automata. A unified view of these methods would hopefully allow us to construct matrix interpretations as efficiently as match bound certificates.

Acknowledgements. We have discussed the matrix method with Jörg Endrullis, Alfons Geser and Hans Zantema. Jörg Endrullis has built an independent implementation [3]. Niklas Sörensson provided information on finite domain constraint solving, and Axel Schüler pointed us to rings with polynomial identities. We are grateful to the referees for careful reading and detailed remarks.

References

1. T. Arts and J. Giesl. Termination of term rewriting using dependency pairs. *Theoret. Comput. Sci.*, 236:133–178, 2000.
2. R. V. Book and F. Otto. *String-Rewriting Systems*. Texts Monogr. Comput. Sci.. Springer-Verlag, New York, 1993.
3. J. Endrullis. Jambox: Automated termination proofs for string rewriting. http://joerg.endrullis.de/
4. J. Endrullis, J. Waldmann, and H. Zantema. Matrix interpretations for proving termination of term rewriting. *Proc. 3rd Int. Conf. Automated Reasoning IJCAR-06*, to appear, 2006.
5. L. Fuchs. *Partially Ordered Algebraic Systems*. Pergamon Press, 1963.
6. N. Dershowitz and R. Treinen. The RTA list of open problems. http://www.lsv.ens-cachan.fr/rtaloop/
7. N. Eén and A. Biere. Effective preprocessing in SAT through variable and clause elimination. In F. Bacchus and T. Walsh, *Proc. 8th Int. Conf. Theory and Applications of Satisfiability Testing SAT-05*, Lecture Notes in Comput. Sci. Vol. 3569, pp. 61–75. Springer-Verlag, 2005.
8. A. Geser. *Relative Termination*. Dissertation, Universität Passau, Germany, 1990.
9. A. Geser, D. Hofbauer and J. Waldmann. Match-bounded string rewriting systems. *Appl. Algebra Engrg. Comm. Comput.*, 15(3-4):149–171, 2004.
10. D. Hofbauer and J. Waldmann. Termination of $\{aa \rightarrow bc, bb \rightarrow ac, cc \rightarrow ab\}$. *Inform. Process. Lett.*, 98:156–158, 2006.
11. A. Kanel-Below and L. H. Rowen. *Computational Aspects of Polynomial Identities*. A. K. Peters, Wellesley MA, 2005.
12. The Termination Problems Data Base, Version 2.0. 2005. http://www.lri.fr/~marche/tpdb/
13. J. Waldmann. Matchbox: a tool for match-bounded string rewriting, In V. van Oostrom (Ed.), *Proc. 15th Int. Conf. Rewriting Techniques and Applications RTA-04*, Lecture Notes in Comp. Sci. Vol. 3091, pp. 85–94. Springer-Verlag, 2004.
14. J. Waldmann. Weighted automata for proving termination of string rewriting. In M. Droste and H. Vogler (Eds.), *Proc. Weighted Automata: Theory and Applications*. Leipzig, 2006.
15. H. Zantema. Termination. In Terese, *Term Rewriting Systems*, pp. 181–259. Cambridge Univ. Press, 2003.
16. H. Zantema. TORPA: Termination of Rewriting Proved Automatically. In V. van Oostrom (Ed.), *Proc. 15th Int. Conf. Rewriting Techniques and Applications RTA-04*, Lecture Notes in Comp. Sci. Vol. 3091, pp. 95–104. Springer, 2004.
17. H. Zantema. Problem #104. In [6], http://www.lsv.ens-cachan.fr/rtaloop/problems/104.html, 2005.

Decidability of Termination for Semi-constructor TRSs, Left-Linear Shallow TRSs and Related Systems

Yi Wang* and Masahiko Sakai

Graduate School of Information Science, Nagoya University
Furo-cho, Chikusa-ku, Nagoya, 464-8603 Japan
{ywang80@trs.cm., sakai@}is.nagoya-u.ac.jp

Abstract. We consider several classes of term rewriting systems and prove that termination is decidable for these classes. By showing the cycling property of infinite dependency chains, we prove that termination is decidable for semi-constructor case, which is a superclass of right-ground TRSs. By analyzing argument propagation cycles in the dependency graph, we show that termination is also decidable for left-linear shallow TRSs. Moreover we extend these by combining these two techniques.

1 Introduction

Termination is one of the central properties of term rewriting systems (TRSs for short). We say a TRS terminates if it does not admit any infinite reduction sequences. Termination guarantees that any expression cannot be infinitely rewritten, and hence the existence of a normal form for it. As we go from simple to more general classes of term rewriting systems, the difficulty of deciding termination increases until it becomes undecidable. It is meaningful to identify the decidability barrier and study decidability issues for some intermediate classes, especially if these classes are expressive enough to capture interesting rules.

As a generalization of the decidable classes of ground TRSs [7] and right-ground TRSs [4], the class of semi-constructor TRSs is studied. A TRS is called semi-constructor if every defined symbol in the right-hand sides of rules takes ground terms as its arguments. By showing the cycling property of infinite dependency chains, we give a positive answer to this problem.

The class of shallow TRSs has been attracting some interests from researchers due to the decidability of reachability and joinability problems for this class [3,8,10,13]. A TRS is called shallow if all variables in l, r occur at positions with depth 0 or 1 for each rule $l \rightarrow r$. In 2005, the affirmative result on termination of TRSs that contains right-linear shallow rules was shown by Godoy and Tiwari [5]. Here we propose a technique based on the analysis of argument propagation in the dependency graph.

Combining the two techniques for semi-constructor case and shallow case, we prove the decidability of termination for the following TRSs:

---

* Presently, with Financial Services Dept., Accenture Japan Ltd.

F. Pfenning (Ed.): RTA 2006, LNCS 4098, pp. 343–356, 2006.
© Springer-Verlag Berlin Heidelberg 2006

1. right-linear reverse-growing TRSs with all the dependency pairs being shallow or right-ground
2. left-linear growing TRSs with all the dependency pairs being shallow or right-ground.

The organization of this paper is as follows. In Section 2, we review preliminary definitions of term rewriting systems and introduce basic definitions and results concerning dependency pair method that will be used in Section 3. In Section 3, we give the definition of loop, head-loop and cycle first, then list our results and give their proofs. In Section 4, we compare our results with some existing results.

2   Preliminaries

We assume the reader is familiar with the standard definitions of term rewriting systems [2] and here we just review the main notations used in this paper.

A *signature* $\mathcal{F}$ is a set of function symbols, where every $f \in \mathcal{F}$ is associated with a non-negative integer by an arity function: $arity\colon \mathcal{F} \to \mathbb{N}(= \{0, 1, 2, \ldots\})$. Function symbols of arity 0 are called *constants*. The set of all *terms* built from a signature $\mathcal{F}$ and a countable infinite set $\mathcal{V}$ of *variables* such that $\mathcal{F} \cap \mathcal{V} = \emptyset$, is represented by $\mathcal{T}(\mathcal{F}, \mathcal{V})$. The set of *ground terms* is denoted by $\mathcal{T}(\mathcal{F}, \emptyset)$ ($\mathcal{T}(\mathcal{F})$ for short). We write $s = t$ when two terms s and t are identical. The *root symbol* of a term t is denoted by $root(t)$.

The set of all *positions* in a term t is denoted by $\mathcal{P}os(t)$ and ε represents the root position. We denote the *subterm ordering* by $\trianglelefteq$, that is, $t \trianglelefteq s$ if t is a subterm of s, and $t \triangleleft s$ if $t \trianglelefteq s$ and $t \neq s$. The *depth* of a position $p \in \mathcal{P}os(t)$ is $|p|$. The *height* of a term t is 0 if t is a variable or a constant, and $1 + max(\{height(s_i) \mid i \in \{1, \ldots, m\}\})$ if $t = f(s_1, \ldots, s_m)$. Let C be a *context* with a hole $\square$. We write $C[t]$ for the term obtained from C by replacing $\square$ with a term t.

A *substitution* θ is a mapping from $\mathcal{V}$ to $\mathcal{T}(\mathcal{F}, \mathcal{V})$ such that the set $\mathrm{Dom}(\theta) = \{x \in \mathcal{V} \mid \theta(x) \neq x\}$ is finite. We usually identify a substitution θ with the set $\{x \mapsto \theta(x) \mid x \in \mathrm{Dom}(\theta)\}$ of variable bindings. In the following, we write $t\theta$ instead of $\theta(t)$.

A *rewrite rule* $l \to r$ is a directed equation which satisfies $l \notin \mathcal{V}$ and $\mathrm{Var}(r) \subseteq \mathrm{Var}(l)$. A *term rewriting system* TRS is a finite set of rewrite rules. If the two conditions $l \notin \mathcal{V}$ and $\mathrm{Var}(r) \subseteq \mathrm{Var}(l)$ are not imposed, then we call it *eTRS*. We use R^{-1} for the reverse eTRS of R; $R^{-1} = \{r \to l \mid l \to r \in R\}$. The *reduction relation* $\to_R \subseteq \mathcal{T}(\mathcal{F}, \mathcal{V}) \times \mathcal{T}(\mathcal{F}, \mathcal{V})$ associated with a TRS R is defined as follows: $s \to_R t$ if there exist a rewrite rule $l \to r \in R$, a substitution θ, and a context C such that $s = C[l\theta]$ and $t = C[r\theta]$. The subterm $l\theta$ of s is called a *redex* and we say that s is reduced to t by contracting redex $l\theta$. The transitive closure of $\to_R$ is denoted by $\to_R^+$. The transitive and reflexive closure of $\to_R$ is denoted by $\to_R^*$. If $s \to_R^* t$, then we say that there is a *reduction sequence* starting from s to t or t is *reachable* from s by R. A term without redexes is called a *normal form*. A rewrite rule $l \to r$ is called *left-linear* (resp. *right-linear*) if no variable occurs twice in l (resp. r). It is called *linear* if it is both left- and right-linear. A TRS is called left-linear (resp. right-linear, resp. linear) if all of its rules are left-linear (resp. right-linear, resp. linear).

For a TRS R, a term $t \in T(\mathcal{F}, \mathcal{V})$ *terminates* if there is no infinite reduction sequence starting from t. We say that R terminates if every term terminates.

For a TRS R, a function symbol $f \in \mathcal{F}$ is a *defined symbol* of R if $f = root(l)$ for some rule $l \to r \in R$. The set of all defined symbols of R is denoted by $D_R = \{root(l) \mid \exists l \to r \in R\}$. We write C_R for the set of all *constructor symbols* of R which is defined as $\mathcal{F} \backslash D_R$. A term t has a *defined root symbol* if $root(t) \in D_R$.

Let R be a TRS over a signature $\mathcal{F}$. $\mathcal{F}^\sharp$ denotes the union of $\mathcal{F}$ and $D_R^\sharp = \{f^\sharp \mid f \in D_R\}$ where $\mathcal{F} \cap D_R^\sharp = \emptyset$ and $f^\sharp$ has the same arity as f. We call these fresh symbols *dependency pair symbols*. Given a term $t = f(t_1, \ldots, t_n) \in T(\mathcal{F}, \mathcal{V})$ with f defined, we write $t^\sharp$ for the term $f^\sharp(t_1, \ldots, t_n)$. If $l \to r \in R$ and u is a subterm of r with a defined root symbol, then the rewrite rule $l^\sharp \to u^\sharp$ is called a *dependency pair* of R. The set of all dependency pairs of R is denoted by $\mathrm{DP}(R)$.

For TRSs R and $\mathcal{C}$, a (possibly infinite) sequence of the elements of $\mathcal{C}$ $s_1^\sharp \to t_1^\sharp$, $s_2^\sharp \to t_2^\sharp$, ... is an $(R, \mathcal{C})$-*chain* if there exist substitutions $\tau_1, \tau_2, \ldots$ such that $t_i^\sharp \tau_i \to_R^* s_{i+1}^\sharp \tau_{i+1}$ holds for every $s_i^\sharp \to t_i^\sharp$ and $s_{i+1}^\sharp \to t_{i+1}^\sharp$ in the sequence. An $(R, \mathrm{DP}(R))$-chain is called *dependency chain*.

Theorem 1 ([6,1]). *For a TRS R, R does not terminate if and only if there exists an infinite dependency chain.*

The nodes of an $(R, \mathcal{C})$-*graph* denoted by $\mathrm{G}(R, \mathcal{C})$ are the elements of $\mathcal{C}$ and there is an edge from a node $s^\sharp \to t^\sharp$ to $u^\sharp \to v^\sharp$ if and only if there exist substitutions σ and τ such that $t^\sharp \sigma \to_R^* u^\sharp \tau$. An $(R, \mathrm{DP}(R))$-graph is called *dependency graph* and denoted by $\mathrm{DG}(R)$. Note that the dependency graph is not computable in general. However, our results will work on any approximation of the dependency graph. We say a graph is an *approximate graph* of a $(R, \mathcal{C})$-graph G if it contains G as a subgraph and $root(t) = root(u)$ for each arrow from a node $s^\sharp \to t^\sharp$ to $u^\sharp \to v^\sharp$. We remark that there exists at least one computable approximate graph for every $(R, \mathcal{C})$-graph.

The special notations "$(R, \mathcal{C})$-chain" and "$(R, \mathcal{C})$-graph" adopted in this paper is for handling left-linear TRSs as right-linear ones. For example, we will use an "$(R^{-1}, \mathrm{DP}(R)^{-1})$-chain".

3   Decidability of Termination Based on Cycle Detection

Infinite reduction sequences are often composed of cycles. A cycle is a reduction sequence where a term is rewritten to the same term. More generally, a loop is a reduction sequence where an instance of the starting term re-occurs as a subterm. It is obvious that a loop gives an infinite reduction sequence. In fact, the usual way to deduce non-termination is to construct a loop.

Definition 2 (Loop, Head-Loop, Cycle).

1. *A reduction sequence* loops *if it contains* $t' \to_R^+ C[t'\theta]$ *for some context C, substitution θ and term t'. Similarly, a reduction sequence* head-loops *if containing* $t' \to_R^+ t'\theta$, *and* cycles *if containing* $t' \to_R^+ t'$.
2. *A term t* loops *(resp.* head-loops, *resp.* cycles*) with respect to R if there is a looping (resp. head-looping, resp. cycling) reduction sequence starting from t.*
3. *A TRS R* admits *a loop (resp. head-loop, resp. cycle) if there is a term t such that t loops (resp. head-loops, resp. cycles) with respect to R.*

Proposition 3. *The following statements hold:*

1. *If t cycles, then t head-loops. If t head-loops, then t loops.*
2. *A TRS does not terminate if it admits a loop or a head-loop or a cycle.*

Example 4. Let $R_1 = \{f(x) \to h(f(g(a))), g(x) \to g(h(x))\}$ and $t = f(x)$. We can construct the following reduction sequence by only applying the former rule: $f(x) \to h(f(g(a))) \to h(h(f(g(a)))) \to \cdots$ which loops with $C = h[\Box]$, $\theta = \{x \mapsto g(a)\}$ and $t' = f(x)$. Notice there are more than one looping reduction sequences for R_1.

Naturally, the observation above inspires us to find some class of TRSs, whose non-termination is *equivalent to* the existence of loops. If we are able to check the existence of loops, then termination of such a class becomes decidable.

The following theorem lists our main results and will be proved in the following subsections.

Theorem 5. *Termination of the following classes of TRSs is decidable:*

1. *semi-constructor TRSs*
2. *right-linear shallow TRSs*
3. *left-linear shallow TRSs*
4. *right-linear rev-growing TRSs with all the dependency pairs being shallow or right-ground.*
5. *left-linear growing TRSs with all the dependency pairs being shallow or right-ground.*

3.1   Semi-constructor TRSs

Definition 6 (Semi-Constructor TRS). *A term $t \in \mathcal{T}(\mathcal{F}, \mathcal{V})$ is a semi-constructor* term *if every term s such that $s \trianglelefteq t$ and $root(s) \in D_R$ is ground. A TRS R is a semi-constructor system* if r is a semi-constructor term for every rule $l \to r \in R$.

Example 7. The TRS $R_2 = \{f(x) \to h(x, f(g(a))), g(x) \to g(h(a, a))\}$ is a semi-constructor system.

Proposition 8. *A TRS R is called* right-ground *if for every $l \to r \in R$, r is ground. The following statements hold:*

1. *Right-ground TRSs are semi-constructor systems.*
2. *For a semi-constructor TRS R, rules in $\mathrm{DP}(R)$ are right-ground.*

For a given TRS, let $\mathcal{T}_\infty$ denote the set of all *minimal non-terminating terms*, here "minimal" is used in the sense that all its proper subterms terminate.

Definition 9 ($\mathcal{C}$-min). *For a TRS R, let $\mathcal{C} \subseteq \mathrm{DP}(R)$. An infinite reduction sequence in $R \cup \mathcal{C}$ in the form $t_1^\sharp \to_R^* t_2^\sharp \to_{\mathcal{C}} t_3^\sharp \to_R^* t_4^\sharp \to_{\mathcal{C}} \cdots$ with $t_i \in \mathcal{T}_\infty$ for all $i \geq 1$ is called a $\mathcal{C}$-min reduction sequence. We use $\mathcal{C}_{min}(t^\sharp)$ to denote the set of all $\mathcal{C}$-min reduction sequences starting from $t^\sharp$.*

Proposition 10 ([1,6]). *Given a TRS R, we have the following statements:*

1. *If there exists an infinite dependency chain, then $\mathcal{C}_{min}(t^\sharp) \neq \emptyset$ for some $\mathcal{C} \subseteq \mathrm{DP}(R)$ and $t \in \mathcal{T}_\infty$.*
2. *For any sequence in $\mathcal{C}_{min}(t^\sharp)$, reduction $\to_R$ takes place below the root while reduction $\to_{\mathcal{C}}$ takes place at the root.*
3. *For any sequence in $\mathcal{C}_{min}(t^\sharp)$, subsequence $s^\sharp \to_{R \cup \mathcal{C}}^* t^\sharp$ implies $s \to_R^* C[t]$ for some context C.*
4. *For any sequence in $\mathcal{C}_{min}(t^\sharp)$, there is at least one rule in $\mathcal{C}$ which is applied infinitely often.*

Lemma 11. *For a TRS R, if $sq \in \mathcal{C}_{min}(t^\sharp)$ loops, then sq head-loops.*

Proof. Let $sq \in \mathcal{C}_{min}(t^\sharp)$ loops, then there is a subsequence $t_k^\sharp \to_{R \cup \mathcal{C}}^+ C[t_k^\sharp \theta]$ in sq. From Prop.10–(2) and the fact that dependency pair symbols appears only in dependency pairs, we have $C[t_k^\sharp \theta] = t_k^\sharp \theta$, which implies that sq head-loops.   □

Lemma 12. *For a TRS R, if $sq \in \mathcal{C}_{min}(t^\sharp)$ loops, then there is a term $t_k^\sharp$ in sq such that t_k loops with respect to R.*

Proof. From Lemma 11 and Prop. 10–(3).   □

Lemma 13. *For a semi-constructor TRS R, the following statements are equivalent:*

1. *R does not terminate.*
2. *There exists $l^\sharp \to u^\sharp \in \mathrm{DP}(R)$ such that sq cycles for some $sq \in \mathcal{C}_{min}(u^\sharp)$.*

Proof. $(2 \Rightarrow 1)$: It is obvious by Lemma 12. $(1 \Rightarrow 2)$: By Theorem 1, there exists an infinite dependency chain. By Prop. 10–(1), there exists a sequence $sq \in \mathcal{C}_{min}(t^\sharp)$. By Prop. 10–(4), there is some rule $l^\sharp \to u^\sharp \in \mathcal{C}$ which is applied at root reduction in sq infinitely often. By Prop. 8–(2), $u^\sharp$ is ground. Thus $u^\sharp$ cycles in the form $u^\sharp \to_{R \cup DP(R)}^* \cdot \to_{\{l^\sharp \to u^\sharp\}} u^\sharp$ in sq.   □

Notice that non-termination of semi-constructor systems depends on the existence of a *cycling dependency chain*, which represents the cycle "$u^\sharp \to_{R \cup DP(R)}^* \cdot \to_{\{l^\sharp \to u^\sharp\}} u^\sharp$ in sq" in the proof of Lemma 13. Here, *cycle* is guaranteed by the fact that $\mathrm{DP}(R)$ is right-ground.

Proof. **(Theorem 5–(1))** The decision procedure for termination of semi-constructor TRS R is as follows: consider all terms $u_1, u_2, \ldots, u_n$ corresponding to the right-hand sides of $\mathrm{DP}(R) = \{ l_i^\sharp \to u_i^\sharp \mid 1 \le i \le n \}$, and simultaneously generate all reduction sequences with respect to R starting from $u_1, u_2, \ldots, u_n$. It terminates if it enumerates all reachable terms exhaustively or it detects a looping reduction sequence $u_i \to_R^+ C[u_i]$ for some i.

Suppose R does not terminate. By Lemma 12, 13 and the groundness of u_i's, we have a looping reduction sequence $u_i \to_R^+ C[u_i]$ for some i and C. Hence we detect non-termination of R. If R terminates, then the execution of the reduction sequence generation stops finally since it is finitely branching. Thus we detect termination of R after finitely many steps. □

Next we make a natural extension by relaxing the condition for assuring *cycling*, which is mainly used in the Subsection 3.3.

Lemma 14. *Let R be a TRS whose termination is equivalent to the non-existence of a dependency chain that contains infinite use of right-ground dependency pairs. Then termination of R is decidable.*

Proof. We apply the above procedure starting from terms $u_1, u_2, \ldots, u_n$, where $u_i^\sharp$'s are all ground right-hand sides of dependency pairs. Suppose R is non-terminating, we have a dependency chain with infinite use of a right-ground dependency pair. Similarly to the semi-constructor case, we have a loop $u_i \to_R^+ C[u_i]$, which can be detected by the procedure. □

Example 15. Let $R_3 = \{ f(a) \to g(b), g(x) \to f(x), h(a, x) \to h(b, x) \}$. We can compute the dependency graph. It has only one cycle, which contains a right ground node. From Lemma 14 we can show termination of R_3 by the procedure starting from $g(b)$.

3.2 Right-Linear Shallow or Left-Linear Shallow TRSs

In this subsection, we show how to analyze cycle of dependency chains that consist only of right-linear shallow dependency pairs and then show the decidability of termination for right-linear shallow TRSs and left-linear shallow TRSs.

Definition 16 (Shallow TRS). *A rewrite rule $l \to r$ is shallow if all variables in $\mathrm{Var}(l) \cup \mathrm{Var}(r)$ occur at positions with depth 0 or 1. An eTRS is shallow if all its rewrite rules are shallow.*

Example 17. TRS $R_4 = \{ f(x, y) \to f(g(a), y), f(g(a), z) \to f(z, b) \}$ and $R_5 = \{ g(x, x) \to f(x, a), f(c, x) \to g(x, b), a \to c, b \to c \}$ are shallow.

We say that T *is joinable to s* if $\forall t \in T.\ t \to_R^* s$ and T *is joinable* if it is joinable to some s. From now on, we assume R in which both of the following properties are decidable.

Ground Reachability: $t \to_R^* s$ for given ground terms t and s.
Ground Joinability: T is joinable for a given set T of ground terms.

For dependency chains composed of shallow dependency pairs, all informations carried by variables are passed to the next dependency pairs in its derivation. For example, consider R_5 and an infinite sequence of dependency pairs:

$$g^\sharp(\underline{x},\underline{x}) \to f^\sharp(\underline{x},a),\ f^\sharp(c,\underline{x}) \to g^\sharp(\underline{x},b),\ g^\sharp(\underline{x},\underline{x}) \to f^\sharp(\underline{x},a),\ \ldots\ .$$

It is a dependency chain because we have a derivation:

$$g^\sharp(\underline{c},\underline{c}) \to f^\sharp(\underline{c},a) = f^\sharp(c,\underline{a}) \to g^\sharp(\underline{a},b) \to \cdot \to g^\sharp(\underline{c},\underline{c}) \to f^\sharp(\underline{c},a) \to \cdots\ .$$

In order to analyze such information flows caused by variables, we introduce some notions.

Definition 18 (Labeling Function). *Let R and $\mathcal{C}$ be eTRSs and AG be an approximate graph of $(R,\mathcal{C})$-graph. Let p be a path $nd_1, nd_2, nd_3 \cdots$ in AG and M be the maximum arity of root symbols of right-hand sides of nodes in AG. A labeling function $L_p\colon \mathbb{N} \to \mathcal{C} \times \mathcal{P}(\mathcal{T}(\mathcal{F}))^M$ is defined as follows:*

1. Let nd_1 be $f^\sharp(t_1,\ldots,t_n) \to g^\sharp(s_1,\ldots,s_m)$. Then $L_p(1) = (nd_1, S_1^1,\ldots,S_M^1)$ where

$$S_i^1 = \begin{cases} \{s_i\} & \text{if } s_i \notin \mathcal{V} \text{ and } 1 \leq i \leq n \\ \emptyset & \text{otherwise.} \end{cases}$$

2. Let $nd_i = h^\sharp(v_1,\ldots,v_k) \to f^\sharp(u_1,\ldots,u_n)$, $L_p(i) = (nd_i, S_1^i,\ldots,S_M^i)$ and $nd_{i+1} = f^\sharp(t_1,\ldots,t_n) \to g^\sharp(s_1,\ldots,s_m)$. Then $L_p(i+1) = (nd_{i+1}, S_1^{i+1}, \ldots, S_M^{i+1})$ where

$$S_j^{i+1} = \begin{cases} \{s_j\} & \text{if } s_j \notin \mathcal{V} \text{ and } 1 \leq j \leq k \\ \displaystyle\bigcup_{l \in \{1,\ldots,n\} \wedge s_j = t_l} S_l^i & \text{if } s_j \in \mathcal{V} \text{ and } 1 \leq j \leq k \\ \emptyset & \text{otherwise.} \end{cases}$$

Example 19. Consider $R = R_5$ and $\mathcal{C} = \{\, g^\sharp(x,x) \to f^\sharp(x,a),\ f^\sharp(c,x) \to g^\sharp(x,b)\,\} \subset \mathrm{DP}(R_5)$. The labeling function for a path $N_1, N_2, N_1, \ldots$ is $L(1) = (N_1, \emptyset, \{a\})$, $L(2) = (N_2, \{a\}, \{b\})$, $L(3) = (N_1, \{a,b\}, \{a\})$, $L(4) = (N_2, \{a\}, \{b\})$, $L(5) = (N_1, \{a,b\}, \{a\}), \ldots$, where $N_1 = g^\sharp(x,x) \to f^\sharp(x,a)$ and $N_2 = f^\sharp(c,x) \to g^\sharp(x,b)$. (See Fig. 1)

Definition 20 (Argument Propagation Cycling). *Let L_p be a labeling function over $p = nd_1, nd_2, nd_3, \ldots$. We say a finite sequence of labels $L_p(I), L_p(I+1), \ldots, L_p(J)$ is an argument propagation cycling (APC for short) if $L_p(I) = L_p(J)$ and the following condition, called smoothness condition, are satisfied for all i $(I \leq i < J)$:*

For all j $(1 \leq j \leq n)$,
1. S_j^i is joinable if t_j is a variable;
2. otherwise S_j^i is joinable to t_j

i		1	2	3	4
$L(i)$	nd_i	$N_1 =$ $g^\sharp$ $f^\sharp$ $\overparen{x\text{-}}\,\overparen{,x}\text{-}\text{-}$ $\underparen{x'}\,\underparen{a}$	$N_2 =$ $f^\sharp$ $g^\sharp$ $\overparen{-c}\,\overparen{,x}\text{-}$ $\underparen{-x'}\,\underparen{b}$	$N_1 =$ $g^\sharp$ $f^\sharp$ $\overparen{-x\text{-}}\,\overparen{,x}\text{-}$ $\underparen{-x'}\,\underparen{a}$	$N_2 =$ $f^\sharp$ $g^\sharp$ $\overparen{-c}\,\overparen{,x}$ $\underparen{-x'}\,\underparen{b}$
	S_1^i	$\emptyset$	$\{a\}$	$\{a,b\}$	$\{a\}$
	S_2^i	$\{a\}$	$\{b\}$	$\{a\}$	$\{b\}$

Fig. 1. Labeling for Example 19

where
$$L_p(i) = (v^\sharp \to u^\sharp, S_1^i, \ldots, S_M^i) \ and$$
$$L_p(i+1) = (f^\sharp(t_1, \ldots, t_n) \to s^\sharp, S_1^{i+1}, \ldots, S_M^{i+1}) \ .$$

We say an APC is minimal *if all its proper subsequences are not APC.*

Example 21. Consider the labeling function L in Example 19. The sequence $L(2), L(3), L(4)$ is a minimal APC.

One may think that every minimal APC contains no repetition of a same node except the edges. However it is not correct in general as shown by the following example.

Example 22. Consider a TRS $R_6 = \{\, g(x,y) \to f(y,b,x,a), \ g(x,y) \to f(y,a,x,b),$ $f(x,x,y,y) \to g(x,y) \,\}$ and $\mathcal{C} = \mathrm{DP}(R_6)$. The minimal APC over a path $N_1, N_3, N_2, N_3, N_1, \ldots$ is the sequence $L(2), L(3), \ldots, L(6)$ as shown in Fig. 2. There is no APC over a path $N_1, N_3, N_1, N_3, \ldots$.

Lemma 23. *For an eTRS R such that ground reachability and ground joinability are decidable and for a shallow eTRS $\mathcal{C}$, the existence of APC is decidable.*

Proof. Firstly we take an approximate graph G of $(R, \mathcal{C})$-graph. The procedure tries searches starting from every node in G. In traversing edges, it quits if an APC is found and backtracks traversal if the path does not satisfy the smoothness condition. The correctness of this procedure is obvious. The range of the labeling function is finite since the possible value in S_k of the labeling function is a ground term at depth 1 that occurs in the right-hand side of nodes. Since the smoothness condition is decidable by the assumption, termination of the procedure is guaranteed. □

We say that a natural extension of $(R, \mathcal{C})$-chain $\cdots nd_{-1}, nd_0, nd_1$ is *backward-infinite*. In order to avoid confusion, we sometimes say that an infinite $(R, \mathcal{C})$-chain is *forward-infinite*. Next lemma will formally express the relation between an APC and an infinite $(R, \mathcal{C})$-chain.

i	1	2	3	4	5	6
nd_i / $L(i)$	$N_1 =$ $g^\sharp$ $f^\sharp$	$N_3 =$ $f^\sharp$ $g^\sharp$	$N_2 =$ $g^\sharp$ $f^\sharp$	$N_3 =$ $f^\sharp$ $g^\sharp$	$N_1 =$ $g^\sharp$ $f^\sharp$	$N_3 =$ $f^\sharp$ $g^\sharp$
S^i_1	$\emptyset$	$\{b\}$	$\{a\}$	$\{a\}$	$\{b\}$	$\{b\}$
S^i_2	$\{b\}$	$\{a\}$	$\{a\}$	$\{b\}$	$\{b\}$	$\{a\}$
S^i_3	$\emptyset$		$\{b\}$		$\{a\}$	
S^i_4	$\{a\}$		$\{b\}$		$\{a\}$	

Fig. 2. Labeling for Example 22

Lemma 24. *Let R be an eTRS and C be a right-linear shallow eTRS. Then,*

1. *there exists a forward and backward-infinite (R, C)-chain if there exists an APC, and*
2. *there exists an APC if there exists a forward or backward-infinite (R, C)-chain.*

Proof. We firstly show the former part. Let $L_p(I), \ldots, L_p(J)$ be an APC over a path $nd_1, \cdots, nd_I, \cdots, nd_J$. In order to construct substitutions $\tau_I, \cdots, \tau_J$ that satisfy the chain condition and $\tau_I = \tau_J$, do the following repeatedly while applicable, starting with empty substitutions $\tau_i = \emptyset$ $(I \le i \le J)$.

- Let $I \le k < J$, $nd_k = v^\sharp \to f^\sharp(u_1, \ldots, u_n)$ and $nd_{k+1} = f^\sharp(t_1, \ldots, t_n) \to s^\sharp$. Set $\tau_k := \tau_k \cup \{u_i \mapsto t_i \tau_{k+1}\}$ if $u_i \in \mathcal{V} - \mathrm{Dom}(\tau_k)$ and $t_i \tau_{k+1} \notin \mathcal{V}$ $(1 \le i \le n)$.
- Set $\tau_J := \tau_I$ if $\tau_J \ne \tau_I$.

Note that the uniqueness of each substitution τ_i is guaranteed by the right-linearity of nodes. This procedure eventually stops and the chain is an (R, C)-chain with $nd_I \tau_I = nd_J \tau_J$ from the smoothness of the APC. Hence the existence of an forward and backward-infinite (R, C)-chain is easily shown.

Next, we argue that there exists an APC over a given forward-infinite (R, C)-chain. Let the (R, C)-chain be $nd_1, nd_2, \ldots$ where $nd_i = t_i^\sharp \to s_i^\sharp \in C$. There exists an APC (with smoothness condition ignored) over the path. Note that it is also possible even if the given chain is backward-infinite one $\cdots, nd_{-1}, nd_0, nd_1$, since we can choose a natural number N small enough such that an APC can be found along the path $nd_N, nd_{N+1}, \cdots, nd_0, nd_1$. Let the APC be $L(I), \cdots, L(J)$. We

have an sequence $t_I^\sharp \tau_I \to_C s_I^\sharp \tau_I \to_R^* t_{I+1}^\sharp \tau_{I+1} \to_C s_{I+1}^\sharp \tau_{I+1} \to_{R\cup C}^* \cdots \to_{R\cup C}^*$
$t_J^\sharp \tau_J \to_C s_J^\sharp \tau_J$, where $t_i^\sharp \to s_i^\sharp$ is a rule in $L(i)$. The satisfaction of the smoothness
condition follows from the traces of the reductions of ground terms at depth 1
in the sequence.   □

Example 25. Consider the APC $L(2), L(3), L(4)$ from a path $nd_1, nd_2, \ldots$ (see
Fig. 1). We show the existence of cycling reduction sequence $t \to_{R\cup C}^+ t$. According
to the procedure in the proof of Lemma 24, we obtain substitutions $\tau_2 = \tau_3 =$
$\tau_4 = \{x \mapsto c\}$. Thus, we have $f^\sharp(c, c) \to_C g^\sharp(c, b) \to_R^* g^\sharp(c, c) \to_C f^\sharp(c, a) \to_R^*$
$f^\sharp(c, c)$ from the sequence $nd_2\tau_2, nd_3\tau_3, nd_4\tau_4$.

Next, based on Lemma 24, we give proofs for Theorem 5–(2) and (3).

Proof. **(Theorem 5–(2))** Let R be a right-linear shallow TRS. Then, $\mathrm{DP}(R)$
is also right-linear shallow. We know ground reachability and ground joinability of right-linear shallow TRSs are decidable [9, 3, 10, 13]. By Lemma 23,
we can decide the existence of APC. Thus we can decide the existence of
a forward-infinite $(R, \mathrm{DP}(R))$-chain by Lemma 24. The theorem follows from
Theorem 1.   □

Proof. **(Theorem 5–(3))** Let R be a left-linear shallow TRS. Then R^{-1} and
$\mathrm{DP}(R)^{-1}$ are right-linear shallow eTRSs We know ground reachability and
ground joinability of right-linear shallow TRSs are decidable [9, 3, 10, 13]. By
Lemma 23, we can decide the existence of APC. If an APC exists, we have
a backward-infinite $(R^{-1}, \mathrm{DP}(R)^{-1})$-chain from the former part of Lemma 24,
which shows the existence of a forward-infinite $(R, \mathrm{DP}(R))$-chain. If no APC
exists, we have no backward-infinite $(R^{-1}, \mathrm{DP}(R)^{-1})$-chain from the latter part
of Lemma 24, which shows the non-existence of a forward-infinite $(R, \mathrm{DP}(R))$-
chain. The theorem follows from Theorem 1.   □

3.3   Combining the Two Techniques

In this subsection, we combine the techniques in the above two subsections and
show the decidability of termination for some larger classes. This is based on the
following lemma.

Proposition 26. *For TRSs R, C and C' such that $C \supseteq C'$, the following statements are equivalent.*

1. *There exists an infinite (R, C)-chain.*
2. *There exists an infinite (R, C')-chain or there exists an infinite (R, C)-chain
   with infinite use of pairs in $C - C'$.*

Proof. Since the latter implies the former trivially, we show the converse. Suppose we have an infinite (R, C)-chain $nd_1, nd_2, \ldots$ with finite use of pairs in
$C - C'$. Letting nd_n is the last use of a pair in $C - C'$, the infinite subsequence
$nd_{n+1}, nd_{n+2}, \ldots$ is a (R, C')-chain.   □

Definition 27 (Growing TRS). *A rewrite rule* $l \to r$ *is* growing *if all variables in* $\mathrm{Var}(l) \cap \mathrm{Var}(r)$ *occur at positions with depth 0 or 1 in* l. *An eTRS R is* growing *if every rewrite rule in R is growing and R is* rev-growing *if* R^{-1} *is growing.*

Example 28. TRS $R_7 = \{ \ f(a,x) \to g(x,b), \ g(x,y) \to h(x,p(x,y)), \ h(c,x) \to f(x,x) \}$ is left-linear growing.

Proof. **(Theorem 5–(4))** Let R be a right-linear rev-growing TRS with $\mathrm{DP}(R)$ being shallow or right-ground. Let $\mathcal{C}_s$ be the set of all shallow pairs in $\mathrm{DP}(R)$. We know ground reachability and ground joinability of right-linear rev-growing TRSs are decidable [10, 13]. Since $\mathcal{C}_s$ is right-linear shallow, we can decide the existence of APC by Lemma 23. If an APC exists then we have an infinite $(R, \mathcal{C}_s)$-chain by Lemma 24, which implies that R is non-terminating. Otherwise, from Prop. 26, it is enough to decide the existence of an infinite $(R, \mathrm{DP}(R))$-chain with infinite use of pairs in $\mathrm{DP}(R) - \mathcal{C}_s$, which is a set of right-ground pairs. This is decidable from Lemma 14. □

Proof. **(Theorem 5–(5))** Let R be a left-linear growing TRS with $\mathrm{DP}(R)$ being shallow or right-ground. Let $\mathcal{C}_s$ be the set of all shallow pairs in $\mathrm{DP}(R)$. Then R^{-1} is right-linear rev-growing and $\mathcal{C}_s^{-1}$ is right-linear shallow. Since we know ground reachability and ground joinability of right-linear rev-growing TRSs are decidable [10, 13], we can decide the existence of APC by Lemma 23. If an APC exists then we have a backward-infinite $(R^{-1}, \mathcal{C}_s^{-1})$-chain by Lemma 24, which implies the existence of an infinite $(R, \mathcal{C}_s)$-chain and hence R is non-terminating. Otherwise, from Prop. 26, it is enough to decide the existence of an infinite $(R, \mathrm{DP}(R))$-chain with infinite use of pairs in $\mathrm{DP}(R) - \mathcal{C}_s$, which is a set of right-ground pairs. This is decidable from Lemma 14. □

4   Comparison

In this section, we compare our results with some existing results.

Lemma 29. *For a semi-constructor TRS R, the following statements are equivalent:*

1. *R does not terminate.*
2. *DG(R) contains a cycle.*

Proof. Suppose R does not terminate. There exists an infinite dependency chain by Theorem 1. Hence the dependency graph must have a cycle, otherwise it causes a contradiction.

   Conversely, for every edge from a node $s^{\sharp} \to t^{\sharp}$ to a node $u^{\sharp} \to v^{\sharp}$ in a cycle, there exists a substitution τ such that $t^{\sharp} \to_R^* u^{\sharp}\tau$. Thus we can easily construct an infinite dependency chain. □

Lemma 30. *The dependency graph of semi-constructor TRSs is not computable.*

Proof. By encoding Post's Correspondence Problem. Let $\{\langle u_i, v_i \rangle \in \Sigma^+ \times \Sigma^+ \mid 1 \leq i \leq n\}$ be a finite set of PCP pairs.

$$TRS\ R_8 = \begin{array}{l} \{\varepsilon \to e_i(\varepsilon) \mid 1 \leq i \leq n\} \cup \\ \{\varepsilon \to f(c, d)\} \cup \\ \{b \to a(b),\ b \to a(\varepsilon) \mid b \in \{c, d\},\ a \in \Sigma\} \cup \\ \{f(x, x) \to g(x, x)\} \cup \\ \{e_i(g(u_i(x), v_i(y))) \to g(x, y) \mid 1 \leq i \leq n\} \cup \\ \{e_i(g(u_i(\varepsilon), v_i(\varepsilon))) \to \varepsilon \mid 1 \leq i \leq n\} \end{array}$$

Defined symbol of R_8 is $\{\varepsilon, c, d, f\} \cup \{e_i \mid 1 \leq i \leq n\}$, R_8 is a semi-constructor TRS and it is a variant of the example in [12]. Notice that the following statement is true: in $DG(R_8)$, there is an edge from node $\varepsilon^\sharp \to e_1^\sharp(\varepsilon)$ to node $e_1^\sharp(g(u_1(\varepsilon), v_1(\varepsilon))) \to \varepsilon^\sharp$ if and only if PCP has a solution.   □

Note that reachability problem is undecidable for linear semi-constructor TRSs [11]. However this fact is not enough to prove the above lemma because the use of reachability in dependency graphs are limited.

In the reference [9], Middeldorp proposed a decision procedure for termination of right-ground TRSs which is dependency graph based. Denoting growing approximation dependency graph by $DG_g(R)$, he showed that for right-ground TRS R, $DG(R) = DG_g(R)$, that is, the dependency graph of the right-ground TRS is computable. Thus, the decision procedure proposed is that: compute the dependency graph of R using the growing approximation and then check the existence of cycles. For semi-constructor case, we also have Lemma 29 to assure that semi-constructor TRS terminates if and only if there is no cycles in the dependency graph. However, the dependency graph based method can not be applied to semi-constructor case, since its dependency graph is not computable by Lemma 30.

The following theorem shown by Godoy and Tiwari [5] is also given as a corollary of Theorem 5–(4) since TRSs in this class satisfy the assumption of our theorem.

Theorem 31 ([5]). *Termination of TRSs that consist of right-linear shallow rules, collapsing rules and right-ground rules is decidable.*

Nagaya and Toyama [10] obtained the decidability result for almost orthogonal growing TRSs. We claim that the applicable classes of Nagaya's method and ours do not cover each other. Considering R_4 in Example 17 and R_7 in Example 28: R_4 is left-linear shallow, but it is not almost orthogonal since there is a non-trivial critical pair $\langle f(g(a), z), f(z, b) \rangle$; $DP(R_7)$ does not fit either of the applicable classes we proposed, but it is orthogonal.

5   Conclusion

One research direction on termination is to find more general classes of TRSs whose termination is decidable. We proposed several positive results listed in Theorem 5.

We propose some conjectures and list them as follows:

Conjecture 32. Termination of right-linear rev-growing TRSs with all the dependency pairs being left-linear is decidable.

Conjecture 33. Termination of left-linear growing TRSs with all the dependency pairs being right-linear is decidable.

Conjecture 34. Termination of shallow TRSs is undecidable.

Acknowledgement

We would like to thank Ashish Tiwari and the anonymous referees for their helpful comments and remarks. We are particularly grateful to one of anonymous referees who indicated the idea for combining our methods (Prop. 26). This work is partly supported by MEXT.KAKENHI #18500011 and #16650005.

References

1. T. Arts and J. Giesl: Termination of term rewriting using dependency pairs. Theoretical Computer Science **236** (2000) 133–178.
2. F. Baader and T. Nipkow: Term rewriting and all that. Cambridge University Press (1998).
3. H. Comon, M. Dauchet, R. Gilleron, F. Jacquemard, D. Lugiez, S. Tison and M. Tommasi: Tree automata techniques and applications (1997). http://www.grappa.univ-lille3.fr/tata/
4. N. Dershowitz: Termination of linear rewriting systems. Proc. of the 8th international colloquium on automata, languages and programming, LNCS **115** Springer-Verlag (1981) 448–458.
5. G. Godoy and A. Tiwari: Termination of rewrite systems with shallow right-linear, collapsing and right-ground rules. Proc. of the 20th international conference on automated deduction, LNAI **3632** (2005) 164–176.
6. N. Hirokawa and A. Middeldorp: Dependency pairs revisited. Proc. of the 15th international conference on rewriting techniques and applications, LNCS **3091** (2004) 249–268.
7. G. Huet and D. Lankford: On the uniform halting problem for term rewriting systems. Technical Report **283** INRIA (1978).
8. F. Jacquemard: Decidable approximations of term rewriting systems. Proc. of the 7th international conference on rewriting techniques and applications, LNCS **1103** (1996) 362–376.
9. A. Middeldorp: Approximating dependency graphs using tree automata techniques. Proc. of the international joint conference on automated reasoning, LNAI **2083** (2001) 593–610.

10. T. Nagaya and Y. Toyama: Decidability for left-linear growing term rewriting systems. Proc. of the 10th international conference on rewriting techniques and applications, LNCS **1631** (1999) 256–270.
11. I. Mitsuhashi, M. Oyamaguchi, Y. Ohta and T. Yamada. The joinability and unification problems for confluent semi-constructor TRSs. Proc. of the 15th international conference on rewriting techniques and applications, LNCS **3091** (2004) 285–300.
12. I. Mitsuhashi, M. Oyamaguchi and T. Yamada: The reachability and related decision problems for monadic and semi-constructor TRSs. Information Processing Letters **98** (2006) 219–224.
13. T. Takai, Y. Kaji and H. Seki: Right-linear finite path overlapping term rewriting systems effectively preserve recognizability. Proc. of the 11th international conference on rewriting techniques and applications, LNCS **1833** (2000) 246–260.

Proving Positive Almost Sure Termination Under Strategies

Olivier Bournez and Florent Garnier

LORIA/INRIA, 615 Rue du Jardin Botanique
54602 Villers lès Nancy Cedex, France

Abstract. In last RTA, we introduced the notion of probabilistic rewrite systems and we gave some conditions entailing termination of those systems within a finite mean number of reduction steps.

Termination was considered under arbitrary unrestricted policies. Policies correspond to strategies for non-probabilistic rewrite systems.

This is often natural or more useful to restrict policies to a subclass. We introduce the notion of positive almost sure termination under strategies, and we provide sufficient criteria to prove termination of a given probabilitic rewrite system under strategies. This is illustrated with several examples.

1 Introduction

As discussed in several papers such as [7,22,15], when specifying probabilistic systems, it is rather natural to consider that the firing of a rewrite rule can be subject to some probabilistic law.

Considering rewrite rules subject to probabilities leads to numerous questions about the underlying notions and results. In [7], we introduced probabilistic abstract reduction systems, and we introduced notions like almost-sure termination or probabilistic confluence, with relations between all these notions. In [6], we proved that, unlike what happens for classical rewriting logic, there is no hope to build a sound and complete proof system with probabilities in the general case [6]. In [5], we argue that positive almost sure termination is a better notion than simple almost sure termination for probabilistic systems and we provide necessary and sufficient criteria entailing positive almost sure termination.

In this paper, we pursue the investigation, by considering positive almost sure termination under strategies. As we show through several examples, it is often natural to restrict strategies to a subset of strategies. Many simple probabilistic rewrite systems do not terminate under arbitrary strategies, whereas they terminate if strategies are restricted to natural strategies.

The idea of adding probabilities to high level models of reactive systems is not new, and has also been explored for models like Petri Nets [3,26], automata based models [10,27], or process algebra [16]. There is now a rather important literature about model-checking techniques for probabilistic systems: see example [21] and the references there. Computer Tools like PRISM [20], APMC [19], do exist. Observe however, that most of the studies and techniques restrict to finite state systems.

F. Pfenning (Ed.): RTA 2006, LNCS 4098, pp. 357–371, 2006.

Termination of probabilistic concurrent programs has already been investigated. In particular, in [18] it has been argued that this is important to restrict to fair schedulers, and techniques for proving termination under fair schedulers have been provided. These techniques have been extended to infinite systems in [17]. Compared to our work, they focus on almost sure termination, whereas we focus on positive almost sure termination. Furthermore, we deal with probabilistic abstract reduction systems or rewrite systems, whereas these two papers are focusing on concurrent programs, where strategies correspond to schedulers.

Several notions of fairness have been introduced for concurrent programs, and in particular for probabilistic concurrent programs. In particular, Pnueli [23], and Pnueli and Zuck have introduced extreme fairness and α-fairness [24]. Hart, Sharir and Pnueli [18] and Vardi [27] consider probabilistic systems in which the choice of actions at the states is subject to fairness requirements, and proposed model checking algorithms. A survey and discussion of several fairness notions for probabilistic systems can be found in chapter 8 of [10].

Probabilistic abstract reduction systems and probabilistic rewrite systems do correspond to classical abstract reduction systems and classical rewrite systems where probabilities can only be 0 or 1 [5]. Therefore, any technique for proving termination of a probabilistic system must have a counterpart for classical systems. In particular, any technique for proving termination of probabilistic rewrite systems under strategies is an extension of a technique for proving termination of classical rewrite systems under strategies. The termination of rewrite systems under strategies has been investigated in e.g. [12,13]. Since the extension to the probabilistic case of very basic techniques already yields several problems discussed in this paper, we do not consider so general strategies.

The paper is organized as follows: in Section 2, we recall probabilistic abstract reduction systems, and probabilistic rewrite systems, as well as several concepts and results from [5]. In Section 3, we introduce positive almost sure termination under strategies, and we discuss several examples of systems that are non positively almost surely terminating but which are positively almost surely terminating under some strategies. In Section 4, we derive some techniques to prove positive almost sure termination under strategies. In Section 5, we discuss several applications of our results.

2   Probabilistic Abstract Reduction Systems and Probabilistic Rewrite Systems

A *stochastic sequence on a set A* is a family $(X_i)_{i \in \mathbb{N}}$, of random variables defined on some fixed probability space (Ω, σ, P) with values on A. It is said to be *Markovian* if its conditional distribution function satisfies the so-called Markov property, that is for all n and $s \in A$,

$$P(X_n = s | X_0 = \pi_0, X_1 = \pi_1, \ldots, X_{n-1} = \pi_{n-1}) = P(X_n = s | X_{n-1} = \pi_{n-1}),$$

and *homogeneous* if furthermore this probability is independent of n.

Probabilistic abstract reduction systems (PARS) were introduced in [5]. In the same way that abstract reduction systems are also called *transition systems* in other contexts, PARS correspond[12] to *Markov Decision Processes* [25].

Definition 1 (PARS). *Given some denumerable set S, we note $Dist(S)$ for the set of probability distributions on S: $\mu \in Dist(S)$ is a function $S \to [0,1]$ that satisfies $\sum_{i \in S} \mu(i) = 1$.*

A probabilistic abstract reduction system (PARS) is a pair $\mathcal{A} = (A, \to)$ consisting of a countable set A and a relation $\to \subset A \times Dist(A)$. A state $a \in A$ with no μ such that $a \to \mu$ is said terminal.

A PARS is said deterministic if, for all a, there is at most one μ with $a \to \mu$. We denote $Dist(\mathcal{A})$ for the set of distributions μ with $a \to \mu$ for some a.

We now need to explain how such systems evolve: a *history* is a finite sequence $a_0 a_1 \cdots a_n$ of elements of the state space A. It is non-terminal if a_n is.

Definition 2 (Deterministic Policy/Strategy). *A (deterministic) policy ϕ, that can also be called a (deterministic) strategy, is a function that maps non-terminal histories to distributions in such a way that $\phi(a_0 a_1 \cdots a_n) = \mu$ is always one (of the possibly many) distribution μ with $a_n \to \mu$. A history is said realizable, if for all $i < n$, if μ_i denotes $\phi(a_0 a_1 \cdots a_i)$, one has $\mu_i(a_{i+1}) > 0$.*

Actually, previous definition assumes that strategies must be deterministic (μ is a deterministic function of the history). If we want to be very general, we can also allow the strategy to be itself random (μ is selected among the possible μ with $a_n \to \mu$ in a random fashion).

Definition 3 (Randomized Policy/Strategy). *A randomized policy ϕ, that can also be called a randomized strategy, is a function that maps non-terminal histories to $Dist(M)$, where M is the set of μ with $a_n \to \mu$.*

Following the classification from [25], one can also distinguish history dependent strategies (the general case) from Markovian strategies (the value of the function on a history $a_0, \cdots, a_n$ depends only on a_n), to get the classes HD, HR, MD, MD, where H is for history dependent, M for Markovian, D for deterministic, R for randomized. In what follows, when we talk about strategies, it may mean a strategy of any of these classes.

A *derivation of $\mathcal{A}$* is then a stochastic sequence where the non-deterministic choices are given by some policy ϕ, and the probabilistic choices are governed by the corresponding distributions.

---

[1] The only true difference with [25] is that here action names are omitted.

[2] We prefer to keep to the terminology of [5], since we think that PARS indeed correspond to a probabilistic extension of Abstract Reduction Systems (ARS), Markov Decision Processes indeed correspond to a probabilistic extension of transition systems, and hence that the question of the best terminology is related to the question of the best terminology for ARS/transition systems, i.e. a cultural question.

Definition 4 (Derivations). *A derivation π of $\mathcal{A}$ over policy ϕ is a stochastic sequence $\pi = (\pi_i)_{i \in \mathbb{N}}$ on set $A \cup \{\bot\}$ (where $\bot$ is a new element: $\bot \notin A$) such that for all n,*

$$P(\pi_{n+1} = \bot | \pi_n = \bot) = 1,$$

$$P(\pi_{n+1} = \bot | \pi_n = s) = 1 \text{ if } s \in A \text{ is terminal,}$$

$$P(\pi_{n+1} = \bot | \pi_n = s) = 0 \text{ if } s \in A \text{ is non-terminal,}$$

and for all $t \in A$.

$$P(\pi_{n+1} = t | \pi_n = a_n, \pi_{n-1} = a_{n-1}, \ldots, \pi_0 = a_0) = \mu(t)$$

whenever $a_0 a_1 \cdots a_n$ is a realizable non-terminal history and $\mu = \phi(a_0 a_1 \cdots a_n)$.

If a derivation is such that $\pi_n = \bot$ for some n, then $\pi_{n'} = \bot$ almost surely for all $n' \geq n$. Such a derivation is said to be *terminating*. In other words, a non-terminating derivation is such that $\pi_n \in A$ ($\pi_n \neq \bot$) for all n.

The following two notions were introduced in [5]:

Definition 5 (Almost Sure Termination). *A PARS $\mathcal{A} = (A, \rightarrow)$ will be said almost surely (a.s) terminating iff for any policy ϕ, the probability that a derivation $\pi = (\pi_i)_{i \in \mathbb{N}}$ under policy ϕ terminates is 1: i.e. for all ϕ, $P(\exists n | \pi_n = \bot) = 1$.*

Definition 6 (Positive Almost Sure Termination). *A PARS $\mathcal{A} = (A, \rightarrow)$ will be said positively almost surely (+a.s.) terminating if for all policies ϕ, for all states $a \in A$, the mean number of reduction steps before termination under policy ϕ starting from a, denoted by $T[a, \phi]$, is finite.*

The following was proved in [5].

Theorem 1. *A PARS $\mathcal{A} = (A, \rightarrow)$ is +a.s. terminating if there exist some function $V : A \rightarrow \mathbb{R}$, with $\inf_{i \in A} V(i) > -\infty$, and some $\epsilon > 0$, such that, for all states $a \in A$, for all μ with $a \rightarrow \mu$, the drift in a according to μ defined by*

$$\Delta_\mu V(a) = \sum_i \mu(i) V(i) - V(a)$$

satisfies

$$\Delta_\mu V(a) \leq -\epsilon.$$

The technique was proved complete for finitely branching systems in [5]: such a function V always exists for +a.s. terminating finitely branching systems.

In [5], we also introduce the following notion, that covers classical (i.e. non-probabilistic) rewrite systems, and also Markov chains over finite spaces. It follows in particular that all examples that have been modeled in literature using finite Markov chains (for e.g. in model-checking contexts [21,20]) can be modeled as probabilistic rewrite systems.

Definition 7 (Probabilistic Rewrite system). *Given a signature Σ and a set of variables X, the set of terms over Σ and X is denoted by $T(\Sigma, X)$.*

A probabilistic rewrite rule is an element of $T(\Sigma, X) \times Dist(T(\Sigma, X))$. A probabilistic rewrite system is a finite set $\mathcal{R}$ of probabilistic rewrite rules.

To a probabilistic rewrite system is associated a probabilistic abstract reduction system $(T(\Sigma, X), \to_{\mathcal{R}})$ over the set of terms $T(\Sigma, X)$ where $\to_{\mathcal{R}}$ is defined as follows: When $t \in T(\Sigma, X)$ is a term, let $Pos(t)$ be the set of its positions. For $\rho \in Pos(t)$, let $t|_\rho$ be the subterm of t at position ρ, and let $t[s]_\rho$ denote the replacement of the subterm at position ρ in t by s. The set of all substitutions is denoted by Sub.

Definition 8 (Reduction relation). *To a probabilistic rewrite system $\mathcal{R}$ is associated the following PARS $(T(\Sigma, X), \to)$ over terms: $t \to_{\mathcal{R}} \mu$ iff there is a rule $(g, M) \in \mathcal{R}$, some position $p \in Pos(t)$, some substitution $\sigma \in Sub$, such that $t|_p = \sigma(g)$, and, for all t', $\mu(t') = \sum_{d | t' = t[\sigma(d)]_p} M(d)$.*

For example, a probabilistic rewrite rule can be $f(x, y) \mapsto \begin{cases} g(a) : 1/2 \\ y \quad : 1/2 \end{cases}$, where right hand side denotes the distribution with value $1/2$ on $g(a)$ and value $1/2$ on y. Then $f(b, c)$ rewrites to $g(a)$ with probability $1/2$, and to c with probability $1/2$. Now, $f(b, g(a))$ rewrites to $g(a)$ with probability 1.

Example 1. Consider[3] the following probabilistic rewrite system, with two rules R_1 and R_2 (of course, we assume $0 \le p_1 \le 1$, $0 \le p_2 \le 1$).

$$X \odot (Y \oplus Z) \quad \to \begin{cases} (X \odot Y) \oplus (X \odot Z) : p_1 \\ X \odot (Y \oplus Z) \qquad : 1 - p_1 \end{cases}$$

$$((X \odot Y) \oplus (X \odot Z)) \oplus X \to \begin{cases} (X \odot (Y \oplus Z)) \oplus X : p_2 \\ X \odot ((Y \oplus Z) \oplus X) : 1 - p_2 \end{cases}$$

Consider the polynomial interpretation of symbols $\{\oplus, \odot\}$ given by $[X \oplus Y] = 2[X] + [Y] + 1$ and $[X \odot Y] = [X] * [Y]$, where $[P] \in \mathbb{N}[X_1, \dots, X_n]$ denotes the polynomial interpretation of a term P of arity n.

Fix some integer $n_0 \ge 2$, yet to be determined. The set of integers $\ge n_0$ is preserved by the polynomials $[P]$. Consider function V that maps any term P to $[P](n_0, \cdots, n_0)$. Denote also $V(P)$ by $\{P\}$.

We have $[X \odot (Y \oplus Z)] = 2[X][Y] + [X][Z] + [X]$, $[(X \odot Y) \oplus (X \odot Z)] = 2[X][Y] + [X][Z] + 1$, and hence $\{X \odot (Y \oplus Z)\} = 2\{X\}\{Y\} + \{X\}\{Z\} + \{X\}$, $\{(X \odot Y) \oplus (X \odot Z)\} = 2\{X\}\{Y\} + \{X\}\{Z\} + 1$, and the drift of the first rule (see [5]) is given by $\Delta_{R_1} V(X \odot (Y \oplus Z)) = p_1 \times \{(X \odot Y) \oplus (X \odot Z)\} + (1 - p_1)\{X \odot (Y \oplus Z)\} - \{X \odot (Y \oplus Z)\} = p_1 \times (1 - \{X\})$. This is negative, and any

---

[3] Example obtained by modifying an example discussed in [2] about polynomial interpretations. As far as we know, this is the first time that a polynomial interpretation is used to prove termination of a probabilistic system (the examples from [5] used only linear interpretation functions).

substitution on X can only decrease it: R_1 is substitution decreasing following the terminology of [5].

Considering the second rule, we have $[((X \odot Y) \oplus (X \odot Z)) \oplus X] = 4[X][Y] + 2[X][Z] + [X] + 3$, $[(X \odot (Y \oplus Z)) \oplus X] = 4[X][Y] + 2[X][Z] + 3[X] + 1$, $[X \odot ((Y \oplus Z) \oplus X)] = 2[X][Y] + 2[X][Z] + X + 2$, and hence $\Delta_{R_2} V(((X \odot Y) \oplus (X \odot Z)) \oplus X) = p_2 \times \{(X \odot (Y \oplus Z)) \oplus X\} + (1 - p_2)\{X \odot ((Y \oplus Z) \oplus X)\} - \{((X \odot Y) \oplus (X \odot Z)) \oplus X\} = 2(p_2 - 1)\{X\}\{Y\} + 2p_2\{X\} - p_2 - 1$. This drift is not necessarily negative: in particular for $p_2 = 1$, it is positive. However, assume $p_2 < 1$. If we take, $n_0 \geq p_2/(1 - p_2)$, we can be sure that it becomes negative, since $2(p_2 - 1)\{X\}\{Y\} \leq -2p_2\{X\}$. For such an n_0, it is substitution decreasing.

Now, observing the form of the interpretation of symbols $\{\oplus, \odot\}$, which are linear in each of their variables with integer positive coefficients, a context can only decrease a drift. We get that the probabilistic rewrite system is $+$a.s. terminating for $p_2 < 1$.

This is a fortiori true for the following system, since the drift of the third rule is $-1 - 2\{X\}$, and hence negative.

Example 2. Consider the following probabilistic rewrite system, with three rules R_1, R_2, R_3:

$$X \odot (Y \oplus Z) \quad \rightarrow \begin{cases} (X \odot Y) \oplus (X \odot Z) : p_1 \\ X \odot (Y \oplus Z) \qquad\quad : 1 - p_1 \end{cases}$$

$$((X \odot Y) \oplus (X \odot Z)) \oplus X \rightarrow \begin{cases} (X \odot (Y \oplus Z)) \oplus X : p_2 \\ X \odot ((Y \oplus Z) \oplus X) : 1 - p_2 \end{cases}$$

$$(X \oplus Y) \oplus Z \quad \rightarrow \{X \oplus (Y \oplus Z) : 1$$

3 Positive Almost Sure Termination Under Strategies

Positive almost sure termination means that for all starting term the mean number of rewrite steps to reach a terminal state is finite under any policy/strategy. In particular, non termination can happen with a single very specific strategy.

In many examples, one is often tempted not to consider arbitrary strategies, but to restrict to a subset of strategies. Whatever the considered class of strategies is, the following notion is rather natural.

Definition 9 (Positive Almost Sure Termination Under Strategies).
Fix a class Φ of strategies (i.e. policies);

A PARS $\mathcal{A} = (A, \rightarrow)$ will be said positively almost surely ($+$a.s.) terminating under Φ if for all strategy (i.e. policy) $\phi \in \Phi$, for all states $a \in A$, the mean number of reduction steps before termination under ϕ starting from a, denoted by $T[a, \phi]$, is finite.

Example 3. Consider the following probabilistic rewrite system, with two rules.

$$a \rightarrow \{a : 1$$
$$a \rightarrow \{b : 1$$

This system is clearly not (almost surely) terminating , since there is the infinite derivation $a \to a \to \cdots a \to \cdots$.

However, it is +a.s. terminating under Markovian non-deterministic[4] randomized strategies: indeed, in state a, such a strategy selects either the first rule with probability p_1, or the second with probability $1 - p_1$, for some fixed $p_1 < 1$. The system is then equivalent to the probabilistic rewrite system

$$a \to \begin{cases} a : p_1 \\ b : 1 - p_1 \end{cases}$$

whose positive almost sure termination can be established easily, for example using previous theorem and $V(a) = 1$, $V(b) = 0$.

Example 4. Consider the following probabilistic rewrite system, with two rules named *red* and *green*: see Figure 1.

$$s(x) \to \begin{cases} x & : p_1 \\ s(s(x)) : 1 - p_1 \end{cases}$$
$$s(x) \to \begin{cases} x & : p_2 \\ s(s(x)) : 1 - p_2 \end{cases}$$

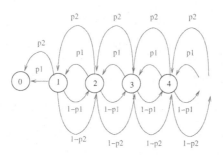

Fig. 1. Example 4

The red (respectively: green) rule[5] is easily shown to be +a.s. terminating iff $p_1 > 1/2$ (resp. $p_2 > 1/2$).

Suppose that $p_1 < 1/2$, $p_2 > 1/2$. The whole system is not +a.s. terminating: consider the strategy that always selects the red rule.

However, it is +a.s. terminating under the strategy that always selects the green rule.

Intuitively, in a more general case, its +a.s. termination depends on the ratio of selection of the red versus green rule. Indeed, if we focus on Markovian randomized strategies that select the red (respectively green) rule with a fixed probability p (resp. $1 - p$), the whole system is equivalent to

$$s(x) \to \begin{cases} x & : p_1 * p + p_2 * (1 - p) \\ s(s(x)) : (1 - p_1) * p + (1 - p_2) * (1 - p) \end{cases}$$

---

[4] We want to avoid $p_1 = 1$.

[5] That is to say: the probabilistic rewrite system made of this rule alone.

which is easily shown to be +a.s. terminating iff $p_1 * p + p_2 * (1 - p) > 1/2$, i.e. $p < (1 - 2p_2)/(2(p_2 - p_1))$.

Example 5. Consider the following probabilistic rewrite system, over signature $\Sigma = \{A, B, C\}$, with four rules, where we assume $p_1 > 0$, $p_2 > 0$.

$$A \to \begin{cases} B : p_1 \\ A : 1 - p_1 \end{cases}$$
$$B \to \begin{cases} A : p_2 \\ B : 1 - p_2 \end{cases}$$
$$A \to \{C \qquad\qquad : 1$$
$$B \to \{C \qquad\qquad : 1$$

We have only states A and B, and in each of these states, a strategy can either select the rule among the two first that applies or the rule among the two last that applies. It is easy to see that with probability one, an infinite derivation is made of a sequence of A and B, each of them appearing infinitely often.

This probabilistic rewrite system is not +a.s. terminating: consider the strategy ϕ_∞ that always excludes the second possibility (i.e. never choose third or fourth rule).

However, it is clearly +a.s. terminating under Φ, for any class Φ that does not contain this specific strategy ϕ_∞.

This example illustrates that one may want to restrict to fair strategies, for some or one's preferred notion of fairness: in this example, since third and fourth rule can fire infinitely often, one may want that they fire at least once (or with positive probability).

In literature, several notions of fairness have been introduced: see [23,24,27,10] and references in the introduction of this paper. Termination of probabilistic systems under fairness constraints has been investigated, in particular in [18] for probabilistic finite state systems, and in [17] for probabilistic infinite state systems.

Next section will be devoted to provide techniques to prove positive almost sure termination of a probabilistic rewrite system under strategies. These results can be applied with classes of strategies constrained by several of these notions of fairness. The following results can also be seen as an extension of the two papers [18,17] to deal with +a.s. termination (and not only almost sure termination).

4   Proving +a.s. Termination Under Strategies

A slight generalization of Theorem 1 yields rather directly:

Theorem 2. *Fix a class of strategies Φ.*

A PARS $\mathcal{A} = (A, \to)$ is +a.s. terminating under Φ if there exist some function $V : A \to \mathbb{R}$, with $\inf_{i \in A} V(i) > -\infty$, and some $\epsilon > 0$, such that, for all realizable non-terminal history $h = a_0 a_1 \cdots a_n$, for all $\phi \in \Phi$, the drift in h according to ϕ defined by

$$\Delta_\phi V(h) = \sum_i \phi(h)(i)V(i) - V(a_n)$$

satisfies

$$\Delta_\phi V(h) \leq -\epsilon.$$

Fortunately, we can do better in many cases.

Consider a PARS $\mathcal{A} = (A, \rightarrow)$. Assume that $Dist(\mathcal{A})$ (see Definition 1) can be partitioned into finitely many subsets $Dist(\mathcal{A}) = D_1 \cup D_2 \ldots \cup D_k$. Intuitively, when $\mathcal{A}$ is corresponding to a PARS associated to some probabilistic rewrite system with k probabilistic rewrite rules $R_1, \ldots, R_k$, each D_i corresponds to rewrite rule R_i: D_i is made of distributions μ obtained by varying position p, and substitution σ in the distribution of rule R_i, according to Definition 8.

We assume that for any strategy $\phi \in \Phi$, $\phi^{-1}(D_i)$ is measurable. The expectation of a random variable X is denoted by $E[X]$.

Definition 10 (Next Selection of a Rule). *Fix some D_i.*

Fix some deterministic policy ϕ and some realizable non-terminal history $h = a_0 a_1 \cdots a_n$. Let $(\pi_i)_{i \in \mathbb{N}}$ be a derivation starting from h: $(\pi_i)_{i \in \mathbb{N}}$ is a stochastic sequence as in Definition 4 with $\pi_0 = a_0, \cdots, \pi_n = a_n$.

Let τ be the random variable denoting the first index greater than n at which D_i is selected, or a terminal state is reached (set $\tau = \infty$ if there is no such index). I.e. $\tau = m$ iff $\phi(\pi_0, \cdots, \pi_m) \in D_i$, and $\phi(\pi_0, \cdots, \pi_{m'}) \notin D_i$ for $n < m' < m$, or $\pi_m = \bot$ and $\pi_{m'} \neq \bot$ for $n < m' < m$.

Let $\tau_{D_i, \pi, \phi, h}$ denote the τ for the corresponding D_i, π, ϕ and h.

Each random variable $\tau_{D_i, \pi, \phi, h}$ is a stopping time with respect to derivation π (see e.g. [9]): it is a random variable taking its value in $\mathbb{N} \cup \{\infty\}$, such that for all integers $m \geq 0$, the event $\{\tau = m\}$ can be expressed in terms of $\pi_0, \pi_1, \ldots, \pi_m$.

Remark 1. One must understand that even if the policy is deterministic, and hence not depending on any random choice, each $\tau_{D_i, \pi, \phi, h}$ is random. Indeed, when $h = a_0 \cdots a_n$ is fixed, the choice of a_{n+1} is made according to distribution $\phi(a_0 \cdots a_n)$, and hence random; the choice of a_{n+2} is then made according to distribution $\phi(a_0 \cdots a_n a_{n+1})$, and hence random. And so on. The event D_i is selected or a terminal state is reached at time n is then random.

Definition 11 (Bounded Mean Selection). *A class of strategy Φ has bounded mean selection $\alpha \in \mathbb{R}$ for D_i, if for any strategy $\phi \in \Phi$, for any history, the expected time to wait before reaching a final state or selecting a rule from D_i is less than α. I.e. for any realizable non-terminal history $h = a_0 \cdots a_n$, for any policy $\phi \in \Phi$, for any derivation π starting from h, $\tau_{D_i, \pi, \phi, h}$ has a finite mean with*

$$E[\tau_{D_i, \pi, \phi, h}] \leq n + \alpha.$$

Observe that a variable taking values in $\mathbb{N} \cup \{\infty\}$ with a finite mean is necessarily almost surely finite: in other words, when the conditions of the previous definition hold, one knows that almost surely starting from any history h, one reaches either a final state, or one selects a rule from D_i.

Definition 12 (Expected Value of V At Time τ). *Let $V : A \to \mathbb{R}$ be some function. Let $\tau \in \mathbb{N} \cup \{\infty\}$ denotes some stopping time with respect to derivation π, which is almost surely finite: $P(\tau < \infty) = 1$. Fix some policy ϕ, and a corresponding derivation $(\pi_i)_{i \in \mathbb{N}}$.*

We denote by $E_\tau V$ the expected value of V at time τ: formally

$$E_\tau V = E[V(\pi_\tau)]$$

when it exists.

We claim:

Theorem 3 (Almost Sure Termination Under Strategy). *Fix a class of strategies Φ.*

A PARS $\mathcal{A} = (A, \to)$ is almost surely terminating under strategies Φ if there exist some function $V : A \to \mathbb{R}$, with $\inf_{i \in A} V(i) > -\infty$, some $\epsilon > 0$, and some D_i such that for all strategy $\phi \in \Phi$, for all realizable non-terminal history $h = a_0 \ldots a_n$, for all derivation π starting from h,

1. *the stopping time $\tau_{D_i, \pi, \phi, h}$ is almost surely finite,*
2. *and*
$$E_{\tau_{D_i, \pi, \phi, h}} V \leq V(a_n) - \epsilon.$$

This follows from the following result from Martingale theory: See [11] for a proof (1_A denotes the characteristic function of a set A).

Proposition 1. *Let $(\Omega, \mathcal{F}, P)$ be a given probability space, and $\{\mathcal{F}_n, n \geq 0\}$ an increasing family of σ-algebra.*

Consider a sequence $(S_i)_{i \in \mathbb{N}}$ of real non-negative random variables, such that S_i is $\mathcal{F}_i$-measurable, for all i. Assume S_0 to be constant, w.l.o.g.

Denote by τ the $\mathcal{F}_n$-stopping time representing the epoch of the first entry into $[0, C]$, for some C: $\tau = \inf\{i \geq 1, S_i \leq C\}$.

Introduce the stopped sequence $S'_i = S_{\min(i, \tau)}$.

Assume $S_0 > C$, and for some $\epsilon > 0$, and for all $n \geq 0$, almost surely

$$E[S'_{i+1} | \mathcal{F}_{n-1}] \leq S'_i - \epsilon 1_{\tau > i}. \tag{1}$$

Then:

- *Almost surely τ is finite.*
- *$E[\tau] < S_0 / \epsilon$.*

Proof (of Theorem 3). Replacing function V by $V + K$ for some constant K if needed, we can assume without loss of generality that $V(a) \geq 2\epsilon$ for all $a \in A$. Extend function V on $A \cup \{\bot\}$ by $V(\bot) = 0$.

Fix a strategy $\phi \in \Phi$, a realizable non-terminal history h, and a derivation $(\pi_i)_{i \in \mathbb{N}}$ starting from h.

From Condition 1., one can build a sequence of random functions $(\psi_n)_{n \in \mathbb{N}}$ such that almost surely, for all $n > 1$, either $\pi_{\psi(n)} = \bot$, or D_i is selected at rank $\psi(n)$. Indeed: Take $\psi(0) = 0$; when $\psi(n)$ is built, build $\psi(n + 1)$ as $\psi(n)$ if $\pi_{\psi(n)} = \bot$ and as $\psi(n) + \tau_{D_i, \pi, \phi, h}$ otherwise.

Consider the increasing family of σ-algebra $\mathcal{F}_n$ where $\mathcal{F}_n$ is the σ-algebra generated by $\pi_0, \cdots, \pi_n$. Condition 2. implies almost surely $E[S'_{n+1} | \mathcal{F}_{n-1}] \leq S'_n - \epsilon 1_{\pi_{\psi(n)} \neq \bot}$, where $S'_n = V(\pi_{\psi(n)})$ for all n. By Proposition 1 above with $C = \epsilon$, almost surely there must exist some n with $\pi_n = \bot$.

In other words, the PARS is almost surely terminating under Φ.

Remark 2. Previous hypotheses yield almost sure termination, but not positive almost sure termination. Indeed, the proof build a subsequence of indexes $\psi(n)$ yielding almost surely to termination. But there is no reason that $\psi(n+1) - \psi(n)$ stay bounded, and hence the original derivation can be non positively almost surely terminating (such an example is easy to build).

Actually, weaker conditions entailing almost sure termination have been derived in [17]: in particular ϵ can be taken as 0. However, for +a.s. almost sure derivation, we claim:

Theorem 4 (+ A.S. Termination Under Strategy). *Fix a class of strategies Φ.*

A PARS $\mathcal{A} = (A, \rightarrow)$ is +a.s. terminating under strategies Φ if there exist some function $V : A \rightarrow \mathbb{R}$, with $\inf_{i \in A} V(i) > -\infty$, some $\epsilon > 0$, and some D_i such that Φ has bounded mean selection for D_i, and such that for all strategy $\phi \in \Phi$, for all realizable non-terminal history $h = a_0 \ldots a_n$, for all derivation π starting from h,

$$E_{\tau_{D_i, \pi, \phi, h}} V \leq V(a_n) - \epsilon.$$

Proof. By previous discussion, the fact that Φ has bounded mean selection for D_i entails Condition 1. of previous theorem, and hence we have almost sure termination. Even if we did not mention it, the application of Proposition 1 in the proof of previous Theorem also yields that the random variable N giving the smallest n with $\pi_n = \bot$ has a finite mean with $E[N] \leq V(a_n)/\epsilon$.

Now, since Φ has bounded mean selection α for some $\alpha > 0$, we can bound $E[\psi(N)]$ by $\alpha V(a_0)/\epsilon$ using following Lemma, whose proof can easily be established (for example by adapting the proof of Wald's Lemma in [9]).

Lemma 1. *Consider a stochastic sequence $(X_i)_{i \in \mathbb{N}}$ taking non-negative values. Let N be an integer-valued random variable, with a finite expectation. Assume there exists some constant M such that $0 \leq X_{n+1} - X_n \leq M$ almost surely for all n. The random variable X_N has an expectation bounded by $E[X_0] + M * E[N]$.*

5 Applications

We first derive one simple case:

Proposition 2. *Consider a PARS $\mathcal{A} = (A, \rightarrow)$ so that there exists $V : A \rightarrow \mathbb{R}$, with $\inf_{i \in A} V(i) > -\infty$, some (possibly positive) α, some $\epsilon > 0$, such that $Dist(\mathcal{A})$ can be partitioned into $Dist(\mathcal{A}) = D_1 \cup D_2$ such that for all $a \rightarrow \mu$, we have*

1. $\Delta_\mu V(a) \leq \alpha$ whenever $\mu \in D_1$.
2. $\Delta_\mu V(a) \leq -\epsilon$ whenever $\mu \in D_2$.

Assume that a rule of the form $a \to \mu$, with $\mu \in D_1$ never lead to a terminal: for all $a \to \mu$, $\mu \in D_1$, for all a' with $\mu(a') > 0$, a' is not a terminal.

Assume that ϕ selects D_2 at least once every k steps for some constant k: for any $h = a_0 \cdots a_n$, for any $\phi \in \Phi$, for any π, we assume that $\tau_{D_2,\pi,\phi,h}$ exists and satisfies $\tau_{D_2,\pi,\phi,h} \leq n + k$.

Assume that $(k-1)\alpha - \epsilon < 0$.

Then $\mathcal{A}$ is $+a.s.$ terminating under strategies Φ.

Proof. It is easy to see that we always have $E_{\tau_{D_2,\pi,\phi,h}} V \leq V(a_n) + (k-1)\alpha - \epsilon$ in this case: Indeed, a derivation starting from h must either reach a terminal or lead to a state where D_2 is selected. In any case, the last applied rule will be a rule from D_2, and hence V will decrease in mean of at least ϵ, after the at most $k-1$ first rules that can make it increase in mean of at most $(k-1)\alpha$. We can then apply previous theorem.

Example 6. Consider the following probabilistic rewrite system, with three rules R_1, R'_2, R_3:

$$
X \odot (Y \oplus Z) \quad \to \begin{cases} (X \odot Y) \oplus (X \odot Z) : p_1 \\ X \odot (Y \oplus Z) \quad\quad\; : 1 - p_1 \end{cases}
$$
$$
((X \odot Y) \oplus (X \odot Z)) \oplus X \to \begin{cases} (X \odot (Y \oplus Z)) \oplus X \quad\quad : p_2 \\ ((X \odot Y) \oplus (X \odot Z)) \oplus X : 1 - p_2 \end{cases}
$$
$$
(X \oplus Y) \oplus Z \quad \to \begin{cases} X \oplus (Y \oplus Z) : 1 \end{cases}
$$

This probabilistic rewrite system is not positively almost surely terminating. Indeed, for the policy which always apply the first two rules and never the third, we have an infinite derivation with terms $((X \odot Y) \oplus (X \odot Z)) \oplus X$ and $(X \odot (Y \oplus Z)) \oplus X$, each of them appearing almost surely infinitely often.

The drift of the rules R_1 and R_3 have been computed in Example 2. Now, the drift of the rule R'_2 is $\Delta_{R'_2} V(((X \odot Y) \oplus (X \odot Z)) \oplus X) = 2p_2 \times (\{X\} - 1)$, and hence positive.

If we choose a policy ϕ with a bounded mean selection for the rewrite rule R_3, and if ϕ always reduce the term of a cycle $((X \odot Y) \oplus (X \odot Z)) \oplus X \to (X \odot (Y \oplus Z)) \oplus X$ until it can be broken by firing rule R_3, then conditions of Theorem 4 are satisfied, because, for all histories $h = a_0, \ldots, a_n$ such that a_n contains a subterm which is an instance of $((X \odot Y) \oplus (X \odot Z)) \oplus X$, then $E_{\tau_{D_3},\pi,\phi,h} V \leq V(a_n) - 2 \times \{X\} - 1$.

Example 7. Let's now consider the following term rewrite system, coming from the model of [8] of a simulator for the CSMA-CA protocol [1]. The rules rewrite lists of couples. Each couple is made of two positive integers. The *sort* operator triggers a rule based sort algorithm, which sorts in increasing order the list in function of the value of the first field. The first rule will take the head of the list, replace the first field by a random value between 1 and p following an uniform law with probability μ and decrease the value of the second field with probability $1 - \mu$.

$$
\begin{aligned}
(\Delta t, n+1), \ldots, (\Delta_k, n_k) &\rightarrow \begin{cases} (U(1, \ldots, p), n+1), \ldots, (\Delta_k, n_k) & : \mu \\ (U(1, \ldots, p), n), \ldots, (\Delta_k - \Delta t, n_k) : 1 - \mu \end{cases} \\
(\Delta t, n+1), \ldots, (\Delta_k, n_k) &\rightarrow sort((\Delta t, n+1), \ldots, (\Delta_k, n_k)) \\
sort((\Delta_t, n_t), X) &\rightarrow sort1((\Delta_t, n_t), nil, X) \\
sort1((\Delta_t, n_t), l, (\Delta'_t, n'_t).X) &\rightarrow sort1(((\Delta_t, n_t), l.(\Delta'_t, n'_t), X)) \text{ If } \Delta_t < \Delta'_t \\
sort1((\Delta_t, n_t), l, (\Delta'_t, n'_t).X) &\rightarrow sort1(((\Delta'_t, n'_t), l.(\Delta_t, n_t), X)) \text{ If } \Delta_t > \Delta'_t \\
sort1((\Delta_t, n_t), l, (\Delta'_t, n'_t).nil) &\rightarrow (\Delta_t, n_t).sort(l.(\Delta'_t, n'_t)) \text{ If } \Delta'_t > \Delta_t \\
sort1((\Delta_t, n_t), l, (\Delta'_t, n'_t).nil) &\rightarrow (\Delta'_t, n'_t).sort(l.(\Delta_t, n_t)) \text{ If } \Delta'_t < \Delta_t
\end{aligned}
$$

where, X, l are some lists of couples of integers, the operator "."denotes the concatenation of lists and nil is the empty list. $U(1, \ldots, p)$ is a random integer variable following an uniform law on $\{1, \ldots, p\}$. n and $n_{i_{i \in 1, \ldots, k}}$ are non negative integers.

This PRS is easily seen not +a.s. terminating: For example the first two rules always apply on every list or sublists.

Now let's build a policy under which the PRS positively almost surely terminates. Let's start with a_0 a list of length n, $\phi(a_0)$ is the rule that rewrites a_0 to $sort(a_0)$. The length of the sorting process is $n(n-1)$, and the policy ϕ chooses only the rules coding the sort algorithm during the sort process. If the first element of the list has a zero second field, there's no rule matching this list and this term is terminal. Otherwise, the policy ϕ will choose again the rule that triggers the sort of the list, and later apply the rule number one, and so one since no terminal state is reached.

To show this system is +a.s. terminating, let's consider the function $V : T(\Sigma, X) \rightarrow \mathbb{N}$ computing the sum of the second field of each element of a list, and apply Proposition 2.

An alternative proof is the following: We can apply Theorem 4, because ϕ has bounded mean selection for the first rule rewrite relation D_1, because such a rule is triggered between two sorts of length $n(n-1)$ and $E_{\tau_{D_1}, \pi, \phi, h} V = V(a_n) + \mu - 1$, because V does not change during the sorting process since the values of the second field are not touched, and the only variation of the mean is induced by the rule D_1 whose drift is $\mu - 1$. V, as the sum of positive value, is lower bounded.

6 Conclusion and Future Work

In this paper, we introduced positive almost sure termination under strategies, and we provide sufficient criteria to prove positive almost sure termination of a given probabilitic rewrite system under strategies.

We plan to apply our techniques on industrially motivated examples of bigger size. It may be possible to weaken the hypotheses of our theorems since they mainly use a special case of Proposition 1. As mentioned in the introduction, any technique to deal with probabilistic systems, must work for classical ones, since probabilities can be 0/1. The classical (non-probabilistic) counterpart of our

framework for proving termination under strategies is very poor: the question
of understanding which of the techniques from literature for non-probabilistic
systems can be extended to deal with probabilistic systems seems fascinating.

References

1. Ieee csma/ca 802.11 working group home page. http://www.ieee802.org/11/.
2. Franz Baader and Tobias Nipkow. *Term Rewriting and All That*. Cambridge University Press, 1998.
3. Gianfranco Balbo. Introduction to stochastic Petri nets. *Lecture Notes in Computer Science*, 2090:84, 2001.
4. P. Borovanský, C. Kirchner, and H. Kirchner. Controlling Rewriting by Rewriting. In J Meseguer, editor, *Proceedings of 1st International Workshop on Rewriting Logic*, volume 4, Asilomar (CA, USA), September 1996. Electronic Notes in Theoretical Computer Science.
5. Olivier Bournez and Florent Garnier. Proving positive almost sure termination. In Jürgen Giesl, editor, *16th International Conference on Rewriting Techniques and Applications (RTA'2005)*, volume 3467 of *Lecture Notes in Computer Science*, page 323, Nara, Japan, 2005. Springer.
6. Olivier Bournez and Mathieu Hoyrup. Rewriting logic and probabilities. In Robert Nieuwenhuis, editor, *Rewriting Techniques and Applications, 14th International Conference, RTA 2003, Valencia, Spain, June 9-11, 2003, Proceedings*, volume 2706 of *Lecture Notes in Computer Science*, pages 61–75. Springer, June 2003.
7. Olivier Bournez and Claude Kirchner. Probabilistic rewrite strategies: Applications to ELAN. In Sophie Tison, editor, *Rewriting Techniques and Applications*, volume 2378 of *Lecture Notes in Computer Science*, pages 252–266. Springer-Verlag, July 22-24 2002.
8. Olivier Bournez Florent Garnier Claude Kirchner. Termination in finite mean time of a csma/ca rule-based model. Technical report, LORIA, Nancy, 2005.
9. Pierre Brémaud. *Markov Chains, Gibbs Fields, Monte Carlo Simulation, and Queues*. Springer-Verlag, New York, 2001.
10. Luca de Alfaro. *Formal Verification of Probabilistic Systems*. PhD thesis, Stanford University, 1998.
11. Guy Fayolle, Vadim A. Malyshev, and Mikhail V. Menshikov. *Topics in constructive theory of countable Markov chains*. Cambridge University Press, 1995.
12. Olivier Fissore, Isabelle Gnaedig, and Hélène Kirchner. Simplification and termination of strategies in rule-based languages. In *PPDP '03: Proceedings of the 5th ACM SIGPLAN international conference on Principles and practice of declaritive programming*, pages 124–135, New York, NY, USA, 2003. ACM Press.
13. Olivier Fissore, Isabelle Gnaedig, and Hélène Kirchner. A proof of weak termination providing the right way to terminate. In *Theoretical Aspects of Computing – ICTAC 2004: First International Colloquium*, Lecture Notes in Computer Science. Springer, 2004.
14. F. G. Foster. On the stochastic matrices associated with certain queuing processes. *The Annals of Mathematical Statistics*, 24:355–360, 1953.
15. Thom Frühwirth, Alexandra Di Pierro, and Herbert Wiklicky. Toward probabilistic constraint handling rules. In Slim Abdennadher and Thom Frühwirth, editors, *Proceedings of the third Workshop on Rule-Based Constraint Reasoning and Programming (RCoRP'01)*, Paphos, Cyprus, December 2001. Under the hospice of the International Conferences in Constraint Programming and Logic Programming.

16. H. Hansson. *Time and Probability in Formal Design of Distributed Systems*. Series in Real-Time Safety Critical Systems. Elsevier, 1994.
17. Sergiu Hart and Micha Sharir. Concurrent probabilistic programs, or: How to schedule if you must. *SIAM Journal on Computing*, 14(4):991–1012, November 1985.
18. Sergiu Hart, Micha Sharir, and Amir Pnueli. Termination of probabilistic concurrent program. *ACM Transactions on Programming Languages and Systems*, 5(3):356–380, July 1983.
19. Thomas Hérault, Richard Lassaigne, Frédéric Magniette, and Sylvain Peyronnet. Approximate probabilistic model checking. In *VMCAI*, pages 73–84, 2004.
20. Marta Kwiatkowska, Gethin Norman, and David Parker. PRISM: Probabilistic symbolic model checker. *Lecture Notes in Computer Science*, 2324:200, 2002.
21. Marta Z. Kwiatkowska. Model checking for probability and time: from theory to practice. In *LICS*, page 351, 2003.
22. Kumar Nirman, Koushik Sen, Jose Meseguer, and Gul Agha. A rewriting based model for probabilistic distributed object systems. In *Proceedings of 6th IFIP International Conference on Formal Methods for Open Object-based Distributed Systems (FMOODS'03)*, volume 2884 of *Lecture Notes in Computer Science*, pages 32–46. Springer, Paris, France, November 2003.
23. Amir Pnueli. On the extremely fair treatment of probabilistic algorithms. In *STOC: ACM Symposium on Theory of Computing (STOC)*, 1983.
24. Amir Pnueli and Lenore D. Zuck. Probabilistic verification. *Information and Computation*, 103(1):1–29, March 1993.
25. Martin L. Puternam. *Markov Decision Processes - Discrete Stochastic Dynamic Programming*. Wiley series in probability and mathematical statistics. John Wiley & Sons, 1994.
26. William H. Sanders and John F. Meyer. Stochastic activity networks: Formal definitions and concepts. *Lecture Notes in Computer Science*, 2090:315, 2001.
27. Moshe Y. Vardi. Automatic verification of probabilistic concurrent finite-state programs. In *26th Annual Symposium on Foundations of Computer Science*, pages 327–338, Portland, Oregon, 21–23 October 1985.

A Proof of Finite Family Developments for Higher-Order Rewriting Using a Prefix Property

H.J. Sander Bruggink

Department of Philosophy, Utrecht University
bruggink@phil.uu.nl
http://www.phil.uu.nl/~bruggink

Abstract. A prefix property is the property that, given a reduction, the ancestor of a prefix of the target is a prefix of the source. In this paper we prove a prefix property for the class of Higher-Order Rewriting Systems with patterns (HRSs), by reducing it to a similar prefix property of a λ-calculus with explicit substitutions. This prefix property is then used to prove that Higher-order Rewriting Systems enjoy Finite Family Developments. This property states, that reductions in which the creation depth of the redexes is bounded are finite, and is a useful tool to prove various properties of HRSs.

1 Introduction

Higher-order Rewriting Systems (HRSs), as introduced by Nipkow in 1991 [12,10], are a powerful tool to study the metatheory of declarative programming languages, like λProlog and Haskell, on the one hand, and theorem provers and proof checkers, like Isabelle, on the other. Also, many (extensions of) λ-calculi can be encoded as instances of HRSs, so that results obtained for HRSs carry over to other interesting domains.

In this paper, we prove two properties of HRSs where left-hand sides of rule are restricted to be patterns. First we prove a prefix property, by reducing this property to a similar prefix property for a λ-calculus with explicit substitutions. The prefix property says that, given a step, the ancestor of a prefix of the target is a prefix of the source. Consider, as an example, the (first-order) rewriting system with the single rule $f(x) \rightarrow g(f(x), x)$ and the step $f(h(a)) \rightarrow g(f(h(a)), h(a))$. Now, $p = g(f(\square), h(\square))$ is a prefix of the target. Intuitively, its ancestor is $f(h(\square))$, because $s = f(h(\square)) \rightarrow g(f(h(\square)), h(\square)) = t$, and p is contained in t. And indeed, $f(\square)$ is a prefix of the source.

Many different prefix properties are possible: we can, e.g., vary in how the notions of prefix and ancestor are formalized, and we may impose additional conditions on the form of the prefixes. Prefix properties are already known for first-order TRSs [2,13] and (a labelling of) the λ-calculus with β-reductions [2], and have many applications, such as (head) needed reductions [13, Chap. 8] and normalization of outermost-fair reductions [13, Chap. 9]. A similar property is proved in Van Daalen's Square Brackets Lemma [14].

The second contribution is that we prove Finiteness of Family Developments (FFD) for HRS, by reducing this property to the prefix property described above.

F. Pfenning (Ed.): RTA 2006, LNCS 4098, pp. 372–386, 2006.

FFD states that each reduction, in which the "creation depth", or *family*, of function symbols is bounded, is finite. The intuition behind the notion of family is that in a step $C[l^\sigma] \to C[r^\sigma]$, the symbols of r depend on the symbols of l, and therefore have a higher creation depth, while the symbols in C and σ do not take part in the step and have the same creation depth in both source and target. For example, consider the following infinite reduction, using the rewrite system above. We label the function symbols with their creation depth.

$$\mathsf{f}^0(\mathsf{a}^0) \to \mathsf{g}^1(\mathsf{f}^1(\mathsf{a}^0), \mathsf{a}^0) \to \mathsf{g}^1(\mathsf{g}^2(\mathsf{f}^2(\mathsf{a}^0), \mathsf{a}^0)) \to \mathsf{g}^1(\mathsf{g}^2(\mathsf{g}^3(\mathsf{f}^3(\mathsf{a}^0), \mathsf{a}^0), \mathsf{a}^0)) \to \cdots$$

Clearly, in this infinite reduction, the creation depth of the f's grows without bound. FFD states that restricting the creation depth to a finite number, yields finite reductions. FFD is a useful tool to prove various properties of rewrite systems, such as termination (e.g. termination of simply typed λ-calculus follows from FFD, cf. [7, page 31]), existence of standard reductions, etc.

Of some lemmas and theorems the proof is omitted or only sketched. Full proofs are made available in the technical report [4].

2   Preliminaries

We presuppose knowledge of the simply typed λ-calculus. Here we give a short overview of Higher-Order Rewrite Systems (HRSs) [10]. In particular, we consider HRSs as HORSs [15] with the simply typed λ-calculus as substitution calculus. We refer to [13, Sect. 11.2] for a good introduction.

Simple types are generated from a set of base types by the type constructor $\to$. Let Σ be a signature of simply typed function symbols. We define a *preterm* to be a simply typed λ-terms over Σ. We want to consider $\beta\eta$-equivalence classes of preterms. Since it is well known that β-reduction, combined with restricted η-expansion ($\overline{\eta}$-reduction), is confluent and terminating, we take $\beta\overline{\eta}$-normal forms as unique representatives of the $\beta\overline{\eta}$-equivalence classes. We define: a *term* is a preterm in $\beta\overline{\eta}$-normal form. In the following, s, t will range over terms (and, whenever indicated, over preterms as well).

A sequence $a_1, \ldots, a_n$ will sometimes be written as $\overline{a_n}$, or just $\overline{a}$ if the length is not important or clear from the context. Juxtaposition of two sequences denotes concatenation.

For terms or preterms $s, t_1, \ldots, t_n$, we write $s(t_1, \ldots, t_n)$ for $st_1 \cdots t_n$, or, in the case of terms, the β-normal form thereof. We also introduce the shorthand $\lambda \overline{x_n}.s$ for $\lambda x_1 \ldots \lambda x_n.s$. With $FV(s)$ we denote the set of free variables of term or preterm s, and with $Sym(s)$ the set of function symbols present in the term or preterm. If $\lambda \overline{x}.a(\overline{s})$ is a term, then a is called the *head* of that term (a is a function symbol or variable).

In the class of HRSs that we consider, the left-hand sides of rules are restricted to be *local patterns*. For patterns, unification is decidable and unique most general unifiers exist [11]. Local patterns, additionally, are linear (each free variable occurs at most once) and fully-extended (free variable have all bound variables in scope as argument). These extra requirements have a similar purpose as the requirement of left-linearity in first-order TRS: they keep matching local. To match a non-linear pattern, it is possible that subterms outside the pattern need to be checked for equality; to match a non fully extended pattern, it is possible that such a subterm must be checked for the non-occurrence of a variable. Because the notion of pattern depends on what the free variables are, we need to parametrize the notion with a context of variables, and obtain the following inductive definition:

Definition 2.1 (Pattern). *Let $\bar{x}$ be a sequence of variables.*

- *(i) A term s is an $\bar{x}$-pattern if:*
  - *$s = a(s_1, \ldots, s_n)$ and either $a \in \bar{x} \cup \Sigma$ and $s_1, \ldots, s_n$ are $\bar{x}$-patterns; or $s_1, \ldots, s_n$ is η-equivalent to a sequence of distinct variables from $\bar{x}$.*
  - *$s = \lambda y.s_0$ and s_0 is an $\bar{x}y$-pattern.*
- *(ii) A term s is* linear outside $\bar{x}$, *if each free variable which is not in $\bar{x}$, occurs in it at most once. A term s is a* fully extended $\bar{x}$-pattern, *if, in the second case of the above definition, $s_1, \ldots, s_n =_\eta \bar{x}$. A term s is a* local $\bar{x}$-pattern, *if s is linear outside $\bar{x}$ and a fully extended $\bar{x}$-pattern.*

Examples of local patterns are $f(x)$, $g(\lambda xy.f(z(x, y)))$ and $h(\lambda x.z(x))$. Examples of non-local patterns are $g(\lambda xy.f(y))$ (not fully-extended) and $h(\lambda x.z(x), \lambda x.z(x))$ (not linear). An example of a non-pattern is $g(z(a))$.

In the following, p, q will range over patterns, and the word pattern (without the sequence of variables) will refer to a $\emptyset$-pattern.

Definition 2.2 (HRS). *An rewrite rule (for a signature Σ) is a pair $\lambda\bar{x}.l_0 \rightarrow \lambda\bar{x}.r_0$ of closed Σ-terms of the same type, such that $l_0 = f(s_1, \ldots, s_n)$ and l_0 is a local pattern not η-equivalent to a variable. An HRS is a tuple $\mathcal{H} = \langle \Sigma, R \rangle$, where Σ is a signature and R a set of rewrite rules for Σ.*

The rewrite relation $\rightarrow_{\mathcal{H}}$ is defined as follows: $s \rightarrow_{\mathcal{H}} t$ if there exist a context C such that $s =_\beta C[l]$ and $t =_\beta C[r]$, for some rule $l \rightarrow r \in R$.

For arbitrary rewrite system $\mathcal{R}$, we denote with $\twoheadrightarrow_{\mathcal{R}}$ the reflexive, transitive closure of $\rightarrow_{\mathcal{R}}$.

Note that there is no substitution in the definition of the rewrite relation, such as in first-order term rewriting systems (but see also Remark 2.4). The leading abstractions of the rules take the role of the substitution, as can be seen in the next example:

Example 2.3. Let the HRS $\mathcal{M}ap$, implementing the higher-order function map, be defined by:

$$\lambda z.\text{map}(\lambda x.z(x), \text{nil}) \rightarrow \lambda z.\text{nil}$$
$$\lambda zuv.\text{map}(\lambda x.z(x), \text{cons}(u, v)) \rightarrow \lambda zuv.\text{cons}(z(\text{e}(u)), \text{map}(\lambda x.z(x), v))$$

Here, cons and nil are the list constructors, viz. list composition and the empty list, respectively. The reason for the symbol e is to make the HRS non-collapsing (see Def. 2.5). A reduction of two $\mathcal{M}ap$-steps is the following:

$\mathsf{map}(\lambda x.\mathsf{f}(x), \mathsf{cons}(\mathsf{a}, \mathsf{nil}))$
$=_\beta \quad (\lambda zuv.\mathsf{map}(\lambda x.z(x), \mathsf{cons}(u, v)))(\lambda x'.\mathsf{f}(x'), \mathsf{a}, \mathsf{nil})$
$\rightarrow_{\mathcal{M}ap} (\underline{\lambda zuv.\mathsf{cons}(z(\mathsf{e}(u)), \mathsf{map}(\lambda x.z(x), v))})(\lambda x'.\mathsf{f}(x'), \mathsf{a}, \mathsf{nil})$
$=_\beta \quad \mathsf{cons}(\mathsf{f}(\mathsf{e}(\mathsf{a})), \mathsf{map}(\lambda x.\mathsf{f}(x), \mathsf{nil}))$
$=_\beta \quad \mathsf{cons}(\mathsf{f}(\mathsf{e}(\mathsf{a})), (\lambda z.\mathsf{map}(\lambda x.z(x), \mathsf{nil}))(\lambda x'.\mathsf{f}(x')))$
$\rightarrow_{\mathcal{M}ap} \mathsf{cons}(\mathsf{f}(\mathsf{e}(\mathsf{a})), (\underline{\lambda z.\mathsf{nil}})(\lambda x'.\mathsf{f}(x')))$
$=_\beta \quad \mathsf{cons}(\mathsf{f}(\mathsf{e}(\mathsf{a})), \mathsf{nil})$

Note how the (underlined) left-hand sides are *literally* replaced by the (also underlined) right-hand sides.

In later examples, the leading abstractions of rewrite rules will be omitted; in other words, we will write $l \to r$ for $\lambda \overline{x}.l \to \lambda \overline{x}.r$.

Substitutions are maps from variables to terms. Application of a substitution $\sigma = [x_1 \mapsto t_1, \ldots, x_n \mapsto t_n]$ to a term s is defined as: $s^\sigma = (\lambda x_1 \ldots x_n.s)t_1 \ldots t_n$ (where this term is, as always, implicitly reduced to $\beta\overline{\eta}$-normal form). In the following, $\rho, \sigma, \tau, \upsilon$ will rangle over substitutions. The composition of substitutions σ and τ is denoted by $\sigma \; ; \; \tau$, where $s^{\sigma;\tau} = (s^\sigma)^\tau$. A substitution is called linear, if each free variable occurs in its codomain at most once, i.e. if all terms of its codomain are linear and have mutually disjoint free variables. A (fully extended) $\overline{x}$-pattern substitution is a substitution of which the codomain consists of (fully extended) $\overline{x}$-patterns.

Remark 2.4. The rewrite relation of Def. 2.2 can alternatively, and more in the fashion of first-order TRSs, be defined in the following way: $s \to_{\mathcal{H}} t$ if $s =_\beta C[l_0^\sigma]$ and $t =_\beta C[r_0^\sigma]$, where $\lambda \overline{x}.l_0 \to \lambda \overline{x}.r_0 \in R$ and σ is a substitution with $Dom(\sigma) = \overline{x}$. This alternative definition, however, requires the notion of substitution to be defined, and therefore we prefer the other one. In the rest of the paper, we will sometimes implicitly switch definitions.

Intuitively, a rewrite rule is collapsing, if it can bring context and subtitution, or different parts of the substitution, together, i.e. if, after the application of the rule, a function symbol of the context can be directly connected to a function symbol of the substitution. This can happen in two specific cases, which we will use as a definition:

Definition 2.5. *A term s is* collapsing, *if one of the following applies:*

- *(context-subst):* $s = x(s_1, \ldots, s_n)$, *where x is a free variable; or*
- *(subst-subst):* $s = C[x(s_1, \ldots, s_n)]$, *and for some k, $s_k = \lambda \overline{z}.y(t_1, \ldots, t_m)$, where C is a context, x is a free variable, and y a free or bound variable.*

A rewrite rule $\lambda \overline{x}.l \to \lambda \overline{x}.r$ is collapsing, *if r is collapsing, and an HRS is collapsing, if at least one of its rules is.*

Example 2.6.

- The rules $\lambda x.\mathsf{f}(x) \to \lambda x.x$ and $\lambda z.\mathsf{mu}(\lambda x.z(x)) \to \lambda z.z(\mathsf{mu}(\lambda x.z(x)))$ are collapsing due to the (context-subst) condition.

- The rule $\lambda yz.\mathsf{g}(\lambda x.z(x), y) \to \lambda yz.\mathsf{f}(z(y))$ is collapsing due to the (subst-subst) condition.
- The rule $\lambda yz.\mathsf{app}(\mathsf{lam}(\lambda x.z(x)), y) \to \lambda yz.z(y)$ is collapsing due to both the (context-subst) and the (subst-subst) conditions.

3   Labelling HRSs with Natural Numbers

Labelling rewriting systems is a well-known method to formalize the notion of redex family; see e.g. [8,9]. In this section, we develop a labelling, in the sense of [17,13], for HRSs, analogous to the labelling for the λ-calculus used by Hyland [6] and Wadsworth [18]. Each function symbol is labelled by a natural number, representing the "creation depth" of the function symbol, and the rules are labelled such that every function symbol of the right-hand side is labelled with the largest label of the left-hand side plus one.

Definition 3.1 (ω-labelling).

(i) *The ω-labelling of a signature Σ is defined as:* $\Sigma^\omega = \{f^\ell \mid f \in \Sigma, \ell \in \mathbb{N}\}$.
(ii) *The family of a term s, denoted* $\mathrm{fam}(s)$, *is the largest label of s, i.e.:*
$\mathrm{fam}(s) = \max\{\ell \mid f^\ell \in Sym(s)\}$
(iii) *Let s be a term, and $\ell \in \mathbb{N}$ a label. Then:*

$$x(s_1, \ldots, s_n)^\ell = x(s_1^\ell, \ldots, s_n^\ell)$$
$$f(s_1, \ldots, s_n)^\ell = f^\ell(s_1^\ell, \ldots, s_n^\ell)$$
$$(\lambda x.s_0)^\ell = \lambda x.s_0^\ell$$

(iv) *The projection operation $|\cdot|_\omega$ is the mapping from Σ^ω to Σ given by* $|f^\ell|_\omega = f$. *The mapping is homomorphically extended to terms.*
(v) *Let $\mathcal{H} = \langle \Sigma, R \rangle$. The ω-labelled version of $\mathcal{H}$ is defined as:* $\mathcal{H}^\omega = \langle \Sigma^\omega, R^\omega \rangle$, *where R^ω consist of all rules $l' \to r^{(\mathrm{fam}(l')+1)}$ such that $l \to r \in R$ and $|l'|_\omega = l$.*

The ω-labelling only labels *function symbols*, not variables, abstractions or applications. The reason for this is that we want the ω-labelling of an HRS to be an HRS itself (otherwise it would not be a labelling in the sense of [17,13]). Labelling variables is impossible, because α-equivalent terms are identified. Labelling abstractions and applications is impossible because we have fixed the (unlabelled) simply typed λ-calculus as substitution calculus.

Example 3.2. The labelled HRS $\mathcal{M}ap^\omega$ consists, among others, of the rules:

$$\mathsf{map}^0(\lambda x.z(x), \mathsf{nil}^0) \to \mathsf{nil}^1$$
$$\mathsf{map}^1(\lambda x.z(x), \mathsf{nil}^1) \to \mathsf{nil}^2$$
$$\mathsf{map}^0(\lambda x.z(x), \mathsf{cons}^0(u, v)) \to \mathsf{cons}^1(z(\mathsf{e}^1(u)), \mathsf{map}^1(\lambda x.z(x), v))$$
$$\mathsf{map}^0(\lambda x.z(x), \mathsf{cons}^1(u, v)) \to \mathsf{cons}^2(z(\mathsf{e}^2(u)), \mathsf{map}^2(\lambda x.z(x), v))$$

A labelled reduction corresponding to the reduction of Ex. 2.3 is the following:

$\mathsf{map}^0(\lambda x.\mathsf{f}^0(x), \mathsf{cons}^0(\mathsf{a}^0, \mathsf{nil}^0))$

$=_\beta \quad (\lambda zuv.\mathsf{map}^0(\lambda x.z(x), \mathsf{cons}^0(u, v)))(\lambda x'.\mathsf{f}^0(x'), \mathsf{a}^0, \mathsf{nil}^0)$

$\to_{Map} \underline{(\lambda zuv.\mathsf{cons}^1(z(\mathsf{e}^1(u)), \mathsf{map}^1(\lambda x.z(x), v)))}(\lambda x'.\mathsf{f}^0(x'), \mathsf{a}^0, \mathsf{nil}^0)$

$=_\beta \quad \underline{\mathsf{cons}^1(\mathsf{f}^0(\mathsf{e}^1(\mathsf{a}^0)), \mathsf{map}^1(\lambda x.\mathsf{f}^0(x), \mathsf{nil}^0))}$

$=_\beta \quad \mathsf{cons}^1(\mathsf{f}^0(\mathsf{e}^1(\mathsf{a}^0)), (\lambda z.\mathsf{map}^1(\lambda x.z(x), \mathsf{nil}^0))(\lambda x'.\mathsf{f}^0(x')))$

$\to_{Map} \mathsf{cons}^1(\mathsf{f}^0(\mathsf{e}^1(\mathsf{a}^0)), \underline{(\lambda z.\mathsf{nil}^2)}(\lambda x'.\mathsf{f}^0(x')))$

$=_\beta \quad \mathsf{cons}^1(\mathsf{f}^0(\mathsf{e}^1(\mathsf{a}^0)), \mathsf{nil}^2)$

Notice how only the labels of function symbols involved in the step (i.e. the underlined ones) are increased.

The following two lemmas provide a correspondence between labelled and unlabelled reductions, and are easily proved by induction:

Lemma 3.3. *Let* $\mathcal{H}$ *be an HRS.* $\mathcal{H}^\omega$ *is orthogonal/collapsing/erasing, if and only if* $\mathcal{H}$ *is.*

Lemma 3.4. *Let* $\mathcal{H}$ *be an HRS.*

(i) *If* $s \to_{\mathcal{H}} t$, *then, for each* s' *such that* $|s'|_\omega = s$, *there is a* t' *such that* $s' \to_{\mathcal{H}^\omega} t'$ *and* $|t'|_\omega = t$.

(ii) *If* $s \to_{\mathcal{H}^\omega} t$, *then* $|s|_\omega \to_{\mathcal{H}} |t|_\omega$.

4   The Prefix Property

We call p a prefix of term t, if it is a pattern, and there exists a substitution σ such that $p^\sigma = t$. Given a step $s \to t$, a subterm q of s is the ancestor of a subterm p of t, if the symbols of t "trace to" the symbols of s. This notion is formalized here using labelling together with the rewrite relation: q is an ancestor of p, if $\mathrm{fam}(p) \geq \mathrm{fam}(q)$ and $q \twoheadrightarrow_{\mathcal{H}^\omega} p^\upsilon$. The substitution υ is necessary because q might reduce to a "bigger" term than p; typically, υ has only function symbols which are also in p. Using these formalizations, we prove in this section the following theorem (proof begins on page 382).

Theorem 4.1 (Prefix Property). *Let* $\mathcal{H}^\omega$ *be the* ω-labelling *of a non-collapsing HRS,* s *a term,* p *a local* $\bar{x}$-pattern *and* σ *a substitution. If* $s \to_{\mathcal{H}^\omega} p^\sigma$, *then there exist a local* $\bar{x}$-pattern q *and a substitution* τ, *such that* $s = q^\tau$, $\mathrm{fam}(p) \geq \mathrm{fam}(q)$, *and either:*

- $q \to_{\mathcal{H}^\omega} p^\upsilon$, *for some substitution* υ *such that* $\upsilon ; \tau = \sigma$; *or* \hfill (trm)
- $q = p$ *and* $\tau \to_{\mathcal{H}^\omega} \sigma$. \hfill (sub)

The theorem states that, given a prefix of the target, its ancestor is a prefix of the source. There are two possibilities: either the prefix takes part in the step, or the step occurred fully in the substitution. Note that, in the first case, we do not only require that its ancestor is a prefix, but also that the suffix stays the same (except for duplicated subterms). In this regard, the lemma is stronger than e.g. the prefix property (for the λ-calculus) proved in [2, Prop. 7.4].

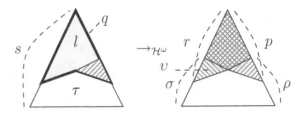

Fig. 1. The interesting case in the proof of the Prefix Property for HRSs

Example 4.2. Consider the following $\mathcal{M}ap^\omega$-step (see page 374):

$$h^1(map^3(\lambda x.f^2(x), cons^2(a^5, nil^1))) \to h^1(cons^4(f^2(e^4(a^5)), map^4(\lambda x.f^2(x), nil^1)))$$

First, let the prefix $p = h^1(cons^4(f^2(y_1), y_2))$ of the target be given. The suffix is then given by $\sigma = [y_1 \mapsto e^4(a^5), y_2 \mapsto map^4(\lambda x.f^2(x), nil^1)]$. Then:

$$q = h^1(map^3(\lambda x.f^2(x), cons^2(y_1, z_2)))$$
$$\upsilon = [y_1 \mapsto e^4(z_1), y_2 \mapsto map^4(\lambda x.f(x), z_2)]$$
$$\tau = [z_1 \mapsto a^5, z_2 \mapsto nil^1]$$

satisfy the conditions of the (trm) case. Second, let $p = h(y)$ and $\sigma = [y \mapsto cons^4(f^2(e^4(a^5)), map^4(\lambda x.f^2(x), nil^1)))]$. Then:

$$q = h^1(y) \quad \text{and} \quad \tau = [y \mapsto map^3(\lambda x.f^2(x), cons^2(a^5, nil^1))]$$

satisfy the conditions of the (sub) case.

The interesting case in the proof of the Prefix Property is the case that the step $s \to_{\mathcal{H}^\omega} p^\sigma$ occurs at the head. In this case we have that $s = l^\rho$ and $p^\sigma = r^\rho$, for some rule $l \to r$ and substitution ρ. This situation is depicted in Fig. 1. We want to construct an ancestor q that satisfies the (trm) case. It makes sense to try to do this by adding to the pattern l the parts of p that are not in r. However, due to the implicit β-conversions, these "parts of p that are not in r" are not easily obtained. The key observation is that the β-reduction from p^σ to normal form is a variable renaming, because p is a pattern and has only bound variables as arguments of free variables. The trick is to translate the prefix and suffix in such a way, that the variable renamings are already carried out (we need variable capturing, first-order substitutions for this, called *graftings*), trace the prefix back over the β-reduction from r^ρ to normal form, and find the prefix's ancestor, which is a prefix of r^ρ. Now, we are dealing with terms that are exactly equal, instead of only equal up to β-equality, and now the problem can be solved by using first-order unification techniques.

The above proof technique suggests that we need to prove a prefix property for β-reductions in the λ-calculus. This is difficult, however, since the λ-calculus does not cope well with graftings, because of the global nature of substitution. For example, let $C = (\lambda x.\Box)a$, $D = \Box$ and $s = x$. Then $C \to_\beta D$, and $C[s] \to_\beta a$, because the x in s is captured by the abstraction in the context and substituted for. However, $D[s] = x$ and thus $C[s] \not\to_\beta D[s]$. To tackle this problem, we use a λ-calculus with explicit substitutions, a variant of the λx-calculus, and prove a prefix property for it. Then, we simulate β-equality with this new calculus. In [5] a similar approach is taken w.r.t. higher-order unification.

4.1   The Prefix Property of the λx-Calculus

We use a variant of the λx-calculus [3], with explicit renamings. The calculus has both object variables (x, y, z) and metavariables (X, Y, Z). In the following, we will refer to it simply by λx-calculus. The terms of the λx-calculus over some signature Σ are first-order terms given by the following grammar:

$$\Lambda_x := x \mid X \mid f \mid \lambda x.\Lambda_x \mid \Lambda_x\Lambda_x \mid \Lambda_x\{x\backslash\Lambda_x\}$$

where $f \in \Sigma$ and the object variables are considered as constants or names. M, N will range over λx-terms. Terms of the form $M\{x\backslash N\}$ will be called *explicit substitutions*, and the $\{x\backslash N\}$ part of an explicit substitution is called a *closure*. With $\text{MV}(M)$ we will denote the set of metavariables of M; and with $Sym(M)$ the set of function symbols of M. The reduction rules of the λx-calculus are:

$$(\lambda x.M)N \to_B M\{x\backslash N\}$$

$$
\begin{array}{ll}
x\{x\backslash N\} \to_x N & (\lambda x.M)\{x\backslash N\} \to_x \lambda x.M \\
y\{x\backslash N\} \to_x y & (\lambda y.M)\{x\backslash N\} \to_x \lambda z.M\{y\backslash z\}\{x\backslash N\} \\
f\{x\backslash N\} \to_x f & (M_1 M_2)\{x\backslash N\} \to_x M_1\{x\backslash N\}M_2\{x\backslash N\}
\end{array}
$$

where $x \neq y$ and z is a fresh object variable. The subcalculus x consists of all rules except the B-rule. The reduction relations $\to_{Bx}$ and $\to_x$ are the contextual closures of the above steps. Note that there is no reduction rule for terms of the form $X\{x\backslash N\}$, where X is a metavariable, and thus x-normal forms are characterized by the fact that sequences of closures are only applied to metavariables.

A λx-term is called *passive* if no metavariable X occurs in a subterm of the form $X\overline{\mu}(M_1, \ldots, M_n)$, where $\overline{\mu}$ is a sequence of closures; it is called *linear*, if every metavariable occurs in it at most once. In the following P, Q will range over linear, passive λx-terms.

Remark 4.3. It is well-known that the λx-calculus is *not* confluent on terms containing metavariables. At first sight, non-confluence seems problematic, because we're trying to use the λx-calculus to simulate the (confluent) λ-calculus. However, the translation to λ-calculus (see page 381) will remove all closures, and will project normal forms of the same λx-term to the same λ-term (modulo α-equivalence).

A *grafting* is a mapping from metavariables to λx-terms. The greek lowercase letters $\zeta, \eta, \theta, \kappa$ will range over graftings. Applying a grafting ζ to a term M, written $M[\zeta]$, is defined exactly as first order substitution, i.e.:

$$
\begin{array}{ll}
x[\zeta] = x & (\lambda x.M)[\zeta] = \lambda x.M[\zeta] \\
X[\zeta] = \zeta'(X) & (M_1 M_2)[\zeta] = M_1[\zeta]M_2[\zeta] \\
f[\zeta] = f & (M\{x\backslash N\})[\zeta] = M[\zeta]\{x\backslash N[\zeta]\}
\end{array}
$$

where $\zeta'(X) = \zeta(X)$, if $X \in Dom(\zeta)$, and $\zeta'(X) = X$, otherwise. A grafting is called *linear*, if every metavariable occurs in its codomain only once, i.e. its codomain consists of linear λx-terms with mutually disjoint metavariables. A grafting is called *passive*, if all the terms of its codomain are passive.

Because λx-terms are first-order terms, unification is decidable. In the proof of the Prefix Property, we need the following property: if two λx-terms are unifiable, there exists a most general unifier (mgu). In fact, if we assume the unifiable terms to be linear and passive, then the mgu applied to one of the terms is a linear, passive λx-term again:

Lemma 4.4. *Let M, N be linear λx-terms, where $MV(M) \cap MV(N) = \emptyset$, and let ζ, η be graftings such that $M[\zeta] = N[\eta]$. There exist graftings ζ_0, η_0, κ such that $M[\zeta_0] = N[\eta_0]$, $\zeta_0 ; \kappa = \zeta$, $\eta_0 ; \kappa = \eta$, $Sym(\zeta_0) \subseteq Sym(N)$, $Sym(\eta_0) \subseteq Sym(M)$. Moreover, if M (N) is passive, then η_0 (ζ_0) consists of passive λx-terms.*

Proof. (Sketch) Since λx-terms are basically first-order terms, we can use first-order unification techniques. Because of disjointness of the metavariables we can consider the two graftings as one unifier, and the linearity assumption is needed for the condition on the symbols. $\qquad\square$

Example 4.5. Let:

$$M = \lambda x.g(f_1(X_1), X_2) \qquad N = \lambda x.g(Y_1, f_2(Y_2))$$
$$\zeta = [X_1 \mapsto a, X_2 \mapsto f_2(a)] \qquad \eta = [Y_1 \mapsto f_1(a), Y_2 \mapsto a]$$

Then $M[\zeta] = \lambda x.g(f_1(a), f_2(a)) = N[\eta]$. We take $\zeta_0 = [X_2 \mapsto f_2(Z_1)]$, $\eta_0 = [Y_1 \mapsto f_1(Z_2)]$ and $\kappa = [Z_i \mapsto a]$ to satisfy the conditions of the lemma.

In the next theorem, we prove the Prefix Property for the λx-calculus. P is a prefix of a λx-term M, if it is a linear, passive λx-term, and there exists a grafing ζ such that $P[\zeta] = M$. The notion of ancestor is again formalized using labelling and the rewrite relation; however, because we do not count creation depth in Bx-reductions, now the labels, or more generally, the function symbols of the prefix must be the same as those of its ancestor. Just like in Theorem 4.1, a prefix can either take part in the step, or not, resulting in two cases. Item (ii) is the extension of the Prefix Property to Bx-*reductions*.

Theorem 4.6 (λx-Prefix Property). *Let M be a closed λx-term, P a linear, passive λx-term and ζ a grafting.*

(i) *If $M \to_{Bx} P[\zeta]$, then there exist a linear, passive λx-term Q and a grafting η such that $M = Q[\eta]$, $Sym(Q) = Sym(P)$ and either:*
   - *$Q \to_{Bx} P[\kappa]$ where κ is some grafting such that $\kappa ; \eta = \zeta$; or* $\qquad$ (trm)
   - *$Q = P$ and $\eta \to_x \zeta$.* $\qquad$ (sub)

(ii) *If $M \twoheadrightarrow_{Bx} P[\zeta]$, then there exist a linear, passive λx-term Q and a grafting η such that: $M = Q[\eta]$, $Sym(Q) = Sym(P)$, $Q \twoheadrightarrow_{Bx} P[\kappa]$ where κ is some grafting such that $\kappa ; \eta \twoheadrightarrow_{Bx} \zeta$.*

Proof. (Sketch) Item (i) is proved by induction on the context of the step $M \to_{\mathrm{Bx}} P[\zeta]$, using a case analysis and Lemma 4.4 in the base case, and (ii) by induction on the length of the reduction. $\qquad\square$

Example 4.7. Consider the Bx-reduction $(\lambda x.\mathsf{g}(x,x))(\mathsf{f}(\mathsf{a})) \twoheadrightarrow_{\mathrm{Bx}} \mathsf{g}(\mathsf{f}(\mathsf{a}),\mathsf{f}(\mathsf{a}))$, and the prefix $P = \mathsf{g}(\mathsf{f}(X),Y)$ of the target. The suffix is $\zeta = [X \mapsto \mathsf{a}, Y \mapsto \mathsf{f}(\mathsf{a})]$. We can take $Q = (\lambda x.\mathsf{g}(x,x))(\mathsf{f}(Y))$, $\kappa = [Y \mapsto \mathsf{f}(X)]$ and $\eta = [X \mapsto \mathsf{a}]$, satisfying the conditions of Theorem 4.6 (ii).

4.2 Translating Between Terms, Preterms and λx-Terms

We are now dealing with three types of terms: terms, preterms and λx-terms. In this section we develop translations between (pre)terms and λx-terms. The "translation" between terms and preterms will be done completely implicitly, here. See [4] for a more detailed approach.

Translating Terms. We introduce the operations $\cdot_{\overline{x}}^{\ominus}$ and $\cdot^{\oplus}$, which map λ-terms to λx-terms, and vice versa, as follows:

$$
\begin{aligned}
& & M^{\oplus} &= (M{\downarrow_{\mathsf{x}}})_{\mathrm{N}}^{\oplus} \\
y_{\overline{x}}^{\ominus} &= Y \text{ if } y \notin \overline{x} & (Y\overline{\sigma})_{\mathrm{N}}^{\oplus} &= y \\
x_{\overline{x}}^{\ominus} &= x \text{ if } x \in \overline{x} & x_{\mathrm{N}}^{\oplus} &= x \\
f_{\overline{x}}^{\ominus} &= f & f_{\mathrm{N}}^{\oplus} &= f \\
(\lambda y.s)_{\overline{x}}^{\ominus} &= \lambda y.s_{\overline{x}y}^{\ominus} & (\lambda y.M)_{\mathrm{N}}^{\oplus} &= \lambda y.M^{\oplus} \\
(s_1 s_2)_{\overline{x}}^{\ominus} &= (s_1)_{\overline{x}}^{\ominus}(s_2)_{\overline{x}}^{\ominus} & (M_1 M_2)_{\mathrm{N}}^{\oplus} &= M_1^{\oplus} M_2^{\oplus}
\end{aligned}
$$

Note that $\cdot^{\oplus}$ also normalizes the term to x-normal form and removes explicit substitutions, and that, for each preterm s and sequence of variables $\overline{x}$, $(s_{\overline{x}}^{\ominus})^{\oplus} = s$. The operations above are naturally generalized to translations between substitutions and graftings.

Lemma 4.8. *Let M, N be λx-terms. $M \twoheadrightarrow_{\mathrm{Bx}} N$ if and only if $M^{\oplus} \twoheadrightarrow_{\beta} N^{\oplus}$.*

Proof. ($\Rightarrow$) and ($\Leftarrow$) are proved by induction on the length of the reductions $M \twoheadrightarrow_{\mathrm{Bx}} N$ and $M^{\oplus} \twoheadrightarrow_{\beta} N^{\oplus}$, respectively. $\qquad\square$

Although the above lemma suggests that Bx-reduction in the λx-calculus can easily simulate β-reduction, there is still a problem: $\cdot^{\oplus}$ does not distribute properly over grafting application. The problem is similar to the problem given on page 378. Consider the λx-term $M := (\lambda x.\mathsf{f}(Y))\mathsf{a}$ and grafting $\zeta := [Y \mapsto x]$. Now $M[\zeta]^{\oplus} = (\lambda x.\mathsf{f}(x))\mathsf{a}$, $M^{\oplus} = (\lambda x.\mathsf{f}(y))\mathsf{a}$. $\zeta^{\oplus} = [y \mapsto x]$. Note that $(M^{\oplus})^{(\zeta^{\oplus})} = \lambda z.\mathsf{f}(x)$, because substitutions are capture-avoiding, and thus $M[\zeta]^{\oplus} \neq_{\beta} (M^{\oplus})^{(\zeta^{\oplus})}$.

The solution is to add as arguments to the free variables of the preterms as many (bound) variables as necessary (or more) to make the distribution work. In the example above we would have $s = (\lambda x.\mathsf{f}(y(x)))\mathsf{a}$ and $\sigma = [y \mapsto \lambda x.x]$. Now, s and σ are, in a way that will be formalized in the next definition, similar to M and ζ, but now $M[\zeta]^{\oplus} =_{\beta} s^{\sigma}$.

Definition 4.9. *Let M be a λx-term and ζ a grafting. A tuple $\langle s, \sigma \rangle$ of preterm and substitution is a λ-extension of $\langle M, \zeta \rangle$ if there are graftings θ_1, θ_2 such that:*

- *$s = M[\theta_1]^\oplus$ and $\sigma = (\theta_2 \, ; \zeta)^\oplus$;*
- *for each $X \in \mathrm{MV}(M)$, $\theta_1(X) = X(\overline{z})$ and $\theta_2(X) = \lambda \overline{z}.X$, where $\overline{z}$ is a list of variables containing at least the bound variables of M in scope that occur in $\zeta(X)$ (in arbitrary order).*

The notion of λ-extension is, again, naturally generalized to graftings and substitutions as the first component of the tuples.

Lemma 4.10. *Let $\langle s, \sigma \rangle$ be a λ-extension of $\langle M, \zeta \rangle$. Then:*

(i) $s^\sigma =_\beta M[\zeta]^\oplus$;
(ii) for each λx-term N such that $M \twoheadrightarrow_{\mathrm{Bx}} N$, $s^\sigma =_\beta N[\zeta]^\oplus$.

The lemma works, because the arguments of the free variables in the term and the leading abstractions in the substitution, take over the role of the explicit substitutions, as can be seen in the following example:

Example 4.11. Let $M = (\lambda x.(\lambda y.Z)\mathsf{b})\mathsf{a}$ be a λx-term, and $\zeta = [Z \mapsto \mathsf{f}(x, y)]$ a grafting. Now, according to Def. 4.9, $\langle s, \sigma \rangle$, where $s = (\lambda x.(\lambda y.z(x, y))\mathsf{b})\mathsf{a}$ and $\sigma = [z \mapsto \lambda xy.\mathsf{f}(x, y)]$ is a λ-extension of $\langle M, \zeta \rangle$, with, $\theta_1 = [Z \mapsto Z(x, y)]$ and $\theta_2 = [\lambda xy.Z]$. We check both cases of Lemma 4.10:

(i) $s^\sigma = (\lambda x.(\lambda y.(\lambda xy.\mathsf{f}(x, y))(x, y))\mathsf{b})\mathsf{a} =_\beta (\lambda x.(\lambda y.\mathsf{f}(x, y))\mathsf{b})\mathsf{a} = M[\zeta]^\oplus$.
(ii) Let $N = Z\{y\backslash\mathsf{b}\}\{x\backslash\mathsf{a}\}$. Then $M \twoheadrightarrow_x N$. Let $t = z(\mathsf{a}, \mathsf{b})$. Now $t^\sigma =_\beta \mathsf{f}(\mathsf{a}, \mathsf{b}) = M[\zeta]^\oplus$. Since $s =_\beta t$, this means that $s^\sigma =_\beta M[\zeta]^\oplus$, as required. (Note that the $\cdot^\oplus$ operation also reduces to x-normal form.)

Translating Patterns. Among the λ-extensions of a pair $\langle P, \zeta \rangle$ of linear, passive λx-term and grafting, there is, for each sequence of variables $\overline{x}$ exactly one λ-extension $\langle p, \sigma \rangle$ where p is a $\overline{x}$-pattern, viz. the one in which in p the free variables have *all* bound variables in scope as arguments. We denote by $\mathbf{P}_{\overline{x}}^+ \langle P, \zeta \rangle$ the function which returns this specific λ-extension, and by $\mathbf{P}_{\overline{x}}^-$ the inverse of $\mathbf{P}_{\overline{x}}^+$. See [4] for a more detailed definition of these operations.

Example 4.12. Consider the linear, local λx-terms $P = \mathsf{f}(\lambda xy.\mathsf{g}(Z, x))$ and $Q = \mathsf{map}(\lambda x.Z, \mathsf{nil})$, and the grafting $\zeta = [Z \mapsto \mathsf{f}(x)]$. Then:

$$\mathbf{P}_\emptyset^+ \langle P, \zeta \rangle = \langle \mathsf{f}(\lambda xy.\mathsf{g}(z(x, y), x)), [z \mapsto \lambda xy.\mathsf{f}(x)] \rangle$$
$$\mathbf{P}_\emptyset^+ \langle Q, \zeta \rangle = \langle \mathsf{map}(\lambda x.Z(x), \mathsf{nil}), [z \mapsto \lambda x.\mathsf{f}(x)] \rangle$$

4.3   Proof of the Prefix Property

Proof (of Theorem 4.1). (Sketch). The interesting case is the case that the step occurs at the head, i.e.: $s = l^\rho$ and $p^\sigma = r^\rho$, for some rule $l \to r \in R$ and substitution ρ. We translate the terms to λx-terms: $\langle P, \zeta \rangle := \mathbf{P}_{\overline{x}}^- \langle p, \sigma \rangle$, $R := r_\emptyset^\ominus$, $L := l_\emptyset^\ominus$ and $\mu := \rho_{\overline{x}}^\ominus$. Because $p^\sigma =_\beta r^\rho$, and $P[\zeta]$ is a Bx-normal form by construction, it is the case that $R[\mu] \twoheadrightarrow_{\mathrm{Bx}} P[\zeta]$ (using Lemma 4.8).

Now, we use the λx-Prefix Property (Theorem 4.1) to find the ancestor P' of P in this reduction. This gives us, among other things, a graftings η, κ_1 such that $P'[\eta] = R[\mu]$, and $P' = P[\kappa]$. Equality here is first-order equality, and thus we apply first-order unification techniques (Lemma 4.4) to find an mgu $\langle \eta_0, \mu_0 \rangle$ for the unifier $\langle \eta, \mu \rangle$, and grafting κ_2 such that $\eta_0 ; \kappa_2 = \eta$ and $\mu_0 ; \kappa_2 = \mu$.

Now we translate everything back to (pre)terms, using the techniques discussed in the previous subsection: v is the translation of κ_1 followed by η_0 (using λ-extensions to make the two composable), τ is the translation of κ_2 and Q is the translation of $L[\eta_0]$. This translation is cumbersome, but not hard in principle. The λ-extensions make sure that Bx-equality can be transformed to β-equality.

The last thing we have to prove is that $\text{fam}(p) \geq \text{fam}(q)$. This holds because $Sym(\eta_0) \subseteq Sym(p)$, because $Sym(\eta_0) \subseteq Sym(r)$, all labels in r are the same and p and r have at least one symbol in common because r is non-collapsing. □

5 Finite Family Developments

In this section we apply the prefix property of the previous section to prove that all family developments of HRSs are finite. We restrict our attention to non-collapsing HRSs first. In the next section, we will describe a way to generalize the result to collapsing HRSs as well.

Families are formalized by labelling all function symbols with natural numbers, as defined in Def. 3.1. We prove that the resulting system is terminating if we restrict the labels to some finite bound. The proof is inspired by the proof by Van Oostrom [16]. The differences between this proof and the one by Van Oostrom are the following:

- We use a different method of labelling. Our labelling has the property that one step of the labelled HRS corresponds exactly to one step in the original. Also, our notion of labelling is an instance of the abstract notion of labelling put forth in [17,13].
- In Van Oostrom's paper, the proof of Lemma 15 is omitted. Here, we give a proof of that lemma (adapted for the different method of labelling) by reducing it to the Prefix Property.

Lemma 5.1. *Let $\mathcal{H}^\omega$ be the labelling of a non-collapsing HRS, s be a term, p a local pattern, $\ell \in \mathbb{N}$ a label and τ and σ substitutions such that for any $x \in Dom(\sigma)$, $\sigma(x)$ has a function symbol labelled with ℓ as head. If $s^\sigma \twoheadrightarrow_{\mathcal{H}^\omega} p^\tau$, then either:*

- $\text{fam}(p) \geq \ell$; or   (int)
- $s \rightarrow_{\mathcal{H}} p^v$, for some v such that $v ; \sigma \twoheadrightarrow_{\mathcal{H}^\omega} \tau$.   (ext)

Proof. By induction on the length of the reduction $s^\sigma \twoheadrightarrow_{\mathcal{H}^\omega} p^\tau$. If the length is 0, the result follows easily. Otherwise, suppose $s^\sigma \twoheadrightarrow_{\mathcal{H}^\omega} s' \twoheadrightarrow_{\mathcal{H}^\omega} p^\tau$. By Theorem 4.1, there exist a local pattern q and substitution σ' such that $s' = q^{\sigma'}$, $\text{fam}(p) \geq \text{fam}(q)$ and either (trm) $q \twoheadrightarrow_{\mathcal{H}^\omega} p^{v'}$ and $v' ; \sigma' = \tau$; or (sub) $p = q$ and $\sigma' \twoheadrightarrow_{\mathcal{H}^\omega} \tau$. Applying the induction hypothesis to $s^\sigma \twoheadrightarrow_{\mathcal{H}^\omega} q^{\sigma'}$ yields that one of the following cases must apply:

- *(int)* $\mathrm{fam}(q) \geq \ell$, but then $\mathrm{fam}(p) \geq \ell$ by transitivity of $\geq$.
- *(ext)* $s \twoheadrightarrow_{\mathcal{H}^\omega} q^\upsilon$ and $\upsilon \,;\, \sigma \twoheadrightarrow_{\mathcal{H}^\omega} \sigma'$, for some substitution υ. We distinguish the following cases:
    - *(trm)* $s \twoheadrightarrow_{\mathcal{H}^\omega} q^\upsilon \twoheadrightarrow_{\mathcal{H}^\omega} p^{\upsilon';\upsilon}$ and $\upsilon' \,;\, \upsilon \,;\, \sigma \twoheadrightarrow_{\mathcal{H}^\omega} \upsilon' \,;\, \sigma' = \tau$.
    - *(sub)* $s \twoheadrightarrow_{\mathcal{H}^\omega} q^\upsilon = p^\upsilon$ and $\upsilon \,;\, \sigma \twoheadrightarrow_{\mathcal{H}^\omega} \sigma' \twoheadrightarrow_{\mathcal{H}^\omega} \tau$. □

Theorem 5.2. *Let $\mathcal{H}^\omega$ be the labelling of a non-collapsing HRS, and let $\mathcal{R} : s_1 \twoheadrightarrow_{\mathcal{H}^\omega} s_2 \twoheadrightarrow_{\mathcal{H}^\omega} \cdots$ be a $\mathcal{H}^\omega$-reduction. $\mathcal{R}$ is finite, if and only if there is a $\ell_{\max} \in \mathbb{N}$ such that $\mathrm{fam}(s_i) \leq \ell_{\max}$ for all s_i.*

Proof. (Sketch) ($\Rightarrow$): Trivial. ($\Leftarrow$): We prove the theorem by showing that $\mathcal{H}^\omega = \langle \Sigma^\omega, R^\omega \rangle$ is terminating if we restrict it to rules $l \to r \in R^\omega$ where $\mathrm{fam}(r) \leq \ell_{\max}$. It suffices to show that r^σ terminates for all right-hand sides r and terminating substitutions σ. We do this by assuming, to the contrary, that a non-terminating term exists. Let $(s^\ell)^\sigma$ be a *minimal* non-terminating term such that s is non-(subst-subst)-collapsing[1], and σ is a terminating substitution. By minimality, this reduction is of the form: $(s^\ell)^\sigma \twoheadrightarrow_{\mathcal{H}^\omega} l^\tau \twoheadrightarrow_{\mathcal{H}^\omega} r^\tau \twoheadrightarrow_{\mathcal{H}^\omega} \cdots$, where $\lambda\overline{x}.l \to \lambda\overline{x}.r \in R^\omega$. We will show, by induction on $(\ell_{\max} - \ell)$, that $(s^\ell)^\sigma$ is terminating, contradicting the assumption that it's not.

The interesting case is that $s = \lambda\overline{x}.y(s_1, \ldots, s_n)$, where $y \in Dom(\sigma)$. Let $t = \sigma(y)$, and $\sigma' = [x_i \mapsto s_i]$. Then $t^{\sigma'} \twoheadrightarrow_{\mathcal{H}} l^\tau$. By the fact that s is non-(subst-subst)-collapsing, the heads of the s_i are function symbols labelled with ℓ, and thus we can apply Lemma 5.1. Again, the interesting case is if $t \twoheadrightarrow_{\mathcal{H}^\omega} l^\upsilon$, and now termination of r^τ follows from the fact that σ is terminating by assumption. □

6   Dealing with Collapsing HRSs

In the previous sections we restricted our attention to non-collapsing HRSs. Both the Prefix Property and FFD do not hold for collapsing HRSs, as is witnessed by the following two counterexamples:

Example 6.1 (Prefix Property). Consider the collapsing HRS $\mathcal{M}u$:

$$\mathsf{mu}(\lambda x.z(x)) \to z(\mathsf{mu}(\lambda x.z(x))$$

and the following $\mathcal{M}u^\omega$-step:

$$\mathsf{mu}^3(\lambda x.\mathsf{f}^2(x)) \to_{\mathcal{M}u^\omega} \mathsf{f}^2(\mathsf{mu}^4(\lambda x.\mathsf{f}^2(x)))$$

It is easy to check that the prefix $p = \mathsf{f}^2(u)$ of the target of the step has no ancestor q that satisfies the requirements of the Prefix Property (Theorem 4.1).

Example 6.2 (FFD). Consider the collapsing HRS $\mathcal{L}am$:

$$\mathsf{app}(\mathsf{lam}(\lambda x.z(x), y)) \to z(y)$$

---

[1] We drop the (context-subst) condition of Def 2.5, because subterms of non (context-subst)-collapsing terms can be (constext-subst)-collapsing, meaning that an infinite reduction from a minimal counter example might not contain a head step.

Then one $\mathcal{L}am^\omega$-step is the following:

$$\mathsf{app}^1(\mathsf{lam}^1(\lambda x.\mathsf{app}^1(x,x)),\mathsf{lam}^1(\lambda x.\mathsf{app}^1(x,x)))$$
$$\rightarrow_{\mathcal{L}am^\omega} \mathsf{app}^1(\mathsf{lam}^1(\lambda x.\mathsf{app}^1(x,x)),\mathsf{lam}^1(\lambda x.\mathsf{app}^1(x,x)))$$

So we see that $\mathcal{L}am^\omega$ has a one-step cycle, and thus an infinite reduction with bounded labels.

The problem in both cases is that, because of applying a collapsing rule, a function symbol can be directly connected to a previously unconnected function symbol from the context or substitution, or to the root of the term, without the rule leaving any trace in between, in the form of a label. This can be remedied by including "empty" function symbols, named ϵ_α, for each base type α, in the right-hand sides of rules, and "saturating" the left-hand sides of rules with those empty function symbols. The same approach is taken for the first-order case in [13, Chap. 8]. We sketch the idea of this "ϵ-lifting", $\mathcal{H}^\epsilon$, by giving two examples; for a formal definition, see [4].

Example 6.3. The ϵ-lifting of $\mathcal{M}u$ (types of ϵ's omitted):

$$\mathsf{mu}(\lambda x.z(x)) \rightarrow \epsilon(z(\epsilon(\mathsf{mu}(\lambda x.\epsilon(z(\epsilon(x))))))))$$

Note that more ϵ's are added than strictly necessary; this is for ease of definition (see [4] for details). A $(\mathcal{M}u^\epsilon)^\omega$ step corresponding to the step of Ex. 6.1 is:

$$\mathsf{mu}^3(\lambda x.\mathsf{f}^2(x)) \rightarrow_{(\mathcal{M}u^\epsilon)^\omega} \epsilon^4(\mathsf{f}^2(\epsilon^4(\mathsf{mu}^4(\lambda x.\epsilon(\mathsf{f}^2(\epsilon(x)))))))$$

Take the corresponging prefix $p = \epsilon^4(\mathsf{f}^2(y))$. Now, the Prefix Property is satisfied with $q = \mathsf{mu}^3(\lambda x.\mathsf{f}^2(x))$, $\tau = \emptyset$ and $\upsilon = [z \mapsto \epsilon^4(\mathsf{mu}^4(\lambda x.\epsilon(\mathsf{f}^2(\epsilon(x)))))]$.

Example 6.4. The ϵ-lifting of $\mathcal{L}am$ consists of (among others) the following rules:

$$\mathsf{app}(\mathsf{lam}(\lambda x.z(x),y)) \rightarrow \epsilon(z(\epsilon(y)))$$
$$\mathsf{app}(\epsilon(\mathsf{lam}(\lambda x.z(x))),y) \rightarrow \epsilon(z(\epsilon(y)))$$
$$\mathsf{app}(\epsilon(\epsilon(\mathsf{lam}(\lambda x.z(x)))),y) \rightarrow \epsilon(z(\epsilon(y)))$$

Then a $(\mathcal{L}am^\epsilon)^\omega$-step corresponding to the step of Ex. 6.2 is the following:

$$\mathsf{app}^1(\mathsf{lam}^1(\lambda x.\mathsf{app}^1(x,x)),\mathsf{lam}^1(\lambda x.\mathsf{app}^1(x,x)))$$
$$\rightarrow_{(\mathcal{L}am^\epsilon)^\omega} \epsilon^2(\mathsf{app}^1(\epsilon^2(\mathsf{lam}^1(\lambda x.\mathsf{app}^1(x,x))),\epsilon^2(\mathsf{lam}^1(\lambda x.\mathsf{app}^1(x,x)))))$$

Now, all redex patterns have a maximum label of 2, instead of 1.

Theorem 6.5 (FFD). *Let $(\mathcal{H}^\epsilon)^\omega$ be the $\epsilon\omega$-labelling of an HRS, and let $\mathcal{R}$: $s_1 \rightarrow_{(\mathcal{H}^\epsilon)^\omega} s_2 \rightarrow_{(\mathcal{H}^\epsilon)^\omega} \cdots$ be a $(\mathcal{H}^\epsilon)^\omega$-reduction. $\mathcal{R}$ is finite, if and only if there is a $\ell_{\max} \in \mathbb{N}$ such that $\mathrm{fam}(s_i) \leq \ell_{\max}$ for all s_i.*

7   Applications and Further Research

The Prefix Property and Finite Family Developments are useful tools for proving various properties of HRSs. For example, an alternative proof of termination of the simply typed λ-calculus (encoded as an HRS) uses FFD. Also, in a work in progress by the author, FFD is used to prove the termination of a higher-order standardization procedure. This result can be used to formalize the notion of equivalence of reductions, in a similar way as is done in [13].

For future research, it might be interesting to further investigate the relation between FFD and the Dependency Pair method [1], both in the higher-order and first-order case. Since FFD and the Dependency Pair method both essentially depend on the same principle, that an infinite reduction must have an unbounded creation depth, it the author's conjecture that FFD, or the Prefix Property, can be used to design a higher-order Dependency Pair method.

Acknowledgements. I wish to thank Vincent van Oostrom, Delia Kesner and the anonymous referees for useful remarks on preliminary versions of this paper.

References

1. Thomas Arts and Jürgen Giesl. Termination of term rewriting using dependency pairs. *Theoretical Computer Science*, 236(1–2):133–178, 2000.
2. Inge Bethke, Jan Willem Klop, and Roel de Vrijer. Descendants and origins in term rewriting. *Information and Computation*, 159(1–2):59–124, 2000.
3. Roel Bloo. *Preservation of Termination for Explicit Substitution*. PhD thesis, Technische Universiteit Eindhoven, 1997.
4. H. J. Sander Bruggink. A proof of finite family developments for higher-order rewriting using a prefix property. Preprint, LGPS 245, Zeno Inst. of Phil., 2006.
5. Gilles Dowek, Thérèse Hardin, and Claude Kirchner. Higher order unification via explicit substitutions. *Information and Computation*, 157(1–2):184–233, 2000.
6. J.M.E. Hyland. A syntactic characterization of the equality in some models of the λ-calculus. *Journal of the London Mathematical Society*, 12(2):361–370, 1976.
7. J. W. Klop. *Combinatory Reduction Systems*. PhD thesis, Utrecht Univ., 1980.
8. Jean-Jacques Lévy. *Réductions correctes et optimales dans le λ-calcus*. PhD thesis, Université Paris VII, 1978.
9. Luc Maranget. Optimal derivations in weak lambda-calculi and in orthogonal term rewriting systems. In *POPL*, 1991.
10. Richard Mayr and Tobias Nipkow. Higher-order rewrite systems and their confluence. *Theoretical Computer Science*, 192:3–29, 1998.
11. Dale Miller. A logic programming language with lambda abstraction, function variables and simple unification. *Journal of Logic and Computation*, 1(4), 1991.
12. Tobias Nipkow. Higher-order critical pairs. In *LICS*, 1991.
13. Terese. *Term Rewriting Systems*. Number 55 in CTTCS. CUP, 2003.
14. D.T. van Daalen. *The language theory of Automath*. PhD thesis, Technische Universiteit Eindhoven, 1980.
15. Vincent van Oostrom. *Confluence for Abstract and Higher-Order Rewriting*. PhD thesis, Vrije Universiteit Amsterdam, 1994.
16. Vincent van Oostrom. Finite family developments. In *RTA*, 1997.
17. Vincent van Oostrom and Roel de Vrijer. Four equivalent equivalences of reductions. *ENTCS*, 70(6), 2002.
18. C. P. Wadsworth. The relation between computational and denotational properties for Scott's D_∞-models of the λ-calculus. *SIAM Journal on Computing*, 5, 1976.

Higher-Order Orderings for Normal Rewriting

Jean-Pierre Jouannaud[1,*] and Albert Rubio[2,**]

[1] LIX, École Polytechnique, 91400 Palaiseau, France
[2] Technical University of Catalonia, Pau Gargallo 5, 08028 Barcelona, Spain

Abstract. We extend the termination proof methods based on reduction orderings to higher-order rewriting systems *à la Nipkow* using higher-order pattern
matching for firing rules, and accommodate for any use of eta, as a reduction,
as an expansion or as an equation. As a main novelty, we provide with a mechanism for transforming any reduction ordering including beta-reduction, such as
the higher-order recursive path ordering, into a reduction ordering for proving
termination of rewriting à la Nipkow. Non-trivial examples are carried out.

1 Introduction

Rewrite rules are used in logical systems to describe computations over lambda-terms
used as a suitable abstract syntax for encoding functional objects like programs or specifications. This approach was pioneered in this context by Nipkow [18] and is available
in Isabelle [21]. Its main feature is the use of higher-order pattern matching for firing
rules. A recent generalization of Nipkow's setting allows one for rewrite rules of polymorphic, higher-order type [15], see also [10]. Besides, it is shown that using the η-rule
as an expansion [20] or as a reduction [15] yields very similar confluence checks based
on higher-order critical pairs.

A first contribution of this paper is a general setting for addressing termination of
all variants of higher-order rewriting à la Nipkow, thanks to the notion of a *normal
higher-order reduction ordering*. While higher-order reduction orderings actually *include* $\beta\eta$-reductions, normal higher-order reduction orderings must be compatible with
$\beta\eta$-equality since higher-order rewriting operates on $\beta\eta$-equivalence classes of terms.
This is done by computing with $\beta\eta$-normal forms as inputs. Since this may destroy
stability under substitution, it becomes necessary to use higher-order reduction orderings enjoying a stronger stability property. Restricting the higher-order recursive path
ordering [12] to achieve this property is our second contribution. Finally, the obtained
ordering is used inside a powerful schema transforming an arbitrary higher-order reduction ordering satisfying the stronger stability property into a normal higher-order
reduction ordering. This is our third contribution. The obtained ordering allows us to
prove all standard examples of higher-order rules processing abstract syntax.

We describe our framework for terms in Section 2, and for higher-order rewriting
in Section 3. The schema is introduced and studied in Section 4. The restricted higher-
order recursive path ordering is given in Section 5. Two complex examples are carried
out in Section 6. Significance of the results is briefly discussed in Section 7.

---

* Project LogiCal, Pôle Commun de Recherche en Informatique du Plateau de Saclay, CNRS,
  École Polytechnique, INRIA, Université Paris-Sud.
** Project LogicTools (TIN2004-03382) of the Spanish Min. of Educ. and Science.

F. Pfenning (Ed.): RTA 2006, LNCS 4098, pp. 387–399, 2006.

Readers are assumed familiar with the basics of term rewriting [9,16] and typed lambda calculi [4,5]. Most ideas presented here originate from [14], an unpublished preliminary draft. A full version is [13].

2 Higher-Order Algebras

Rewrite rules of polymorphic higher type are our target. To define them precisely, we need to recall the framework of higher-order algebras [12]. We will consider the non-polymorphic case for simplicity. The general case of polymorphic higher-order rewrite rules is carried out in [13].

2.1 Types

Given a set S of *sort symbols* of a fixed arity, denoted by $s : *^n \to *$, the set of *types* is generated by the constructor $\to$ for *functional types*:

$$\mathcal{T}_S := s(\mathcal{T}_S^n) \mid (\mathcal{T}_S \to \mathcal{T}_S) \qquad \text{for } s : *^n \to * \in S$$

Types are *functional* when headed by the $\to$ symbol, and *data types* otherwise. $\to$ associates to the right. We use $\sigma, \tau, \rho, \theta$ for arbitrary types. The type $\sigma = \tau_1 \to \dots \to \tau_n \to \tau$, τ not functional, has *arity* $ar(\sigma) = n$.

2.2 Signatures

Function symbols are meant to be algebraic operators equipped with a fixed number n of arguments (called the *arity*) of respective types $\sigma_1 \in \mathcal{T}_S, \dots, \sigma_n \in \mathcal{T}_S$, and an *output type* $\sigma \in \mathcal{T}_S$. Let

$$\mathcal{F} = \biguplus_{\sigma_1, \dots, \sigma_n, \sigma} \mathcal{F}_{\sigma_1 \times \dots \times \sigma_n \to \sigma}$$

The membership of a given function symbol f to $\mathcal{F}_{\sigma_1 \times \dots \times \sigma_n \to \sigma}$ is called a *type declaration* and written $f : \sigma_1 \times \dots \times \sigma_n \to \sigma$. A type declaration is *first-order* if it uses only sorts, and *higher-order* otherwise.

2.3 Terms

The set $\mathcal{T}(\mathcal{F}, \mathcal{X})$ of *raw algebraic λ-terms* is generated from the signature $\mathcal{F}$ and a denumerable set $\mathcal{X}$ of variables according to the grammar:

$$\mathcal{T} := \mathcal{X} \mid (\lambda \mathcal{X} : \mathcal{T}_S.\mathcal{T}) \mid @(\mathcal{T}, \mathcal{T}) \mid \mathcal{F}(\mathcal{T}, \dots, \mathcal{T}).$$

The raw term $\lambda x : \sigma.u$ is an *abstraction* and $@(u, v)$ is an *application*. We may omit σ in $\lambda x : \sigma.u$ and write $@(u, v_1, \dots, v_n)$ or $u(v_1, \dots, v_n)$, $n > 0$, omitting applications. The raw term $@(u, \overline{v})$ is a (partial) *left-flattening* u being possibly an application. $\mathcal{V}ar(t)$ is the set of free variables of t. A term t is *ground* if $\mathcal{V}ar(t) = \emptyset$. The notation $\overline{s}$ shall be ambiguously used to for a list, a multiset, or a set of raw terms $s_1, \dots, s_n$.

Raw terms are identified with finite labeled trees by considering $\lambda x : \sigma.u$, for each variable x and type σ, as a unary function symbol taking u as argument to construct the raw term $\lambda x : \sigma.u$. *Positions* are strings of positive integers. Λ and $\cdot$ denote respectively the empty string (root position) and string concatenation. $\mathcal{P}os(t)$ is the set of positions in t. $t|_p$ denotes the *subterm* of t at position p. We use $t \trianglerighteq t|_p$ for the subterm relationship. The result of replacing $t|_p$ at position p in t by u is written $t[u]_p$. A raw term $t[x : \sigma]_p$ with a hole of type σ at position p is a *context*.

Given a binary relation $\longrightarrow$ on raw terms, a raw term s such that $s|_p \longrightarrow t$ for some position $p \in \mathcal{P}os(s)$ is called *reducible*. Irreducible raw terms are in *normal form*. A raw term s is *strongly normalizable* if there is no infinite sequence of $\longrightarrow$-steps issuing from s. The relation $\longrightarrow$ is *strongly normalizing*, or *terminating* or *well-founded*, if all raw terms are strongly normalizable. We denote by $\longleftrightarrow$ the symmetric closure of the relation $\longrightarrow$, by $\xrightarrow{*}$ its reflexive, transitive closure, and by $\xleftrightarrow{*}$ its reflexive, symmetric, transitive closure. The relation $\longrightarrow$ is *confluent* (resp. *Church-Rosser*) if $s \longrightarrow^* u$ and $s \longrightarrow^* v$ (resp. $u \longleftrightarrow^* v$) implies $u \longrightarrow^* t$ and $v \longrightarrow^* t$ for some t.

2.4 Typing Rules

Definition 1. *An* environment Γ *is a finite set of pairs written as* $\{x_1 : \sigma_1, \ldots, x_n : \sigma_n\}$, *where* x_i *is a variable,* σ_i *is a type, and* $x_i \neq x_j$ *for* $i \neq j$. $Var(\Gamma) = \{x_1, \ldots, x_n\}$ *is the set of variables of* Γ. *Given two environments* Γ *and* Γ', *their* composition *is the environment* $\Gamma \cdot \Gamma' = \Gamma' \cup \{x : \sigma \in \Gamma \mid x \notin Var(\Gamma')\}$. *Two environments* Γ *and* Γ' *are* compatible *if* $\Gamma \cdot \Gamma' = \Gamma \cup \Gamma'$.

Our typing judgements are written as $\Gamma \vdash_{\mathcal{F}} s : \sigma$. A raw term s has type σ in the environment Γ if the judgement $\Gamma \vdash_{\mathcal{F}} s : \sigma$ is provable in our inference system given at Figure 1. Given an environment Γ, a raw term s is *typable* if there exists a type σ such that $\Gamma \vdash_{\mathcal{F}} s : \sigma$. Typable raw terms are called *terms*. An important property of our simple type system is that a raw term typable in a given environment has a unique type.

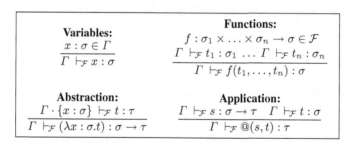

Fig. 1. The type system for monomorphic higher-order algebras

Because variables are typed, they must be replaced by typable terms:

Definition 2. *A* substitution $\gamma = \{(x_1 : \sigma_1) \mapsto (\Gamma_1, t_1), \ldots, (x_n : \sigma_n) \mapsto (\Gamma_n, t_n)\}$, *is a finite set of quadruples made of a variable symbol, a type, an environment and a term, such that*

(i) $\forall i \in [1..n]$, $t_i \neq x_i$ *and* $\Gamma_i \vdash_{\mathcal{F}} t_i : \sigma_i$,
(ii) $\forall i \neq j \in [1..n]$, $x_i \neq x_j$, *and*

(iii) $\forall i \neq j \in [1..n]$, Γ_i *and* Γ_j *are compatible environments.*
We may omit the type σ_i *and environment* Γ_i *in* $(x_i : \sigma_i) \mapsto (\Gamma_i, t_i)$.
The set of (input) variables of γ *is* $Var(\gamma) = \{x_1, \ldots, x_n\}$, *its* domain *is the environment* $Dom(\gamma) = \{x_1 : \sigma_1, \ldots, x_n : \sigma_n\}$ *while its* range *is the environment* $Ran(\gamma) = \bigcup_{i \in [1..n]} \Gamma_i$.

Definition 3. *A substitution* γ *is* compatible *with the judgement* $\Gamma \vdash_{\mathcal{F}} s : \sigma$ *(or simply, with* Γ*) if   (i)* $Dom(\gamma)$ *is compatible with* Γ, *and*
(ii) $Ran(\gamma)$ *is compatible with* $\Gamma \setminus Dom(\gamma)$.

A substitution γ compatible with a judgement $\Gamma \vdash_{\mathcal{F}} s : \sigma$ operates classically as an endomorphism on s, resulting in a term denoted by $s\gamma$.

Lemma 1. *Given a signature* $\mathcal{F}$ *and a substitution* γ *compatible with the judgement* $\Gamma \vdash_{\mathcal{F}} s : \sigma$, *then* $\Gamma \cdot Ran(\gamma) \vdash_{\mathcal{F}} s\gamma : \sigma$.

2.5   Conversions

We consider α-convertible terms as identical, and hence α-conversions are omitted. The congruence generated by the β- and η-equalities
$$(\lambda x.u, v) =_\beta u\{x \mapsto v\} \qquad \lambda x.@(u, x) =_\eta u \text{ if } x \notin Var(u)$$
is written $=_{\beta\eta}$. An important property, *subject reduction*, is that typable terms u, v such that $u =_{\beta\eta} v$ have the same type. Both equalities can be oriented as rewrite rules. There are two possible choices for rewriting with η, either as a reduction or as an expansion, in which case termination is ensured by restricting its use to positions other than the first argument of an application. Typed lambda-calculi have all termination and confluence properties one may need, with respect to: $\beta\eta$-reductions; β-reductions and η-expansions; β-reductions modulo η-equality. Using the notations $u \longrightarrow_\beta v$ for one β-rewrite step, $u \longrightarrow_\beta^* v$ for its transitive closure, $u \downarrow_\beta$ ($u\downarrow$ for short) for the β-normal form of u, and $\longleftrightarrow_\eta^*$ or $=_\eta$ for η-equality, the Church-Rosser property of β-reductions modulo η-equality for typable terms can be phrased as

$$s =_{\beta\eta} t \text{ iff } s\downarrow_\beta =_\eta t\downarrow_\beta$$

3   Normal Higher-Order Rewriting of Higher Type

Normal higher-order rewriting [20,18] allows defining computations on λ-terms used as a suitable abstract syntax for encoding functional objects like programs or specifications. Nipkow's framework assumes that rules are of basic type, and that left-hand sides of rules are patterns in the sense of Miller [19], assumptions which are not made here, nor in [15].

Nipkow's normal higher-order rewriting uses $\beta\eta$-equalities in two different ways: given a term s to be rewritten with a set R of rules, s is first normalized, using η-expanded β-normal forms, before to be searched for left-hand sides of rules in R via higher-order pattern matching, that is, matching modulo $=_{\beta\eta}$. In this section, we define higher-order rewriting so as to capture the different ways in which a term can be $\beta\eta$-normalized before pattern matching its subterm with a left-hand side of rule.

Definition 4. *A* normal rewrite rule *is a rewrite rule* $\Gamma \vdash l \rightarrow r : \sigma$ *such that l and r are higher-order terms in β-normal form satisfying $\Gamma \vdash_{\mathcal{F}} l : \sigma$ and $\Gamma \vdash_{\mathcal{F}} r : \sigma$. A* normal term rewriting system *is a set of normal rewrite rules.*

Given a normal term rewriting system R, an environment Γ, two β-normal terms s and t, and a type σ such that $\Gamma \vdash_{\mathcal{F}} s : \sigma$, we say that s rewrites to t at position p with the normal rule $\Gamma_i \vdash l_i \rightarrow r_i : \sigma_i$ and the term substitution γ, written $\Gamma \vdash s \longrightarrow^p_{R_{\beta\eta}} t$, or $s \longrightarrow^p_{R_{\beta\eta}} t$ assuming the environment Γ, if the following conditions hold:

$$\begin{array}{ll} \text{(i) } \mathcal{D}om(\gamma) \subseteq \Gamma_i & \text{(iii) } s|_p =_{\beta\eta} l_i\gamma \\ \text{(ii) } \Gamma_i \cdot \mathcal{R}an(\gamma) \subseteq \Gamma_{s|_p} & \text{(iv) } t =_\eta s[r_i\gamma]_p{\downarrow}_\beta \end{array}$$

where $\Gamma_{s|_p}$ is the environment of the judgement $\Gamma_{s|_p} \vdash_{\mathcal{F}} s|_p : \sigma_i$, obtained as a subterm of the proof of the judgement $\Gamma \vdash_{\mathcal{F}} s : \sigma$.

Note that t is any term in the η-equivalence class of $s[r_i\gamma]_p{\downarrow}_\beta$. Higher-order rewriting is therefore defined up to η-equivalence of target terms. By providing a method for proving termination of this relation, we do provide a termination method for all variants of higher-order rewriting based on higher-order pattern matching. A key observation is this:

Lemma 2. *Assume $\Gamma \vdash_{\mathcal{F}} s : \sigma$ and $\Gamma \vdash s \rightarrow_{R_{\beta\eta}} t$. Then $\Gamma \vdash_{\mathcal{F}} t : \sigma$.*

Example 1. We present here an encoding of symbolic derivation in which functions are represented by λ-terms of a functional type. We give two typical rules of higher type. Both rules have the same environment $\Gamma = \{F : real \rightarrow real\}$, and x, y stand for real values. Let $\mathcal{S} = \{real\}$, and

$$\begin{aligned} \mathcal{F} = \{ & \mathsf{sin, cos} : real \rightarrow real; \mathsf{diff} : (real \rightarrow real) \rightarrow real \rightarrow real \\ & +, \times : (real \rightarrow real) \rightarrow (real \rightarrow real) \rightarrow real \rightarrow real\} \end{aligned}$$

$$\mathsf{diff}(\lambda\mathsf{x}.\,\mathsf{sin}(@(F,x))) \rightarrow \lambda\mathsf{x}.\,\mathsf{cos}(@(F,x)) \times \mathsf{diff}(\lambda\mathsf{x}.@(F,x))$$
$$\mathsf{diff}(\lambda\mathsf{x}.@(F,x) \times \lambda\mathsf{y}.@(F,y)) \rightarrow (\mathsf{diff}(\lambda\mathsf{x}.@(F,x)) \times \lambda\mathsf{y}.@(F,y))+$$
$$(\lambda\mathsf{x}.@(F,x) \times \mathsf{diff}(\lambda\mathsf{y}.@(F,y)))$$

This example makes sense when using normal higher-order rewriting, because using plain pattern matching instead would not allow to compute the derivative of all expressions: $\mathsf{diff}(\lambda\mathsf{x}.\mathsf{sin}(x)) =_\beta \mathsf{diff}(\lambda\mathsf{x}.\mathsf{sin}((\lambda\mathsf{y}.y)\,x))$ does require higher-order pattern matching. We shall give a mechanical termination proof of both rules in Section 5.

3.1 Normal Higher-Order Reduction Orderings

We shall use well-founded relations for proving strong normalization properties. For our purpose, these relations may not be transitive, but their transitive closures will be well-founded orderings, justifying some abuse of terminology. Reduction orderings operating on judgements turn out to be an adequate tool for showing termination of normal rewriting. We consider two classes of reduction orderings called *higher-order reduction ordering* when they include β-reductions and *normal higher-order reduction ordering* when they are compatible with $=_{\beta\eta}$.

Definition 5. *A binary relation* $\succ$ *on the set of judgements is*

- coherent *iff for all terms* s, t *such that* $(\Gamma \vdash_{\mathcal{F}} s : \sigma) \succ (\Gamma \vdash_{\mathcal{F}} t : \sigma)$, *and for all environment* Γ' *such that* Γ *and* Γ' *are compatible,* $\Gamma' \vdash_{\mathcal{F}} s : \sigma$ *and* $\Gamma' \vdash_{\mathcal{F}} t : \sigma$, *then* $(\Gamma' \vdash_{\mathcal{F}} s : \sigma) \succ (\Gamma' \vdash_{\mathcal{F}} t : \sigma)$;
- stable *iff for all terms* s, t *such that* $(\Gamma \vdash_{\mathcal{F}} s : \sigma) \succ (\Gamma \vdash_{\mathcal{F}} t : \sigma)$, *and all substitution* γ *whose domain is compatible with* Γ, *then* $(\Gamma \cdot \mathcal{R}an(\gamma) \vdash_{\mathcal{F}} s\gamma : \sigma) \succ (\Gamma \cdot \mathcal{R}an(\gamma) \vdash_{\mathcal{F}} t\gamma : \sigma)$;
- monotonic *iff for all terms* s, t *and type* σ *such that* $(\Gamma \vdash_{\mathcal{F}} s : \sigma) \succ (\Gamma \vdash_{\mathcal{F}} t : \sigma)$, *for all* Γ' *compatible with* Γ *and ground context* $u[]$ *such that* $\Gamma' \vdash_{\mathcal{F}} u[x : \sigma] : \tau$, *then* $(\Gamma \cdot \Gamma' \vdash_{\mathcal{F}} u[s] : \tau) \succ (\Gamma \cdot \Gamma' \vdash_{\mathcal{F}} u[t] : \tau)$ *(note the assumption that* $u[]$ *is ground)*;
- normal-monotonic *iff for all terms* s *and* t *such that* $(\Gamma \vdash_{\mathcal{F}} s : \sigma) \succ (\Gamma \vdash_{\mathcal{F}} t : \sigma)$, *for all* Γ' *compatible with* Γ *and for all ground context* $u[]$ *such that* $\Gamma' \vdash_{\mathcal{F}} u[x : \sigma] : \tau$ *and* $u[s]$ *is in* β-*normal form, then* $(\Gamma \cdot \Gamma' \vdash_{\mathcal{F}} u[s] : \tau) \succ (\Gamma \cdot \Gamma' \vdash_{\mathcal{F}} u[t] : \tau)$;
- functional *iff for all terms* s, t *such that* $(\Gamma \vdash_{\mathcal{F}} s : \sigma \longrightarrow_{\beta} t : \sigma)$, *then* $(\Gamma \vdash_{\mathcal{F}} s : \sigma) \succ (\Gamma \vdash_{\mathcal{F}} t : \sigma)$;
- compatible *iff for all terms* s', s, t, t' *such that* $(\Gamma \vdash_{\mathcal{F}} s' : \sigma =_{\beta\eta} s : \sigma)$, $(\Gamma \vdash_{\mathcal{F}} t : \sigma =_{\beta\eta} t' : \sigma)$ *and* $(\Gamma \vdash_{\mathcal{F}} s : \sigma) \succ (\Gamma \vdash_{\mathcal{F}} t : \sigma)$ *then* $(\Gamma \vdash_{\mathcal{F}} s' : \sigma) \succ (\Gamma \vdash_{\mathcal{F}} t' : \sigma)$.

A higher-order reduction ordering $\succ$ is a well-founded ordering of the set of judgements satisfying coherence, stability, monotonicity and functionality.

A normal higher-order reduction ordering $\succ_n$ is a well-founded ordering of the set of judgements satisfying coherence, stability, normal-monotonicity and compatibility.

Let us show that no ordering $\succ$ can satisfy monotonicity, stability, compatibility and well-foundedness, therefore explaining the need for the weaker notion of normal-monotonicity. Assume $s : \sigma \succ t : \sigma$ (omitting judgements), where $s : \sigma$ is in β-normal form. Consider the term $\lambda y.a : \sigma \to \tau$ where $a : \tau$ is a constant. Then, by monotonicity, $@(\lambda y.a, s) : \tau \succ @(\lambda y.a, t) : \tau$ and by compatibility, $a : \tau \succ a : \tau$, contradicting well-foundedness. Normal-monotonicity removes the problem since $@(\lambda y.a, s)$ is not in β-normal form. As a consequence, we cannot have $@(X, s) \succ @(X, t)$ when $s \succ t$, but only $@(X, s) = @(X, t)$.

Theorem 1. *Let* $R = \{\Gamma_i \vdash l_i \to r_i : \sigma_i\}_i$ *be a higher-order rewrite system and* $\succ$ *a normal higher-order reduction ordering s.t.* $(\Gamma_i \vdash_{\mathcal{F}} l_i) \succ (\Gamma_i \vdash_{\mathcal{F}} r_i) \ \forall i$. *Then the relation* $\longrightarrow_{R_{\beta\eta}}$ *is strongly normalizing.*

Proof. Without loss of generality, let s be a ground normal term such that $\Gamma \vdash_{\mathcal{F}} s \xrightarrow[\Gamma_i \vdash l_i \to r_i : \sigma_i]{p} t$. By definition 4, t is a ground normal term. It therefore suffices to show that $\Gamma \vdash_{\mathcal{F}} s \succ t$, which we proceed to do now. By assumption, $\Gamma_i \vdash_{\mathcal{F}} l_i \succ r_i$. By stability, $\Gamma_i \cdot \mathcal{R}an(\gamma) \vdash_{\mathcal{F}} l_i\gamma \succ r_i\gamma$, therefore, by coherence, $\Gamma_{s|_p} \vdash_{\mathcal{F}} l_i\gamma \succ r_i\gamma$. By definition, $s|_p =_{\beta\eta} l_i\gamma$, hence, by compatibility, $\Gamma_{s|_p} \vdash_{\mathcal{F}} s|_p \succ r_i\gamma$. By monotonicity of $\succ$ for normal ground terms (of equal type), $\Gamma_{s|_p} \cdot \Gamma \vdash_{\mathcal{F}} s \succ s[r_i\gamma]_p$. By coherence $\Gamma \vdash_{\mathcal{F}} s \succ s[r_i\gamma]_p$, hence $\Gamma \vdash_{\mathcal{F}} s \succ t$ by compatibility. $\qquad\square$

By lemma 2, higher-order rewriting can be seen as a type preserving relation on terms in a given environment Γ typing the term originating the sequence of rewrites. We can therefore simplify our notations by omitting the typing judgements unless they are really necessary.

4 Building Normal Higher-Order Reduction Orderings

In this section, we assume given a new function symbol $\perp_\sigma$ for every type σ and a function f_{new} for some of the function symbols in $\mathcal{F}$. We denote by $\mathcal{F}_{new}$ the augmented signature. We write $\perp_\sigma$ for $\perp_\sigma()$. The higher-order rules we want to prove terminating are built from terms in $\mathcal{T}(\mathcal{F}, \mathcal{X})$, not in $\mathcal{T}(\mathcal{F}_{new}, \mathcal{X})$. We successively introduce neutralization, and the neutralized ordering schema. Neutralization replaces a term of functional type by its application to a term headed by a $\perp$-operator seen as a container for its arguments. Neutralizing an abstraction creates a redex which will be later eliminated by a β-normalization step.

Definition 6. *The* neutralization of level i *(i-neutralization in short) of a term $t : \tau \in \mathcal{T}(\mathcal{F}_{new}, \mathcal{X})$ w.r.t. a list of (typable) terms $\langle u_1 : \theta_1, \ldots, u_n : \theta_n \rangle$ in $\mathcal{T}(\mathcal{F}_{new}, \mathcal{X})$, is the term $\mathcal{N}_i(t, \langle u_1, \ldots, u_n \rangle)$ defined as follows:*

1. $\mathcal{N}_0(t, \langle u_1, \ldots, u_n \rangle) = t$;
2. $\mathcal{N}_{i+1}(t, \langle u_1, \ldots, u_n \rangle) = t$ *if τ is a data type;*
3. $\mathcal{N}_{i+1}(t, \langle u_1, \ldots, u_n \rangle) =$
   $\mathcal{N}_i(@(t, \perp_{\theta_1 \to \ldots \to \theta_n \to \sigma}(u_1, \ldots, u_n)), \langle u_1, \ldots, u_n \rangle)$ *if $\tau = \sigma \to \rho$.*

From now one, we shall very precisely control for each function symbol which of its arguments of a functional type are neutralized:

Definition 7. *A signature $\mathcal{F}$ is* neutralized *if each symbol $f : \sigma_1 \times \ldots \times \sigma_n \to \sigma \in \mathcal{F}$ comes along with, for each argument position $j \in [1..n]$:*

- *a natural number $\mathcal{L}_f^j \leq ar(\sigma_j)$, called* neutralization level *of f at position j. We call* neutralized *those positions j for which $\mathcal{L}_f^j > 0$.*
- *a subset $\mathcal{A}_f^j \subseteq [1..n]$ of argument positions of f used to filter out the list $\bar{t}$ of arguments of f by defining $\bar{t}_f^j = \langle t_k \mid k \in \mathcal{A}_f^j \rangle$.*

The role of full neutralization is to neutralize terms of functional type recursively from arguments of function symbols up to a given depth depending on the function symbol itself and its selected argument. This will allow us to eventually eliminate undesirable abstractions. This huge flexibility provided by levels and argument positions allows us to tune our coming normal higher-order ordering and carry out difficult and important examples taken from the literature. In most of them, the chosen level is 1, implying that neutralization applies to the top of the arguments only, and the set of argument positions is empty, implying that $\perp$ is a constant.

To neutralize terms recursively, we need to introduce new function symbols in the signature : for every declaration $f : \sigma_1 \times \ldots \times \sigma_n \to \sigma$, we assume given a new symbol

$f_{new} : \sigma'_1 \times \ldots \times \sigma'_n \to \sigma$ whose type declaration depends upon the neutralization level of its argument positions: if $\sigma_i = \tau_1 \to \ldots \to \tau_k \to \tau$ and $\mathcal{L}^i_f = q \leq k$, then $\sigma'_i = \tau_{q+1} \to \ldots \to \tau_k \to \tau$.

Definition 8. *The* full neutralization *of a term* t *is the term* $\mathcal{FN}(t)$ *s.t.*

1. *if* $t \in \mathcal{X}$, *then* $\mathcal{FN}(t) = t$;
2. *if* $t = \lambda x.u$, *then* $\mathcal{FN}(t) = \lambda x.\mathcal{FN}(u)$;
3. *if* $t = @(t_1, t_2)$, *then* $\mathcal{FN}(t) = @(\mathcal{FN}(t_1), \mathcal{FN}(t_2))$;
4. *if* $t = f(t_1, \ldots, t_n)$ *with* $f \in \mathcal{F}$, *then*
   $$\mathcal{FN}(t) = f_{new}(\mathcal{N}_{\mathcal{L}^1_f}(\mathcal{FN}(t_1), \bar{t}^1_f), \ldots, \mathcal{N}_{\mathcal{L}^n_f}(\mathcal{FN}(t_n), \bar{t}^n_f)).$$

Our definition makes sense since, in all cases, $\mathcal{FN}(t)$ is typable with the same type as t. Note also that using Case 3 repeatedly for flattened applications yields $\mathcal{FN}(@(t_1, \ldots, t_n)) = @(\mathcal{FN}(t_1), \ldots, \mathcal{FN}(t_n))$.

Example 1 (continued). We show here the full neutralization of the left-hand and right-hand of the rules of Example 1 after β-normalizing. To this end, we choose a neutralization level for each function symbol and argument position. The associated subsets of argument positions are all chosen empty. As a consequence, $\perp_{real}$ is a constant abbreviated as $\perp$.

$\mathcal{L}^1_{diff} = 1$	$\mathcal{L}^1_{sin} = 0$	$\mathcal{L}^1_{cos} = 0$	
$\mathcal{A}^1_{diff} = \{\}$	$\mathcal{A}^1_{sin} = \{\}$	$\mathcal{A}^1_{cos} = \{\}$	
$\mathcal{L}^1_{\times} = 1$	$\mathcal{L}^2_{\times} = 1$	$\mathcal{L}^1_{+} = 1$	$\mathcal{L}^2_{+} = 1$
$\mathcal{A}^1_{\times} = \{\}$	$\mathcal{A}^2_{\times} = \{\}$	$\mathcal{A}^1_{+} = \{\}$	$\mathcal{A}^2_{+} = \{\}$

We now compute the β-normalization of the full neutralization of both sides of the first rule:

$$\mathcal{FN}(\quad \text{diff} \quad (\ \lambda x.\, \sin(@(F, x))\ )\ ) \!\downarrow$$
$$= \quad \text{diff}_{new} (\quad \sin(@(F, \perp))\quad)$$

$$\mathcal{FN}(\ \lambda x.\, \cos(@(F, x)) \quad \times \qquad \text{diff}\ (\ \lambda x.@(F, x)\ ) \qquad)\!\downarrow$$
$$= \quad \cos(@(F, \perp)) \quad \times_{new} @(\text{diff}_{new}(\quad @(F, \perp)\ )\ , \perp)$$

and of both sides of the second rule:

$$\mathcal{FN}(\quad \text{diff} \quad (\qquad \lambda x.@(F, x) \quad \times \quad \lambda y.@(F, y) \qquad)\ )\!\downarrow$$
$$= \quad \text{diff}_{new} (\ @(\quad @(F, \perp) \quad \times_{new} \quad @(F, \perp)\quad , \perp)\ )$$

$$\mathcal{FN}((\text{diff}(\lambda x.@(F, x)) \times \lambda y.@(F, y)) + (\lambda x.@(F, x) \times \text{diff}(\lambda y.@(F, y))))\!\downarrow =$$
$$@(@(\text{diff}_{new}(@(F, \perp)), \perp) \times_{new} @(F, \perp), \perp) +_{new} @(@(F, \perp) \times_{new} @(\text{diff}_{new}(@(F, \perp)), \perp), \perp)$$

Definition 9. *Given a neutralized signature, two typable terms* s, t *and a higher-order ordering* $\succ$, *we define the* neutralized ordering $\succ_n$ *as:*

$$s \succ_n t \text{ if and only if } \mathcal{FN}(s)\!\downarrow \succ \mathcal{FN}(t)\!\downarrow$$

Note that normalization applies after neutralization: we will actually see that these two operations commute, therefore implying compatibility of $\succ_n$. Well-foundedness follows from well-foundedness of $\succ$. Stability and normal-monotonicity depend upon the particular ordering $\succ$ used in the construction, which must satisfy two stronger properties:

Definition 10. *An ordering $\succ$ on higher-order terms satisfies*

(i) schema-stability *if for all $\beta\eta$-normal terms s, t and substitutions γ, $s \succ t$ implies $t\gamma \longrightarrow^*_{\beta\eta} t'\gamma$ for some term t' such that $s\gamma\!\downarrow\, \succ t'\gamma$.*

(ii) schema-monotonicity *if for all $\beta\eta$-normal terms $\lambda x.v : \sigma \to \rho \succ t : \sigma \to \rho$, and for all sequences of $\beta\eta$-normal terms $\langle u_1, \ldots, u_n \rangle$,*

- *if $t = \lambda x.w$, then $v\{x \mapsto \perp_\sigma(u_1, \ldots, u_n)\} \succ w\{x \mapsto \perp_\sigma(u_1, \ldots, u_n)\}$*
- *otherwise, $v\{x \mapsto \perp_\sigma(u_1, \ldots, u_n)\} \succ @(t, \perp_\sigma(u_1, \ldots, u_n))$.*

Theorem 2. *Let $\succ$ be a higher-order reduction ordering fulfiling the schema-stability and schema-monotonicity properties. Then $\succ_n$ is a normal higher-order reduction ordering.*

The proof of this theorem requires several preliminary technical lemmas stating properties of neutralization with respect to normalization, before to start proving stability and normal-monotonicity of $\succ_n$.

5   Normal Higher-Order Recursive Path Orderings

While the higher-order recursive path ordering satisfies schema-monotonicity, it does not satisfy schema-stability. Fortunately, a simple natural restriction suffices in case of an application on left (Cases 5 and 7 of the coming definition). In order to ease the presentation, we present a simple version of the (restricted) higher-order recursive path ordering, which will be sufficient for all examples to come. We assume given:

1. a partition $Mul \uplus Lex$ of $\mathcal{F}$ and a quasi-ordering $\geq_{\mathcal{F}}$ on $\mathcal{F}$, called the *precedence*, such that $>_{\mathcal{F}}$ is well-founded;
2. a quasi-ordering $\geq_{T_S}$ on types such that $>_{T_S}$ is well-founded and preserves the functional structure of functional types [12].

Because of type comparisons, the higher-order recursive path ordering enjoys but a weak subterm property A used in its definition:

Definition 11. *Given $s : \sigma$ and $t : \tau$, $s \underset{rhorpo}{\succ} t$ iff $\sigma \geq_{T_S} \tau$ and*

1. $s = f(\overline{s})$ with $f \in \mathcal{F}$, and $u \underset{rhorpo}{\succeq} t$ for some $u \in \overline{s}$

2. $s = f(\overline{s})$ and $t = g(\overline{t})$ with $f >_{\mathcal{F}} g$, and A

3. $s = f(\overline{s})$ and $t = g(\overline{t})$ with $f =_{\mathcal{F}} g \in Mul$ and $\overline{s}(\underset{rhorpo}{\succ})_{mul}\overline{t}$

4. $s = f(\overline{s})$ and $t = g(\overline{t})$ with $f =_{\mathcal{F}} g \in Lex$ and $\overline{s}(\underset{rhorpo}{\succ})_{lex}\overline{t}$, and A

5. $s = @(s_1, s_2)$, s_1 is not of the form $@(X, \overline{w})$ with $X \in \mathcal{X}$ and $u\succeq_{rhorpo}t$ for some $u \in \{s_1, s_2\}$

6. $s = f(\overline{s})$, $@(\overline{t})$ is an arbitrary left-flattening of t, and A

7. $s = @(s_1, s_2)$, s_1 is not of the form $@(X, \overline{w})$ with $X \in \mathcal{X}$, $@(\overline{t})$ is an arbitrary left-flattening of t and $\{s_1, s_2\}(\succ_{rhorpo})_{mul} \overline{t}$

8. $s = \lambda x : \alpha.u$, $t = \lambda x : \beta.v$, $\alpha =_{T_S} \beta$ and $u \underset{rhorpo}{\succ} v$

9. $s = @(\lambda x.u, v)$ and $u\{x \mapsto v\} \underset{rhorpo}{\succeq} t$

$$where \begin{cases} s \succeq_{rhorpo} t \text{ iff } s \succ_{rhorpo} t \text{ or } s =_\alpha t \text{ or} \\ \qquad\qquad s = @(X, u), t = @(X, v) \text{ and } u \succ_{rhorpo} v \\ A = \forall v \in \bar{t} \ s \succ_{rhorpo} v \text{ or } u \succeq_{rhorpo} v \text{ for some } u \in \bar{s} \end{cases}$$

Of course, making bound variables fit may need renaming in Case 8, and as usual, $\succ_{mul}$ and $\succ_{lex}$ denote respectively the multiset and lexicographic extensions of the relation $\succ$.

Theorem 3. $(\succ_{rhorpo})^+$ *is a higher-order reduction ordering satisfying schema-stability and schema-monotonicity.*

The relation $\succeq_{rhorpo}$ being non-transitive in general because of case 9, taking its transitive closure is needed to make it into an ordering.

The property of being a higher-order reduction ordering is inherited from the non-restricted version of the ordering, for which Cases 5 and 7 do not restrict the form of s_1 [12]. Without this restriction, we run into the aforementioned problem that any ordering satisfying monotonicity, compatibility and well-foundedness must violate stability. Note that the pairs which cause this violation do not become incomparable in the (quasi-) ordering: they are used to enrich its equality part.

Schema-monotonicity is straightforward, while schema-stability is by induction on the term structure. As a consequence of Theorems 2 and 3:

Theorem 4. $(\succ_{rhorpo})^*_n$ *is a normal higher-order reduction ordering.*

We will approximate $(\succ_{rhorpo})^*_n$ by $(\succ_{rhorpo})_n$ in all coming examples. As can be guessed, we need to define the precedence on the extended signature. In practice, we always make $\perp_\sigma$-function symbols small.

Example 1 (end). Let diff$_{new} >_\mathcal{F} \{\times_{new}, +_{new}, \cos, \perp\}$ and diff$_{new} \in$ Mul.
First rule: $s =$ diff$_{new}(\sin(@(F, \perp))) \succ_{rhorpo} \cos(@(F, \perp)) \times_{new} @(\text{diff}_{new}(@(F, \perp)), \perp)$
Applying first case 2, we recursively obtain two subgoals:
(i) $s \succ_{rhorpo} \cos(@(F, \perp))$ and (ii) $s \succ_{rhorpo} @(\text{diff}_{new}(@(F, \perp)), \perp)$.
(i): applying Case 2 yields $s \succ_{rhorpo} @(F, \perp)$ shown by Case 1 twice.
(ii): applying Case 6 generates two new subgoals
(iii) $s \succ_{rhorpo}$ diff$_{new}(@(F, \perp))$, which holds by case 3, then case 1.
(iv) diff$_{new}(\sin(@(F, \perp))) \succ_{rhorpo} \perp$, which holds by case 2.
Second rule: $s =$ diff$_{new}(@(@(F, \perp) \times_{new} @(F, \perp), \perp)) \succ_{rhorpo}$
$@(@(\text{diff}_{new}(@(F,\perp)), \perp) \times_{new} @(F,\perp), \perp) +_{new} @(@(F,\perp) \times_{new} @(\text{diff}_{new}(@(F,\perp)), \perp), \perp)$
Case 2 generates two subgoals:
(i) $s \succ_{rhorpo} @(@(\text{diff}_{new}(@(F, \perp)), \perp) \times_{new} @(F, \perp), \perp)$
(ii) $s \succ_{rhorpo} @(@(F, \perp) \times_{new} @(\text{diff}_{new}(@(F, \perp)), \perp), \perp)$.
By Case 6, (i) generates two new subgoals:
(iii) $s \succ_{rhorpo} @(\text{diff}_{new}(@(F, \perp)), \perp) \times_{new} @(F, \perp)$ and (iv) $s \succ_{rhorpo} \perp$.
The latter holds by case 1 and then case 5. By Case 2, (iii) yields two subgoals:
(v) $s \succ_{rhorpo} @(\text{diff}_{new}(@(F, \perp)), \perp)$ and (vi) $s \succ_{rhorpo} @(F, \perp)$.
By Case 6, (v) generates (vii) $s \succ_{rhorpo}$ diff$_{new}(@(F, \perp))$ and (viii) $s \succ_{rhorpo} \perp$
(vii) is solved by Case 3, 5, and 1 successively, and (viii) is solved by cases 1 and 5.

6   Examples

We present two complex examples proven terminating with $(\succ_{rhorpo})^*_n$. For all of them, we give the necessary ingredients for computing the appropriate neutralizations and comparisons. The precise computations can be found in the full version of the paper available from the web. An implementation is available for the original version of the ordering which will be extended to the present one in a near future.

As a convention, missing neutralization levels are equal to 0, in which case the corresponding subset of argument positions will be empty. In all examples, we use a simple type ordering $\geq_{T_s}$ equating all data types, which satisfies the requirements given in Section 5. Precedence on function symbols and statuses will be given in full.

In all examples we write $F(X)$ instead of $@(F, X)$ to ease the reading.

Example 2. The coming encoding of first-order prenex normal forms is adapted from [20], where its local confluence is proved via the computation of its (higher-order) critical pairs. Formulas are represented as λ-terms with sort *form*. The idea is that quantifiers are higher-order constructors binding a variable via the use of a functional argument.

$$S = \{form\}, \quad \mathcal{F} = \{ \wedge, \vee : form \times form \to form; \neg : form \to form; \\ \forall, \exists : (form \to form) \to form\}.$$

$$
\begin{array}{ll}
P \wedge \forall(\lambda x.Q(x)) \to \forall(\lambda x.(P \wedge Q(x))) & P \wedge \exists(\lambda x.Q(x)) \to \exists(\lambda x.(P \wedge Q(x))) \\
\forall(\lambda x.Q(x)) \wedge P \to \forall(\lambda x.(Q(x) \wedge P)) & \exists(\lambda x.Q(x)) \wedge P \to \exists(\lambda x.(Q(x) \wedge P)) \\
P \vee \forall(\lambda x.Q(x)) \to \forall(\lambda x.(P \vee Q(x))) & P \vee \exists(\lambda x.Q(x)) \to \exists(\lambda x.(P \vee Q(x))) \\
\forall(\lambda x.Q(x)) \vee P \to \forall(\lambda x.(Q(x) \vee P)) & \exists(\lambda x.Q(x)) \vee P \to \exists(\lambda x.(Q(x) \vee P)) \\
\neg(\forall(\lambda x.Q(x))) \to \exists(\lambda x.\neg(Q(x))) & \neg(\exists(\lambda x.Q(x))) \to \forall(\lambda x.\neg(Q(x)))
\end{array}
$$

Ingredients for neutralization: $\mathcal{L}^1_\forall = 1, \mathcal{L}^1_\exists = 1, \mathcal{A}^1_\forall = \{\}, \mathcal{A}^1_\exists = \{\}$.
Precedence: $\wedge >_{\mathcal{F}} \{\forall_{new}, \exists_{new}\}, \vee >_{\mathcal{F}} \{\forall_{new}, \exists_{new}\}, \neg >_{\mathcal{F}} \{\forall_{new}, \exists_{new}\}$.
Statuses: $\forall_{new}, \exists_{new} \in \mathsf{Mul}$ □

Example 3. Encoding of natural deduction, taken from [6].
Let $S = \{o, c : * \times * \to *\}$. Because we did not consider polymorphism, the following signature and rules is parameterized by all possible types $\sigma, \tau, \rho \in T_S$.

$$
\mathcal{F} = \{ app_{\sigma,\tau} : (\sigma \to \tau) \times \sigma \to \tau; \; abs_{\sigma,\tau} : (\sigma \to \tau) \to (\sigma \to \tau); \\
\Pi_{\sigma,\tau} : \sigma \times \tau \to c(\sigma,\tau); \; \Pi^0_{\sigma,\tau} : c(\sigma,\tau) \to \sigma; \; \Pi^1_{\sigma,\tau} : c(\sigma,\tau) \to \tau; \\
\exists^+_\sigma : o \times \sigma \to c(o,\sigma); \; \exists^-_{\sigma,\tau} : c(o,\sigma) \times (o \to \sigma \to \tau) \to \tau \; \}.
$$

$$
\mathcal{X} = \{ X : \sigma; \; Y : \tau; \; Z : o; \; T : c(o,\rho); \; F : \sigma \to \tau; \; G : o \to \sigma \to \tau, \\
H : o \to \rho \to (\sigma \to \tau), I : o \to \rho \to c(\sigma,\tau), J : o \to \rho \to c(o,\sigma)\}.
$$

$$
\begin{array}{c}
app_{\sigma,\tau}(abs_{\sigma,\tau}(F), X) \to F(X) \\
\Pi^0_{\sigma,\tau}(\Pi_{\sigma,\tau}(X,Y)) \to X \\
\Pi^1_{\sigma,\tau}(\Pi_{\sigma,\tau}(X,Y)) \to Y \\
\exists^-_{\sigma,\tau}(\exists^+_\sigma(Z,X), G) \to G(Z,X)
\end{array}
$$

$$app_{\sigma,\tau}(\exists^{-}_{\rho,\sigma\to\tau}(T,H),X) \to \exists^{-}_{\rho,\tau}(T,\lambda x : o\,y : \rho.app_{\sigma,\tau}(H(x,y),X))$$
$$\Pi^0_{\sigma,\tau}(\exists^{-}_{\rho,c(\sigma,\tau)}(T,I)) \to \exists^{-}_{\rho,\tau}(T,\lambda x : o\,y : \rho.\Pi^0_{\sigma,\tau}(I(x,y)))$$
$$\Pi^1_{\sigma,\tau}(\exists^{-}_{\rho,c(\sigma,\tau)}(T,I)) \to \exists^{-}_{\rho,\tau}(T,\lambda x : o\,y : \rho.\Pi^1_{\sigma,\tau}(I(x,y)))$$
$$\exists^{-}_{\sigma,\tau}(\exists^{-}_{\rho,c(o,\sigma)}(T,J),G) \to \exists^{-}_{\rho,\tau}(T,\lambda x : o\,y : \rho.\exists^{-}_{\sigma,\tau}(J(x,y),G))$$

Neutralization: $\mathcal{L}^2_{\exists^{-}_{\sigma,\tau}} = 2$ and $\mathcal{A}^2_{\exists^{-}_{\sigma,\tau}} = \{1\}$ for all possible types σ and τ.

Precedence: $\{app_{\sigma,\tau}, \Pi^0_{\sigma,\tau}, \Pi^1_{\sigma,\tau}\} >_{\mathcal{F}} \exists^{-}_{new\ \rho,\tau}$ and $\exists^{-}_{new\ \rho,\tau} = \exists^{-}_{new\ \sigma,\tau}$ for all possible types ρ, σ and τ.

Statuses: $\exists^{-}_{new\ \sigma,\tau} \in$ Lex and $app_{\sigma,\tau}, \Pi^0_{\sigma,\tau}, \Pi^1_{\sigma,\tau} \in$ Mul for all $\rho\,\sigma$ and τ;   □

7 Conclusion

Proving termination properties of Nipkow's rewriting was considered in [8] and [2]. The former yields a *methodology* needing important user-interaction to prove that the constructed ordering has the required properties. Here, our method does provide with an ordering having automatically all desired properties. The user has to provide with a precedence and statuses as usual with the recursive path ordering. He or she must also provide with neutralization levels together with filters selecting appropriate arguments for each function symbols. This requires of course some expertise, but can be implemented by searching non-deterministically for appropriate neutralization levels and filters, as done in many implementations of the recursive path ordering for the precedence and statuses.

The higher-order recursive path ordering generalizes the notion of general schema as formulated in [3] where the notion of computability closure was introduced. However, what can be done with the schema can be done with the higher-order recursive path ordering when using the computability closure of $f(\overline{t})$ in the subterm case, instead of simply the set of subterms $\overline{t}$ itself. The general definition of the higher-order recursive path ordering with closure is given in [12]. It is however interesting to notice that the neutralization mechanism is powerful enough so as to dispense us with using the closure for all these complex examples taken from the literature that we have considered here and in [13]. It remains to be seen whether the closure plays in the context of normal higher-order rewriting, a role as important as for proving termination of recursor rules for inductive types for which plain pattern matching is used instead of higher-order pattern matching.

References

1. F. Blanqui, J.-P. Jouannaud, and M. Okada. The Calculus of Algebraic Constructions. In Narendran and Rusinowitch, Proc. RTA'99, 1999.
2. F. Blanqui. Termination and Confluence of Higher-Order Rewriting Systems. In Proc. RTA'00, 2000.
3. F. Blanqui, J.-P. Jouannaud, and M. Okada. Inductive Data Type Systems. *Theoretical Computer Science*, 272(1-2):41–68. 2002.
4. H. Barendregt. Functional Programming and Lambda Calculus. In [22], pages 321–364.

5. H. Barendregt. *Handbook of Logic in Computer Science*, chapter Typed lambda calculi. Oxford Univ. Press, 1993. eds. Abramsky et al.
6. J. van de Pol. Termination of Higher-Order Rewrite Systems. PhD thesis, Department of Philosophy, Utrecht University, 1996.
7. N. Dershowitz. Orderings for term rewriting systems. *Theoretical Computer Science*, 17(3):279–301, March 1982.
8. J. van de Pol and H. Schwichtenberg. Strict functional for termination proofs. In *Typed Lambda Calculi and Applications, Edinburgh*. Springer-Verlag, 1995.
9. N. Dershowitz and J.-P. Jouannaud. Rewrite systems. In [22], pages 321–364.
10. J.-P. Jouannaud. Higher-Order rewriting: Framework, Confluence and termination. In A. Middeldorp, V. van Oostrom, F. van Raamsdonk and R. de Vrijer eds., *Processes, Terms and Cycles: Steps on the road to infinity*. Essays Dedicated to Jan Willem Klop on the occasion of his 60th Birthday. LNCS 3838. Springer Verlag, 2005.
11. J.-P. Jouannaud and A. Rubio. The higher-order recursive path ordering. In Giuseppe Longo, editor, *Fourteenth Annual IEEE Symposium on Logic in Computer Science*, Trento, Italy, July 1999. IEEE Comp. Soc. Press.
12. J.-P. Jouannaud and A. Rubio. Polymorphic Higher-Order Recursive Path Orderings. 2005. Submitted to JACM. http://www.lix.polytechnique.fr/Labo/Jean-Pierre.Jouannaud.
13. J.-P. Jouannaud and A. Rubio. Higher-Order Orderings for Normal Rewriting. 2005. Full version. http://www.lix.polytechnique.fr/Labo/Jean-Pierre.Jouannaud.
14. J.-P. Jouannaud and A. Rubio. Higher-Order Recursive Path Orderings à la carte. 2001. http://www.lix.polytechnique.fr/Labo/Jean-Pierre.Jouannaud.
15. J.-P. Jouannaud, F. van Raamsdonk and A. Rubio Higher-order rewriting with types and arities. 2005. http://www.lix.polytechnique.fr/Labo/Jean-Pierre.Jouannaud.
16. J. W. Klop. Combinatory Reduction Relations. Mathematical Centre Tracts 127. Mathematisch Centrum, Amsterdam, 1980.
17. J. W. Klop. Term Rewriting Systems. In S. Abramsky, D.M. Gabbay, and T.S.E. Maibaum eds., *Handbook of Logic in Computer Science*, vol. 2:2–116. Oxford University Press, 1992.
18. R. Mayr and T. Nipkow. Higher-order rewrite systems and their confluence. *Theoretical Computer Science*, 192(1):3–29, February 1998.
19. D. Miller. A Logic Programming Language with Lambda-Abstraction, Function Variables, and Simple Unification. In *Journal and Logic and Computation* 1(4):497–536, 1991.
20. T. Nipkow. Higher-order critical pairs. In *6th IEEE Symp. on Logic in Computer Science*, pages 342–349. IEEE Computer Society Press, 1991.
21. L. C. Paulson. Isabelle: the next 700 theorem provers. In P. Odifreddi, editor, *Logic and Computer Science*. Academic Press, 1990.
22. J. van Leeuwen, ed. Handbook of Theoretical Computer Science, vol. B. North-Holland, 1990.

Bounded Second-Order Unification
Is NP-Complete[*]

Jordi Levy[1], Manfred Schmidt-Schauß[2], and Mateu Villaret[3]

[1] IIIA, CSIC, Campus de la UAB, Barcelona, Spain
http://www.iiia.csic.es/~levy
[2] Institut für Informatik, Johann Wolfgang Goethe-Universität,
Postfach 11 19 32, D-60054 Frankfurt, Germany
http://www.ki.informatik.uni-frankfurt.de/persons/schauss/schauss.html
[3] IMA, UdG, Campus de Montilivi, Girona, Spain
http://ima.udg.es/~villaret

Abstract. Bounded Second-Order Unification is the problem of deciding, for a given second-order equation $t \stackrel{?}{=} u$ and a positive integer m,
whether there exists a unifier σ such that, for every second-order variable F, the terms instantiated for F have at most m occurrences of every
bound variable.

It is already known that Bounded Second-Order Unification is decidable and NP-hard, whereas general Second-Order Unification is undecidable. We prove that Bounded Second-Order Unification is NP-complete,
provided that m is given in unary encoding, by proving that a size-minimal
solution can be represented in polynomial space, and then applying a
generalization of Plandowski's polynomial algorithm that compares com-
pacted terms in polynomial time.

1 Introduction

Second-order unification (SOU) is a generalization of first-order unification,
where variables are permitted also at the position of function symbols, hence
they may have arguments. These variables are also called second-order variables. When solving an equation, the second-order variables can stand for an
arbitrary first-order term with holes for plugging in the arguments, which must
be terms. In lambda-notation, a second-order variable may be instantiated by a
term $\lambda x_1 . \cdots \lambda x_n . t$, where t is a first-order term, and the variables x_i also stand
for first-order terms. SOU extends the expressivity of first-order unification, and
is a restriction of higher-order unification (see [6,3]). It is known that SOU is
undecidable [5], even under severe syntactic restrictions [4,20,10,11].

A decidable variant is bounded second-order unification (BSOU) [17], which
restricts the possible instantiations of second-order variables by limiting the
number of occurrences of bound variables. However, the described algorithm
for BSOU has non-elementary complexity. Recently, we described an improved

---

[*] This research has been partially founded by the CICYT research projects iDEAS
(TIN2004-04343) and Mulog (TIN2004-07933-C03-01).

F. Pfenning (Ed.): RTA 2006, LNCS 4098, pp. 400–414, 2006.

algorithm for monadic SOU [9] —which is BSOU where only unary function symbols and constants are permitted— and determined its complexity to be NP-complete.

In this paper we apply and extend methods used in [9] for monadic SOU to improve the BSOU algorithm by compressing the computed solution, and as a main result we prove that BSOU is in NP, which means that it is NP-complete. To obtain this result requires compression techniques and, as a basis, the BSOU-algorithm in [17]. This result shows that BSOU may become a practically useful restriction of higher-order unification, perhaps via using a SAT-solver.

To illustrate the difficulties in proving the membership of BSOU in NP, we will compare this problem with other unification problems. Most general first-order unifiers σ have a very nice property: for every solvable equation E and variable X_i there exists a subterm t_i of the equation E such that σ can be decomposed in the form $\sigma = [X_1 \mapsto t_1] \circ \cdots \circ [X_n \mapsto t_n]$. This representation is polynomial and ensures that the problem is in NP. In [13] it is proved that the search of these subterms can be done very efficiently and the problem is in fact linear. Well-nested context unifiers [8] —this is, context unifiers where instances of variables do not overlap— have the same property, but replacing subterms t_i by subcontexts c_i of the equation. This property is used to prove that well-nested context unification is in NP. However, in this case the search of these subcontexts cannot be done efficiently. The property held by these two problems suggested us to represent substitutions as compositions of instantiations to save space. In monadic SOU we have a weaker property: instead of just one subcontext, we need to compose a bounded number of subcontexts, and in some cases raise them to an exponent. Moreover, we get the instance of *only one* of the variables $[X_1 \mapsto t_1]$. This means that we have to use the same property applied to $[X_1 \mapsto t_1] E$ to find the instance of another variable. Notice that $[X_1 \mapsto t_1] E$ may be bigger than E, and the size of the instance of X_n could be exponential in n. In [9] it is proved that this is not the case, if we represent such an instance using a context free grammar (CFG). If we have a CFG generating E, to represent a subcontext of E, we have to enlarge the grammar, and in the worst case to duplicate the size, obtaining an exponential representation. To avoid this problem in [9] we propose the conjugation of size and *depth* of the grammar —the depth of the parse tree—, which has an effect similar to balancing conditions. Then, the representation of a subcontext preserves the depth and requires to increase the size of the grammar only on the depth (see Lemma 4). Showing PSPACE [16] as an upper complexity bound for stratified context unification used an ad-hoc compression technique composed of sharing and exponentiation. The algorithm given there does not look for a polynomial-sized solution, and erases partial solution as early as possible to keep the PSPACE-bound.

Compared with monadic SOU, the situation in BSOU is even worse. Given an equation E we can only find a "partial" instance of some variable. This means that we remove a variable, say X, but we have to introduce new variables, say X', by instantiating $[X \mapsto c[X'[\bullet]]]$ where c is a context composed by a bounded number of subcontexts of E. Moreover, this context is not ground, contrarily to the monadic SOU case. Fortunately, we have a well-founded ordering where $[X \mapsto c[X'[\bullet]]] E$ is smaller than E.

This paper proceeds as follows. After some preliminaries, we define an extension of singleton CFG for trees in Section 3. In Section 4 we define an order on equations and show a polynomial upper bound for the length of decreasing sequences. Then, in Section 5 we prove that given an equation E, and a size-minimal solution σ, we can find a polynomial-sized partial description ρ of σ, such that $\rho(E)$ is strictly smaller than E and $\sigma = \sigma' \circ \rho$. Finally, in Section 6 we show how we can get a compact representation of these partial instantiations, and represent σ in polynomial size. Using an extension of Plandowski's [14,15] result for CFG, we can check in polynomial time if a substitution in such a representation is a solution, proving that BSOU is in NP.

2   Preliminary Definitions

We consider one base (first-order) type o, and second-order types described by the syntax $\tau ::= o \to o \mid o \to \tau$, with the usual convention that $\to$ is associative to the right. We deal with a *signature* $\Sigma = \bigcup_{i \geq 0} \Sigma_i$, where *constants* of Σ_i are i-ary, and a set of *variables* $\mathcal{X} = \bigcup_{i \geq 0} \mathcal{X}_i$, where variables of $\mathcal{X}_i$ are also i-ary. Variables of $\mathcal{X}_0$ are therefore first-order typed and those of $\mathcal{X}_i$, with $i \geq 1$, are second-order typed, and similarly for Σ. We use the convention that X, Y (possibly with primes and subindexes) mean free first-order or second-order variables (unknowns), while constants are denoted by lower-case letters $a, b, \ldots$, for first-order, and $f, g, \ldots$, for second-order ones.

Terms are built as usual in simply typed λ-calculus. We assume that they are in $\beta\eta$-long normal form, or are immediately reduced, so we will use a first-order notation, if possible. We denote terms with lower case letters like $t, u, \ldots$.

Contexts are first-order typed terms with *one hole* at some position, notated as $\bullet$. We call *Z-contexts* to the union of first-order terms and contexts, hence they may contain zero or one hole. We denote contexts and Z-contexts by lower case letters: $c, d, \ldots$ for contexts and $c, d, \ldots, t, u, \ldots$ for Z-contexts. If the Z-context d is plugged into the hole of a Z-context c, we denote the result as the Z-context $c[d]$. (In the special case that c is a term, $c[d] = c$). We sometimes abbreviate $c_1[c_2[c_3 \ldots]]$ as $c_1 c_2 c_3 \ldots$ and $c[c[c \cdot^n \cdot]]$ as c^n. For any pair of Z-contexts c_1 and c_2, if for some Z-context d we have $c_1 = c_2[d]$, then c_2 is said to be a *prefix* of c_1 (notated $c_2 \preceq c_1$ and $c_2 \prec c_1$ for strict prefixes), and if for some context d (with hole) we have $c_1 = d[c_2]$, then c_2 is said to be a *suffix* of c_1. Notice that, if c_2 is a suffix of c_1, then c_1 contains a hole iff c_2 contains a hole. On the contrary, a *subterm* u of a context c does not need to contain a hole. This distinguishes a suffix from a subterm. If c is a prefix of a subterm of d, then c is called a *subcontext* of d. The *size* of a Z-context c is denoted $|c|$, and defined as its number of symbols (including the hole).

We use *positions* in terms, noted p, q, as sequence of positive integers following Dewey's notation. The empty word is notated ϵ, $p \prec q$ notates the prefix relation, $p \cdot q$ the concatenation, and $t|_p$ the subterm at position p of t. For a context c, its *main path* is the position of the hole. A position p is in the main path of c if p is a prefix of the main path of c.

Second-order *substitutions* are functions from terms to terms, defined as usual. The application of a substitution σ to a term t is written $\sigma(t)$. An instance of the bounded second-order problem (BSOU) is an *equation* $t \overset{?}{=} u$, where t and u are first-order terms, and a number m given in unary encoding. The set of variables (unknowns) occurring in an equation E is denoted by $Var(E)$, and the notational size by $|E|$. We assume that equations are symmetric. A substitution σ is said to be a *unifier* of $(t \overset{?}{=} u, m)$, iff $\sigma(t) = \sigma(u)$, and for all $X \in Var(E)$ every bound variable in $\sigma(X)$ occurs at most m times. A unifier σ is said to be a *solution* of (E, m), iff $\sigma(t)$ and $\sigma(u)$ are ground (do not contain free variables).

It is easy to see that it suffices to consider only unifiers and solutions built from constant, and function symbols that occur in E. A solution σ of $(t \overset{?}{=} u, m)$ is said to be *size-minimal* if it minimizes $|\sigma(t)|$ among all solutions of $(t \overset{?}{=} u, m)$.

As already shown in [17], there is an NP-reduction of BSOU to the specialized problem, where $m = 1$, and every second-order variable is unary. Hence in the following, we will only treat this case. In the simplification of the problem we go a step further by considering only second-order variables. To do so we can replace all occurrences of the first-order variable X by the term $X'(a)$ where X' is a fresh (unary) second-order variable and a is any 0-ary constant. This transformation allow us to P-reduce BSOU to BSOU without first-order variables. Therefore, from now on, all variables will have type $o \to o$, and all terms type o, or $o \to o$. Moreover, we will represent second-order typed terms $\lambda y . t$ as the Z-context resulting from replacing in t the occurrence of y (if any) by the hole. Thus, from now on, we will only deal with Z-contexts, and terms will be assumed to be first-order typed.

We know that size-minimal solutions of a BSOU equation satisfy the exponent of periodicity lemma [12,7,19,17]. However, since we have a slightly different definition of size-minimality, after some encoding by enlarging E, we have a quadratic dependency on $|E|$:

Lemma 1 ([17], Lemma 4.1). *There exists a constant $\alpha \in \mathbb{R}$ such that, for every BSOU-problem E, every size-minimal unifier σ, and every variable X, if d^n is a nonempty subcontext of $\sigma(X)$, then $n \le 2^{\alpha|E|^2}$.*

3   Singleton Tree Grammars

We generalize singleton context free grammars (SCFG) to trees, since we require a device for a compressed representation of solutions. We extend the expressivity of SCFGs by permitting terms and contexts. The definition is a special case of the context free tree grammars defined in [2].

Definition 1. *A singleton tree grammar (STG) is a tree grammar, i.e. a 4-tuple $(\mathcal{TN}, \mathcal{CN}, \Sigma, R)$, where $\mathcal{TN}$ are tree nonterminals, and $\mathcal{CN}$ are context nonterminals, and Σ is a signature of terminals symbols (variables and constants), such that the sets $\mathcal{TN}, \mathcal{CN}, \Sigma$ are pairwise disjoint. The set of nonterminals $\mathcal{N}$ is defined as $\mathcal{N} = \mathcal{TN} \cup \mathcal{CN}$. The rules in R may be of the form:*

- $A ::= f(A_1, \ldots, A_n)$, where $A, A_i \in \mathcal{TN}$, and $f \in \Sigma$ is an n-ary terminal symbol.
- $A_1 ::= C[A_2]$ where $A_1, A_2 \in \mathcal{TN}$, and $C \in \mathcal{CN}$.
- $C_1 ::= C_2 C_3$, where $C_i \in \mathcal{CN}$.
- $C ::= f(A_1, \ldots, A_{i-1}, [\bullet], A_{i+1}, \ldots, A_n)$, where $A_i \in \mathcal{TN}$, $C \in \mathcal{CN}$, $[\bullet]$ is the hole, and $f \in \Sigma$ an n-ary terminal symbol.

The tree grammar must be non-recursive (the relation $\xrightarrow{+}$ has no cycles). Furthermore, for every non-terminal N there is exactly one rule having N as left hand side. Give a term t where nonterminals may occur, the derivation by G is an exhaustive iterated replacement of the nonterminals by the corresponding right hand sides.

Definition 2. The size of a grammar (STG) G is the number of its rules and denoted as $|G|$.
The depth of a nonterminal D is defined as the maximal number of $\rightarrow_G$-steps from D, where $D' \rightarrow_G D''$ for two nonterminals D', D'', iff $D' ::= T$ is a rule of G, and D'' occurs in T.
The depth of a grammar is the maximum of the depths of all nonterminals.
When a grammar G generates a Z-context t from a non-terminal symbol D (and the grammar is clear from the context) we write $depth(t)$ to denote $depth(D)$.

The following theorem is a generalization to trees of Plandowski's one in [14,15].

Theorem 1 ([1,18]). Given an STG G, and two tree nonterminals from G, it is decidable in polynomial time depending on $|G|$ whether they generate the same tree or not.

The following lemmas state how the size and the depth of a grammar are increased by extending it with concatenations, exponentiation, prefixes and suffixes of Z-contexts. Proofs may be adapted from the extended version of [9].

Lemma 2. Let G be an STG defining the Z-contexts $d_1, \ldots, d_n$ for $n \geq 1$. Then there exists an STG $G' \supseteq G$ that defines the Z-context $d_1 \cdots d_n$ and satisfies $|G'| \leq |G| + n - 1$ and $depth(d_1 \cdots d_n) \leq \max\{depth(d_1), \ldots, depth(d_n)\} + \lceil \log n \rceil$.

Lemma 3. Let G be an STG defining the context d. For any $n \geq 1$, there exists an STG $G' \supseteq G$ that defines the context d^n and satisfies $|G'| \leq |G| + 2 \lfloor \log n \rfloor$ and $depth(d^n) \leq depth(d) + \lceil \log n \rceil$.

Lemma 4. Let G be an STG defining the context d. For any nontrivial prefix or suffix context d' of d, there exists an STG $G' \supseteq G$ that defines d' and satisfies $|G'| \leq |G| + depth(d)$ and $depth(d') \leq depth(d)$.
Similarly if d is a Z-context and d' is a subterm of d.

Lemma 5. Let G be an STG defining the term t. For any nontrivial prefix context d of the term t, there exists an STG $G' \supseteq G$ that defines d and satisfies $|G'| \leq |G| + 2 \, depth(t) \, (\log(depth(t)) + 1)$ and $depth(d) \leq depth(t) + 2 + \log(depth(t))$.

Notice that for prefixes of contexts we get better bounds than for prefixes of terms.

4   A Well-Founded Ordering on Equations

In this section we define an ordering on the equations. This order is similar to the one proposed in [17] to prove the decidability of BSOU. However, in our case, the order is not only well-founded: we prove that the length of any strictly decreasing sequence is polynomially bounded on the size of the first element.

Definition 3. *We say that p is a* surface position *of t if there are no variable occurrences strictly above p.*

Given an equation $E = (t \overset{?}{=} u)$, the relation $\approx_E \subseteq Var(E) \times Var(E)$ is the reflexive-transitive closure of the relation defined by: if X occurs at the surface position p of t and Y occurs at the same surface position p of u, then $X \approx_E Y$.

The relation $\succ_E \subseteq Var(E) \times Var(E)$ is the relation defined by: if X occurs at the surface position p of t and, for some nonempty sequence q, Y occurs at the surface position $p \cdot q$ of u, then $X \succ_E Y$. We extend this relation to classes of equivalences with if $X \succ_E Y$ then $\overline{X} \succ_E \overline{Y}$.

If p is a surface position of t and of u, then $t|_p \overset{?}{=} u|_p$ is called a subequation *of $t \overset{?}{=} u$.*

In first-order unification all variable occurrences are at surface positions. Moreover, if $\succ_E^+$ is not irreflexive then there is occur-check and the equation is unsolvable. In second-order unification this is not the case, $\succ_E^+$ may be not irreflexive and E solvable.

Definition 4. *A* cycle *in an equation $E = (t \overset{?}{=} u)$ is a sequence of variables $X_1, \ldots, X_n$ and pairs of positions $\langle p_1, p_1 \cdot q_1 \rangle, \ldots, \langle p_n, p_n \cdot q_n \rangle$, such that, for $i = 1, \ldots, n$, X_i is at the surface position p_i of t, and X_{i+1} is at the surface position $p_i \cdot q_i$ of u, and there is at least one nonempty q_i.[1]*

The length *of the cycle is n.*

Notice that an equation E contains a cycle iff the relation $\prec_E^+$ for classes of equivalences is not irreflexive. The shortest cycle in an equation E is shorter than $|Var(E)|$.

Definition 5. *Given an equation E, the* measure *$\mu(E)$ is a lexicographic combination $\langle \mu_1(E), \mu_2(E), \mu_3(E) \rangle$ of the following components:*

1. *$\mu_1(E) = |Var(E)|$ is the number of variables occurring in E.*
2. *$\mu_2(E)$ is the length of the shortest cycle in E, or ∞ if there are no cycles.*
3. *$\mu_3(E)$ is zero, if E contain cycles, otherwise*

$$\mu_3(E) = |Var(E)| - |Var(E)/\approx_E| + 2|\succ_E| = \sum_{C \,\in\, Var(E)/\approx_E} (|C| - 1) \;+\; \sum_{\substack{X, Y \in Var(E) \\ X \succ_E Y}} 2$$

---

[1] When the length n of the cycle is clear from the context, all indexes i greater than n are replaced by $((i - 1) \mod n) + 1$.

Lemma 6. *Any decreasing sequence of equations $\{E_i\}_{i\geq 1}$, i.e. where $\mu(E_i) > \mu(E_{i+1})$, terminates in at most $2\,|\,Var(E_1)|^3$ steps.*

PROOF: Let $n = |Var(E_1)|$. The first component of $\mu(E_i)$ can have values from $j = n, \ldots, 1$. When the first component is j, the second component can have values from $\infty, j, \ldots, 1$. When there are no cycles, the third component is maximal when all the equivalence classes are singletons, and is $j(j-1)$. Therefore, the set of possible values of $\mu(E_i)$ is smaller than $\sum_{j=1}^{n} j + j\,(j-1) + 1 = 1/3n^3 + 1/2n^2 + 7/6n \leq 2n^3$. ∎

5   Finding the Partial Instance of Some Variable

In this section we show how, given an equation E and a minimal solution σ, we can find an instantiation $[X \mapsto t]$ or a partial instantiation $[X \mapsto c[X'(\bullet)]]$ for every variable $X \in Var(E)$ such that the composition ρ of all them satisfies $\sigma = \sigma' \circ \rho$, where σ' is a size-minimal solution of $\rho(E)$, and the new equation $\rho(E)$ is smaller than E w.r.t. μ. Moreover the (partial) instantiation can be built up from a linear number of pieces (subcontexts) of E, which as we show in the next section, ensures that it can be efficiently represented.

Lemma 7 (Partial instance). *Given an equation E and a size-minimal solution σ, with exponent of periodicity bounded by e, there exist substitutions $\rho = \rho_2 \circ \rho_1$ such that the ρ_i's have the form*

$$[X_1 \mapsto c_1[X_1'(\bullet)]\,, \ldots,\ X_n \mapsto c_n[X_n'(\bullet)]]$$

such that:

1. *$n \leq |\,Var(E)|$,*
2. *$X_1', \ldots, X_n'$ are fresh variables not occurring in E,*
3. *the Z-contexts c_i can be constructed taking $\mathcal{O}(n)$-many subcontexts of E [or of $\rho_1(E)$ in the case of ρ_2], composing them, raising the result to some exponent smaller than e and taking a prefix,*
4. *ρ is coherent with σ, i.e. σ decomposes as $\sigma = \sigma' \circ \rho$, for some σ', and*
5. *$\mu(E) > \mu(\rho(E))$.*

Remark 1. Notice that Lemma 7 and 6 allow us to decompose $\sigma = \rho_m \circ \cdots \circ \rho_1$, where m is polynomial on the size of the original equation E, and ρ_i can be represented polynomially on the size of $\rho_{i-1} \circ \cdots \circ \rho_1(E)$ using singleton tree grammars. From this we can only conclude that σ has a representation bounded by a composition of a polynomial number of polynomials, i.e. that σ has an exponential-size representation. Obviously, this is not enough for proving the NP-completeness of BSOU. We need an important result that will be proved in Section 6.

Lemma 7 is proved in the following subsections. We also need the following Lemma.

Lemma 8. *If σ is a size-minimal solution of E, and σ decomposes as $\sigma = \sigma' \circ \rho$, then σ' is a size-minimal solution of $\rho(E)$.*

5.1   There Are Cycles in the Set of Equations

If $E = (t \overset{?}{=} u)$ contains a cycle defined by $X_1, \ldots, X_n$ and $\langle p_1, p_1 \cdot q_1 \rangle, \ldots, \langle p_n, p_n \cdot q_n \rangle$, then, for every $i = 1, \ldots, n$, we have a subequation $t|_{p_i} \overset{?}{=} u|_{p_i}$ of the form

$$X_i(v_i) \overset{?}{=} c_i[X_{i+1}(w_i)]$$

for some terms v_i and w_i, and some context c_i that has its hole at position q_i and has no variables in its main path. Note that there is at least one context c_i different from $\bullet$. The unifier σ of $t \overset{?}{=} u$ has to solve all these subequations.

Now we find how long each variable "stays" in the cycle: For $i = 1, \ldots, n$, let r_i be the longest prefix of $(q_i \cdot \ldots \cdot q_n \cdot q_1 \cdot \ldots \cdot q_{i-1})^\infty$ such that, if $\sigma(X_i)$ has no hole, then r_i is a position inside the term $\sigma(X_i)$ and, if $\sigma(X_i)$ has a hole, then this hole must be below or at position r_i.

We select a minimal r_i: Let $minlength = \min_{i \in \{1, \ldots, n\}} |r_i|$, and assume w.l.o.g. that r_1 is minimal, i.e. $minlength = |r_1|$.

We make all variables copy along this distance: For $i = 1, \ldots, n$, let s_i be the prefix of $(q_i \cdot \ldots \cdot q_n \cdot q_1 \cdot \ldots \cdot q_{i-1})^\infty$ of length $minlength$, and let d_i be the context resulting from putting a hole at position s_i of $(c_i \ldots c_n c_1 \ldots c_{i-1})^\infty$. Note that, since the exponent of periodicity of σ does not exceed e, then d_i has the form $(c_i \ldots c_n c_1 \ldots c_{i-1})^{e_i} d_i'$ where $e_i \leq e$ and the context d_i' is a prefix of $c_i \ldots c_n c_1 \ldots c_{i-1}$.

Since d_i is a prefix of $\sigma(X_i)$, the substitution $\rho_1 = [X_1 \mapsto d_1[X_1'(\bullet)], \ldots, X_n \mapsto d_n[X_n'(\bullet)]]$ is coherent with σ. Moreover, the sequences $X_1', \ldots, X_n'$ and $\langle p_1 \cdot s_1, p_1 \cdot q_1 \cdot s_2 \rangle$, $\langle p_2 \cdot s_2, p_2 \cdot q_2 \cdot s_3 \rangle$, $\ldots$, $\langle p_n \cdot s_n, p_n \cdot q_n \cdot s_1 \rangle$ define a cycle in $\rho_1(E)$ of the same length as the original cycle. Now, we define a new substitution ρ_2 such that $\rho = \rho_2 \circ \rho_1$ is coherent with σ and $\mu(\rho(E)) < \mu(E)$. There are three cases:

Case 1: If $\sigma(X_1)$ does not contain any hole, then r_1 corresponds to the position of a first-order constant in $\sigma(X_1)$. Since r_1 and $q_1 \cdot r_2$ are both prefixes of $(q_1 \cdot \ldots \cdot q_n)^\infty$ and $|r_1| \leq |r_2|$, r_1 is a prefix of $q_1 \cdot r_2$. Since σ solves $X_1(v_1) = c_1[X_2(w_1)]$, and r_2 is a position inside $\sigma(X_2)$, $\sigma(X_2)$ has a first-order constant at position r_2 and $r_1 = q_1 \cdot r_2$. Therefore, since $|r_1| \leq |r_2|$, we have $q_1 = \epsilon$ and $c_1 = \bullet$. Thus, $|r_2| = minlength$ and $\sigma(X_2)$ also has a constant at position r_2. Repeating this argument we would get $c_i = \bullet$, for every $i = 1, \ldots, n$, which contradicts the assumption that we have a cycle (for some i, $c_i \neq \bullet$). Therefore this situation is not possible.

Case 2: If for some $i = 1, \ldots, n$, s_i corresponds to the position of the hole in $\sigma(X_i)$ then take $\rho_2 = [X_i' \mapsto \bullet]$. The variable X_i is completely instantiated, and the first component of μ decreases. This situation corresponds to some variable that "finishes inside the cycle, i.e. it is completely instantiated".

Case 3: Otherwise, r_1 corresponds to some *proper* prefix of the hole position of $\sigma(X_1)$. Let m be the minimal index such that $c_1 = \cdots = c_{m-1} = \bullet$ and $c_m \neq \bullet$. Notice that $q_1, \ldots, q_{m-1}$ are empty, $r_1 = s_1 = \cdots = s_{m-1}$, and, for $j = 2, \ldots m - 1$, $r_1 \prec r_j$ strictly. For $i = 1, \ldots, m - 1$, let $l_i \in \mathbb{N}$ satisfy: the hole of $\sigma(X_i)$ is below or at $r_1 \cdot l_i$, if $\sigma(X_i)$ has a hole, or $l_i = 1$, otherwise.

Let $l \in \mathbb{N}$ satisfies $r_1 \cdot l \preceq q_m \cdot s_{m+1}$. The equation $\rho_1(E)$ contains as subequations $\{X_1'(\rho_1(v_1)) \overset{?}{=} X_2'(\rho_1(w_1)), \cdots, X_{m-1}'(\rho_1(v_{m-1})) \overset{?}{=} X_m'(\rho_1(w_{m-1})),$ $X_m'(\rho_1(v_m)) \overset{?}{=} \rho_1(c_m[d_{m+1}[X_{m+1}'(w_m)]]|_{r_1})\}$ where r_1 is a proper prefix of the main path of $c_m[d_{m+1}[\bullet]]$, i.e. $r_i \prec q_m \cdot s_{m+1}$. Let f be the constant at the root of $c_m[d_{m+1}[X_{m+1}'(w_m)]]|_{r_1}$. We take

$$\rho_2 = [X_i' \mapsto f(w_i^1, \ldots, w_i^{l_i-1}, X_i''(\bullet), w_i^{l_i+1}, \ldots, w_i^{arity(f)})]_{i \in \{1,\ldots,m\}}$$

where, for $k \neq l_m$, $w_m^k = \rho_1(c_m[d_{m+1}[X_{m+1}'(w_m)]]|_{r_1 \cdot k})$. And, for every $i = m-1, \ldots, 1$, let $X_i'(\rho_1(v_i)) \overset{?}{=} X_{i+1}'(\rho_1(w_i))$ be the subequation of $\rho_1(E)$. For every $k \neq l_i, l_{i+1}$, we have $w_i^k = w_{i+1}^k$ and, if $l_i \neq l_{i+1}$, $w_i^{l_{i+1}} = X_{i+1}'(\rho_1(w_i))$. The new equation $\rho_2 \circ \rho_1(E)$ contains a cycle defined by the variables $X_{m+1}', \ldots, X_n'$ and the variables X_i'' that do not leave the cycle, i.e. that satisfy $l_i = l$. Among the pairs of positions we have $\langle p_n \cdot s_n, p_n \cdot q_m \cdot s_{m+1} \rangle$. This cycle is shorter than the original one because $l_1 \neq l$. This situation corresponds to some variable that "leaves the cycle". Notice that in the special situation where $n = 1$, we always fall into case 2.

5.2   There Are No Cycles

If the surface positions of variables in t are the same as in u, then either $\sigma(X) = a$ or $\sigma(X) = \bullet$, for every $X \in Var(E)$. Therefore if we take $\rho = \sigma$ we fulfil the requirement of the lemma. Notice that the size-minimality of σ is only needed in this point and in the exponent of periodicity lemma.

Otherwise, there exists a $\prec_E^*$-maximal $\approx_E$-equivalence class $\{X_1, \ldots, X_n\}$ such that, there exists a variable (assume w.l.o.g. that it is X_1) and a surface position q of X_1 in t, satisfying $u|_q$ has not a variable in the root. Let $v = u|_q$, then $X_1(\ldots) \overset{?}{=} v$ is a subequation of E. We consider two cases:

Case 1. If, for all $i = 1, \ldots, n$, $\sigma(X_i)$ does not contain the hole, then take $\rho = [X_1 \mapsto v, \ldots, X_n \mapsto v]$. It is easy to prove that ρ is coherent with σ, and since it completely instantiates some variable, $\mu(\rho(E)) < \mu(E)$.

Case 2. Otherwise, let p be the largest sequence such that, for all $i = 1, \ldots, n$

1. if $\sigma(X_i)$ contains a hole, then p is a prefix of this hole occurrence,
2. if $\sigma(X_i)$ does not contain a hole, then p is inside $\sigma(X_i)$, and
3. for any q and r, if q is a surface occurrence of X_i in t, and $u|_q$ has not a variable on the root, and $q \cdot r$ is a surface occurrence of a variable in u, then $r \not\prec p$.

Notice that p is a position of v, and there are not variables in v above or at p. Roughly speaking, p is the result of following the main path of the Z-contexts $\sigma(X_i)$ until they split, or someone finish, or we find another variable below. Let c be the context resulting of putting a hole at position p of v. Then $\rho_1 = [X_1 \mapsto c[X_1'(\bullet)], \ldots, X_n \mapsto c[X_n'(\bullet)]]$ is coherent with σ. Moreover $X_1', \ldots, X_n'$ belong to the same equivalence class of $\rho_1(E)$, and $X_1'(\ldots) \overset{?}{=} \rho_1(v|_p)$ is a subequation of $\rho_1(E)$. Now there are three possibilities:

Case 2a. For some $i = 1, \ldots, n$, the hole of $\sigma(X_i)$ is at position p. Then take $\rho = [X_i' \mapsto \bullet] \circ \rho_1$. The first component of μ decreases. This situation corresponds to the case when one of the main paths finish.

Case 2b. If there exists a surface position q of some variable X_i in t, and $q \cdot p$ is a surface position of some variable Y in u (hence $X_i \succ_E Y$) then take $\rho = \rho_1$. This situation corresponds to the case when we have found a variable Y (belonging to a smaller equivalence class) before two main paths split or some one finishes.

In this situation $\rho(E)$ either contains a cycle, or the equivalence class $C = \{X_1, \ldots, X_n\}$ is merged getting $C' = \{X_1', \ldots, X_n'\} \cup \overline{Y} \cup \ldots$ In the second case, $|C'| > |C|$, but we pass from $X_i \succ_E Y$ to $X_i' \not\succ_{\rho(E)} Y$, and no new $\succ$ related pairs are added. The increasing in the first term of μ_3 is strictly compensated by the decreasing in the second term of μ_3.

Case 2c. For every $i = 1, \ldots, n$, let $l_i \in \mathbb{N}$ satisfy: if $\sigma(X_i)$ has a hole, then it is below or at $p \cdot l_i$, otherwise, $l_i = 1$. If cases 2a and 2b does not apply, then there exists at least two distinct l_i's. This situation corresponds to the case when two main paths of variable instantiations split.

Let f be the constant symbol at the root of $v|_p$. We take $\rho = \rho_2 \circ \rho_1$ where

$$\rho_2 = [X_i' \mapsto f(w_i^1, \ldots, w_i^{l_i-1}, X_i''(\bullet), w_i^{l_i+1}, \ldots, w_i^{arity(f)})]_{i \in \{1, \ldots, n\}}$$

Now we show how the subterms w_i^k are chosen. First $w_1^j = \rho_1(v|_{p \cdot j})$, for every $j \neq l_1$, being $X_1(\ldots) \stackrel{?}{=} v$ a subequation of E. Then, for every $i, j = 1, \ldots, n$, if $X_i(w') \stackrel{?}{=} X_j(w'')$ is a subequation of E, then $w_i^k = w_j^k$, for any $k \neq l_i, l_j$, and, if $l_i \neq l_j$, then $w_i^{l_j} = X_j''(w'')$ and $w_j^{l_i} = X_i''(w')$. The existence of a connection between any pair of variables of the same equivalence class ensures that we define all the w_i^k's. We can prove that ρ is coherent with σ. Moreover ρ_2 can be built up from a linear number of pieces of $\rho_1(E)$, and ρ_1 from a linear number of pieces of E.

If we compare $\approx_E$ and $\succ_E$ with $\approx_{\rho(E)}$ and $\succ_{\rho(E)}$, we see that the equivalence class $C = \{X_1, \ldots, X_n\}$ has been split into $arity(f)$ (possibly empty) subsets $C_k = \{X_i'' \mid l_i = k\}$. The existence of two distinct l_i's ensures that there are at least two nonempty of such equivalence classes, and the first term of μ_3 has decreased. There can also be merges between equivalence classes, but then the second term of μ_3 decreases and compensates the increasing in the first term of μ_3. There can also appear cycles, but then μ_2 decreases.

6   Compacting the Solutions

One of the key ideas to compact the representation of a unifier is notating it as a composition of instantiations $[X_1 \mapsto v_1] \circ \cdots \circ [X_n \mapsto v_n]$. Another key idea is representing the Z-contexts v_i using a STG. Finally, the representation of the instance of a variable may involve the computation of subcontext of a term t represented as $t = [X_1 \mapsto v_1] \circ \cdots \circ [X_n \mapsto v_n] \, u$. In this section we show how this can be done efficiently without increasing very much the depth of the grammar.

To understand the main ideas, assume that the v_i's, t and u are words, and we have a grammar G that generates $A_i \to^* v_i$ and $A_0 \to^* u$. We can get a grammar G' that generates $B \to^* t$ replacing in G the variables X_i by the nonterminals A_i. This preserves the size of the grammar, but not the depth: in the worst case $depth(B) = \sum_{i=0}^{n} depth(A_i)$. This means that, to represent a prefix t' of t, we have to enlarge G' in $depth(B)$. A less expensive solution is finding a prefix v'_i of v_i and u' of u such that $t' = [X_1 \mapsto v_1] \circ \cdots \circ [X_n \mapsto v_n] u' v'_n \ldots v'_1$, and enlarge G in order to generate $B \to^* A'_0 A'_n \cdots A'_1 \to^* u' v'_n \ldots v'_1$. Then, in the worst case the depth is only $depth(B) = \log n + \max_{i=0}^{n} \{depth(A_i)\}$.

Definition 6. *We say that a term t is compactable as $t = [X_1 \mapsto v_1] \circ \cdots \circ [X_n \mapsto v_n] u$ with a grammar G, if*

1. *$X_i \neq X_j$, when $i \neq j$,*
2. *X_i does not occur in $v_1, \ldots, v_{i-1}$,*
3. *G generates v_i, for $i = 1, \ldots, n$, and u.*

Similarly when t and u are equations.

The following is a technical lemma used to handle the proof by induction of Lemma 10.

Lemma 9. *Let $\sigma = [X_1 \mapsto v_1] \circ \cdots \circ [X_n \mapsto v_n]$. For any context t compactable as $t = \sigma(X_i(u))$ with a grammar G, and any prefix $c \preceq t$ satisfying $\sigma(X_i) \not\prec c$, c is compactable as $c = \sigma(d)$ with a grammar $G' \supseteq G$ satisfying*

$$depth(d) \leq 3\,i + M$$
$$|G'| \leq |G| + i^2 + 3\,i + 2\,i\,M$$

where $M = \max\{depth(u), depth(v_1), \ldots, depth(v_n)\}$.

PROOF: We proceed by induction on i.

In the base case $i = 1$ we have $c \preceq t = \sigma(X_1(u)) = v_1[\sigma(u)]$. The position of the hole of c must correspond to some position inside v_1 (otherwise $\sigma(X_1) = v_1 \prec c$, contrarily to the assumptions). Therefore, either c does not contain any part of $\sigma(u)$ or contains it completely. So, there exists a prefix d of $v_1[u]$ such that $c = \sigma(d)$. Now, by Lemmas 2 and 4 we can generate any prefix d of $v_1[u]$ with $depth(d) \leq depth(v_1[u]) = 1 + \max\{depth(v_1), depth(u)\}$ using a grammar G' with size $|G'| \leq |G| + 2 + \max\{depth(v_1), depth(u)\}$.

In the induction case $i > 1$ we have $c \preceq \sigma(X_i(u)) = \sigma(v_i[u])$. Let d_i be the largest prefix of $v_i[u]$ such that $\sigma(d_i) \preceq c$. This prefix is uniquely defined.

By Lemmas 2 and 4, since $d_i \preceq v_i[u]$, we can generate d_i with $depth(d_i) \leq 1 + \max\{depth(v_i), depth(u)\}$ using a grammar G'' of size $|G''| \leq |G| + 2 + \max\{depth(v_i), depth(u)\}$.

If $\sigma(d_i) = c$, taking $d = d_i$ and $G' = G''$ we fulfil the requirements of the lemma.

Otherwise $\sigma(d_i) \prec c$, and the hole of c fall inside the instance of some variable X_k occurring in v_i, with $k < i$ (remember that $\sigma(X_i) \not\prec c$). This position may be in the main path of v_i or not. In the first case, we have $v_i[u] = d_i[X_k[v_i'[u]]]$, for some suffix Z-context v_i' of v_i, i.e. d_i does not contain any part of u. In the second case, we have $v_i[u] = d_i[X_k[v_i']]$, for some subterm v_i' of v_i, i.e. d_i completely contains u.

In the first case, we can decompose $c = \sigma(d_i)[\hat{c}]$, for some Z-context $\hat{c}$ satisfying $\hat{c} \prec \sigma(X_k[v_i'[u]])$ and $\sigma(X_k) \not\prec \hat{c}$. Using Lemmas 2 and 4, we see that there exists a grammar $\hat{G}$ deriving $X_k[v_i'[u]]$ with $depth(X_k[v_i'[u]]) \leq 2 + \max\{depth(v_i), depth(u)\}$ and satisfying $|\hat{G}| \leq |G''| + 2 + depth(v_i)$. Moreover, $\sigma(X_k[v_i'[u]])$ is compactable with $\hat{G}$. Since $k < i$, by induction hypothesis, we can compact $\hat{c} = \sigma(\hat{d})$ with a grammar $\hat{G}'$ that generates $\hat{d}$ with

$$depth(\hat{d}) \leq 3k + \max\{depth(X_k[v_i'[u]]), depth(v_1), \ldots, depth(v_n)\}$$
$$\leq 3k + 2 + M$$

and has size

$$|\hat{G}'| \leq |\hat{G}| + k^2 + 5k + + 2k \max\{depth(v_1), \ldots, depth(v_n), depth(v_i'[u])\}$$
$$\leq |\hat{G}| + k^2 + 3k + 2k(1 + M)$$

Now since $c = \sigma(d_i)[\hat{c}]$ and $\hat{c} = \sigma(\hat{d})$, we have $c = \sigma(d_i[\hat{d}])$. Therefore, by Lemmas 2 and 4, we can find a grammar G' with $|G'| \leq |\hat{G}'| + 1$ that generates $d = d_i[\hat{d}]$ with $depth(d) = 1 + \max\{depth(d_i), depth(\hat{d})\}$ and allow us to compact c.

In the second case we obtain lower bounds. Finally, all the inequalities allow us to conclude

$$depth(d) = 1 + \max\{depth(d_i), depth(\hat{d})\}$$
$$\leq 1 + \max\{1 + \max\{depth(v_i), depth(u)\},$$
$$3k + \max\{depth(X_k[v_i'[u]]), depth(v_1), \ldots, depth(v_n)\}\}$$
$$\leq 1 + \max\{1 + M, 3k + \max\{2 + M, M\}\}$$
$$= 3(k + 1) + M \leq 3i + M$$

$$|G'| \leq |\hat{G}'| + 1$$
$$\leq |\hat{G}| + k^2 + 3k + 2k(M + 1) + 1$$
$$\leq |G''| + 2 + M + k^2 + 3k + 2k(M + 1) + 1$$
$$\leq |G| + 2 + M + 2 + M + k^2 + 3k + 2k(M + 1) + 1$$
$$= |G| + (k + 1)^2 + 3(k + 1) + 1 + 2(k + 1)M \leq |G| + i^2 + 3i + 2iM$$

∎

Lemma 10. *For any Z-context t compactable as $t = [X_1 \mapsto v_1] \circ \cdots \circ [X_n \mapsto v_n]u$ with a grammar G, any prefix, subterm or subcontext t' of t, is also compactable as $t' = [X_1 \mapsto v_1] \circ \cdots \circ [X_n \mapsto v_n]u'$, for some Z-context u', with a grammar $G' \supseteq G$ satisfying*

$$depth(u') \leq M + \mathcal{O}(n)$$
$$|G'| \leq |G| + \mathcal{O}(nM)$$

where $M = \max\{depth(u), depth(v_1), \ldots depth(v_n)\}$.

PROOF: We only show the proof when t' is a prefix of t, and t is a context. We can write $t = [X_1 \mapsto v_1] \circ \cdots \circ [X_n \mapsto v_n]u$ as $t = [X_1 \mapsto v_1] \circ \cdots \circ [X_n \mapsto v_n] \circ [X_{n+1} \mapsto u]X_{n+1}(\bullet)$ for any fresh variable X_{n+1}. Then we can apply Lemma 9.

For subterms we need a variant of Lemma 9, and for subcontexts the application of a subterm and then a prefix. For prefixes of terms we need a variant of Lemma 9 based on Lemma 5. These proofs exceeds the length of this paper. ∎

Lemma 11. *For any equation E, and any substitution $\tau = [X \mapsto c[X'(\bullet)]]$, where c is a Z-context not containing X, and built up using $\mathcal{O}(|Var(E)|)$ subcontexts of E, and one exponentiation to e, if E is compactable as*

$$E = [X_1 \mapsto v_1] \circ \cdots \circ [X_n \mapsto v_n]E'$$

with a grammar G, then, for some Z-context d, some $m \in \{0, \ldots, n\}$, and some permutation π, $\tau(E)$ is also compactable as

$$\tau(E) = [X_{\pi(1)} \mapsto v_{\pi(1)}] \circ \cdots \circ [X_{\pi(m)} \mapsto v_{\pi(m)}] \circ [X \mapsto d]$$
$$\circ [X_{\pi(m+1)} \mapsto v_{\pi(m+1)}] \circ \cdots \circ [X_{\pi(n)} \mapsto v_{\pi(n)}] E'$$

with a grammar $G' \supseteq G$ deriving d and satisfying

$$depth(d) \le M + \mathcal{O}(|Var(E)| \, n + \log e)$$
$$|G'| \le |G| + \mathcal{O}(|Var(E)| \, n \, M + \log e)$$

where $M = \max\{depth(E), depth(v_1), \ldots, depth(v_n)\}$.

PROOF: By Lemma 10, we can compact each one of the $\mathcal{O}(Var(|E|))$ subcontexts c_i of E that compound c as $c_i = \sigma(d_i)$ with the same grammar G' increasing the size of G in $\mathcal{O}(Var\,|E|)\,\mathcal{O}(n\,M)$ and the depth of the symbols generating d_i being at most $M + \mathcal{O}(Var\,|E|)\,\mathcal{O}(n)$. Let d be constructed from the pieces d_i as c is constructed from the pieces c_i.

By Lemmas 3 and 2, applied as many-times as pieces we have to assemble, we can prove that there exists a grammar $G'' \supseteq G'$ that generates d with depth $M + \mathcal{O}(|Var(E)| \, n + \log e)$, and satisfying $|G''| = |G'| + \mathcal{O}(\log e)$. Using this grammar G'', we can compact $\tau(E)$ as

$$\tau(E) = \big[X \mapsto [X_1 \mapsto v_1] \circ \cdots \circ [X_n \mapsto v_n]d\big]E$$
$$= \big[X \mapsto [X_1 \mapsto v_1] \circ \cdots \circ [X_n \mapsto v_n]d\big] \circ [X_1 \mapsto v_1] \circ \cdots \circ [X_n \mapsto v_n]E'$$
$$= [X_1 \mapsto v_1] \circ \cdots \circ [X_n \mapsto v_n] \circ [X \mapsto d] \circ [X_1 \mapsto v_1] \circ \cdots \circ [X_n \mapsto v_n]E'$$

Let $\sqsubset$ be the transitive closure of the relation: if X_i occurs in v_j then $X_i \sqsubset X_j$. By definition of compaction this relation is irreflexive. Extend this relation considering $X \sqsubset X_i$ when X occurs in v_i and and $X_i \sqsubset X$ when X_i occurs in d. Then, for every $i = 1, \ldots, n$, either $X_i \not\sqsubset X$ or $X \not\sqsubset X_i$. (Otherwise we would get $X \sqsubset X$ and either c_1 or c_2 would contain X, contrarily to our assumption). Now, for every i, if $X_i \not\sqsubset X$ we can remove $[X_i \mapsto v_i]$ from the left of $[X \mapsto d]$, and if $X \not\sqsubset X_i$ we can remove $[X_i \mapsto v_i]$ from the right of $[X \mapsto d]$. In this way we obtain the desired compaction. Notice that we have to re-order the X_i according to the extension of $\sqsubseteq$, i.e. $X_{\pi(1)} < \cdots < X_{\pi(m)} < X < X_{\pi(m+1)} < \cdots < X_{\pi(n)}$ is a total ordering of the variables compatible with the partial ordering $\sqsubseteq$. ∎

Theorem 2. *If σ is a size-minimal solution of $E = (t \overset{?}{=} u)$, then $\sigma(t)$ is compactable as $\sigma(t) = [X_1 \mapsto v_1] \circ \cdots \circ [X_m \mapsto v_m] \, t'$ with a grammar of depth $\mathcal{O}(|E|^9)$ and size $\mathcal{O}(|E|^{18})$, where $m = \mathcal{O}(|E|^4)$.*
Similarly for u.

PROOF: Using Lemmas 7 and 8 inductively, we can get a decomposition $\sigma = \rho_n \circ \cdots \circ \rho_1$ such that $\mu(\rho_i \circ \cdots \circ \rho_1(E)) < \mu(\rho_{i-1} \circ \cdots \circ \rho_1(E))$. Therefore, by Lemma 6, we have $n = \mathcal{O}(|E|^3)$. Moreover, each one of the ρ_i's is the composition of at most $|Var(E)|$ many (partial) instantiations of just one variable. So, there are $m = \mathcal{O}(|E|^4)$ of these instantiations.

Each one of these partial instantiations τ_j fulfill the requirements of Lemma 11. So, using this Lemma 11 inductively, we can prove that $\tau_i \circ \cdots \circ \tau_1(E)$ is compactable with a grammar G_i such that the maximal depth d_i of a Z-context derived by G_i is $d_i \leq d_{i-1} + \mathcal{O}(|Var(E)| i + \log e)$, i.e. $d_i = \mathcal{O}(|Var(E)| i^2 + i \log e)$, and for the size $|G_i| \leq |G_{i-1}| + \mathcal{O}(|Var(E)| i \, d_i + \log e) = |G_{i-1}| + \mathcal{O}(|Var(E)|^2 i^3 + |Var(E)| i^2 \log e)$, i.e. $|G_i| = \mathcal{O}(|Var(E)|^2 i^4 + |Var(E)| i^3 \log e)$.

We have $i \leq \mathcal{O}(|E|^4)$. The exponent of periodicity lemma ensures that $\log e = \mathcal{O}(|E|^2)$. We have also $|Var(E)| = \mathcal{O}(|E|)$.

Finally, composing all the bounds we get the polynomial bounds stated in the Theorem. ∎

Corollary 1. *Bounded Second-Order Unification is NP-complete.*

PROOF: For any equation E, and any size-minimal solution σ, there exists a STG of polynomial size in $|E|$ that generates $\sigma(X)$, for every $X \in Var(E)$. Notice that we represent σ as a composition of substitutions, and the grammar can generate each one of the compositions, but replacing variables by non-terminal symbols of the grammar, we can (increasing the depth, but without increasing the size) generate σ. A small enlargement of the grammar allow us to generate $\sigma(t)$ and $\sigma(u)$.

Now, a nondeterministic algorithm, guessing a representation of the substitution σ not exceeding the polynomial bound, and using Theorem 1 to check if $\sigma(t) = \sigma(u)$ can decide if $t \overset{?}{=} u$ is solvable or not.

NP-hardness is proved in [17]. ∎

References

1. G. Busatto, M. Lohrey, and S. Maneth. Efficient memory representation of XML documents. In *DBPL'05*, volume 3774 of *LNCS*, pages 199–216, 2005.
2. H. Comon, M. Dauchet, R. Gilleron, F. Jacquemard, D. Lugiez, S. Tison, and M. Tommasi. Tree automata techniques and applications. Available on: http://www.grappa.univ-lille3.fr/tata, 1997. release 1.10.2002.

3. G. Dowek. Higher-order unification and matching. In A. Robinson and A. Voronkov, editors, *Handbook of Automated Reasoning*, volume II, chapter 16, pages 1009–1062. Elsevier Science, 2001.
4. W. M. Farmer. Simple second-order languages for wich unification is undecidable. *Theoretical Computer Science*, 87:173–214, 1991.
5. W. D. Goldfarb. The undecidability of the second-order unification problem. *Theoretical Computer Science*, 13:225–230, 1981.
6. G. Huet. A unification algorithm for typed λ-calculus. *Theoretical Computer Science*, 1:27–57, 1975.
7. A. Kościelski and L. Pacholski. Complexity of Makanin's algorithm. *J. ACM*, 43(4):670–684, 1996.
8. J. Levy, J. Niehren, and M. Villaret. Well-nested context unification. In *CADE'05*, volume 3632 of *LNCS*, pages 149–163, 2005.
9. J. Levy, M. Schmidt-Schauß, and M. Villaret. Monadic second-order unification is NP-complete. In *RTA'04*, volume 3091 of *LNCS*, pages 55–69, 2004.
10. J. Levy and M. Veanes. On the undecidability of second-order unification. *Information and Computation*, 159:125–150, 2000.
11. J. Levy and M. Villaret. Currying second-order unification problems. In *RTA'02*, volume 2378 of *LNCS*, pages 326–339, 2002.
12. G. S. Makanin. The problem of solvability of equations in a free semigroup. *Math. USSR Sbornik*, 32(2):129–198, 1977.
13. A. Martelli and U. Montanari. An efficient unification algorithm. *ACM TOPLAS*, 4(2):258–282, 1982.
14. W. Plandowski. Testing equivalence of morphisms in context-free languages. In *ESA'94*, volume 855 of *LNCS*, pages 460–470, 1994.
15. W. Plandowski. *The Complexity of the Morphism Equivalence Problem for Context-Free Languages*. PhD thesis, Department of Mathematics, Informatics and Mechanics, Warsaw University, 1995.
16. M. Schmidt-Schauß. Stratified context unification is in PSPACE. In *CSL'01*, volume 2142 of *LNCS*, pages 498–512, 2001.
17. M. Schmidt-Schauß. Decidability of bounded second order unification. *Information and Computation*, 188(2):143–178, 2004.
18. M. Schmidt-Schauß. Polynomial equality testing for terms with shared substructures. Frank report 21, Institut für Informatik. FB Informatik und Mathematik. J. W. Goethe-Universität Frankfurt am Main, November 2005.
19. M. Schmidt-Schauß and K. U. Schulz. On the exponent of periodicity of minimal solutions of context equations. In *RTA'98*, volume 1379 of *LNCS*, pages 61–75, 1998.
20. A. Schubert. Second-order unification and type inference for church-style polymorphism. In *POPL'98*, pages 279–288, 1998.

Author Index

Lecture Notes in Computer Science

For information about Vols. 1–3992

please contact your bookseller or Springer